Encyclopedia of

Plant Physiology

New Series Volume 10

Editors

A. Pirson, Göttingen
M.H. Zimmermann, Harvard

Hormonal Regulation of Development II

The Functions of Hormones from the Level of the Cell to the Whole Plant

Edited by Tom K. Scott

Contributors

M.L. Evans W. P. Jacobs H. Kaldewey A.C. Leopold
A.G. Matthysse A.W. Naylor L.D. Noodén J.A. Romberger
T.K. Scott T.J. Wodzicki S. Zajączkowski

With 42 Figures

Springer-Verlag
Berlin Heidelberg NewYork Tokyo 1984

Editor:
Professor TOM K. SCOTT
Department of Biology
The University of North Carolina
Chapel Hill, North Carolina 27514/USA

Planning Volumes 9–11:
Professor N.P. KEFFORD
College of Tropical Agriculture
and Human Resources
University of Hawaii at Manoa
Honolulu, Hawaii 96822/USA

ISBN-13: 978-3-642-67733-5 e-ISBN-13: 978-3-642-67731-1
DOI: 10.1007/978-3-642-67731-1

Library of Congress Cataloging in Publication Data. Main entry under title: The functions of hormones from the level of the cell to the whole plant. (Hormonal regulation of development; 2) (Encyclopedia of plant physiology; new ser., v. 10). Includes indexes. 1. Plant hormones. I. Scott, Tom K. II. Evans, M.L. III. Series. IV. Series: Encyclopedia of plant physiology; new ser., v. 10. QK711.2.E5 vol. 10 581.ls [581.1'4] 83-27185 [QK731].

Softcover reprint of the hardcover 1st edition 1984

2131/3130-543210

In memory of

Martin H. Zimmermann

whose dedication and enthusiasm
for Developmental Botany
was an inspiration to all who knew him

T.K. Scott

In Memoriam

With the death of Martin Huldrych Zimmermann on March 7, 1984, at the age of 58, the international community of plant physiology has lost not only a colleague whose own practical scientific work has played a decisive role in the development of this field, but also the author and editor of numerous scientific works that have greatly served the dissemination of scientific knowledge.

We mourn for a friend, who as long ago as 1974 entrusted a book to us, (Zimmermann and Brown: "Trees"). On the recent founding of a new book series ("Springer Series in Wood Science"), Martin H. Zimmermann willingly contributed his own work as the initial volume ("Xylem Structure and the Ascent of Sap"). This was to be his last book; the moving epilog – placed at his own wish at the last moment at the conclusion of the volume – became also his farewell to active publishing.

Martin H. Zimmermann's main work in the long years of cooperation with Springer-Verlag was, however, the handbook series "Encyclopedia of Plant Physiology, New Series". Over more than 12 years of intensive and often laborious effort he achieved, together with his editorial colleague, the realization of a complete survey of plant physiology. All those who experienced personally with what patience, efficiency, and diplomacy Martin H. Zimmermann not only planned each volume, but also indefatigably supported the numerous editors and authors with good counsel and help, will feel, with us, a great debt of gratitude to him.

The first volume of this series was edited by him jointly with J.A. Milburn. Tragically, it was no longer granted to him to see the series brought to its conclusion.

Springer-Verlag
Spring 1984

André Pirson

Foreword

This is the second of the set of three volumes in the Encyclopedia of Plant Physiology, New Series, that will cover the area of the hormonal regulation of plant growth and development. The overall plan for the set assumes that this area of plant physiology is sufficiently mature for a review of current knowledge to be organized in terms of unifying principles and processes. Reviews in the past have generally treated each class of hormone individually, but this set of volumes is subdivided according to the properties common to all classes. Such an organization permits the examination of the hypothesis that differing classes of hormones, acting according to common principles, are determinants of processes and phases in plant development. Also in keeping with this theme, a plant hormone is defined as a compound with the properties held in common by the native members of the recognized classes of hormone.

Current knowledge of the hormonal regulation of plant development is grouped so that the three volumes consider advancing levels of organizational complexity, viz: molecular and subcellular; cells, tissues, organs, and the plant as an organized whole; and the plant in relation to its environment.

The first volume, entitled *Molecular Aspects of Plant Hormones* and edited by Professor J. MACMILLAN, was published in 1980. It treated the molecular and subcellular aspects of hormones and the processes they regulate. Although it dealt with chemically distinct classes of hormone, the first volume stressed properties and modes of studying them that were common to all classes.

In the present volume, the roles of hormones at levels of organization from the cell up to the whole plant are traced. The cellular processes of increase and change, and the interrelations of cells in tissues, of tissues in organs, and of organs in the whole plant, are considered in turn. During this progressive treatment of levels of organization, the relevant basic properties of hormones are introduced and illustrated.

A third volume addresses the interrelationships of hormones with factors in the environments of the tissues, the organs and the whole plants, within which the hormones are functioning. When this volume touches upon wide-reaching topics such as photomorphogenesis or plant movements, only those aspects that relate to principles of hormonal regulation are treated. Separate volumes of the Encyclopedia of Plant Physiology, New Series, provide comprehensive treatment of topics such as photomorphogenesis and plant movements.

My role in the preparation of these volumes has been to propose a theme and prepare a plan to cover the current status of the field of hormonal regulation, then to circumscribe the portions of the plan that form logical volumes. Thereafter, the editors of the individual volumes have determined the manner in which the domain for which they accepted responsibility was treated. The editor of the present volume is Professor TOM K. SCOTT and, in the Introduction to his volume, he outlines his approach and that of his authors.

The base from which these volumes have developed is the old series of the Encyclopedia of Plant Physiology. The volumes in the New Series of the Encyclopedia may, therefore, concentrate on principles that may be derived from the mass of older information and on the findings of the past 20 years. The length of each volume has been deliberately restricted, but effective organization of topics and their succinct treatment assures the reader of a concise but comprehensive statement of current knowledge and thought in the field.

I thank Professor Kenneth V. Thimann for reviewing the theme and initial plan for these volumes with me.

May 1984 N.P. KEFFORD

Contents

Introduction

T.K. SCOTT 1

1 Hormonal Regulatory Systems in Plants
A.C. LEOPOLD and L.D. NOODÉN

1.1 Introduction 4
1.1.1 The Concept of the Plant Hormone as a Chemical Messenger 4
1.1.2 Types of Hormones 5
1.2 Hormone Messenger Systems 6
1.2.1 Common Difficulties with Single Hormonal Controls 6
1.2.2 Is a Hormone Always a Hormone? 7
1.2.3 Further Complications in the Hormone Concept 8
1.2.4 Regulatory Sites in Hormonal Controls 9
1.2.5 Patterns in Multiple Hormonal Control 10
1.2.6 Complications Arising from Compartmentation 11
1.3 Analysis of Hormonal Controls 11
1.3.1 Two Major Approaches from the Past 12
1.3.2 Analyses of Hormone Systems 13
1.4 Conclusions 15
1.4.1 The Complexity of Hormonal Controls 15
1.4.2 Nonhormonal Correlative Controls 15
1.4.3 Evolutionary Plasticity of Hormonal Controls 16
1.4.4 The Future 16
References 17

2 Functions of Hormones at the Cellular Level of Organization
M.L. EVANS (With 1 Figure)

2.1 Introduction 23
2.2 Promotive and Inhibitory Classes of Plant Hormones 23
2.3 Modes of Hormone Interaction 24
2.3.1 Interaction and Independence in Hormone Action 24
2.3.2 Synergism and Antagonism in Hormone Action 25
2.3.2.1 Synergism 25
2.3.2.2 Antagonism 25
2.3.3 Hormone Effects on Endogenous Hormone Levels 27
2.3.3.1 Hormonal Modification of Endogenous Auxin 27
2.3.3.2 Hormonal Modification of Endogenous Gibberellin 28
2.3.3.3 Hormonal Modification of Endogenous Ethylene 29
2.3.3.4 Hormonal Modification of Endogenous Abscisic Acid . . . 29
2.3.3.5 Hormonal Modification of Endogenous Cytokinin 30
2.3.3.6 Summary 30
2.3.4 Influence of Environmental Parameters on Endogenous Hormones 31
2.3.4.1 Light 31
2.3.4.2 Water Stress 33

2.3.4.3 Temperature 34
2.3.4.4 Mechanical Stimulation 34
2.3.4.5 Summary 34
2.4 Hormonal Regulation 35
2.4.1 The Cell Cycle 35
2.4.1.1 Effect of Cytokinins 36
2.4.1.2 Effect of Auxins 39
2.4.1.3 Effects of Other Growth Regulators 40
2.4.2 Endomitosis 42
2.4.3 Meiosis 43
2.4.4 Summary 44
2.5 Hormone Effects on the Plane of Cell Division 44
2.6 Relationship of Cell Multiplication to Cell Expansion 46
2.7 Hormonal Regulation of Cell Expansion 48
2.7.1 Stem Cells 48
2.7.1.1 Effects of Auxin 48
2.7.1.2 Effects of Gibberellin 49
2.7.1.3 Effects of Cytokinin, Ethylene, and Abscisic Acid 49
2.7.2 Root Cells 50
2.7.2.1 Effects of Auxin 51
2.7.2.2 Effects of Gibberellin 52
2.7.2.3 Effects of Ethylene 53
2.7.2.4 Effects of Abscisic Acid 54
2.7.2.5 Effects of Cytokinin 54
2.7.3 Leaf Cells 55
2.7.3.1 Effects of Auxin 55
2.7.3.2 Effects of Gibberellin 56
2.7.3.3 Effects of Ethylene 56
2.7.3.4 Effects of Abscisic Acid 56
2.7.3.5 Effects of Cytokinin 56
2.7.4 Other Cell Types 57
2.7.4.1 Pollen Tubes 57
2.7.4.2 Epidermal Hairs 57
2.7.5 Lower Plants 57
2.7.5.1 Algae 57
2.7.5.2 Fungi 59
2.7.6 Summary 61
References 62

3 Transport and Other Modes of Movement of Hormones (Mainly Auxins)
H. Kaldewey (With 7 Figures)

3.1 Introduction 80
3.2 Historical Sketch 80
3.2.1 Indirect Evidence for Hormone Translocation 80
3.2.2 Direct Evidence for Hormone Translocation 81
3.2.3 Hormone Transport Experiments 81
3.2.3.1 The Early Period 81
3.2.3.2 The Period of Confirmation, Extension, and Criticism 82
3.2.3.3 The "Modern" Period 83
3.3 General Remarks on Hormone Movement 86
3.3.1 Terminology and the Transport Equation 86
3.3.2 Methods to Estimate Hormone Translocation Characteristics 88
3.3.2.1 Intercept Method of van der Weij 88
3.3.2.2 Short-Term Collecting Method 92
3.3.2.3 Pulse Experiments 94

3.3.2.4 Penetration-Depletion Method 98
3.3.2.5 Electric Potential Wave Method 101
3.3.3 Evaluation of Estimation and Interpretation of Hormone Movement 103
3.3.3.1 Does Auxin Move in a Stream? 103
3.3.3.2 Loading of the Transport System 105
3.3.3.3 Validity of Velocity Estimations 106
3.3.3.4 Fluctuations of Transport Velocity 108
3.3.3.5 Flexibility of the Transport System 109
3.3.4 Models for Transport Mechanisms 111
3.3.4.1 Conveyor Belt Model 112
3.3.4.2 Electrical Polarity Hypothesis 112
3.3.4.3 Polar Secretion Theory 112
3.3.4.4 Theory of Chemiosmotic Polar Diffusion 115
3.3.5 Binding Sites 116
3.3.6 Pathways of Hormone Movement 117
3.3.6.1 Shoots 117
3.3.6.2 Roots 118
3.4 Evidence of Hormone Translocation 119
3.4.1 Hormone Movement in Vascular Systems 119
3.4.2 Hormone Translocation in Shoot Sections 122
3.4.2.1 Auxins 122
3.4.2.2 Hormones Other than Auxins 123
3.4.3 Hormone Translocation in Root Sections 125
3.4.3.1 Auxins 125
3.4.3.2 Hormones Other than Auxins 126
References 126

4 Functions of Hormones at Tissue Level of Organization
W.P. Jacobs (With 12 Figures)

4.1 Introduction 149
4.2 Role of Hormones in Differentiation of Vascular Tissues 149
4.2.1 Regeneration in Shoots 149
4.2.1.1 Xylem 149
4.2.1.2 Sieve-Tube Elements 154
4.2.1.3 Fibers 159
4.2.2 Differentiation During Normal Development in Shoots 159
4.2.2.1 Xylem 159
4.2.2.2 Sieve-Tube Elements 160
4.2.2.3 Fibers 160
4.2.2.4 Cambium 162
4.2.3 Control of Differentiation in Roots 165
4.2.4 Evidence of Differentiation from Tissue Cultures 165
4.3 General Discussion 166
References 168

5 Functions of Hormones at the Organ Level of Organization
A.W. Naylor (With 8 Figures)

5.1 Introduction 172
5.2 Embryogeny — Establishment of Polarity 174
5.3 Hormonal Relations of the Developing Embryo 177
5.4 Vascular Development — Spatial Arrangement 178
5.5 Regeneration 180
5.6 Flower Initiation 186
5.7 Flower Development 192

5.8 Fruit Setting and Development 195
5.9 Seed Development 200
5.10 Seed Dormancy—Germination Inhibitors 204
5.11 Conclusions 207
References 208

6 Functions of Hormones at the Whole Plant Level of Organization
A.G. Matthysse and T.K. Scott (With 8 Figures)

6.1 Introduction 219
6.2 Some Perspectives on the Nature of Hormones 220
6.3 A General View of the Role of Hormones in the Whole Plant 221
6.4 Some Considerations Regarding Experimental Design 224
6.5 Some Considerations Regarding Research Strategies 224
6.6 Conclusion 226
References 226

7 Auxin Waves and Plant Morphogenesis
S. Zajączkowski, T.J. Wodzicki, and J.A. Romberger (With 6 Figures)

7.1 Introduction 244
7.2 Polar Transport of Auxin and Positional Information 244
7.3 The Origin of a New Concept 245
7.4 Some Experimental Evidence 246
7.5 The Proposed Model 252
7.6 Further Implications 258
References 259

Author Index 263

Subject Index 289

List of Contributors

M.L. Evans
Department of Botany
Ohio State University
1735 Neil Avenue
Columbus, Ohio 43210/USA

W.P. Jacobs
Department of Biology
Princeton University
Princeton, New Jersey 08544/USA

H. Kaldewey
Fachrichtung Botanik
Universität des Saarlandes
6600 Saarbrücken/FRG

A.C. Leopold
Boyce Thompson Institute
Ithaca, New York 14853/USA

A.G. Matthysse
Department of Biology
University of North Carolina
Chapel Hill, North Carolina 27514/USA

A.W. Naylor
Department of Botany
Duke University
Durham, North Carolina 27706/USA

L.D. Noodén
Botany Department
University of Michigan
Ann Arbor, Michigan 48109-1048/USA

J.A. Romberger
Light and Plant Growth Laboratory
Institute of Plant Physiology
BARC-West, Bldg. 046 A
Beltsville, Maryland 20705/USA

T.K. Scott
Department of Biology
The University of North Carolina
Chapel Hill, North Carolina 27514/USA

T.J. Wodzicki
Department of Forest Botany
Warsaw Agricultural University
26/30 Rakowiecka Street
02-528 Warsaw/Poland

S. Zajączkowski
Department of Forest Botany
Warsaw Agricultural University
26/30 Rakowiecka Street
02-528 Warsaw-Poland

Introduction

T.K. Scott

The field of hormonal control of the development of higher plants continues to attract the interests of a variety of researchers and continues to make steady progress. Increasingly, this field unites the molecular and cell biologists with the plant biochemists, plant geneticists, and plant physiologists, who, in turn, interact with those researchers in the practical fields of agronomy, horticulture, and phytopathology. The fascinating challenge related to understanding a regulation of plant growth and development has recently gained new urgency in the attempt to find answers to pressing demands for increased production of food and other renewable resources. Clearly, at no time has the understanding of higher plant development found more application or relevancy and, as is made clear in these following pages, at no time has this field been more stimulating or provocative.

This volume, number 10 of the Encyclopedia of Plant Physiology, New Series, deals with our most current understanding of plant hormones and how they regulate or influence developmental events at the level of the cell up to and including the whole plant. Volume 10 is the second in a subset of three in a series. It commences where Volume 9, *Molecular and Subcellular Aspects of Hormonal Regulation in Plants* ended. In turn, this volume leads naturally to the consideration of the relationship between environmental factors and their effects on hormone-regulated processes in plants – this being the domain of consideration to be found in Volume 11.

The present volume has been organized and written with the following goals in mind. First and foremost, the authors have brought the field up to date. However, because hormonal control mechanisms in higher plants have, in general, been so extensively reviewed since the publication of Volume XIV of the original Encyclopedia series, some 20 years ago, the authors have been deliberately selective in their discussions and have highlighted, in the main, those contemporary discoveries and advances which are clearly the most significant and which hold the most promise for the encouragement and furthering of future work. As with any such endeavor, much specific information will necessarily have been omitted. However, these omissions are more than offset by the ample referencing made to recent and copious numbers of review works which are devoted directly or indirectly to the topics contained in this volume. The effort to condense the encyclopedia somewhat was agreed to in order to make volume accessible, attractive, and easily manageable. Further, we wish to emphasize our hope that the subject material will attract the attention of a variety of readers. It is hoped in particular that the volume will be used broadly by the teacher. In addition, this volume is directed toward those researchers in need of updating their own knowledge of the field, to those wishing a new and balanced overview of the field, and to those not now in the field but who would like to become acquainted with it. Finally, all the authors have made it their goal to generate incentive as well as enthusiasm for future research by carefully suggesting new (and sometimes old) directions in which the field could profitably progress. Invariably these suggestions are drawn from a combination of experienced vantages and individual philosophical prospectives. Many are proposals for truly new "thought" and consequently will both provoke and challenge.

The seven chapters generally lead the reader through known and suggested roles of hormones at advancing levels of organization in the plant. Chapter 1 (LEOPOLD and NOODEN) is an ideal introduction to the volume since it provides a balanced, rational, and modern overview of the hormonal concept of the regulation of plant development. The result is a provocative and rewarding critique which generates some new concepts of varied roles and activities of hormones acting singly or cooperatively as controlling factors in development. In Chapter 2 (EVANS) the functions of hormones at the level of the plant cell are detailed. Known biological and physical factors which alter endogenous hormone production and levels in the cell are discussed and individual and multiple controls of mitosis and meiosis, as well as cell enlargement, are compared for a variety of plant cell types. These examples are provided against a backdrop of a thorough definition and comparison of all known plant hormonal interactions. In addition, there is a brief account of these same hormonal activities where they are known to occur in the algae and the fungi. Chapter 3 (KALDEWEY) deals with hormone transport which, if not the most, is certainly one of the most critical aspects of hormonal regulation. The volume thus progresses or "moves" from the cell to more highly integrated levels of organization in the plant body by way of this chapter. A great deal of emphasis on technique is provided in order to place proper focus on an aspect of plant hormonal research known for its difficulties, bewilderment, and frustrating aspects. The unique properties of the auxin transport system, including an in-depth perspective as to how and why various methodologies came to be developed, highlights this chapter and the discussion emphasizes the slow and scant progress made over the last half century. The inescapable conclusion emerges that the major components of hormone transport, including secretion, long-distance movement, and uptake, of auxin and the other known plant hormones indeed remain obscure or are only just now becoming understood.

The functions of plant hormones at the level of tissues, organs, and the whole plant are discussed in Chapters 4 (JACOBS), 5 (NAYLOR), and 6 (MATTHYSSE and SCOTT), respectively. The hormonal control of vascular tissue differentiation in shoots and roots is described in relation to what is known regarding the role of hormones in various regeneration phenomenon of the same tissues in herbaceous shoots. Reference is also made to other aspects of hormonal control of tissue differentiation, including results from recent studies involved with the use of tissue culture techniques. The topic of organ development is dealt with methodically and precisely by tracing it from the beginning to the end of the life cycle of the plant. In this chapter, various known hormonal control points are described, starting first with the organization of the embryo itself, continuing with considerations of tissue patterning, leading to vegetative organ formation, and ending with events concerning reproductive phenomena. Flower initiation and formation, fruit formation, as well as seed formation and metabolism, are treated as interlocking regulatory events under a variety of hormonal control mechanisms. This chapter emphasizes the interactive nature of hormones and urges further attention to the promising and yet largely unknown area of development modulation. Finally, the treatment of functions of hormones in the whole intact plant emphasizes integration and coordination within the organism. General rules of hormones for the "typical" higher plant in the regulation of the organism during growth and maturation, in physiological homeostasis, and in its responses to various elements in the external environment are discussed and described and also presented in graphic chart form. Again, as elsewhere in the book, considerations regarding experimental design and research strategies are stressed.

Chapter 7 (ZAJACZKOWSKI, WODZIEKI, and ROMBERGER), the last chapter, provides an ideal capstone to the volume. It takes one hormone and one concept and brings them together, utilizing unifying principles put forward in the preceding chapters. Here, the authors propose a new concept, that of information which is generated by the polar transport of auxin and of how this information is later translated. This chapter includes recent data to substantiate the model; one which emphasizes vectoral waves of hormone (in this case, auxin), rather than emphasizing mere changes in molecular concentration and as such it suggests an elegant precept of the spatial and temporal control of higher plant morphogenesis. The conclusion of the volume, therefore, is far from encyclopedic. Rather, it is a summary of research which holds promise of being a beginning of a systematic means by which to analyze the functions of hormones in the regulation and integration of the developing higher plant.

The Functions of Hormones from the Level of the Cell to the Whole Plant has been an effort which combines ambition, diligence, and very considerable patience. Its origins date back further than one would wish and inevitably there was a lack of synchrony in the completion of the chapters. Although the circumstances leading to the various delays are regrettable, such delays, while obviously undesirable, are understandable. However, all of the authors, in holding to their topic content, have also brought the topic up to the forefront of our knowledge. I wish to thank them for their consideration, for their thorough execution, and for their promptness.

I am further indebted to Mrs. Henrietta S. Brandt for her efforts in organizing the various details associated with all aspects of the production of such a volume, which she has painstakingly done from its very beginning to its very end.

The editors and publishers thank Professor Martin Bopp for his advice and assistance in the publication of this volume.

1 Hormonal Regulatory Systems in Plants

A.C. LEOPOLD and L.D. NOODÉN

1.1 Introduction

1.1.1 The Concept of the Plant Hormone as a Chemical Messenger

Over two millenia, observers of plants have noticed that one part of a plant may influence or control the activities of another part (see, e.g., WIESNER 1871, DOSTAL 1967). There are diverse examples, such as axillary buds growing out when the main bud has been removed (GOEBEL 1900, SNOW 1925), excision of seeds from fleshy fruit promoting their germination (ALBERTUS MAGNUS, thirteenth century, see WAREING 1965), cutting off the coleoptile tip preventing coleoptile tropisms (DARWIN and DARWIN 1880), removal of the embryo blocking starch degradation in grains (BROWN and MORRIS 1890), partial defoliation altering bud growth (LOEB 1918) and removal of reproductive structures delaying the senescence of leaves and stems (MOLISCH 1928).

It has been recognized that these interactions between plant parts might be mediated by chemicals, later known as hormones, which act as messengers. Solid progress in this field had to wait for developments in natural product and organic chemistry; thus, after Darwin completed his studies on the regulation of coleoptile growth and bending, decades passed before indoleacetic acid was shown to be the main regulatory hormone in this system (see WENT and THIMANN 1937, AUDUS 1959, HESLOP-HARRISON 1963).

The classic idea of a hormone as a chemical messenger is expressed in the definition given by WENT and THIMANN (1937):

"a hormone is a substance which, being produced in any one part of the organism, is transferred to another part and there influences a specific physiological process."

Subsequently, this concept has been slightly refined by limiting the term to organic substances acting in small quantities (THIMANN 1960).

Under this conceptual framework, hormone systems may be viewed in terms of three primary functional components: (1) a site of synthesis or source of the hormone, (2) movement from the source to the site of action (this may be passive as in the diffusion of ethylene or active as in polar transport of auxin) and (3) action by the hormone upon the target site. Through such a system, hormones may initiate a new path of development at the target site or simply control an established pathway.

A current trend toward calling these substances plant growth substances may lead to confusion, especially since growth is only one of many aspects of development which can be regulated. The central theme of this review is to provide an overview of the present state of knowledge of hormonal regulation, showing the inadequacy of the concept of single hormonal controls, pointing out the risk in assuming that the presence of a hormone allows inferences about its participation in regulation, and emphasizing the evidence that regulation is achieved by the collective interactions of multiple hormonal systems.

1.1.2 Types of Hormones

The first plant hormone to be identified was auxin or "growth hormone". Until 1950, the hormone concept in plants centered on this one hormone, and attempts were made to explain regulation of most plant developmental processes in terms of auxin. As other hormones (gibberellin, cytokinin, ethylene, and abscisic acid) were discovered, physiologists initially considered them as agents altering the actions of auxin (see PILET 1957, PHILLIPS et al. 1959), but it soon became evident that each could be a hormone in its own right. Nonetheless, numerous examples of overlap in the functions and regulatory actions of these five hormones are best explained with a concept of regulatory control involving an integrated system of multiple hormones. In addition, there is evidence for several minor hormones such as wound hormone (BONNER and ENGLISH 1938, ZIMMERMAN and COUDRON 1979), thiamine (BONNER and BONNER 1948, AUDUS 1972), calorigen (CHEN and MEEUSE 1975), D-pinitol (SCHILDKNECHT and IYENGAR 1975) and gentisic acid glycosides (SCHILDKNECHT et al. 1978), which seem less notable as regulators. There is also a promising array of naturally occurring, physiologically active substances, which may be elevated to hormone status as information on their correlative effects is obtained (see Vol. 9 of this encyclopedia). These include brassins (GROVE et al. 1979, MANDAVA et al. 1981), dihydroconiferol (SHIBATA et al. 1974), triacontanol (RIES et al. 1977), methyljasmonate (UEDA and KATO 1980), a large number of phenolic compounds which alter the effects of known hormones (see THIMANN 1977) and a wide range of inhibitory compounds which may be involved in dormancy (see AUDUS 1972, LETHAM 1978, NOODÉN and WEBER 1978, MOORE 1979, MANDAVA 1980, EVANS and TRAMONTANO 1981).

It is highly probable that there are hormones waiting to be "discovered." Furthermore, there are several correlative control systems in which the known hormones do not seem to account for the observed functions, again suggesting that other hormones are yet to be discovered. Definitely included here are the florigen(s) of some short-day plants (see CLELAND 1978) and possibly antiflorigen (LANG et al. 1977, CLELAND 1978). The continuing uncertainty about the role of ABA in the induction of dormant bud development (PHILLIPS et al. 1980) also suggests that some yet unknown factors may be involved in this process. On the other hand, it should not be expected that all physiologically active substances, even naturally occurring ones, are necessarily hormones.

It is also possible that the five known major hormone classes can be subdivided. Particularly, the plethora of different gibberellins now known may in the future be grouped into different subsets functioning in the same plant. Furthermore, it is probable that indoleacetic acid, zeatin, abscisic acid and ethylene are not the only active hormones in their classes.

In its elementary form, then, the hormonal concept of regulation in plants attempts to account for growth or developmental phenomena as being controlled primarily by any of five major hormone systems, the auxins, gibberellins, cytokinins, abscisic acid or ethylene; each of these is presumed to be formed in one part of the plant and moved to another, where it plays a regulatory role in some physiological process (presumably through a reaction involving its attachment to some stereo-specific site).

There are several shortcomings with this type of regulatory concept. In the first place, almost never has a phase of growth or development been found to be under the exclusive control of a single hormone. Conversely, none of the hormones has been found to have exclusive control over any phase of growth or development. Further, the implications in the definition of a plant hormone as a chemical messenger having a site of synthesis, translocation, and then a site of action have become increasingly complex. In addition, the very definitions of the five hormones have always been empirical, based on the observed effects of members of a family of related chemicals on some bioassay rather than the natural roles of such chemicals in controlling a specific reaction or process. These various shortcomings and complications of the hormonal regulation concept will be discussed in the following sections.

1.2 Hormone Messenger Systems

1.2.1 Common Difficulties with Single Hormonal Controls

Each of the five major hormones has been implicated in regulatory control of specific processes in the plant at the cellular, tissue, or plant level; with time, nearly all of these specific functions have proved to be very complex. Auxin was first described as the regulator of cell enlargement, gibberellin as the regulator of overall stem growth, cytokinin as the regulator of cell division, abscisic acid as the regulator of abscission and dormancy, and ethylene as the regulator of fruit ripening. Now it is evident that each of these aspects of plant development is inadequately explained as a response to a single hormone.

Every one of these assignations of hormonal functions has major exceptions and qualifications. Auxins alone may not account for cell enlargement, and gibberellin alone does not control stem growth (CLELAND 1964). Cytokinin alone does not appear to induce cell division even in the standard cell division tests; the assay for cytokinins requires the presence of another hormone, auxin (SKOOG and MILLER 1957, STEWARD and SHANTZ 1959); abscisic acid itself does not adequately account for the control of either abscission (CRAKER and ABELES 1969, NOODÉN and LEOPOLD 1978) or dormancy (LENTON et al. 1972, LOVEYS et al. 1974, NOODÉN and WEBER 1978, PHILLIPS et al. 1980); and ethylene alone may not account for the ripening of fruits (BIALE 1960, MATSON and JARVIS 1970, FRENKEL 1972).

The shortcomings of the single hormone control concept can be illustrated in another way; if a single hormone was responsible for a particular function in growth or development, then one could use an adaptation of Koch's postulates to test whether the hormone was in fact responsible for the regulatory function. This was attempted by JACOBS et al. (1959) in examining the role of auxin in apical dominance; auxin clearly failed the test of Koch's postulates. We should not be surprised, knowing as we do that several regulatory agents in plants can influence apical dominance. Failing the test, of course, should not lead us to discard the concept that auxin has a role to play in apical dominance, but it might fairly lead us to consider a nonexclusive type of regulation by the hormone, auxin.

A second type of difficulty with the single hormone concept is the frequency with which a hormone fails to show quantitative correlation with the physiological function which it is presumed to regulate. While successful correlations make good subjects for publication, unsuccessful correlations are published only by braver researchers; thus the literature does not always reflect fairly the frequency of cases in which hormones fail to show quantitative correlations with physiological functions.

Auxin has been presumed to be the hormone in control of the elongation of etiolated stems and coleoptiles. What fraction of auxin may serve as the regulatory control? WENT (1942) noted that diffusible auxin correlated with coleoptile growth rate, but SCOTT and BRIGGS (1960) found that neither diffusible nor extractable auxin correlated well with stem growth rate. In another case, gibberellin is recognized as a major regulator in the growth of corn. Although some dwarf corn mutants respond to gibberellin applications by changing from dwarf to standard growth characteristics, certain of these dwarf varieties have endogenous gibberellin contents which were not unlike those in the standard varieties (JONES and LANG 1968). Similarly, cytokinins are considered to have an important role in deferring senescence of leaves, and yet as leaves enter senescence, they may show an increase in cytokinin content (LORENZI et al. 1975, DAVEY and VAN STADEN 1978, ILAN and GOREN 1979). Abscisic acid is presumed to act as an inhibitor of shoot growth, but cases have been reported in which the most rapidly growing shoot regions and young leaves have the highest endogenous content of abscisic acid (POWELL 1975, RASCHKE and ZEEVAART 1976, NOODÉN 1980). In addition, ABA was considered the principal inducer of bud dormancy; yet buds have been found to emerge from dormancy without any measurable decrease in ABA activity (FRANKLAND and WAREING 1962). These are only a few samples of the data which illustrate the lack of correlation between the content of a single hormone and the growth or development process it is presumed to regulate. One reasonable interpretation of these discrepancies is that other hormones are participating in the regulatory controls. Other explanations involving cellular heterogeneity within a developing tissue are discussed in Section 1.2.6.

1.2.2 Is a Hormone Always a Hormone?

In both plant and animal biology, a tacit assumption has been made that where hormones are present in an organism, they must be serving some regulatory function. This assumption has been drastically altered, however, by the recent findings that many animal hormones occur in tissues where they do not appear to serve their usual hormonal function, and in fact they are also synthesized in body parts remote from the usual synthetic sites in an endocrine gland (Roth et al. 1982). Not only are hormone molecules present in tissues in which they are apparently not functioning as hormones, but stereo-specific receptor sites for these "hormones" can also be found in such tissues. For example, biologically active forms of several animal hormones such as insulin, ACTH, and endorphin are found in primitive organisms in which it is exceedingly unlikely that they serve any regulatory function or at least not the same function as in higher animals (Roth et al. 1982). From an evolutionary point of view, then, the compounds which are adapted to serve regulatory roles in plant and animal organisms may be synthesized in a wide array of cells and organisms; with the progress of specialization, organisms have developed systems in which the synthesis, translocation, physiological function, and disposal of these compounds have become adapted to involve them in regulatory roles. A suggested central conclusion from this point of view, then, is that the extraction and identification of plant hormones from various plant tissues does not automatically mean that the hormonal compound is serving regulatory functions. The same can be said for stereo-specific attachment sites which can bind these compounds. Thus, the lack of correlation between the content of auxin, gibberellin, cytokinin, ABA, or ethylene with particular processes of growth or development should be accepted as evidence that the hormone pool being sampled is probably not serving in a regulatory fashion or that its functions are more complex as explained in Sections 1.2.5 and 6.

1.2.3 Further Complications in the Hormone Concept

The simple model of a hormone system implies that the chemical species will induce the regulatory response in some place remote from the site of its synthesis. This model is extremely well adapted to the auxin regulation of coleoptile growth where the indoleacetic acid was found to be produced in the coleoptile tip and moved in a directional transport system to the elongating portion of the coleoptile where it stimulated growth.

One of the complications with this model is that the compound which is synthesized and translocated may not be the same as the one acting at the target site. In the case of the interactions between the seed and the coleoptile of corn, it is the myo-inositol ester of indoleacetic acid which is formed in the seed endosperm and is translocated to the coleoptile tip; subsequent conversion to indoleacetic acid occurs at the tip, and the indoleacetic acid then travels down the coleptile where it stimulates growth (Nowacki and Bandurski 1980). In another instance, environmental stress experienced by the roots is found to cause wilting of the leaves; the roots synthesize ACC (1-aminocyclopropane-1-carboxylic acid), and this chemical is transported to the leaves where it is converted into ethylene, which is the active principal causing the wilting response (Bradford and Yang 1980). In short, then, the messenger function may not be restricted to the chemical species which is physiologically active.

Although movement (active or passive) is required, it is uncertain how far the substance must move to qualify as a hormone. Some influences such as wound hormone (AUDUS 1959) and others (LANG 1966, STEEVES and SUSSEX 1972) span only a few cells. Would it be minimally sufficient for the hormone to move only from one cell to the next? Or, like the neurohumors of animal systems, might it be sufficient for the hormone to move only across a membrane?

In order to exert its effect on the target tissue, the hormone must bind (at least briefly) to a receptor site, which may or may not be a protein. The resultant hormone-binding site complex would alter some biochemical process (primary effect) which in turn may produce secondary effects. It should be noted that the primary effect might be modulating a chemical reaction or it might be a physical change such as an alteration of membrane permeability. It is an uneasy likelihood that most of the observed hormone effects on development are probably not primary (direct) effects, but rather secondary. For example, a basic part of auxin promotion of cell enlargement is a change in plasticity of the cell wall, but even this change is probably a secondary effect of auxin. Obviously, the initial action triggered by the combination of hormone and receptor has been very elusive.

1.2.4 Regulatory Sites in Hormonal Controls

An individual hormone system may have numerous sites for control, and regulation may be accomplished at more than one step. In some cases, the synthesis of the hormone is the regulatory step; for example, α-amylase production by the embryo is controlled by gibberellin synthesis in barley embryos (see LANG 1966, HO 1979). Formation of disorganized tumorous growth appears to be due in part to activation of auxin and cytokinin synthesis thereby lifting the normal restraints on the cells (see BRAUN 1969). Transport allows many opportunities to control the rate of flow of the hormone and its distribution. The classic example is the altered distribution of auxin in geotropism and phototropism of coleoptiles (see LEOPOLD 1960, AUDUS 1972, GOLDSMITH 1977, THIMANN 1977). As another example, light may inhibit corn mesocotyl growth by limiting the supply of auxin transported from the coleoptile (VANDERHOEF and BRIGGS 1978). In addition to control by changes in biosynthesis or translocation, regulation may be achieved through influences on the rate of hormone degradation. Rice coleoptile growth is increased by submergence which decreases auxin degradation (because less O_2 is available for the auxin-destroying enzymes) and increases the endogenous auxin levels (WADA and NAGAO 1960, WADA 1961). Secondary substances such as phenolic compounds may control the activity of auxin degradative enzymes (see THIMANN 1977, LETHAM 1978). Alternatively, nonhormones may influence hormone activity through antagonism or synergism. Also, the activity of the receptors at the target site could be regulated in various ways (VENIS 1977). Thus, light may lower the sensitivity of pea stems to GA_5 (KENDE and LANG 1964). Numerous other examples of differing hormone sensitivity between different tissues or developmental stages exist (CLELAND 1983). If plant hormone actions are expressed through second messengers comparable to the cyclic AMP in animal cells, this would allow yet another level of regulation; however, we should note that cyclic AMP has not been implicated as a component of higher plant regulation (AMRHEIN 1977, LETHAM 1978). In any case, understanding a hormonal system requires not only knowing the hormone, translocation, target and degradation, but also the regulatory sites within the system.

1.2.5 Patterns in Multiple Hormonal Control

Existing examples in the literature on growth and development suggest four general types of interaction which may commonly occur between hormones: regulation may be controlled by a balance or ratio between hormones, by opposing effects between hormones, by alterations of the effective concentrations of one hormone by another, and by sequential actions of different hormones. Illustrative examples of each of these integrative types of actions might be useful to mention here (see also LEOPOLD 1982).

a) Ratios between hormones were first noted to be important in the case of cytokinins and auxins in the regulation of bud differentiation in tobacco callus cultures (SKOOG and TSUI 1948, SKOOG and MILLER 1957). Similarly, in regeneration of chicory cuttings, the ratio of cytokinins to auxins changes from a high ratio during bud differentiation to low during root regeneration (VARDJAN and NITSCH 1961).

b) Opposing effects between pairs of hormones have been noted in several instances. The first suggestion of opposing actions by pairs of hormones was made by HALL (1952) for ethylene and auxin, with ethylene stimulating leaf abscission and auxin inhibiting it. Again, opposing actions have been suggested for auxins and cytokinins in apical dominance (WICKSON and THIMANN 1958), and for growth inhibitors and growth stimulators (gibberellins) in the regulation of seed dormancy (NAYLOR and SIMPSON 1961). Another opposing set of actions has been suggested for gibberellin and ethylene, with each having the ability to inhibit plant responses to the other in several instances (SCOTT and LEOPOLD 1967).

c) Alteration of One Hormone Concentration by Another. Examples may be found where one hormone may *alter the concentration of another* by various means, including alteration of hormone biosynthesis, translocation or inactivation. The first clear example of an alteration of biosynthesis of one hormone by another was described by BURG and BURG (1968); here elevated levels of auxin caused a stimulation of ethylene biosynthesis. In the converse direction, the biosynthesis of ethylene can be inhibited by abscisic acid (LIEBERMAN and KUNISHI 1972). Hormonal influences on hormone biosynthesis can become fairly complex; for example, the stimulation of ethylene production by auxin can be amplified by cytokinins (FUCHS and LIEBERMAN 1968). The ability of one hormone to inhibit transport of another hormone is illustrated by the ethylene inhibition of auxin transport (MORGAN and GAUSMAN 1966); on the other side of the ledger, there is evidence that auxin movement may be stimulated by cytokinins (SETH et al. 1966) or by gibberellin (JACOBS and CASE 1965). Regulation of the degradation of one hormone by another hormone is illustrated by the enhancement of auxin degradation by ethylene (BEYER and MORGAN 1969).

d) Sequential regulation by several hormones was first described by WRIGHT (1961) for the regulation of the growth of oat coleoptiles; in this case, the coleoptile tissue was shown to pass through a stage in which growth was stimulated by gibberellin, followed by a stage of cytokinin stimulation, and then the final growth phase controlled by auxin. An analogous sequential regulation was subsequently shown for the growth of cells in the bean hypocotyl hook (GOTO and ESASHI 1974). No doubt, many more examples of sequential regulation will be unraveled as more developmental sequences are studied.

The complexity of the controls of specific processes by several hormones probably has been an important factor in the slow progress in the understanding of regulatory controls.

1.2.6 Complications Arising from Compartmentation

Developing organs and tissues are made up of mixes of different cell types, which may well mean that several different hormone control systems could be operating within a tissue. For example, auxin can induce channels of xylem differentiation within a mass of parenchymatous cells (JACOBS 1979, WETMORE and RIER 1963). In this case (and others where the responding cells are buried within a larger mass of nonresponding tissue), hormone analysis using the entire tissue mass might not give a good correlation of auxin with xylem differentiation. Clearly, there may be different cellular compartments within a tissue.

There is no doubt that subcellular compartmentation of hormones exists. Evidence for subcellular compartmentation of plant hormones is now being reported (e.g., DE LA FUENTE and LEOPOLD 1972, LOVEYS 1977, OHLROGGE et al. 1980). Water stress induces a large rise in extrachloroplastic ABA as opposed to that inside the chloroplasts (LOVEYS 1977, WALTON 1980). Undoubtedly, more evidence for compartmentation will emerge as hormone target sites are analyzed. Compartmentation of hormones should be viewed as a possible complication in the analysis of hormones extractable or diffusible from tissues.

1.3 Analysis of Hormonal Controls

Having explored some of the difficulties and limitations in the concept of hormone systems and the complications introduced by multiple hormonal controls, it seems particularly important to examine the experimental approaches to working out these systems.

1.3.1 Two Major Approaches from the Past

Most of our ideas on hormonal controls are based on two experimental approaches: (1) application of plant hormones and (2) measurement of changes in endogenous hormone levels. The first approach, often cynically termed "spray and pray," is in fact a logical and inexpensive first test. If a growth process is significantly modified by the applied substance, a possible regulatory role is implied for that hormone. The implication is only valid where the supply of a hormone limits a process and the exogenous hormone corrects that deficiency. These requirements were nicely met in the discovery of growth regulation by gibberellin as well as in subsequent unraveling of its regulatory functions (STOWE and YAMAKI 1959). A useful variation of this approach is to remove the suspected source of the hormone, thus making the hormone level more limiting, and then replace it with preparations containing that hormone. This procedure has provided the basic type of evidence for each of the known hormones as regulators of coleoptile elongation (see LEOPOLD 1960, AUDUS 1972), leaf abscission (see NOODÉN and LEOPOLD 1978, JACOBS 1979), and digestion of barley endosperm (see MAYER and SHAIN 1974, JACOBS 1979). Unfortunately, there is some uncertainty in this kind of experiment since applied hormones can sometimes cause responses which are not apparently regulated by that hormone in normal development.

The second approach, utilizing measurements of changes in endogenous levels of hormones, is still a popular one and has been expanded by major changes in the methods used to measure the hormones. Bioassays were employed in the earlier studies, but when it was realized that (a) bioassays are generally subject to interference from other compounds in the preparations being tested and (b) they are generally less than specific, bioassays fell into disfavor. Still, bioassays combined with improved purification procedures may well provide reliable and inexpensive methods for measuring hormones. Fortunately, the advent of chemical and physical methods provides more precise methods of measuring hormones. With the realization that putative hormone peaks in the eluant profile from a gas chromatograph (GC) might contain more than just the hormone, this method too has been suspect. The need to derivatize most hormones for use on the GC also discouraged its use. Spectacular developments in high pressure liquid chromatography (HPLC) technology opened opportunities for its application in hormone analysis. Eventually, however, because of the relative nonspecificity of most detectors used on HPLC and the relatively low resolving power of HPLC (compared for example with capillary gas chromatography), the same questions will arise concerning the identity (purity) of peaks in the eluant profiles from HPLC. Without a doubt, the most rigorous procedures now available employ mass spectroscopy, MS (e.g., GC coupled with MS). This procedure allows verification of the identity of the hormone as well as its quantitative measurement. Unfortunately, the expense of this procedure will limit its use by most investigators. Although some of the current methods for hormone analysis seem elaborate and expensive, a trend toward simplification (and decrease in cost) may be possible once the procedures and problems are better understood (HILLMAN 1978). Bioassays cannot be completely discarded, and they, in fact, seem to be the most practical way to search for new hormones. The technology of plant hormone purification and measurement is still undergoing development; however, simplification (coupled with a reduction in costs) would be a very great service to this field.

One of the most unnecessary failings in hormone analysis has been a paucity of effort to correlate endogenous hormone levels with quantitative measures of development to establish a relationship or lack of one. Thus, an important aspect in the analysis of hormonal control carries some interpretative difficulties, yet the two approaches outlined plus bits and pieces of other types of evidence described below have given enough data to understand at least parts of some hormonal control systems.

1.3.2 Analyses of Hormone Systems

Knowledge of the location of synthesis (source) of a hormone may be a crucial aspect of understanding a hormonal control system, especially since synthesis may be a controlling step, yet this area is one of the most deficient. A good starting point would be understanding the correlative control of the process in question. This approach has been exploited in an elegantly simple manner to develop a clear picture of auxin control of coleoptile growth (see AUDUS 1972). Hormone synthesis sites have been analyzed by excision and measurement of their ability to produce diffusible hormone, or by their ability to convert radioactively labeled precursors to the hormone. A fairly laborious and relatively little used approach would be to determine whether key biosynthetic enzymes are present or absent. This method is limited by some uncertainties over the biosynthetic enzymes, pathways and even key precursors for hormones (see e.g. WINTER 1966, THIMANN and GROWCHOWSKA 1968, DAVIES et al. 1975). Obviously, individual hormones are synthesized in more than one site during the life cycle of an organism and some, e.g., ethylene, may be produced by almost any cell (ABELES 1973). In apples, GA_{4-7} appears to be synthesized in the seeds, but GA_3 is synthesized elsewhere, possibly in the roots (HAYASHI et al. 1968). Thus, different sites of synthesis may exist for different gibberellins.

Transport offers major opportunities for regulation through control of either the direction(s) or the rate of hormone flux. Of course, auxin with its definitive polar transport influences tropisms in a manner which is the quintessence of the chemical messenger role. Other hormones are not so selectively transported; gibberellins, cytokinin and abscisic acid do occur in the phloem system (HOAD and BOWEN 1968, CARR and REID 1968, HOAD 1973, WALTON 1980), and also in the transpiration stream (KENDE 1965, CARR and REID 1968, DAVISON and YOUNG 1973, WALTON 1980). Thus, the same hormone may move simultaneously in opposite directions through different channels and may perform different simultaneous regulatory functions. Somewhat more removed from the definition of a hormone being "transferred to another part" is the case of ethylene; in this instance, the hormone may move from one part of the plant to another by diffusion processes and sometimes may even travel outside the plant (BURG and BURG 1965). Thus, the "transfer" of hormones from one part of the organism to another may occur by an active transport, by getting carried along with the flow in the vascular system, or even by passive diffusion.

The ability of transfer cells to move solutes in and out of the vascular system and their positioning at strategic locations (GUNNING et al. 1968, PATE and GUNNING 1972, GUNNING and PATE 1974) suggests that transfer cells may play a role in controlling the flux and distribution of plant hormones via the xylem and phloem.

Identification of the responding tissue would seem to be the simplest evidence about the site (target tissue) of the regulatory response to a hormone, but this is not infallible. It is becoming increasingly evident that hormones may indirectly regulate physiological processes even without being moved to the site at which regulation is ultimately achieved. The first example of such a remote regulation to be recognized was the mobilizing effect of cytokinin (MOTHES 1959). This hormone can serve as a mobilizing agent in the growing point of the stem without moving out of the apex; from this locus of action, it can exert regulatory effects on growth of lateral buds (SETH and WAREING 1964). Similarly, cytokinin localized in one leaf can stimulate the senescence of other leaves in the same plant (LEOPOLD and KAWASE 1964). Experimental evidence indicates that this remote type of regulatory effect by cytokinin takes place through directed translocation of mobilizable materials from remote parts to the parts of the plant in which the hormone is located (MÜLLER and LEOPOLD 1966). Another interesting case exists in wheat seeds where gibberellin promotes early leaf and coleoptile growth directly, while it stimulates root growth indirectly, possibly through release of nutrients from the endosperm (KEFFORD and RIJVEN 1966). In the case of ethylene stimulation of leaf abscission, one regulatory function may be the suppression of auxin transport through the petiole, thereby indirectly inducing maturation in the abscission layer (BEYER 1973). The question of what is a primary and what is a secondary effect is an old and difficult one. The problem might be easier if one could use hormone receptors to determine which tissue is the hormone target (VENIS 1977).

One complication in the use of receptors to define a target tissue is the probability that some nontarget tissues may contain receptors. It has already been mentioned that stereo-specific attachment sites have been found in numerous animals which do not apparently utilize the hormone in question in a regulatory role (ROTH et al. 1982). It is probable, therefore, that plants, too, may have receptor sites which are nonfunctional in terms of hormonal regulation. Some "receptors" are no doubt involved in transport of the hormones. Since plant hormones appear to be compartmentalized, it seems probable that hormone receptors are also localized in the specific compartments. Autoradiography will be a useful tool in analyzing the locations of the receptors and cell fractionations can play an important role (HERTEL et al. 1972). Ultimately, some combination of these methods with genetic alterations, such as HO et al. (1980, 1981) are using or subcellular tagging with fluorescent monoclonal antibodies as described by JACOBS and GILBERT (1983), may prove to be particularly powerful.

An equally challenging problem will be the identification of the function of the hormone/receptor complex.

1.4 Conclusions

1.4.1 The Complexity of Hormonal Controls

The various categories of hormone interaction in the regulation of plant processes as suggested here probably represent a rather primitive set of schemes, but they may serve at least to provide a basis for a reconsideration of regulatory systems.

Potentials for multiple hormonal regulation of growth and development may be indicated from the responses to applications of hormones to plants. It has been pointed out earlier that applications of each of the hormones have been found to alter nearly every phase of plant growth and development (LEOPOLD 1972).

A whole plant is a complex, multicellular organism, and no one should expect the controls which govern its formation or its function to be simple. This complexity is evident at the whole plant level in the correlative controls of the maturation phase (pod development and monocarpic senescence) of soybean plants (NOODÉN 1984). Even a single organ consists of different tissues and within each of these an orderly assortment of cell types develops under separate but integrated controls.

In view of the complexity of growth regulation in plants, and the seeming lack of exclusive control of any one step of development by any one of the known hormones, it is hardly surprising that doubts have been expressed as to whether the plant hormones are in fact regulators of development (TREWAVAS 1980, 1983a). The interesting suggestion has been made that instances of poor correlation between concentrations of plant hormones found and changes in development may be due to changes in the sensitivity to the hormones (KENDE and LANG 1964, TREWAVAS 1983b). There are well-known instances of changes in sensitivity to plant hormones (CLELAND 1983), but the existence of this sort of regulatory action still does not clarify on the overall uncertainties in the problem of regulation.

1.4.2 Nonhormonal Correlative Controls

Control mechanisms other than hormones are an interesting possibility. Although hormones may well be the most important mediators of transmitted influences and correlative controls, nutrient supplies, e.g., minerals from the roots or photosynthate from leaves, offer many possibilities for regulation. Nutrient supplies apparently play some role in whole plant senescence (see MOLISCH 1928, NOODÉN 1980), apical dominance (see GREGORY and VEALE 1957, PHILLIPS 1975), root–shoot interactions, flower induction (KRAUS and KRAYBILL 1918) and other processes. Control on a grand scale, such as branching patterns in trees (apical control), apparently differs from apical dominance (BROWN et al. 1967) and may involve regulation of nutrient distribution. For short-range influences such as those which control the initiation of organogenesis in the shoot apex, physical forces such as directional stresses may play a role (GREEN 1980, LINTILHAC and VESECKY 1980). Electrical gradients offer another possibility for correlative controls (JAFFE 1968).

1.4.3 Evolutionary Plasticity of Hormonal Controls

It is highly probable that the evolution of hormonal regulators in plants and animals has involved the utilization of chemicals which can be synthesized by many types of cells, and has involved the gradual specialization of these materials into regulatory roles. Higher plant and animal hormones have been isolated from a startling array of organisms including bacteria, algae, sponges and protozoans. The presence of these chemicals which we associate with regulatory functions in higher animals and higher plants does not automatically implicate them in regulatory functions. The important conclusion which we suggest is that the presence in plants of hormonal compounds which do not show quantitative correlation with growth or development processes may be simply interpreted as probable nonhormonal pools.

The complexity of hormonal regulation is compounded by evolutionary plasticity which may have led to substantial differences even between related plants. One need only look at the tremendous differences in fruit structures between species to realize that development (and its underlying hormonal controls) may differ greatly even within middle level taxonomic groups (see ESAU 1965, COOMBE 1976). Likewise, a wide range of structural differences and controls of dormancy exist in seeds (ESAU 1965, MAYER and SHAIN 1974). It is probable that developmental events such as fruit ripening, dormancy, and monocarpic senescence evolved independently in more than one taxonomic group. Quite likely, the control systems produced by these presumedly independent lines of evolution are also different. For example, monocarpic senescence must have evolved independently in several groups, and there appear to be substantial differences in the physiology of this process in different taxonomic groups (NOODÉN and LINDOO 1978, NOODÉN 1980). Likewise, foliar senescence in different species appears to involve very different antisenescence hormones, cytokinin, gibberellin or auxin (see NOODÉN and LEOPOLD 1978). As pointed out earlier, even a seemingly simple process like cell enlargement may be regulated by different hormones in different tissues (HUMPHRIES and WHEELER 1963, CLELAND 1964). One hormone can even be adapted for opposing roles in different tissues, e.g., auxin induces softening of the cell wall in coleoptiles (HEYN 1931, THIMANN 1977) but inhibits softening of the cell walls in leaf abscission zones (ADDICOTT and LYNCH 1955, KOZLOWSKI 1973). Development is at the forefront of an organism's phylogenetic and ontogenetic adaptation to the environment. One should not be surprised, then, to find that regulatory systems controling development may be very diverse in both their biochemistry and their hormonal mediators. It may be difficult to make generalizations which extend through all plants, let alone all seed plants. Recognizing and accepting this diversity in development and in hormonal controls will avert much confusion and frustration.

1.4.4 The Future

Our analysis of the control of plant development, including hormonal controls, is probably closer to its beginning than its completion, and a clearer picture of the generalizations as well as the diversities may be hoped for in the future. In fact, given our incomplete knowledge of the operation of the best-studied hormonal regulatory systems, the hormone concept may be more of a working hypothesis (though a very plausible one) than a final statement on how plants achieve internal coordination. Since many new tools are now available to analyze hormonal control systems, and improvements are being made all the time, this task should move with greater facility in the future.

Growth and development are of course central to crop productivity. Important as photosynthesis, nitrogen metabolism, etc. are, improvements in these processes may not increase plant productivity unless their products are utilized in growth and development. Understanding hormonal controls should offer opportunities to modify plant growth for enhanced agricultural production. The most conspicuous forces participating in the regulation of growth and development are the hormones, and progress in understanding the dynamics of interacting hormone systems should ultimately expand the potentials for crop productivity. Moreover, the study of plant hormones and their analogs has a clear record of contributing to agriculture (WEAVER 1972, NICKELL 1982).

Perhaps most intriguing are the evidences that developmental capabilities exist in plants which are not normally expressed: capabilities for the production of totally new structures can be brought out by pathogens or symbionts. Incursions by insects and other organisms can induce new and highly organized structures, such as galls (KÜSTER 1925, BRAUN 1969), presumably at least partly through regulatory chemicals. A better understanding of hormonal regulatory systems may reveal some real surprises in the range of possibilities for the manipulation of plant growth.

References

Abeles FB (1973) Ethylene in plant biology. Academic Press, London New York

Addicott FT, Lynch RS (1955) Physiology of abscission. Annu Rev Plant Physiol 6:211–238

Amrhein N (1977) The current status of cyclic AMP in higher plants. Annu Rev Plant Physiol 28:123–132

Audus LJ (1959) Plant growth substances. Interscience, New York

Audus LJ (1972) Plant growth substances Vol I. Chemistry and physiology. Barnes and Noble, New York

Beyer EM Jr (1973) Abscission. Support for a role of ethylene modification of auxin transport. Plant Physiol 52:1–5

Beyer EM, Morgan PW (1969) Time sequence of the effect of ethylene on transport, uptake and decarboxylation of auxin. Plant Cell Physiol 10:787–799

Biale J (1960) The postharvest biochemistry of tropical and subtropical fruits. Adv Food Res 10:293–354

Bonner J, Bonner H (1948) The B vitamins as plant hormones. In: Harris R, Thimann KV (eds) Vitamins and hormones Vol 6. Academic Press, London New York, pp 225–275

Bonner J, English J Jr (1938) A chemical and physiological study of traumatin, a plant wound hormone. Plant Physiol 13:331–348

Bradford KJ, Yang SF (1980) Xylem transport of l-aminocyclopropane-l-carboxylic acid, an ethylene precursor, in waterlogged tomato plants. Plant Physiol 65:322–326

Braun AC (1969) Abnormal growth in plants. In: Steward FC (ed) Plant physiology Vol V B. Academic Press, London New York, pp 379–420

Brown CL, McAlpine RG, Kormanik PP (1967) Apical dominance and form in woody plants: A reappraisal. Am J Bot 54:153–162

Brown HT, Morris GH (1890) Researches on the germination of some of the Gramineae. J Chem Soc 57:458–528

Burg SP, Burg EA (1965) Gas exchange in fruits. Physiol Plant 18:870–884

Burg SP, Burg EA (1968) Auxin-stimulated ethylene formation: its relationship to auxin-inhibited growth, root geotropism and plant processes. In: Wightman FW, Setterfield G (eds) Biochemistry and physiology of plant growth substances. Runge, Ottawa, pp 1275–1294

Carr DJ, Reid DM (1968) The physiological significance of the synthesis of hormones in roots and of their export to the shoot system. In: Wightman F, Setterfield G (eds) Biochemistry and physiology of plant growth substances. Runge, Ottawa, pp 1169–1185

Chen J, Meeuse BJD (1975) Purification and partial characterization of the two biologically active compounds from the inflorescence of *Sauromatum guttatum* Schott (Araceae). Plant Cell Physiol 16:1–11

Cleland CF (1978) The flowering enigma. BioSci 28:265–269

Cleland R (1964) Role of endogenous auxin in the elongation of *Avena* leaf sections. Physiol Plant 17:126–135

Cleland RE (1983) Changes in hormone concentration are important. Trends in Bioch Sci 1983:354–357

Coombe BG (1976) The development of fleshy fruits. Annu Rev Plant Physiol 27:507–528

Craker LE, Abeles FB (1969) Abscission: quantitative measurement with a recording abscissor. Plant Physiol 44:1139–1143

Darwin C, Darwin F (1880) The power of movement in plants. John Murray, London

Davey JE, van Staden J (1978) Cytokinin activity in *Lupinus albus*. II. Distribution in fruiting plants. Physiol Plant 43:82–86

Davies FS, Mau SC, Nooden LD (1975) Auxin synthesis in crown gall tumor tissue: A comparison of three putative precursors. Physiol Plant 33:39–41

Davison RM, Young H (1973) Abscisic acid content of xylem sap. Planta 109:95–98

Dostal R (1967) On integration in plants. Harvard Univ Press, Cambridge Mass

Esau K (1965) Plant anatomy, 2nd edn. Wiley and Sons, New York

Evans LS, Tramontano WA (1981) Is trigonelline a plant hormone? Am J Bot 68:1282–1289

Frankland B, Wareing PF (1962) Changes in endogenous gibberellins in relation to chilling of dormant seeds. Nature 194:313–314

Frenkel C (1972) Involvement of peroxidase and indoleacetic acid oxidase isozymes from pear, tomato and blueberry fruit in ripening. Plant Physiol 49:757–763

Fuchs Y, Lieberman M (1968) Effects of kinetin, IAA and gibberellin on ethylene production and their interactions in growth of seedlings. Plant Physiol 43:2029–2036

Fuente RK de la, Leopold AC (1972) Two components of auxin transport. Plant Physiol 50:491–495

Goebel K (1900) Organography of plants especially of the Archegoniatae and Spermatophyta. Part 1. General organography. Clarendon, Oxford

Goldsmith MHM (1977) The polar transport of auxin. Annu Rev Plant Physiol 28:439–478

Goto N, Esashi Y (1974) Regulation of hypocotyl growth by ethylene. In: Plant Growth Substances 1973. Hirokawa, Tokyo, pp 853–863

Green PR (1980) Organogenesis – a biophysical view. Annu Rev Plant Physiol 31:51–82

Gregory FG, Veale JA (1957) A reassessment of the problem of apical dominance. Soc Exp Biol 11:1–20

Grove MD, Spencer GF, Rohwedder WK, Mandava N, Worley JF, Warthen JD Jr Steffens GL, Flippen-Anderson JL, Cook JC Jr (1979) Brassinolide, a plant growth-promoting steroid isolated from *Brassica napus* pollen. Nature 281:216–217

Gunning BES, Pate JS (1974) Transfer cells. In: Robards Aw (ed) Dynamic aspects of plant ultrastructure. McGraw-Hill, London, pp 441–480

Gunning BES, Pate JS, Briarty LG (1968) Specialized "transfer cells" in minor veins of leaves and their possible significance in phloem translocation. J Cell Biol 37:147–171
Hall WC (1952) Evidence on the auxin-ethylene balance hypothesis of foliar abscission. Bot Gaz 113:310–322
Hayashi F, Naito R, Bukovac MJ, Sell HM (1968) Occurrence of gibberellin A_3 in parthenocarpic apple fruit. Plant Physiol 43:448–450
Hertel R, Thomson KS, Russo VEA (1972) In vitro auxin binding to particulate cell fractions from corn coleoptiles. Planta 107:325–340
Heslop-Harrison J (1963) Plant growth substances. Vistas in Botany III:104–194
Heyn ANJ (1931) Der Mechanismus der Zellstreckung. Recl Trav Bot Neerl 28:113–244
Hillman JR (ed) (1978) Isolation of plant growth substances. Cambridge Univ Press, London
Ho THD (1979) Hormonal control of enzyme formation in barley aleurone layers. In: Rubenstein I (ed) Plant molecular biology. Academic Press, London New York, pp 217–240
Ho THD, Shih S, Kleinhofs A (1980) Screening for barley mutants with altered hormone sensitivity in their aleurone layers. Plant Physiol 66:153–157
Ho THD, Nolan RC, Shute DE (1981) Characterization of a gibberellin-insensitive dwarf wheat, D 6899: Evidence for a regulatory step common to many diverse responses to gibberellins. Plant Physiol 67:1026–1031
Hoad GV (1973) Effect of moisture stress on abscisic acid levels in *Ricinus communis* L. with particular reference to phloem exudate. Planta 113:367–372
Hoad GV, Bowen MR (1968) Evidence for gibberellin-like substances in phloem exudate of higher plants. Planta 82:22–32
Humphries EC, Wheeler AW (1963) The physiology of leaf growth. Annu Rev Plant Physiol 14:385–410
Ilan I, Goren R (1979) Cytokinins and senescence in lemon leaves. Physiol Plant 45:93–95
Jacobs M, Gilbert SF (1983) Basal localization of the presumptive auxin transport carrier in pea stem cells. Science 220:1297–1300
Jacobs WP (1979) Plant hormones and plant development. Cambridge Univ Press, Cambridge
Jacobs WP, Case DB (1965) Auxin transport, gibberellin and apical dominance. Science 148:1729–1731
Jacobs WP, Danielson J, Hurst V, Adams P (1959) What substance normally controls a given biological process? II. The relation of auxin to apical dominance. Dev Biol 1:534–554
Jaffe LF (1968) Localization in the developing *Fucus* egg and the general role of localizing currents. Adv Morphog 1:295–328
Jones RL, Lang A (1968) Extractable and diffusible gibberellins from light- and dark-grown pea seedlings. Plant Physiol 43:629–634
Kefford NP, Rijven MGC (1966) Gibberellin and growth in isolated wheat embryos. Science 151:104–105
Kende H (1965) Kinetin-like factors in the root exudate of sunflowers. Proc Natl Acad Sci USA 53:1302–1307
Kende H, Lang A (1964) Gibberellins and light inhibition of stem growth in peas. Plant Physiol 39:435–440
Kozlowski TT (ed) (1973) Shedding of plant parts. Academic Press, London New York
Kraus EJ, Kraybill HR (1918) Vegetation and reproduction with special reference to the tomato. Oregon State Coll Agric Exp Stn Bull
Küster E (1925) Pathologische Pflanzenanatomie, 3rd edn. Fischer, Jena
Lang A (1966) Intercellular regulation in plants. In: Locke M (ed) Major problems in developmental biology. Academic Press, London New York, pp 251–287
Lang A, Chailakhyan MKh, Frolova IA (1977) Promotion and inhibition of flower formation in a day neutral plant in grafts with a short-day and a long-day plant. Proc Natl Acad Sci USA 74:2412–2416
Lenton JR, Perry VM, Saunders PF (1972) Endogenous abscisic acid in relation to photoperiodically induced bud dormancy. Planta 106:13–22

Leopold AC (1960) Auxins and plant growth. Univ California Press, Berkeley
Leopold AC (1972) Ethylene as a plant hormone. In: Kaldewey H, Vardar Y (eds) Hormonal regulation in plant growth and development. Verlag Chemie, Weinheim, pp 245–262
Leopold AC (1982) Hormonal regulatory systems in plants. In: Sen, SP (ed) Recent developments in plant science. Today and Tomorrow Publ, New Delhi, pp 33–41
Leopold AC, Kawase M (1964) Benzyladenine effects on bean leaf growth and senescence. Am J Bot 51:294–298
Letham DS (1978) Naturally occurring plant growth regulators other than the principle hormones of higher plants. In: Letham DS, Goodwin PB, Higgins TJV (eds) Phytohormones and related compounds: A comprehensive treatise Vol I. Elsevier/North-Holland Biomedical Press, Amsterdam, pp 349–417
Lieberman M, Kunishi AT (1972) Thoughts on the role of ethylene in plant growth and development. In: Carr DJ (ed) Plant growth substances 1970. Springer, Berlin Heidelberg New York, pp 549–560
Linthilhac PM, Vesecky TB (1980) Mechanical stress and cell wall orientation. I. Photoelastic derivation of principal stresses. With a discussion of axillarity and the significance of the "arcuate shell zone". Am J Bot 67:1477–1483
Loeb J (1918) Chemical basis of correlation. Bot Gaz 65:150–174
Lorenzi R, Horgan R, Wareing PF (1975) Cytokinins in *Picea sitchensis*. Biochem Physiol Pflanz 168:333–339
Loveys BR (1977) The intracellular location of abscisic acid in stressed and nonstressed leaf tissue. Physiol Plant 40:6–10
Loveys BR, Leopold AC, Kriedemann PE (1974) Abscisic acid metabolism and stomatal physiology in *Betula lutea* following alteration in photoperiod. Ann Bot 38:85–92
Mandava NB (1980) Natural products in plant growth regulation. In: Mandova NB, Page SW, Wheeler J (eds) ACS Symp Ser. Dekker, New York
Mandava NB, Thompson MJ, Meudt WJ (1981) Brassinosteroids — New plant growth substances. In: Mandava NB, Page SW, Wheeler J (eds) ACS Symp Ser. Dekker, New York
Matson DT, Jarvis WR (1970) Post-harvest ripening of strawberries. Hortic Res 10:125–132
Mayer AM, Shain Y (1974) Control of seed germination. Annu Rev Plant Physiol 25:167–193
Molisch H (1928) Die Lebensdauer der Pflanze. Fischer, Jena (Transl 1938, Fulling EH, Science Press, Lancaster, PA)
Moore TC (1979) Biochemistry and physiology of plant hormones. Springer, New York Berlin Heidelberg
Morgan PW, Gausman HW (1966) Effects of ethylene on auxin transport. Plant Physiol 41:45–52
Mothes K (1959) Bemerkungen über isolierte Blätter. Colloq Ges Physiol Chem 10:72–81
Müller K, Leopold AC (1966) The mechanism of kinetin-induced transport in corn leaves. Planta 68:186–205
Naylor JM, Simpson GM (1961) Dormancy studies in seed of *Avena fatua*. 2. A gibberellin sensitive inhibitory mechanism in the embryo. Can J Bot 39:281–295
Nickell LG (1982) Plant growth regulators. Agricultural uses. Springer, Berlin Heidelberg New York
Noodén LD (1980) Senescence in the whole plant. In: Thimann KV (ed) Senescence in plants. CRC Press, Boca Raton, pp 219–258
Noodén LD (1984) Integration of soybean pod development and monocarpic senescence. A minireview. Physiol Plant (in press)
Noodén LD, Leopold AC (1978) Phytohormones and the endogenous regulation of senescence and abscission. In: Letham DS, Goodwin PB, Higgins TJV (eds) Phytohormones and related compounds — a comprehensive treatise Vol II. Elsevier, Amsterdam, pp 329–369
Noodén LD, Lindoo SJ (1978) Monocarpic senescence. What's New Plant Physiol 9:25–28
Noodén LD, Weber JA (1978) Environmental and hormonal control of dormancy in

terminal buds of plants. In: Clutter ME (ed) Dormancy and developmental arrest. Academic Press, London New York, pp 221–268

Nowacki J, Bandurski RS (1980) Myo-insositol esters of indole-3-acetic acid as seed auxin precursors of *Zea mays*. L. Plant Physiol 65:422–427

Ohlrogge JB, Garcia-Martinez JL, Adams D, Rappaport L (1980) Uptake and subcellular compartmentation of gibberellin A_1 applied to leaves of barley and cowpea. Plant Physiol 66:422–427

Pate JS, Gunning BES (1972) Transfer cells. Annu Rev Plant Physiol 23:173–196

Phillips IDJ (1975) Apical dominance. Annu Rev Plant Physiol 26:341–367

Phillips IDJ, Vlitos AJ, Cutler H (1959) The influence of gibberellic acid upon the endogenous growth substances of the Alaska pea. Contrib Boyce Thompson Inst 20:111–120

Phillips IDJ, Miners J, Roddick JG (1980) Effects of light and photoperiodic conditions on abscisic acid in leaves and roots of *Acer pseudoplatanus* L. Planta 149:118–122

Pilet PE (1957) Action des gibberellins sur l'activite auxines-oxydasique de tissues cultives in vitro. C R Acad Sci Paris 245:1327–1328

Powell LE (1975) Some abscisic acid relationships in apple. Riv Ortoflorofrutt It 59:424–432

Raschke K, Zeevaart JAF (1976) Abscisic acid content, transpiration, and stomatal conductance as related to leaf age in plants of *Xanthium strumarium* L. Plant Physiol 58:169–174

Ries SK, Wert V, Sweeley CC, Leavitt RA (1977) Triacontanol: A new naturally occurring plant growth regulator. Science 195:1339–1341

Roth J, Le Roith D, Shiloach J, Rosenzweig JL, Lesniak MA, Havsankova J (1982) The evolutionary origin of hormones. New Engl J Med 306:523–527

Schildknecht H, Iyengar DSP (1975) On the biological chemistry of Mimosaceae I. D-pinitol from *Mimosa pudica*. Naturwissenschaften 62:533

Schildknecht H, Tausher B, Moeschler H, Edelmann J (1978) Detection and structure elucidation of leaf movement factors from Mimosaceae. Proc 11th Int Symp Chem Nat Prods Part I Vol 4. Bulgarian Acad Sci, Sofia, pp 97–111

Scott PC, Leopold AC (1967) Opposing effects of gibberellin and ethylene. Plant Physiol 42:1021–1022

Scott TK, Briggs WR (1960) Auxin relationships in the Alaska pea. Am J Bot 47:492–498

Seth A, Wareing PF (1964) Interaction between auxins, gibberellins and kinins in hormone-directed transport. Life Sci 3:1483–1486

Seth AK, Davies CR, Wareing PF (1966) Auxin effects on the mobility of kinetin in the plant. Science 151:587–588

Shibata K, Kubota T, Kamisaka S (1974) Isolation and chemical identification of a lettuce cotyledon factor, a synergist of the gibberellin action in inducing hypocotyl elongation. Plant Cell Physiol 15:191–194

Skoog F, Miller CO (1957) Chemical regulation of growth and organ formation in plant tissues cultured in vitro. Symp Soc Exp Biol 11:118–131

Skoog F, Tsui C (1948) Chemical control of growth and bud formation in tobacco stem segments and callus cultured in vitro. Am J Bot 35:782–787

Snow R (1925) The correlative inhibition of the growth of axillary buds. Ann Bot 39:841–859

Steeves TA, Sussex IM (1972) Patterns in plant development. Prentice Hall, Englewood Cliffs

Stewart FC, Shantz EM (1959) The chemical regulation of growth. Annu Rev Plant Physiol 10:379–404

Stowe B, Yamaki T (1959) Gibberellins: Stimulants of plant growth. Science 129:807–816

Thimann KV (1960) Plant growth. In: Nowinski WW (ed) Fundamental aspects of normal and malignant growth. Elsevier, Amsterdam, pp 748–822

Thimann KV (1977) Hormone action in the life of whole plants. Univ Massachusetts Press, Amherst

Thimann KV, Grochowska M (1968) The role of tryptophan and tryptamine as IAA precursors. In: Wightman F, Setterfield G (eds) Biochemistry and physiology of plant growth substances. Runge Press, Ottawa, pp 231–242

Trewavas AJ (1980) Plant growth substances: what is the molecular basis of their action? What's New In Plant Physiol 10:33–36

Trewavas AJ (1983a) Plant growth substances – metabolic flywheels for plant development. Cell Biol Internat Rep 7:569–575

Trewavas AJ (1983b) Is plant development regulated by changes in the concentration of growth substances? Trends in Bioch Sci 1983:354–356

Ueda J, Kato J (1980) Isolation and identification of a senescence-promoting substance from wormwood (*Artemisia absinthium* L.). Plant Physiol 66:246–249

Vanderhoef LN, Briggs WR (1978) Red light-inhibited mesocotyl elongation in maize seedlings I. The auxin hypothesis. Plant Physiol 61:534–537

Vardjan M, Nitsch JP (1961) La régeneration chez *Cichorium endivia* L.: étude des auxines et "kinines" endogènes. Bull Soc Bot Fr 108:363–374

Venis MA (1977) Receptors for plant hormones. Adv Bot Res 5:53–88

Wada S (1961) Growth patterns of rice coleoptiles grown on water and under water. Sci Rep Tohoku Univ Foruth Ser Biol 27:199–207

Wada S, Nagao M (1960) Effect of guaiacol on the auxin-induced growth of rice coleoptile sections. Sci Rep Tohoku Univ Fourth Ser Biol 26:181–188

Walton DC (1980) Biochemistry and physiology of abscisic acid. Annu Rev Plant Physiol 31:453–489

Wareing PF (1965) Endogenous inhibitors in seed germination and dormancy. In: Ruhland W (ed) Encyclopedia of Plant Physiology Vol XV (2). Springer, Berlin Göttingen Heidelberg, pp 909–924

Weaver RJ (1972) Plant growth substances in agriculture. Freeman, San Francisco

Went FW (1942) Growth, auxin and tropism in decapitated *Avena* coleoptiles. Plant Physiol 17:236–249

Went F, Thimann KV (1937) Phytohormones. MacMillan, New York

Wetmore RH, Rier JP (1963) Experimental induction of vascular tissues in callus of angiosperms. Am J Bot 50:418–430

Wickson M, Thimann KV (1958) The antagonism of auxin and kinetin in apical dominance. Physiol Plant 11:62–74

Wiesner J (1871) Untersuchungen über die herbstliche Entlaubung der Holzgewächse. Sitzungsber Akad Wiss Wien 64:465–510

Winter A (1966) A hypothetical route for the biogenesis of IAA. Planta 71:229–239

Wright STC (1961) A sequential growth response to gibberellic acid, kinetin, and indolylacetic acid in the wheat coleoptile. Nature 190:699–700

Zimmerman DC, Coudron CA (1979) Identification of traumatin, a wound hormone, as 12-oxo-trans-10-dodecanoic acid. Plant Physiol 63:536–541

2 Functions of Hormones at the Cellular Level of Organization

M.L. EVANS

2.1 Introduction

With the mounting evidence for subcellular hormone receptor sites and evidence for tissue and sometimes cellular specificity of certain plant hormone effects, it seems clear that the resolution of fundamental problems associated with plant growth and development will require an understanding of the processes of expansion, division, and differentiation at the cellular level. Nevertheless, it is often necessary and sometimes easier to study these processes as they occur in complex tissues. In this chapter the emphasis will be on the hormonal regulation of cell development, while recognizing that the processes involved are strongly influenced by both the physical constraints and metabolic activities of neighboring cells in complex tissues.

2.2 Promotive and Inhibitory Classes of Plant Hormones

Although the effects of hormones on cell expansion or division often depend on the type or state of maturation of the organ being studied, we tend to categorize plant hormones as either stimulatory or inhibitory in their broad spectrum of growth effects. For example, moderate levels of auxin are inhibitory to cell expansion in roots and, in rare instances, shoots (KAUFMAN and JONES 1974). However, the typical effect of auxin is to promote cell elongation, especially in segments of shoot or coleoptile tissue. At low concentrations the auxins also promote cell elongation in roots (SCOTT 1972) so that, in general, they are viewed as stimulators of cell expansion at physiological concentrations.

Gibberellins also promote cell elongation (KAUFMAN and JONES 1974). GA promotes cell elongation in either excised or intact internode tissue, while endogenous levels of GA correlate well with growth rates in certain dwarf (PHINNEY 1956) and normal plants (DURLEY et al. 1976). In roots GA is usually either without effect or somewhat inhibitory to elongation, though there are a few reports of stimulation of root growth by GA (SCOTT 1972, LOW 1975).

In contrast to the auxins and gibberellins, abscisic acid (ABA) and ethylene usually retard cell elongation. With some exceptions (TAKAHASHI 1972, GAITHER et al. 1975, MALIK and MEHAN 1975a, VAN STADEN and BORNMAN 1970) ABA is inhibitory to growth (ADDICOTT and LYON 1969, MILBORROW 1974). ABA inhibits stem and root elongation and counteracts the promotive effects of other substances on these organs (REHM and CLINE 1973, KAUFMAN and JONES 1974,

PILET 1975a, b). Ethylene inhibits cell elongation in roots (CHADWICK and BURG 1967, ANDREAE et al. 1968) and stems (SARGENT et al. 1973, BURG 1973). Although ethylene stimulates elongation in some cell types (SUGE 1971b, PALMER 1972, KU et al. 1970, MUSGRAVE et al. 1972, TAKAHASHI 1973, GOTO and ESASHI 1974b) it usually inhibits elongation with or without concomitant swelling (SARGENT et al. 1973).

It is more difficult to categorize cytokinins as either generally inhibitory or generally promotive. Cytokinins stimulate cell expansion in cotyledons and in leaf discs (LETHAM 1969). They are reported to stimulate cell elongation in coleoptile sections (HEMBERG and LARSSON 1972), but the effect is slight. In contrast to these promotive effects, cytokinin inhibits auxin-induced cell elongation in stem (VANDERHOEF et al. 1973) and root sections (GASPAR and XHAUFFLAIRE 1967).

In general, then, we may view the auxins and GA's as promoters of cell elongation and ABA and ethylene as inhibitors. The cytokinins act as promoters of leaf cell expansion and inhibitors of stem elongation. This is somewhat oversimplified in light of the aforementioned exceptions, and the evidence that effects of applied hormones are often modified by or mediated by other hormones.

2.3 Modes of Hormone Interaction

2.3.1 Interaction and Independence in Hormone Action

Two hormones which affect the same process may act independently or may interact in some fashion. If they act independently, then under appropriate experimental conditions simultaneous application of the two hormones will elicit a response comparable to the sum of the responses induced by each hormone given separately (MILBORROW 1970). If two hormones are interacting, i.e., not acting independently, the response to their simultaneous application should be greater than or less than the sum of the responses induced by each hormone alone (MILBORROW 1970). Accordingly, the expression hormone interaction should be used only for hormones acting interdependently. However, the term interaction is commonly used in reference to any type of interplay of hormones or growth factors (KAUFMAN and JONES 1974, KONDO et al. 1975, EVANS 1976b, BALASIMHA et al. 1977). In order to avoid ambiguity, MILBORROW (1970) has suggested the terms potentiation for nonadditive interplay of hormones and independence for additive interplay. This avoids the ambiguity of the term interaction but introduces the difficulty of relating the term potentiation to cases in which the two factors have opposite effects.

It is proposed here that the term interaction be used in the nonrigorous sense of any type of interplay of two factors in influencing a common parameter. The modes of interaction may then be classified as either *independent interaction* (additive effect) or *interdependent interaction* (nonadditive effects). This introduces no bias as to the sign of the effect and legitimizes the nonrigorous use of the term interaction.

2.3.2 Synergism and Antagonism in Hormone Action

2.3.2.1 Synergism

The term synergism (Gk. synergos = working together) refers to the special case of interdependent interaction in which the effect of two promotive factors is more than the additive effect of each given separately. Examples of apparent synergistic promotive interaction between hormones include (1) the interaction of ethylene and ABA in the promotion or rice mesocotyl elongation (TAKAHASHI 1973), (2) kinetin enhancement of auxin-induced ethylene production in mung bean hypocotyls (LAU et al. 1977a) and (3) GA_3 enhancement of auxin-promoted elongation in cucumber hypocotyl (KATSUMI and KAZAMA 1974) and pea stem segments (OKERSE and GALSTON 1967). These are examples in which two factors cooperate to enhance a particular response and hence the term synergistic promotive interaction is appropriate. Although LOCKHART (1965) has recommended restricting the use of the term synergism to instances of positive interdependent interaction, the term seems to apply to negative interdependent interaction as well. Instances in which a more than additive inhibitory effect is achieved with two inhibitors are common to both plant physiology and classical enzymology (ROSENBERRY and BERNHARD 1972, SALMINEN et al. 1972, WHITE et al. 1974). However, synergism as applied to positive interdependent interaction is not so appropriate for inhibitory effects. If the effect of hormone A, for example, is to reduce growth by 60% and the effect of hormone B is to reduce growth by 70%, it cannot be expected that the combination A + B will reduce growth by 130% (or more than 130% if acting synergistically, according to the classical definition)! Even if A reduces growth by 30% and B reduces growth by 35%, it is not necessary to obtain a combined inhibitory effect greater than 65% to prove synergism. In the presence of B there is less growth to be inhibited by A and vice versa. These considerations have led to the following operational definition of *synergistic inhibitory interaction* of two factors (COLBY 1967)

$$Q < E = (X_1 \cdot Y_1)/100 \qquad (1)$$

where Q is the magnitude of the response expressed (e.g., growth as a percent of control growth) in the presence of inhibitors A and B, E is the magnitude of the response (as percent of control) expected in the presence of A and B assuming there is no synergism, X_1 is the percent of control growth in plants treated with A alone and Y_1 is the percent of control growth in plants treated with B alone.

On occasion, the term synergism has been applied to inhibitor interaction based simply on the observation that inhibition by combined application of A and B exceeds that obtained by either A or B alone. The term synergism is also occasionally used in describing growth promoter interactions in which promotion by A plus B is greater than that obtained by A or B alone. According to accepted definitions, these would not qualify as synergistic interactions.

2.3.2.2 Antagonism

When A suppresses a response promoted by B, A and B are said to have opposite effects. Hormone A inhibits the parameter promoted by B. This does

Table 2.1. Summary of terms applied to modes of hormone interaction

Term	Definition
Interaction	Hormones acting in any fashion to influence the same parameter
Independent interaction	Hormones influencing the same parameter in an additive fashion (action at different sites)
Interdependent interaction	Hormones influencing the same parameter in a non-additive fashion (acting at same site)
Synergistic promotive interaction	Hormones interacting to promote a process such that their combined effect is more than the sum of the promotive effects of each given separately
Synergistic inhibitory interaction	Hormones interacting to inhibit a physiological process such that the combined inhibitory effect is greater than that predicted by the inhibitory effect of each hormone given separately
Opposite interaction	Interaction of two hormones such that one promotes while the other inhibits the same physiological process
Antagonistic interaction	The special case of opposite interaction in which the promotive action of hormone A can be overcome by raising the concentration of the inhibitory hormone (B) and vice versa

not introduce bias concerning the relationship between the modes of action of A and B. The term antagonism may be applied to the special case of opposite interaction of A and B where the action of A can be overcome by increasing the concentration of B and vice versa. The term "negative synergism" has sometimes been used synonymously with antagonism (RUMMENS 1975). However, in view of the original meaning of synergism as "working together", one should avoid the use of the term synergism in describing opposing actions of hormones, especially since "negative synergism" as used in this context may be confused with synergistic inhibitory interaction.

Examples of antagonistic interaction include the effects of ABA and auxin on root tip surface charge (TANADA 1972) the effects of ABA and kinetin on coleoptile elongation (KHAN and DOWNING 1968), the effects of GA and tannins on enzyme biosynthesis (JACOBSON and CORCORAN 1977) and the effects of ethylene and either GA_3 or cytokinin on fruit senescence (GOLDSCHMIDT et al. 1977).

ABA is antagonistic to the chlorophyll-sparing effect of kinetin in detached leaves, i.e., kinetin retards chlorophyll degradation and, at high concentrations, overcomes the promotive effect of ABA (BACK and RICHMOND 1971). The effect of GA on chlorophyll retention is also opposite to that of ABA. In this case, however, the hormones appear not to act antagonistically, i.e., the ABA effects are not overcome be elevated levels of GA (BACK and RICHMOND 1971). Other examples of nonantagonistic opposite effects of hormones include the interaction of ABA and IAA in coleoptile segment elongation (PHILIPSON et al. 1973, ANKER 1975), the interaction of IAA and cytokinin in the regulation of soybean hypocotyl segment elongation (VANDERHOEF et al. 1973), and the inhibition of IAA-induced ethylene production by ABA (KONDO et al. 1975). A summary of definitions of the terms used in describing hormone interactions is presented in Table 2.1.

2.3.3 Hormone Effects on Endogenous Hormone Levels

Much of our information on hormonal regulation has been from experiments in which hormone is applied to tissue segments. While these studies have been informative, they have made us aware of precautions necessary in applying results to functions of hormones in the intact plant. It has become clear that the effects of a particular hormone are dependent on other hormones, and that the applied hormone may trigger changes in the endogenous levels of other hormones.

2.3.3.1 Hormonal Modification of Endogenous Auxin

ABA inhibits the enhancement of elongation by auxin, and there is evidence that ABA can inhibit auxin biosynthesis. ABA appears to reduce IAA levels during the regeneration of a physiological tip in decapitated coleoptiles (ANKER 1975), and there is evidence for ABA-enhanced auxin degradation in the presence of supraoptimal concentrations of IAA (MILBORROW 1966).

Cytokinins may also alter the level of auxin in some tissues. HEMBERG (1972b) reported that kinetin increases the auxin content of *Coleus blumei*, though the major increase was in bound auxin. Kinetin increases the free auxin content of mung bean hypocotyl segments exposed to IAA both by enhancing IAA uptake and suppressing the conversion of IAA to indoleacetylaspartic acid (LAU and YANG 1973). Kinetin enhancement of auxin levels may also explain the ability of kinetin to enhance growth promotion by low levels of auxin (HEMBERG and LARSSON 1972, ANKER 1977). Kinetin may also induce localized increases in auxin by enhancement of auxin movement toward the treated area (LAGERSTEDT and LANGSTON 1967), while cytokinins appear to enhance auxin production in coleoptile tips (JORDAN and SKOOG 1971) and in tobacco callus (DRAVNIEKS et al. 1969). It is suspected that cytokinins alter rates of auxin degradation in some tissues through a concentration-dependent promotion (GASPAR and XHAUFFLAIRE 1967, LEE 1971) or inhibition (LEE 1971) of IAA-oxidase activity.

There is extensive evidence that GA increases auxin levels either by enhancement of auxin biosynthesis (JINDAL and HEMBERG 1976, SASTRY and MUIR 1965) or by retardation of auxin destruction (KÖGL and ELEMA 1960). Nevertheless, there is convincing evidence that GA can also stimulate elongation independently of auxin (KAZAMA and KATSUMI 1974, CLELAND et al. 1968, KAUFMAN et al. 1969). In excised *Avena* stem internodes, for example, GA stimulates cell elongation, while auxin is inhibitory (KAUFMAN et al. 1969). Similarly, stimulation of lettuce hypocotyl elongation has been used as a bioassay for GA, while auxin is ineffective (FRANKLAND and WAREING 1960). In cucumber hypocotyl segments both auxin and GA stimulate cell elongation, but stimulation by auxin is accompanied by increased wall extensibility, while GA-enhanced growth occurs with little or no effect on wall extensibility (CLELAND et al. 1968).

Ethylene may modify the metabolism, uptake, and distribution of auxin (MORGAN and GAUSMAN 1966, MORGAN et al. 1968, BURG and BURG 1966). Ethylene reduces the uptake of auxin in stem sections of cowpea (MORGAN and GAUSMAN 1966) and cotton (MORGAN et al. 1968) and it enhances auxin destruction or conjugation in some types of stem tissue (BEYER and MORGAN 1970, ERNEST and VALDOVINOS 1971, MORGAN et al. 1968). These effects have not been found in all cases (GOREN and BUKOVAC 1973) while in some cases (BURG and BURG 1967) prolonged pretreatment with ethylene is required before alteration of auxin uptake, transport or metabolism occurs. Perhaps the most consistent and dramatic effect of ethylene on auxin physiology is its ability to reduce the capacity of the polar auxin transport system (MORGAN and GAUSMAN 1966, BURG and BURG 1966, 1967) and to reduce lateral auxin movement (BURG and BURG 1966).

One of the most potent factors in the regulation of IAA biosynthesis may be IAA itself. IAA appears to suppress its own biosynthesis in decapitated coleoptiles (ANKER 1973), and certain auxin-induced modifications of growth patterns have been attributed in part to effects on endogenous auxin production (EVANS and SCHMITT 1975, EVANS et al. 1977). Red light-induced suppression of auxin production in coleoptile tips may result from a suppression of auxin movement out of the tip leading to feedback inhibition of auxin biosynthesis (HUISINGA 1976).

2.3.3.2 Hormonal Modification of Endogenous Gibberellin

Cytokinins can increase endogenous GA content or prevent reductions in GA content normally associated with certain stages of development (LOVEYS and WAREING 1971, REID and RAILTON 1974, CHIN and BEEVERS 1970). Enhanced GA production seems to be involved in certain responses to cytokinin. Cytokinin stimulation of leaf unrolling, for example, is accompanied by enhanced GA production. Since GA itself strongly stimulates leaf unrolling and since the effect of cytokinin on leaf unrolling is largely canceled by inhibitors of GA biosynthesis, GA may mediate the activity of cytokinin in this instance (LOVEYS and WAREING 1971).

ABA enhances both the uptake and metabolism of GA_1 in barley half-seeds (NADEAU et al. 1972) with the enhancement of metabolism apparently resulting from enhanced uptake (STOLP et al. 1973). In leaves of potato (*Solanum andigena*) ABA causes as much as a 100-fold increase in GA activity (RAILTON and WAREING 1973a), while in *Solanum tuberosum*, sprouts from ethylene-treated tubers contain high levels of GA (DIMALLA and VAN STADEN 1977). In maize shoots ABA causes a reduction in GA (WAREING et al. 1968) while β-inhibitor preparations exert a similar effect in apices of birch (THOMAS et al. 1965). Auxin causes an increase in the GA content of *Chenopodium rubrum* (TELTSCHEROVA 1970), an effect which may be related to flowering.

2.3.3.3 Hormonal Modification of Endogenous Ethylene

One of the most thoroughly studied examples of modification of hormone production by applied hormone is the enhancement of ethylene production by auxin. Auxin stimulates ethylene formation in stems (BURG and BURG 1966, ABELES and RUBINSTEIN 1964), roots (CHADWICK and BURG 1967, ANDREAE et al. 1968, MCBRIDE and EVANS 1977), leaves (LOVEYS and WAREING 1971, ABELES and RUBINSTEIN 1964), flowers (BURG and DIJKMAN 1967) and fruits (ABELES and RUBINSTEIN 1964) and there is evidence (LAU et al. 1977b) that this is a direct effect rather than an effect caused by oxidation products of auxin. ADAMS and YANG (1979) have determined that ethylene biosynthesis occurs via the pathway: methionine → S-adenosylmethionine (SAM) → aminocyclopropylcarboxylic acid (ACC) → ethylene. Additional evidence obtained by YANG and coworkers (YU et al. 1979, YANG 1980) indicates that auxin enhancement of ethylene production occurs via an auxin-induced increase in the level of ACC synthase, the enzyme catalyzing the conversion of SAM to ACC. This is consistent with earlier observations that auxin-induced ethylene production occurs only after a latent period of one to several hours in some cells (BURG and BURG 1966, ABELES 1966, SAKAI and IMASEKI 1971) and that auxin-induced ethylene production can be largely prevented by inhibitors of RNA and protein synthesis (SAKAI and IMASEKI 1971, KANG et al. 1971).

Cytokinins stimulate ethylene production in wheat leaves (LOVEYS and WAREING 1971), pea stems (FUCHS and LIEBERMANN 1968) and mungbean hypocotyls (LAU and YANG 1976, IMASEKI et al. 1975). Since cytokinins increase the levels of free IAA, it is possible that the cytokinin effect on ethylene production is mediated by IAA.

As noted earlier, ABA can interfere with auxin-induced ethylene biosynthesis in mungbean. ABA also inhibits ethylene production in rose petals (MAYAK and HALEVY 1972) and pea seedlings (GERTMAN and FUCHS 1972, LIEBERMANN and KUNISHI 1971). In contrast to these observations, low concentrations of ABA cause a small stimulation of ethylene production in mungbean (KONDO et al. 1975), while high concentrations stimulate ethylene production in orange peel plugs (GERTMAN and FUCHS 1972) and in carnations (MAYAK and DILLEY 1976).

GA stimulates ethylene production in first internodes of *Phaseolus vulgaris* (ABELES and RUBINSTEIN 1964), but is ineffective in enhancing ethylene production in pea seedlings, at least during the first 6 days of growth (FUCHS and LIEBERMANN 1968).

Since ethylene production is known to be autocatalytic (ABELES 1973), ethylene itself should be added to the list of hormonal factors capable of altering ethylene production.

2.3.3.4 Hormonal Modification of Endogenous Abscisic Acid

ELIASSON (1969, 1975) has shown a correlation between promoter and inhibitor content in growing tissues. He noted that IAA maintains a high level of an ABA-like inhibitor in stem sections of aspen or pea and suggested that IAA plays a role in regulating levels of ABA. In contrast to IAA, both kinetin and GA suppress the accumulation of ABA-like inhibitors, at least in leaves of nasturtium (CHIN and BEEVERS 1970).

Ethylene enhances ABA levels in citrus fruits (GOLDSCHMIDT et al. 1973) and in rose petals (MAYAK and HALEVY 1972). In the latter case the interaction between ethylene and ABA was viewed as an indication that ethylene may initiate senescence by enhancing ABA production and that ABA may, in turn, limit ethylene production by feedback inhibition of its biosynthesis.

MILBORROW (1972) has noted that high levels of ABA can prevent wilt-induced biosynthesis of ABA and he interpreted this as evidence for ABA-mediated feedback control of ABA biosynthesis.

2.3.3.5 Hormonal Modification of Endogenous Cytokinin

The growth retardant CCC increases cytokinins in the bleeding sap of grape (SKENE 1970). SKENE has interpreted this as an indication that CCC acts directly on the root meristem to increase cytokinin production. However, since the effect of CCC is counteracted by GA, it is possible that the enhancement of cytokinin production by CCC is mediated by a CCC-induced reduction in GA. This indicates that GA may act to suppress cytokinin production in roots (WOOLLEY and WAREING 1972a), a possibility supported by TELTSCHEROVA (1970) who noted that CCC causes an increase in the cytokinin content of apical buds of *Chenopodium rubrum* and that GA can overcome this effect. IAA also reduces cytokinins in buds of *Chenopodium*, indicating that both GA and IAA may be involved in the regulation of cytokinin production in this material.

Since cytokinins enhance GA levels and since GA appears to suppress cytokinin levels, the possibility of GA-mediated feedback inhibition of cytokinin production must be recognized. Also, the evidence that IAA can increase GA levels in apical buds of *Chenopodium rubrum* (TELTSCHEROVA 1970) indicates that the suppressing effect of IAA on cytokinin content may be a secondary effect mediated by GA.

In *Solanum andigena*, combinations of IAA and GA reduce cytokinin movement toward the apex of decapitated plants and enhance the conversion of ^{14}C-benzylaminopurine to an unidentified product (WOOLLEY and WAREING 1972b). The ability of combinations of IAA and GA to reduce cytokinin movement toward the apex may also be related to the stimulatory effect of IAA on the basipetal transport of cytokinin (SETH et al. 1966).

2.3.3.6 Summary

The influence of plant hormones on endogenous levels of other hormones is indicative of the complexity of hormone interaction in the regulation of plant growth and development. When one applies auxin to plant tissue in vivo or in vitro, the interpretation of the effects should take into account the evidence that the uptake, metabolism, range of secondary effects and magnitude of direct effects of the hormone are dependent on the presence of other hormones. Furthermore, *each* of these factors may be modified by a variety of environmental parameters. We know that factors such as light, temperature, and moisture stress modify the hormone balance of cells from all types of plant organs. Thus the response of particular cell types to hormones is strongly influenced by the recent history of exposure of the plant to environmental parameters.

2.3.4 Influence of Environmental Parameters on Endogenous Hormones

Light, temperature, moisture content, etc. have strong effects on all phases of plant growth and development, and it is not surprising that these parameters influence the content of endogenous hormones. Some examples of environmental modification of hormone content are described below.

2.3.4.1 Light

Substantial evidence indicates that light modifies endogenous levels of hormones or alters the sensitivity of cells to hormones. Many of these effects appear to be phytochrome-mediated (BLACK and VLITOS 1972).

Effects on Auxin. Red light enhances the growth of seedling tissue near putative endogenous sources of auxin, e.g., near coleoptile tips or cotyledons, but reduces elongation in tissues at a distance from these areas (HUISINGA 1967, GOTO and ESASHI 1974a, LAWSON and WEINTRAUB 1975). This appears to be due, at least in part, to a red light-mediated reduction in auxin transport away from auxin sources (BRIGGS 1963, NAQVI and GORDON 1967, KONDO et al. 1969). The uptake of auxin may also be enhanced by red light (MUIR and CHANG 1974). Auxin production in coleoptile tips is reduced by red light (HUISINGA 1976), an effect that can be largely overcome by GA (MUIR and CHANG 1974).

In some cases light inhibition of hypocotyl or mesocotyl extension can be overcome by IAA (ACTON and MURRAY 1974, VANDERHOEF and BRIGGS 1977). This is the case in lettuce hypocotyls, where the data indicate a correlation between light-induced growth inhibition and both reduced auxin content and apparent reduced sensitivity of the cells to auxin.

Exposure to red light inhibits growth and causes a 40% reduction in extractable IAA in bean seedlings (FLETCHER and ZALIK 1964). Light-induced reductions in free IAA may also arise, in part, by photo-enhancement of conversion of IAA to a bound form (BANDURSKI and SCHULZE 1977). In contrast to these reports, SCOTT and BRIGGS (1960) found an increase in free auxin in light-grown pea seedlings compared with etiolated seedlings. Similar results were reported by TILLBERG (1974) using bean seedlings.

Effects on Gibberellin. In barley and wheat leaf sections or homogenates, red light causes a rapid increase in the GA-like material associated with etioplasts (COOKE et al. 1975, Cooke and SAUNDERS 1975). That phytochrome is involved in the photoregulation of GA metabolism and/or release is indicated by the observations that this effect is far red-reversible (COOKE and KENDRICK 1976) and that night interruption with red light prevents the decline in GA content which normally occurs in leaves of potato subjected to short days (RAILTON and WAREING 1973b).

White light inhibits hypocotyl elongation in a variety of seedlings. In some cases this can be overcome by GA. In hypocotyls of *Lupinus albus*, IAA largely overcomes white light inhibition of hypocotyl growth while GA is ineffective (ACTON and MURRAY 1974).

The inhibitory effect of white light on hypocotyl extension seems not to be due simply to a reduction in GA. JONES and LANG (1968) found no difference in the GA content of light-grown and dark-grown pea stem tissue while in *Lupinus albus*, the content of free GA was found to *increase* during exposure to white light (ACTON and MURRAY 1974). This indicates that white light inhibition of elongation may be more closely related to a photoinduced decrease in the ability of the tissue to respond to GA (JONES and LANG 1968, MUSGRAVE and KENDE 1970, ACTON and MURRAY 1974) than to changes in GA content.

Photoperiodic modification of GA content has also been observed. Low light intensity increases GA in leaves of *Solanum andigena* exposed to short days but not to long days (WOOLLEY and WAREING 1972a, RAILTON and WAREING 1973b) while long-day promotion of growth in spinach appears to be correlated with changes in GA biosynthesis as well as an increased sensitivity to GA (METZGER and ZEEVAART 1980).

Effects on Ethylene. Red light inhibits ethylene production in stems, coleoptiles, and leaves (GOESCHL and PRATT 1968, KANG and RAY 1969, IMASEKI et al. 1971, LOVEYS and WAREING 1971), but not in germinating seeds (ABELES and LONSKI 1969). Red-induced inhibition of ethylene production appears to be a phytochrome-mediated, far red-reversible effect (IMASEKI et al. 1971) which plays a role in the control of hook opening in many plants. Lettuce seedlings are a notable exception since, in lettuce, red light enhances hook closure (MOHR and NOBLÉ 1960) without altering ethylene production (JANES et al. 1976).

A high irradiance blue or far red enhancement of ethylene production has also been found (CRAKER et al. 1973), indicating that the effect of illumination on ethylene production depends strongly on both wavelength and intensity. GEPSTEIN and THIMANN (1980) found that high-intensity white light reduces ethylene production in leaves and suggested that the consequent rise in ethylene production in the dark accounts for certain photoperiodic effects such as nyctinasty. Photoinhibition of ethylene production is also strong in callus tissue cultures. In tobacco callus, dark-grown cultures produce 10 to 30 times more ethylene than light-grown cultures (HUXTER et al. 1977).

Effects on Abscisic Acid and Cytokinin. The ABA content of birch seedlings (LOVEYS et al. 1974) and spinach leaves (ZEEVAART 1971) is increased under long days. The increase in ABA in buds, seeds, and tubers might be controlled, in part, by photoperiod (CORGAN and PEYTON 1970). Light also increases ABA in bean seedlings (TILLBERG 1974) and in pea roots (TIETZ 1974), while red light is postulated to increase ABA in mung bean root tips (TANADA 1973). Light, especially far red-enriched light, accelerates the glucosylation of ABA and the metabolism of ABA to phaseic acid (LOVEYS 1979, PHILLIPS et al. 1980).

Low light intensity decreases the cytokinin content of roots of *Solanum andigena* exposed to short days (WOOLLEY and WAREING 1972a). Similarly, a single short day cycle reduces the cytokinin content of leaves of *Xanthium* raised under long days (HENSON and WAREING 1974), though this may be due to reduced cytokinin import from the roots rather than enhanced turnover in the leaves (WAREING et al. 1977). Since light can increase GA and since GA suppresses cytokinin production in roots (WOOLEY and WAREING 1972a) light-induced reductions in cytokinin content may be mediated by gibberellin.

Light-induced seed germination is often accompanied by an increase in cytokinin (VAN STADEN et al. 1973). In seeds of *Rumex obtusifolius* there is evidence for phytochrome control of cytokinin levels (VAN STADEN and WAREING 1972), while cytokinin levels in leaves of poplar show a large but transient increase following short exposure to red light (HEWETT and WAREING 1973). In *Amaranthus* seedlings blue light induces a large increase in extractable cytokinin-like materials (KÖHLER et al. 1980).

2.3.4.2 Water Stress

Plants exposed to water stress exhibit a variety of changes in hormone levels (see review by LIVNE and VAADIA 1972).

Effects on Abscisic Acid. One of the most clearly defined effects of water stress (decreased turgor) on hormones is the enhancement of ABA biosynthesis in leaves (WRIGHT and HIRON 1972, MILBORROW 1974, ZEEVAART 1980). The response is rapid and the increase in ABA content is large (KRIEDEMANN et al. 1972, MILBORROW 1974, WRIGHT and HIRON 1972, HIRON and WRIGHT 1973). Water stress-induced enhancement of ABA content also occurs in roots (WALTON et al. 1976) but the effect is small in some cases (MILBORROW and ROBINSON 1973).

Effects on Ethylene. Direct desiccation of detached leaves causes a rapid increase in ethylene production. Since the response is larger in leaves with an abscission zone, this stress-induced ethylene production may be related to stress-induced abscission (AHARONI 1977).

Flooding also enhances ethylene production, and some effects of flooding (epinasty, decreased growth, adventitious root formation) have been attributed to elevated levels of the gas. Since the conversion of 1-aminocyclopropane-1-carboxylic acid (ACC) to ethylene is an oxygen-dependent process, it may be that enhanced ethylene production in shoots of flooded plants (BRADFORD and DILLEY 1977) arises from ACC transported to the shoot from the anaerobic root environment.

Effects on Cytokinin, Gibberellin, and Auxin. Desiccation causes a reduction in the cytokinin activity of the root exudate of sunflower (ITAI and VAADIA 1965) as well as the leaf tissue and xylem exudate of *Nicotiana rustica* (ITAI and VAADIA 1971). In the experiments with *Nicotiana* there is evidence for a reversible chemical transformation of ^{14}C-kinetin as the means of altering cytokinin activity in leaves while, in the roots, water stress appears to reduce cytokinin biosynthesis.

Flooding of the roots results in a large but transient increase in the auxin content of sunflower shoots (PHILLIPS 1964) while in tomato, flooding decreases the level of GA in the xylem sap and shoots (REID et al. 1969; REID and CROZIER 1971).

2.3.4.3 Temperature

Effects on Gibberellins. Exposure to cold enhances subsequent GA biosynthesis in certain seeds (ROSS and BRADBEER 1971). The increase appears to arise from de novo GA biosynthesis and cannot be inhibited by concentrations of ABA which inhibit germination (ARIAS et al. 1976) and suppress apparent GA biosynthesis in maize shoots (WAREING et al. 1968). Cold increases GA in potato tubers (THOMAS and WURR 1976), while low temperature during tuber germination increases both GA and auxin activity in leaves of potato (ROMANOVA et al. 1975). In corn roots, on the other hand gibberellin production or export is severely inhibited by prolonged exposure to cold (ATKIN et al. 1973).

Effects on Abscisic Acid and Cytokinins. LEPAGE-DIGIVRY (1973) reported an increase in bound ABA at the expense of free ABA in seeds of *Taxus baccata* during stratification. ENU-KWESI and DUMBROFF (1977) also reported a cold-induced reduction in detectable ABA in *Acer* fruits but, in most cases, changes in ABA levels have not shown a close correlation with thermodormancy (BERRIE and ROBERTSON 1976).

Low temperature strongly retards apparent cytokinin biosynthesis in corn roots (ATKIN et al. 1973). Temperature affects the cytokinin content of the root exudate of Thompson Seedless grape vines in both a qualitative and quantitative manner. So-called Fraction 2 cytokinins are present in exudate of vines grown at 20° C but not in vines grown at 30° C (SKENE and KERRIDGE 1967).

2.3.4.4 Mechanical Stimulation

Mechanical stimulation (rubbing, bending by wind, etc.) causes retardation of growth in a wide variety of vascular plants (JAFFE 1980). Although the physiological basis for this effect is not understood, there is evidence for a correlation between stimulation-induced retardation of growth and enhancement of ethylene production (HIRAKI and OTA 1975, BIRO et al. 1980; see also M.J. JAFFE: Wind and other mechanical effects in development and behaviour, Vol. 11 of this series and P.S. NOBEL: Wind as an ecological factor, Vol. 12A of this series).

2.3.4.5 Summary

The evidence is overwhelming that the balance of endogenous hormones is affected by environmental parameters. It is, therefore, not surprising that the literature contains many apparently conflicting reports of the effects of applied hormones, especially when intact plants are used. The changes in hormone balance and, in many cases, the changes in hormone sensitivity per se (see review by TREWAVAS 1981) under different environmental conditions should emphasize the importance of taking sometimes slight differences in these parameters into account when comparing results from different laboratories.

2.4 Hormonal Regulation

2.4.1 The Cell Cycle

The stability and reliability of meristems is one of the most remarkable aspects of plant function. Meristems produce orderly arrangements of new cells with little or no error under a variety of conditions. An operation of such precision clearly cannot function without inherent points of control. Speculation and experimentation concerning the likely points of control within the cell division cycle have been assisted by the descriptive mitotic cycle outline provided by HOWARD and PELC (1953). They view the cell cycle as comprised of four phases designated G_1, S, G_2, and M. S refers to the period of DNA synthesis that occurs prior to and is necessary for mitosis (M). After mitosis there is a period (G_1) before the beginning of S. This may be viewed as a period of preparation for the DNA synthesis that occurs during S. Following S there is a period (G_2) before the beginning of M. During this period DNA is prepared for organization into chromosomes in anticipation of mitosis. For a discussion of the cell cycle in plants, see BRYANT (1976).

An example of the time devoted to each phase of the cell cycle in the apical meristem of *Vicia faba* roots is shown in Fig. 1, though these values vary somewhat in reports from different laboratories (MACLEOD 1971). Cells which are not committed to nuclear division are sometimes viewed as entering a state G_0 (GOULD et al. 1974, PEAUD-LENOËL 1977), different from those shown in Fig. 1. Cells which begin differentiation leave the cell cycle and initiate a specialized pathway of development.

While the linear sequence shown in Fig. 2.1 represents the simplest relationship of the phases of the cell cycle, an alternative model with more than one independent pathway between two successive steps has also been proposed (PEAUD-LENOËL 1977).

What factors are involved in the initiation and termination of each phase of the cell cycle? What signals lead to the termination of cell division and the initiation of cell differentiation? There is evidence that G_1 and G_2 serve as principal control periods. In most meristems studied, the arrest of cell division, e.g., by starvation or gamma radiation, leaves nearly all of the cells in either G_1 or G_2 indicating that energy-requiring events in G_1 and G_2 are necessary for and therefore control entry into S and M respectively. Similarly, the initiation of differentiation in cells of the root meristem occurs in cells arrested in either G_1 or G_2 (VAN'T HOF 1974, ROBERTS 1976), an indication that the signal causing a switch from cell cycling to cell differentiation acts in G_1 or G_2.

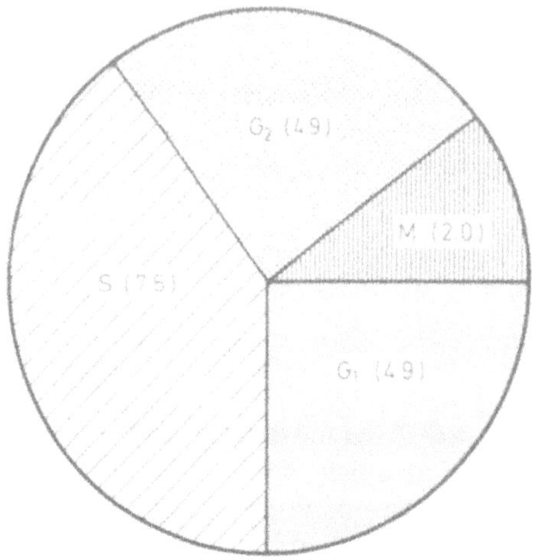

Fig. 2.1. Duration of phases of the cell cycle in the apical meristem of *Vicia faba* at 19°. Values *in parentheses* indicate duration of each phase in h. *M* mitosis; G_1 period between division and the onset of *S*; *S* period of DNA synthesis (DNA content doubles); G_2 period between termination of DNA synthesis and initiation of mitosis (organization of newly synthesized DNA into chromosomes). Total cycle time = 19.3 h. (After EVANS and SCOTT 1964)

The nature of the signals controlling the cell cycle or initiating differentiation is unknown. Possibilities include: physical factors, e.g., position; relative nuclear/cytoplasmic volume, etc. (ROBERTS 1976), genetic factors, e.g., de novo synthesis or activation of specific enzymes (VAN'T HOF 1974, MITCHISON and CREANOR 1969); and chemical factors, e.g., hormones (NAGL 1972).

2.4.1.1 Effect of Cytokinins

The cytokinins seem most likely to play a direct role in the regulation of cell division in vivo since they promote cell division in cultured plant tissues. This, plus reports that kinetin stimulates division in the protozoan, *Paramecium caudatum* (GUTTMAN and BACK 1958), and in yeast (BAREA et al. 1974), indicate a fairly widely occurring cytokinin-sensitive control site.

The effect of cytokinin on division is dependent upon the state of differentiation of the treated cells. In nondividing differentiated cells, for example, cytokinin usually triggers cell proliferation, while in rapidly dividing meristematic cells it retards division (VAN'T HOF 1968). This indicates that sites controlling division change during differentiation, or at least that the hormonal sensitivity of these sites changes.

Cytokinin deprivation of tobacco or sycamore cells in suspension culture temporarily induces partial synchronization of division beginning about 18 h after re-addition of the hormone (JOUANNEAU 1971, PEAUD-LENOËL and JOUANNEAU 1971, ROBERTS and NORTHCOTE 1970). This indicates that cytokinins may influence a specific regulatory point within the cell cycle.

Perhaps the most likely point of cytokinin control of the cell cycle is in G_2 or the entry into mitosis from G_2. Transfer of cultured soybean cells to cytokinin-free medium leads to cessation of division but not to DNA synthesis, so the amount of DNA per cell approximately doubles (SHORT et al. 1974). This would not be expected if the cells became blocked at some point preceding S (i.e., at G_1).

Kinetin enhances the rate of DNA synthesis during S, and a pulse of cytokinin can induce a long-lived enhancement of the ability of the DNA synthesizing machinery to respond to a second exposure to cytokinin (SIMPSON 1977). Under some conditions a 35% decrease in the duration of S accompanies kinetin-enhanced DNA synthesis (MACLEOD 1968). This implies an inverse relationship between the rate of DNA synthesis and the duration of S. Although this is consistent with the observation that mitotic cycle time increases with DNA content in some cases (VAN'T HOF 1965), this has not held in all instances (FRIEDBERG and DAVIDSON 1970). The bulk of the evidence indicates that the effects of cytokinin on the cell cycle, though indirectly dependent on DNA synthesis, are not mediated by modification of DNA synthesis (SIMARD 1971, FOSKET and SHORT 1973). In pea root cells, for example, kinetin has no effect on the rate of travel of the DNA replication fork (VAN'T HOF and BJERKNES 1977). In a cytokinin-requiring strain of soybean tissue, FOSKET and SHORT (1973) noted that cell division and the rate of DNA synthesis were not closely related. In the absence of cytokinin there was greater synthesis of DNA than in cytokinin-supplemented cultures even though there was no cell division in the former and active cell division in the latter. Apparently the DNA synthesis in the absence of cytokinin was associated with endoreduplication (see Sect. 4.2) while in cytokinin-treated cells, division accompanied DNA synthesis. This indicates that, at least in this material, cytokinin is more important as a trigger of mitosis than as a regulator of DNA synthesis, and that the triggering effect of cytokinin on cell division is not mediated by a cytokinin-induced alteration in DNA synthesis. This is supported by the finding of JOUANNEAU and TANDEAU DE MARSAC (1973) that in tobacco cells DNA synthesis proceeds normally in auxin-supplemented media through at least one cell cycle whether or not cytokinin is added. Furthermore, they found that cytokinin induction of synchronous cell division occurs even when DNA synthesis is blocked during cytokinin treatment. The presence of cytokinin in cells whose DNA synthesis is blocked by 5-fluorodeoxyuridine leads to induction of synchrony upon withdrawal of the cytokinin and restoration of DNA synthesis through the addition of thymidine. These data indicate that cytokinin is not required during mitosis itself but is essential for one or more events leading up to mitosis. Although this event(s) appears to be independent of DNA synthesis, it is dependent on continued protein synthesis since inhibition of protein synthesis during exposure to cytokinin prevents the division-enhancing effect of the hormone (JOUANNEAU 1975). Since cytokinins have no early effect on bulk protein synthesis (JOUANNEAU 1970, FOSKET and SHORT 1973), the division-inducing effect of the hormone may be mediated by the synthesis of a specific protein (JOUANNEAU 1970) during the last half of the cell cycle and before mitosis (PEAUD-LENOËL 1977). This is consistent with the observed increase in activity of a number of enzymes during G_1 or early S (AITCHISON and YEOMAN 1974).

Cytokinins are reported to affect both the G_1 and G_2 phases of the cell cycle of root meristems, causing a reduction in G_1 (MACLEOD 1968) and a lengthening of G_2 (MACLEOD 1968, VAN'T HOF 1968). This indicates that even though cytokinins may act most directly on a particular step of the cell cycle

(JOUANNEAU 1971, FOSKET and SHORT 1973), they may have secondary effects on biochemical pathways important to other phases of the cycle (BERGMANN 1965).

Cytokinins and the Quiescent Center of Roots. The experiments mentioned so far involve exogenous cytokinins. The effects therefore do not necessarily reflect functions of cytokinins in vivo. Nevertheless, there is evidence for the occurrence of cytokinins in meristematic regions of intact plants and strong though not conclusive evidence that they are synthesized there (SKENE 1975, FELDMAN 1975, 1979). Of particular interest are investigations of the involvement of endogenous hormones in the determination of variable cell division activity within the root meristem itself. In the quiescent center (QC) of roots, cells proliferate slowly in comparison to surrounding meristematic cells (CLOWES 1954, 1976, PHILLIPS and TORREY 1972). Since supraoptimal levels of cytokinin are inhibitory to cell division (SKOOG et al. 1967, MACKENZIE et al. 1972) it has been proposed (TORREY 1972, CLOWES 1969, SHORT and TORREY 1972) that the QC is the site of cytokinin production and that the high level of cytokinin diffusing into the surrounding meristematic cells maintains active division there. There is at least one report that ABA, which is often antagonistic to cytokinin, can enhance cell division and that this effect can be overcome by cytokinin (MINOCHA 1979). That cell division in the QC is inversely related to cytokinin content is supported by the studies of FELDMAN (1975), who noted that decapping *Zea* roots causes a transient increase in division within the QC, and that this correlates with a transient decrease in cytokinin. Others have also noted a stimulation of cell division in the QC upon removal of all or part of the root cap (CLOWES 1972), and it cannot be ruled out that a noncytokinin inhibitor, e.g., abscisic acid, produced in the cap may suppress activity in the QC.

The major difference in the cell cycle in the QC relative to meristematic cells is the greatly prolonged G_1 phase in the QC (BARLOW and MACDONALD 1973, CLOWES 1976). This is consistent with the suggestion that the QC is high in cytokinin since prolonged exposure of root meristems to cytokinin extends G_1 (VAN'T HOF 1967, 1968, MACLEOD 1968). Reduced levels of auxin may also contribute to slow mitotic cycling in the QC. Application of IAA to root apices speeds up the mitotic cycle in the QC (BARLOW 1976a) and the G_1 phase is essentially omitted in cells at the margin of the meristem (CLOWES 1975) a region postulated to be high in auxin content (BARLOW 1976a).

WEBSTER and LANGENAUER (1973) proposed an alternative to the concept that cytokinin or some other substance produced in the QC maintains active division in meristematic cells. They noted that, after release of cultured root tips from starvation-induced inhibition of division, DNA synthesis and division occur in the QC as well as in the rest of the meristem. After resumption of division in the meristem, however, the quiescence of the QC is re-established, indicating that the activity of the meristem may control the QC rather than vice versa. Other methods of temporarily suppressing meristematic activity also allow resumed DNA synthesis and division in the QC until division activity in the remainder of the meristem recovers (WEBSTER and LANGENAUER 1973, MACLEOD 1976).

Whether the control of cell division in the QC arises internally or from adjacent cells, it is unlikely that this control is by a single promoter or inhibitor (BARLOW 1973, CLOWES 1976). Cell division in culture is dependent on both cytokinin and auxins, while in excised pea root meristems combinations of IAA and kinetin extend G_1 and retard both the progress of cells through S, and their entry from G_2 into M (VAN'T HOF 1968). In root meristems of *Allium cepa*, kinetin alone has no effect on the duration of interphase, but shortens it when given with IAA (GONZALEZ-FERNANDEZ et al. 1972). GUTTMAN (1956), on the other hand, reported an interphase-shortening effect of kinetin in *Allium* root meristems. These conflicting reports indicate that the effect of kinetin on interphase-duration is dependent on cytokinin concentration and probably the level of auxin as well (also see BARLOW 1976a).

2.4.1.2 Effect of Auxins

Auxins modify mitotic activity in cell types ranging from yeast (SHARMA and BARUAH 1975) to bovine muscle cells (BASRUR et al. 1976). The high levels of auxin in human cancer tissue have even led to the suggestion that IAA plays some role in the acceleration of division in animal cancer cells (YAMAKI et al. 1974).

Auxin is essential for cell division in higher plants, but the evidence for direct control of the cell cycle is weaker for auxin than for cytokinin. In tobacco cell suspensions, for example, withdrawal of either cytokinin or auxin leads to cessation of division but, unlike cytokinin, replenishment of auxin has no synchronizing effect on the resumed cell division (JOUANNEAU 1971). However, auxin does play a role in the timing of cytokinin-induced synchronous division; cultures pretreated with auxin before cytokinin restoration begin synchronous cell division earlier than those in which auxin and cytokinin are replenished simultaneously (PEAUD-LENOËL and JOUANNEAU 1971). This indicates that auxin may be necessary, perhaps indirectly, for one or more phases of the cell cycle, possibly the RNA and DNA synthesis associated with the S phase (NITSCH and LANCE-NOUGAREDE 1967).

In cultured carrot cells, temporary auxin deprivation induces at least partial synchronization of cell division (NISHI et al. 1977), a response not shown by cultured tobacco cells. Cell arrest during auxin deprivation appears to be in G_1, suggesting that auxin is needed to initiate the S phase of the cycle.

In some cases, changes in auxin levels show a positive correlation with changes in mitotic activity. NISHINARI and YAMAKI (1976) noted that, in synchronized tobacco cell cultures, there is an increase in the activity of IAA-synthesizing enzymes preceding an increase in the level of auxin, and that these events occur just prior to the increase in the number of dividing cells. The authors concluded that synthesis of free IAA may initiate mitosis. A dependence of cell division on internal auxin concentration in *Acer pseudoplatanus* was also shown by LEQUAY and GUERN (1977).

GAMBURG and OSHAROVA (1973) reported that NAA caused a sharp increase in both mitotic activity and DNA synthesis in tobacco cells. Although DNA synthesis was induced by brief exposure to NAA, mitosis did not follow without either prolonged exposure to NAA or a second addition of NAA after the completion of DNA synthesis. The data indicate that auxin may be needed to initiate progress from G_1 to S and also from G_2 to M. Auxin also accelerates mitosis in antheridial filaments of *Chara*, primarily by decreasing the duration of G_2 and M (GODLEWSKI 1980). An auxin-sensitive transition from G_2 to M was also noted in onion root by NAGL (1972), but in this case the effect was arrest in G_2 after prolonged exposure to 10^{-5} M IAA.

Although auxin is required for cell division, it increases the duration of the mitotic cycle in the root meristem of *Vicia faba* (MACLEOD and DAVIDSON 1966, DAVIDSON and MACLEOD 1966) and at higher concentrations suppresses mitosis completely in onion root meristem (NAGL 1972). In the liverwort, *Riella helicophylla*, anti-auxin treatment accelerates division, presumably by reducing the effective endogenous level of auxin (GROTHA 1976).

In apical meristems of *Vicia faba* roots, IAA increases the duration of S with no change in the initial rate of DNA synthesis (WEBSTER 1967). This indicates that the effects of IAA on S, as with the previously-mentioned effects of cytokinins on S, are not by an overall modification of DNA synthesis. This is supported by the data of NAGL (1972) showing that total suppression of mitosis by IAA in onion root tips occurs with no inhibition of DNA replication.

2.4.1.3 Effects of Other Growth Regulators

There are few specific studies of the effects of gibberellins, ethylene, abscisic acid, and other growth regulators on the cell cycle. While these substances have effects on division in some instances, evidence connecting these effects to specific steps of the cell cycle is generally lacking.

Gibberellins. Although the gibberellins inhibit cell division in some instances (SETTERFIELD 1963, KAUFMAN et al. 1969) more often they are stimulatory (BOEKEN and VAN OOSTVELDT 1977, KÖHLER 1977, SHININGER 1975) especially in subapical tissue (WARDLAW 1965, LIU and LOY 1976). GA usually has less effect on mitotic activity in the shoot apical meristem itself (see review by LOY 1977), an indication that it may be more effective in initiating cell division than in accelerating division in cells already mitotically active. In instances where GA acts to shorten the cell cycle in the shoot apical meristem, it increases the proportion of cells in S and M, an indication that GA may act at the points of entry into S and M (LYNDON 1976).

It is difficult to relate the effects of GA on division to a direct role of GA in the control of the cell cycle or even to prove that the enhancement of cell division by GA is not an indirect effect arising from enhancement of cell elongation (JONES 1973). GA increases the mitotic index in meristematic cells of the leaf and root of barley and in both cases a simultaneous increase in DNA synthesis occurs (SVARINSKAYA and GAVRILOVA 1976). GA may have some selective effect on the cell cycle in the barley root in this case since it reduces the duration of G_1, G_2, and M. GA decreases the cell cycle time in the subapical meristem of dwarf watermelon seedlings (LIU and LOY 1976) where it decreases time spent in S. The primary effect of GA in apical buds of *Rud-*

beckia bicolor (JACQMARD 1968) has been postulated to be on cell progression from G_1 to S. This is consistent with the report (ELLIOTT et al. 1977) that, under appropriate conditions, GA_3 can release sycamore cells from arrest in G_1, GA may be involved in the regulation of division in these cells since the levels of GA fluctuate during the cell cycle, decreasing during the stationary phase and increasing prior to division. Further, inhibitors of GA biosynthesis were found to prevent division in cultures that were normally dividing, and this could be relieved through addition of GA. These results are in contrast to the studies of RAPPAPORT (1979), who found GA_3 to have no effect on cell division in carrot or tobacco cells in culture and noted that GA_3 was unable to overcome inhibition of cell division by growth retardants. He concluded that GA_3 is not required for cell division in tissue and cell suspension cultures and that the effect of GA on cell division in intact plants may be indirect.

Ethylene. The overriding effect of ethylene on cell division is inhibitory. Ethylene inhibits cell division in fern gametophytes (EDWARDS and MILLER 1972) and shoot and root meristems of pea (APELBAUM and BURG 1972a, b, BURG 1973). Ethylene inhibition of cell division in meristems is accompanied by inhibition of DNA synthesis with most of the cells arrested in G_1. The correlation between suppression of DNA synthesis and reduction in cell division has led to the suggestion (KANG and BURG 1973) that arrest in G_1 is caused by ethylene-induced inhibition of DNA synthesis, perhaps by reducing the level of DNA polymerase activity. Since the level of ethylene needed to reduce the frequency of cell divisions in the apical hook of the pea seedling is comparable to the endogenous level of the gas, the authors conclude that ethylene plays a role in the control of cell division in such tissues, a possibility also considered by PRATT and GOESCHL (1969). BARLOW (1976b) also reported that ethylene decreases the number of cells undergoing division in primary root meristems of pea and corn but, in contrast to APELBAUM and BURG (1972a, b), found no effect of ethylene on DNA biosynthesis. He suggested that the apparent ethylene-induced inhibition of DNA synthesis reported by BURG and coworkers may be an artifact. BURG and coworkers measured DNA synthesis (incorporation of ^{3}H-thymidine) of the entire apex but, since ethylene decreases the size of the meristem and increases the size of the quiescent center, one would expect a decrease in total incorporation of ^{3}H-thymidine per apex, even if there were no change in the rate of DNA synthesis in the S phase of each mitotically active cell. Except for a slight increase in the duration of G_1, BARLOW found no effect of ethylene on the duration of particular phases of the cell cycle in dividing cells of the meristem. In pea roots the effect of ethylene on the cell cycle was greater in cells of the stele 600 μm back from the cap-root junction than in cells of the apical meristem. In these cells, ethylene treatment increased the duration of the cell cycle. Ethylene also decreases the length of the meristematic region in roots of pea and corn, apparently by causing cells to switch prematurely from a mitotic to a nonmitotic state.

In suspension cultures of soybean cells, CONSTABEL et al. (1977) succeeded in enhancing growth and partially synchronizing cell division by periodically flushing with 3% ethylene. In lettuce pith tissue cultures, ethylene at high concentration suppresses callus development, while at low concentration it interferes with tracheid development without affecting callus proliferation (ZOBEL and ROBERTS 1978).

Abscisic Acid. ABA is a potent inhibitor of cell division in the developing roots of *Lemna minor* (NEWTON 1977). This effect, plus an apparent direct inhibitory effect on cell expansion in *Lemna* roots, contributes to the overall inhibitory action of ABA on root elongation in this aquatic plant.

Although ABA is generally inhibitory to growth, it acts synergistically with kinetin to promote soybean callus growth (BLUMENFELD and GAZIT 1970) and to stimulate callus formation in the leaf abscission zone of orange (ALTMAN and GOREN 1971) and apple (PIENIAZEK 1971). IAA and to a lesser extent, benzyladenine, act synergistically with ABA to enhance callus formation in this material.

In tuber explants of Jerusalem artichoke, ABA, if added in the presence of auxin, causes a large increase in the number of dividing cells (MINOCHA 1979). This is counteracted by cytokinin and occurs with no change in the duration of different phases of the cell cycle.

ABA stimulates the rooting of stem cuttings of mung beans and English ivy (CHIN et al. 1969). In mung beans this effect is largely overcome by kinetin. Although these effects involve increased cell proliferation, there is no evidence that either the promotion or the inhibition of cell proliferation by ABA is brought about by a direct effect of the hormone on the cell cycle.

A more specific effect of ABA on cell division may be involved in the suppression of root nodulation in pea (PHILLIPS 1971). Nodule tissue develops primarily from the division of polyploid cells, and the withdrawal of ABA from roots after transient exposure to the hormone has been reported to stimulate mature polyploid cells to divide (NAGL 1972).

Trigonelline. It has been known for some time that starvation-induced mitotic cycle arrest in cultured root meristems is nonrandom. Cells are arrested almost exclusively in either G_1 or G_2. In 1973 EVANS and VAN'T HOF reported that a factor present in the cotyledons of pea seedlings enhances the proportion of cells that arrest in G_2 (EVANS and VAN'T HOF 1973). This factor which has been referred to as the G_2 factor has been isolated and identified as N-methylnicotinic acid by NAKANISHI and coworkers at Columbia University (ANONYMOUS 1978). N-methylnicotinic acid (trigonelline) has been known to exist in plants since its discovery some 80 years ago in the fenugreek (*Trigonella foenum-graecum*). As a readily transportable naturally occurring organic product, trigonelline may join the list of plant hormones, if it can be shown to regulate cell division in other plants in which it is found. It may be that other, as yet undiscovered, plant hormones play an important role in the control of cell division.

2.4.2 Endomitosis

Endomitosis (endoreduplication) refers to continued chromosome replication within a nucleus which does not divide. DNA content per cell increases and the cells become polyploid, commonly tetraploid or octaploid. Treatments which suppress nuclear division and cytokinesis with no effect or less effect on chromosomal replication lead to an increase in endomitotic "cycling" relative to normal mitotic cell cycling. Since IAA appears to support or maintain nuclear DNA synthesis (NITSCH and LANCE-NOUGAREDE 1967) while cytokinin appears to favor mitosis and cytokinesis, it has been proposed (BARLOW 1976a) that a high ratio of cytokinin to auxin favors normal mitotic cycling, while a low ratio favors endomitotic cycling (also see FOSKET and SHORT 1973).

While this may be true, under some conditions cytokinins can increase the number of polyploid cells in cultured plant tissues (NAGL and RÜCKER 1974, LIBBENGA and TORREY 1973, FOSKET and SHORT 1973). This effect may be due, in part, to induction of endomitosis by cytokinin (LIBBENGA and TORREY 1973) and, in part, to stimulation of cell division in existing polyploid cells but not in diploid cells (NAGL and RÜCKER 1974). The stimulation of endomitosis by cytokinin may be indicative of the onset of differentiation since endomitotic cycling is commonly associated with (TORREY 1961), but apparently not necessary for cellular maturation (EVANS and VAN'T HOF 1975). Cytokinin stimulation of division in polyploid cells of the root has been suggested as a mechanism for bacterial induction of root nodule formation (LIBBENGA and TORREY 1973). A correlation between cytokinin content and cell division activity has been reported in nodule tissue (SYONO et al. 1976). Stimulation of division in differentiated polyploid cells can also be induced by transient exposure to ABA (NAGL 1972) and in some cases by treatment with auxin (AVANZI 1951). There is also a report that GA can induce more highly polyploid nuclei in already polyploid endosperm nuclei of *Phaseolus vulgaris* (NAGL 1971).

2.4.3 Meiosis

There have been fewer studies of hormonal effects on meiotic cell division than mitotic cell division. Both processes involve DNA synthesis, and mitotic DNA synthesis may be competitive with premeiotic DNA synthesis. In yeast, the ability to undergo meiotic spore production decreases as budding and its associated karyokinesis begin (TSUBOI and YANAGISHIMA 1974). There is evidence for a preferential entry into meiotic development from G_1. Yeast cells receiving a meiosis-inducing signal at any point in the cell cycle other than late G_1 or early S appear to proceed to G_1 before entering meiosis. Cells receiving a meiosis-inducing signal during late G_1 or early S appear to proceed mitotically to a point between the completion of S and the initiation of nuclear division, at which point meiotic development begins (CROES et al. 1976).

Cytokinin enhances meiotic cell division in megaspore mother cells of grape (NEGI and OLMO 1972). Enhancement of meiosis by cytokinin has also been reported in microspore mother cells of *Tradescantia paludosa* (WALKER and DIETRICH 1961). In this case kinetin stimulates the passage of meiotic cells through prophase and, under some conditions, reduces the time between the second meiotic division and the following mitotic division. Prolonged exposure to kinetin, however, causes cessation of meiosis ("meiotic stasis").

The effect of ethylene on meiotic chromosomes in microspore mother cells of rye was examined by BOYLE and EVANS (1974). No information on the effect of ethylene on the timing of the events of meiosis was presented, but the hormone was shown to increase the frequency of chromosomal aberrations.

2.4.4 Summary

Cell division in higher plants requires both cytokinin and auxin. While the cytokinins exhibit the most dramatic effects on cell division they are ineffective in the absence of auxin. The effects of cytokinin and auxin on cell division vary with concentration and with the state of differentiation of the cells. Nondividing differentiated cells are induced to divide when exposed to cytokinin and auxin, while division in rapidly dividing meristematic cells is more likely to be retarded in their presence. This indicates that the control sites for cell cycling may be altered during differentiation, perhaps by a shift in sensitivity to auxins and cytokinins.

Although there is evidence that hormones play a role in the control of the cell cycle, this control may be indirect, i.e., the hormones may play a permissive role in division or they may influence the cell cycle through effects on expansion or metabolism. It is curious, however, that cells arrested experimentally normally arrest in either G_1 or G_2 and that gibberellin and cytokinin are important for progression from G_1 to S and G_2 to M, respectively. Thus the apparent principal control periods of the cell cycle are those most noticeably responsive to hormones; a relationship probably more than coincidental.

Auxin appears to play more of a permissive role than a site-specific control function in cell division. Cell division requires auxin but transient auxin deprivation shows little or no synchronizing effect on cell proliferation in most cases. Auxin may be necessary for DNA synthesis, but it is clear that auxin acts at one or more points in the cell cycle independently of its effects on DNA synthesis.

There is some evidence for correlated changes in hormone levels and cell division activity. This is not proof of a cause/effect relationship, nor does it provide clues to the mechanism of putative cell cycle control by hormones. This will depend on more detailed biochemical analyses relating, for example, GA promotion of entry into S to specific GA effects on DNA synthesizing machinery.

2.5 Hormone Effects on the Plane of Cell Division

There is little evidence that hormones play a direct role in determining the plane of orientation of the cell plate during division. Indeed it is difficult to imagine how subcellular gradients of hormone concentration could be established or maintained. It is clear, on the other hand, that the plane of division during development is not random. Changes in the plane of division are often precisely coordinated with changes in the direction of expansion, as in the periclinal divisions associated with the initiation of lateral roots or leaf primordia (LYNDON 1976), the formation of generative and tube nuclei in pollen grain division, and the deposition of cell plates perpendicular to the axis of greatest relative elongation in rapidly growing cells.

The orientation of the plane of division in a multicellular organ probably does not play a major *direct* role in determining its final shape. This is determined by the relative elemental growth rates of individual cells along the x, y, and z coordinates. HABER (1962), for example, showed that the relation between leaf length and width is the same in leaves growing with normal cell division as in gamma-irradiated leaves with negligible cell divisions. He concluded that "the orientation of cell divisions, whether concurrent with or prior to the growth period studied, plays no role in the basic mechanisms which determine the manner in which leaves change shape."

In contrast to the concept that the plane of cell division determines the direction of cell expansion, STEBBINS (1965) suggested that the direction of expansion may determine the orientation of division. He noted that the direction of spindle orientation is commonly parallel to the long axis of cells dividing rapidly enough that expansion does not keep pace. Since the direction of this axis changes with each mitosis, the direction of spindle orientation will change accordingly, resulting in a series of mitoses in various directions and producing a solid three-dimensional structure, such as occurs in many primary meristems. In meristematic regions where there is sufficient time between divisions and sufficient cell expansion to allow increased cell size in spite of continuing division, the orientation of the spindle tends to follow the direction in which expansion is occurring. This is not always the same as the direction of the long axis of a cell. DYER (1976) points out that the long axis of the spindle tends to parallel the direction of greatest active expansion, regardless of cell shape. According to these ideas, hormones which alter the rate or direction of cell expansion might indirectly determine the plane of cell division accompanying expansion. High concentrations of IAA increase lateral expansion in elongate epidermal cells of *Avena* and also induce occasional longitudinal and oblique cell divisions where normally all divisions would be transverse (KAUFMAN et al. 1969). High levels of IAA also shift the plane of division in guard cell mother cells from the longitudinal direction to transverse or oblique division. In barley leaves, high levels of GA when applied with mercaptoethanol have a similar effect on guard mother cell division (STEBBINS et al. 1967).

In tomato root meristems, auxin causes an increase in the number of longitudinal and a decrease in the number of transverse divisions (HUGHS and STREET 1960), while in intercalary meristem segments from *Avena* stems, kinetin enhances transverse cell division and causes a shift of division within the intercalary meristem from the transverse to the longitudinal direction (JONES and KAUFMAN 1971). These effects were obtained only with high levels of kinetin.

Some of these hormonal effects on cell plate orientation may be caused by alterations in cell expansion patterns. NISHI and SUGANO (1970), showed that carrot cells cultured in 2,4-D exhibit two-dimensional growth and tend to form tetrad-like structures after two divisions. The orientation of cell growth changes after the first division, and the cell plate of the subsequent division is formed at an oblique angle to the plane of the first division. When these cells are cultured in IAA instead of 2,4-D, the cells show one-dimensional growth, and cell division is at a right angle to the direction of elongation, resulting in a filamentous form.

It seems likely that a change in the principal direction of expansion of these cells and of cells in vivo would cause a shift in the distribution of the pattern of mechanical stress born by the cells. This may play a role in determining the plane of cell divisions. YEOMAN and BROWN (1971) and LINTILHAC (1974) have offered the intriguing suggestion that the primary factor in determining the orientation of the plane of cell division is the mechanical stress on the cell at the time of division. LINTILHAC suggests that the cell plate is formed in the shear-free plane perpendicular to the largest nonzero stress passing through the wall. The model is consistent with cell division patterns along free edges, epidermal surfaces and some simple apices in plants.

LINTILHAC (1977) supported this hypothesis by showing that in *Coleus* stems an imposed force induces a pattern of cell wall orientation corresponding to the principal stresses generated by the force. Stress-induced alterations in the plane of cell division were also demonstrated in Jerusalem artichoke callus tissue by YEOMAN and BROWN (1971). LINTILHAC points out that this model may explain the selective development of a single cell among a mass of apparently identical cells. The cell in the very center of the nucellus, for example, is unique in its exposure to isotropic stress, and it is this cell which divides and develops into the megasporangium.

The shear stress model does not readily explain the effects of light on the orientation of cell division. In fern gametophytes at the protonema stage, the orientation of the cell plate during division is sensitive to light quality and direction. In *Adiantum* the first two divisions are normally transverse while the third is longitudinal. WADA and FURUYA (1970) showed that exposure to red light can cause nearly complete alteration of the third division from longitudinal to transverse orientation. They also showed that the orientation of the cell plate can be controlled by the direction of an incident beam of white light (WADA and FURUYA 1971). It is not obvious how light quality or orientation might shift the direction of zero shear stress, although a photoeffect on the mechanical properties of the wall mediated by a shift in wall deposition patterns would seem possible.

2.6 Relationship of Cell Multiplication to Cell Expansion

The necessity of auxin for cell division raises the question of the relationship between auxin-induced cell expansion and the process of cell division. While it is clear that expansion normally accompanies cell division, there are instances in which this is not the case (STEBBINS 1965, DIGBY and WAREING 1966). It cannot be maintained therefore, that cell expansion itself is the trigger for cell division. Conversely, cell division is not necessary for continued cell expansion (HABER 1962, BARLOW 1969, DURNAM et al. 1977). In wheat roots, for example, kinetin severely inhibits cell division while slightly promoting cell elongation (BURSTRÖM 1960), while in wheat leaves cell expansion continues in spite of inhibition of mitosis (HABER 1962). MILTHORPE and NEWTON (1963) noted that, under appropriate conditions, leaves of cucumber plants grown at high light intensity had more cells but less area than leaves grown at lower light intensity. Similarly, SHEN-MILLER et al. (1978) found that in geotropically stimulated corn roots, apparent cell division is greater in the lower half of the root even though there is less elongation there than in the upper half of the root. It seems necessary to conclude that the controls of cell division and cell expansion are not obligatorily coupled. Nevertheless, there is some indirect evidence for regulation of one process by the other. MEYER and BOYER (1972) found that artificially imposed low water potential in hypocotyl cells of soybean led to approximately equal inhibition of division and expansion. A positive relationship between maximum cell length or volume and cellular DNA content occurs in some cells. BOEKEN and VAN OOSTVELDT (1977) showed that the octaploid cells of pea epicotyls reach a greater maximum length than comparable tetraploid cells. Other instances of correlations between DNA content and cell expansion are known (BROUGHTON 1969, VAN OOSTVELDT et al. 1976) but the evidence relating DNA content itself to final cell length is indirect in all of these.

It has been suggested that a "proper balance" between cell division and expansion may favor maximal increase in fresh weight in tissue culture. HAGEN and MARCUS (1975) noted that cytokinin "reduced cell expansion" (as measured by final size of individual cells) in pith cell cultures of cytokinin-requiring lines of tobacco but that this is accompanied by (or more appropriately the reduced final cell size is a result of) enhanced cell division, so that the net effect of "inhibition of cell expansion" is an increase in tissue growth as measured by fresh weight increase. While the data indicate that the increase in fresh weight was maximal when the rate of cell division was high, it is more appropriate to view the increase in fresh weight as resulting from an *increase*, not a decrease, in cell expansion. The fact that cell division was enhanced more than cell expansion is a more acceptable explanation of the decreased final cell size. This avoids the misleading conclusion that decreased cell expansion can lead to enhanced growth.

GREEN (1976) noted the importance of recognizing that increased volume or fresh weight cannot arise from increased cell division. He emphasized that expansion and division together determine individual cell length but that only expansion determines total length. It is not valid to view division and elongation as alternative modes of generating volume. Shoots treated with GA frequently exhibit large increases in cell number but no increase in cell size. However, LOY (1977) has pointed out the fallacy of concluding that GA exerts its effect by stimulating cell division with no effect on cell elongation. The GA-treated shoot clearly increases in length and this can occur only by enhanced cell elongation, even though enhanced cell division may maintain a relatively constant or in some cases a diminished cell size.

2.7 Hormonal Regulation of Cell Expansion

The literature pertaining to the interaction of hormones in the regulation of plant cell expansion is extensive. Representative illustrations of hormone interaction will therefore be emphasized and results of studies with synthetic plant growth regulators will not be considered except where they contribute to our understanding of plant hormones.

2.7.1 Stem Cells

2.7.1.1 Effects of Auxin

The characteristic feature of auxins is their ability to promote cell elongation in coleoptile and stem tissue. For a detailed discussion of the auxins the reader is referred to AUDUS (1972). The emphasis here will be on the interaction of auxin with other plant hormones in the regulation of stem and coleoptile cell elongation. Coleoptiles are included here rather than in the section on leaf cells, since the hormonal sensitivity of coleoptiles is more stem-like than leaf-like.

Auxins stimulate the elongation of stem and coleoptile cells within 10 to 20 min of application (EVANS 1974). This is accompanied by enhancement of wall biosynthesis, wall loosening, and H^+ ion secretion (EVANS 1974). Stimulation of elongation by auxin occurs in hypocotyl, epicotyl, and stem cells of a wide variety of dicots (CLELAND et al. 1968, BARKLEY and EVANS 1970, AUDUS 1972, KAZAMA and KATSUMI 1973), and in the coleoptiles of all of the grasses that have been examined (AUDUS 1972, EVANS 1974). The epidermal cells of seedling stems and hypocotyls are most responsive to auxin and may, in large measure, control the growth response to this hormone (MASUDA and YAMAMOTO 1972).

Auxin has little or no stimulatory effect on the elongation of stem segments from *Avena* (KAUFMAN et al. 1969) or internode elongation in most genetic dwarfs (BRIAN and HEMMING 1955). In fact, in *Avena* stem sections auxin appears to act as a natural suppressor of GA-regulated growth since inhibition of IAA movement toward the internodal intercalary meristem of stem segments promotes elongation relative to the control (DAYANANDAN et al. 1977).

The response of many types of stem and hypocotyl cells to auxin is biphasic (VANDERHOEF and STAHL 1975). The initial phase of the response is similar to growth which can be induced by low pH, while the second phase has properties identical to the thoroughly studied long-term auxin response. Cytokinins inhibit the second phase of the auxin response, but have no effect on the early phase. The response of coleoptile segments to auxin is not so distinctly biphasic, and treatment with cytokinins cannot be used to inhibit selectively the initial phase of the auxin response (EVANS 1976a). This indicates a difference in the nature of the auxin response in dicots and monocots, or at least a difference in the nature of auxin/cytokinin interaction in these two groups.

ABA inhibits auxin-induced cell elongation (HEMBERG 1972a, PHILIPSON et al. 1973, REHM and CLINE 1973, ANKER 1975) but, whereas many inhibitory effects of ABA can be overcome by application of cytokinin, inhibition of coleoptile cell elongation by ABA cannot (HEMBERG 1972a).

2.7.1.2 Effects of Gibberellin

As with the auxins, there is a vast literature dealing with the effects of gibberellins on cell expansion. For detailed discussions the reader is referred to LOW (1975) and JONES (1973). Effects of GA on cell expansion in stems were reviewed by PHINNEY and WEST (1961).

GA promotes cell elongation in internodes of both normal (KAUFMAN et al. 1969, Mondal 1975) and dwarf (PHINNEY and WEST 1960) plants. GA also enhances cell elongation in rice coleoptile sections (SONI and KAUFMAN 1972) and in excised lettuce hypocotyls in light (DURNAM et al. 1977). Perhaps the most striking effect of GA on cell expansion is its enhancement of internode cell elongation in *Avena* stem segments. KAUFMAN and co-workers (KAUFMAN et al. 1969, KAUFMAN and JONES 1974) have shown that, in the presence of sucrose, GA can cause a 20-fold increase in *Avena* stem segment elongation over a 66-h period. This occurs without enhanced cell division. The stimulatory effect of GA on *Avena* stem elongation is inhibited by abscisic acid (KAUFMAN and JONES 1974), IAA (KAUFMAN et al. 1969), or cytokinin (JONES and KAUFMAN 1971) and in each instance the inhibition appears to be noncompetitive.

GA also promotes elongation in intact hypocotyls and hypocotyl segments of dicots (KATSUMI et al. 1965, KAMISAKA et al. 1972). This depends in large measure on some, as yet unidentified, factor produced in the cotyledons (KATSUMI et al. 1965, KAMISAKA 1973). The factor is not IAA or cytokinin since neither of these can substitute for the cotyledon factor. IAA is also capable of promoting growth in hypocotyl cells and, depending on the region of the hypocotyl and its age, the effect of IAA can be as great or greater than the effect of GA (KAZAMA and KATSUMI 1973). There are enough differences in the stimulation of hypocotyl growth by IAA and GA, however, to indicate a fundamental difference in the action of these two substances. Pretreatment of cucumber hypocotyl segments with GA, for example, increases their sensitivity to subsequent exposure to IAA (KATSUMI and KAZAMA 1974). Furthermore, stimulation of cucumber hypocotyl cell elongation by auxin is accompanied by a large increase in wall loosening (CLELAND et al. 1968), while GA does not substantially increase wall loosening. GA does enhance wall loosening in *Avena* stem segments (ADAMS et al. 1975, STUART and JONES 1977), but in *Avena* IAA acts to retard GA-promoted growth (KAUFMAN et al. 1969). In cucumber hypocotyl cells in which both IAA and GA stimulate elongation, sucrose inhibits IAA-induced elongation but acts synergistically with GA in the promotion of elongation. Also, in cucumber hypocotyl cells, 1% mannitol inhibits IAA-induced elongation but does not affect GA-induced elongation (KAZAMA and KATSUMI 1973).

2.7.1.3 Effects of Cytokinin, Ethylene, and Abscisic Acid

The most common effects of cytokinin, ethylene, and abscisic acid on stem and coleoptile cell elongation are inhibitory. They inhibit both auxin-enhanced and GA-enhanced elongation in stem and coleoptile tissue, as described in the preceding sections. However, in some instances, these "inhibitors" stimulate cell elongation.

Kinetin stimulates the elongation of *Avena* coleoptile segments, but the effect is slight compared with effect of IAA (HEMBERG and LARSSON 1972). The effect of kinetin may arise from enhancement of auxin levels rather than from a direct effect of the cytokinin (ANKER 1977). Cytokinins also increase the fresh weight of cultured soybean hypocotyl sections, an effect which has been used as a cytokinin bioassay (MANOS and GOLDTHWAITE 1976). This contrasts with the striking inhibition of auxin-enhanced cell elongation in soybean hypocotyl sections (VANDERHOEF et al. 1973). In cultured soybean hypocotyl segments the major effect of cytokinin is to enhance cell division. The increase in fresh weight results from expansion of the newly formed cells. The hormone appears to enhance division more than expansion, however, since the weight per cell decreases. In isolated soybean hypocotyl segments the primary effect of cytokinin is to interfere selectively with auxin-enhanced cell elongation (VANDERHOEF et al. 1973).

In light-grown hypocotyls of dwarf watermelon seedlings benzyladenine (BA) enhances ethylene production and promotes cell elongation (LOY and POLLARD 1977). Prevention of BA-induced ethylene production prevents BA-induced elongation, an indication that ethylene is necessary for BA-induced growth promotion.

Ethylene stimulates cell elongation in both intact (KU et al. 1970) and excised coleoptiles of rice (IMASEKI and PIJON 1970). This contrasts with the inhibitory effect of ethylene on coleoptile elongation in other grasses (SUGE 1971 a) and is probably a reflection of the adaptation of rice seedling growth to an aquatic environment. Ethylene also stimulates the growth of rice mesocotyls (SUGE 1971 b) and stems of the aquatic plant, *Callitriche*, as well as petioles of the floating leaves of *Ranunculus sceleratus* (MUSGRAVE et al. 1972). The stimulation of internode elongation in *Callitriche* by ethylene (MUSGRAVE et al. 1972) is of interest since it depends strongly on the presence of GA. Ethylene appears to augment the sensitivity of the tissue to GA. Ethylene is also reported to stimulate growth in hypocotyls of bean and cocklebur (GOTO and ESASHI 1974 b) and to enhance the response of *Avena* coleoptile cells to a subsequent treatment with auxin (MARINOS 1960).

ABA stimulates elongation in mesocotyls of dark-grown rice seedlings during long-term experiments (TAKAHASHI 1972). GA also promotes rice mesocotyl elongation, and the elongation in response to GA and ABA is greater than in response to either hormone alone (TAKAHASHI 1972).

2.7.2 Root Cells

The hormonal regulation of root elongation in the intact plant is poorly understood. This is partly because all the major plant hormones can inhibit root cell elongation at physiological concentrations. Since roots produce or at least contain each of these "inhibitors", it seems clear that root elongation must be subject to delicate controls involving the interaction of multiple factors. In the intact root the picture is further complicated because the effects of hormonal factors on elongation can be mediated indirectly by effects on meristematic activity or by modification of the time of initiation or cessation of cell elongation in addition to direct effects on rates of elongation.

2.7.2.1 Effects of Auxin

Detailed discussions of the effects of auxin on root elongation are included in STREET (1966), WHITTINGTON (1969), and SCOTT (1972). IAA remains the most likely candidate for the endogenous promoter of root elongation, though its effect is clearly dependent on other hormones and inhibitors (SCOTT 1972, ELLIOTT 1977). A major source of IAA in the root is that transported there from the shoot. There is apparently also some synthesis of IAA within the root itself (MORRIS et al. 1969, SCOTT 1972, BATRA et al. 1975, ELLIOTT 1977). There is disagreement concerning the supply of auxin to the elongation zone of the root. Auxin can move acropetally toward the elongation zone of the root (FELDMAN 1981), but there are conflicting opinions as to whether or not the auxin present in root caps (RIVIER and PILET 1974) can move basipetally toward the elongation zone. PERNET and PILET (1976) suggest that it cannot, while DAVIES et al. (1976) have shown that at least exogenously supplied ^{3}H-IAA can move from the cap to the elongation zone in corn roots.

IAA can either stimulate or inhibit root cell elongation, depending on concentration. BURSTRÖM (1969) reported that one and the same concentration of auxin can increase the rate of root elongation but cause a net reduction in final length by shortening the duration of the elongation phase. In some short-term experiments (LIST 1969) pulse application of IAA elicits a series of oscillations between growth promotion and inhibition.

Low concentrations of auxin sometimes enhance cell elongation in either isolated roots (EDWARDS and SCOTT 1974, 1976, 1977, ELLIOTT 1977) or intact roots (THIMANN 1936, BURSTRÖM 1969), but the effect is usually small. Inhibition by higher concentrations of auxin, on the other hand, is large and readily demonstrable in either excised or intact roots (THIMANN 1936, EVANS 1976b). Since auxin antagonists can enhance root growth (BURSTRÖM 1950, KOPCEWICZ and KHAN 1977), endogenous levels of auxin may be inhibitory to intact root elongation in some cases. According to THIMANN (1937) the effect of auxin on root cell elongation is the same as its effect on stem or coleoptile cell elongation except that root cells are much more sensitive to auxin. This is consistent with the dose-response relationships of auxin effects on root and stem cell elongation (THIMANN 1937). Recent experiments by EVANS and coworkers have provided strong evidence for the hypothesis that auxin action is similar in roots and shoots, with roots being more sensitive. In roots pretreated to suppress ethylene production (see Sect. 2.7.2.3), low concentrations of auxin (e.g., 10^{-10} M) consistently cause strong promotion of elongation and H^{+} efflux (MULKEY et al. 1982a, b). This is comparable to the action of moderate concentrations of auxin (e.g., 10^{-6} M) on shoot cells.

BURSTRÖM (1969) has questioned the hypothesis that auxin action is similar in shoots and roots, noting that in wheat roots, IAA from 3×10^{-8} to 3×10^{-7} M increases the rate of cell elongation but causes an increasing reduction in the duration of cell elongation. He suggested that auxin stimulates cell elongation in both roots and shoots and that the difference between the two is that, in root cells, auxin shortens the period during which cell elongation continues. According to this model the level of auxin in the root might be suboptimal with respect to the rate of cell elongation, but supraoptimal with respect to the duration of the cell elongation phase of development.

The suggestion that growth inhibitory levels of auxin initially increase the rate of cell elongation in root cells is inconsistent with short-term experiments demonstrating nearly immediate inhibition by auxin of the rate of root elongation (LIST 1969, EVANS 1976c). In recognition of this, BURSTRÖM (1969) recommends focusing on root growth that occurs beginning 24 h after hormone application and making long-term measurements since "what happens the 1st day after the addition of an inhibitor is so irregular that it is difficult to arrive at a meaningful interpretation of growth effects... ". It could be argued, however, that auxin effects occurring soon after application, more accurately reflect the auxin dependence of the tissue than effects observed a day or more later, especially in view of the capacity of roots to adapt to applied auxin (GOUGLER and EVANS 1981). In any case it is clear that the rate of root cell elongation is reduced, at least initially, by concentrations of auxin which promote the elongation of stem cells, while concentrations of auxin too low to affect shoot growth can promote root growth. This difference in sensitivity remains a fundamental difference in the auxin response of root cells and shoot cells.

The kinetics of inhibition of coleoptile cell elongation by supraoptimal concentrations of auxin also differ from the kinetics of inhibition of root cell elongation. In coleoptiles the initial effect of supraoptimal auxin is to promote cell elongation. This is followed by inhibition. In roots the initial effect of high concentrations of auxin is to inhibit elongation (MCBRIDE and EVANS 1977).

GASPAR et al. (1972) suggested that the inhibition of root elongation by IAA may be caused by a metabolite of IAA, most probably 3-methyleneoxindole (3-MO). However, it has since been shown that purified 3-MO does not inhibit root elongation (EVANS 1976c). These data in conjunction with the rapidity (15–20 min) with which intact root growth is inhibited by low concentrations of IAA, indicate that auxin inhibition of root cell elongation is a direct effect.

2.7.2.2 Effects of Gibberellin

Gibberellins can be found in roots and, they appear to be synthesized there (SCOTT 1972). However, there are conflicting reports dealing with the effects of exogenous GA on root elongation. Inhibition of root growth by GA has been reported by BURSTRÖM (1960), BUTCHER and STREET (1960a, b), and by MANOS (1961). Others have found no effect of physiological concentrations of GA on root elongation (BLAKELY et al. 1972, SVENSSON 1972) and, in a few cases (PALEG 1965, MERTZ 1966, FONNESBECH 1972), enhancement of root growth by GA has been reported. That the effect of GA on root elongation is variable is an indication that its action is dependent on other factors such as light, auxin content, inhibitor levels, etc. MERTZ (1966) reported that GA can overcome the inhibitory effects of Amo 1618 and CCC on root elongation in corn. Since IAA also overcomes inhibition by Amo 1618 and CCC, MERTZ suggested that GA acts by enhancement of endogenous auxin. Since root elongation is very sensitive to auxin, GA enhancement of auxin formation in roots may explain the inconsistencies in the effects of GA on root elongation. Application of GA to auxin-deficient roots might be expected to stimulate elongation, while application to roots containing optimal or supraoptimal levels of auxin would be expected to result in retardation of elongation (GASPAR et al. 1965).

The effect of GA on root elongation shows some light dependence. BURSTRÖM (1960) reported that GA reverses the inhibition of wheat root elongation by light while, in the dark, GA acts to inhibit root elongation, apparently by shortening the cell elongation phase of development.

2.7.2.3 Effects of Ethylene

Ethylene rapidly inhibits root elongation (ANDREAE et al. 1968, RADIN and LOOMIS 1969, CHADWICK and BURG 1970, RAUSER and HORTON 1975, MCBRIDE and EVANS 1977) and there is evidence that both roots (RADIN and LOOMIS 1969, CHADWICK and BURG 1970, MCBRIDE and EVANS 1977) and soil microorganism in the vicinity of roots (ILAG and CURTIS 1968, ARCHER 1976) produce ethylene. It therefore seems likely that ethylene plays a role in the control of root elongation.

Auxin enhances ethylene production in roots (CHADWICK and BURG 1967, 1970, ANDREAE et al. 1968, MCBRIDE and EVANS 1977) and this indicates that ethylene may mediate the inhibitory action of auxin on root elongation (CHADWICK and BURG 1967). Arguments against this view (ANDREAE et al. 1968, SVENSSON 1972, RAUSER and HORTON 1975, MCBRIDE and EVANS 1977) are based on the observations that: (1) saturating concentrations of ethylene inhibit root growth less than low concentrations of IAA and (2) in auxin-treated lentil roots detection of exogenously accumulating ethylene was possible only after auxin-induced growth inhibition. With the recent availability of fairly specific inhibitors of ethylene biosynthesis [aminoethoxyvinylglycine (AVG), aminooxyacetic acid (AOA)] and with improved methods for collecting and measuring ethylene, it is now possible to test more directly the involvement of ethylene biosynthesis in auxin action on root growth. MULKEY et al. (1982a) have studied the effects of auxin on the elongation of corn roots which have been treated with AVG and cobalt to suppress ethylene biosynthesis and action. In these pretreated roots, auxin in concentrations from 10^{-10} M through 10^{-8} M strongly (ca. 100%) promotes elongation. Under these conditions, auxin also induces rapid hydrogen ion secretion from roots (MULKEY et al. 1982b). Higher concentrations (e.g., 10^{-6} M) strongly inhibit growth and reverse H^+ secretion, leading to rapid H^+ uptake. In roots that are not pretreated to suppress ethylene production, it is difficult to obtain promotion of elongation by auxin at any concentration. The data indicate that the promotion of root growth by applied auxin may normally be obscured by rapid auxin-induced ethylene production, with the inhibitory effect of ethylene masking potential growth promotion by auxin. The data are also significant in showing that, under the proper conditions of measurement, concentrations of auxins which promote root growth strongly stimulate H^+ secretion from the roots, while concentrations of auxin which inhibit root growth cause apparent H^+ uptake by the roots. This indicates that regulation of H^+ movement by auxin may play a key role in auxin action on root growth just as it appears to in the regulation of shoot growth by auxin (see Sect. 2.7.1.1).

2.7.2.4 Effects of Abscisic Acid

There is evidence that the growth of root cells is controlled, in part, by one or more inhibitors produced in the root cap and moving basipetally toward the elongation zone (PILET 1974). Removal of the cap from the corn root is reported to increase the growth rate of the root. Roots with the caps partially excised or with caps removed but replaced asymmetrically show reduced growth in the region subjacent to the cap (GIBBONS and WILKINS 1970, PILET 1974). Since ABA has been found in root caps and since ABA shows transport and growth inhibition properties similar to those of the inhibitors found in root caps, ABA may constitute a portion of the naturally occurring inhibitors in the cap (KUNDU and AUDUS 1974, WILKINS and WAIN 1974, PILET 1975a, b).

ABA is reported to inhibit the elongation of roots and root segments in certain dicots (PILET 1970, EL-ANTABLY and LARSEN 1974, GAITHER et al. 1975). However, since inhibition by high concentrations of ABA is not as strong as inhibition by low levels of IAA (PILET 1970), there remains some doubt as to the importance of ABA itself as a natural inhibitor of root growth[1].

In seedlings of *Lemna minor*, exogenously applied ABA causes rapid inhibition of root elongation (NEWTON 1977 and references cited there). However, the inhibition by ABA (10 mg l^{-1}) is only on the order of 60%, much less than the usual 90% inhibition of elongation by much lower concentrations of auxin (ca. 0.02 mg l^{-1}) in roots (EVANS 1976b, c). ABA has little or no inhibitory effect on the growth of seedling corn roots in the varieties Bear Hybrid WF 9 × 38 (GOUGLER and EVANS 1979), Golden Cross Bantam 70 (SUZUKI et al. 1979), or Velox (SCHURZMANN and HILD 1980). ABA is reported to promote root growth (GAITHER et al. 1975) in some instances or to reduce the inhibitory effect of auxin (PILET 1970).

2.7.2.5 Effects of Cytokinin

Cytokinins occur in roots and there is evidence that they are synthesized there (SKENE 1975, FELDMAN 1979, WAREING et al. 1977). Cytokinin can inhibit root elongation (DARIMONT et al. 1971, SCOTT 1972, SVENSSON 1972) especially when applied to roots growing in the dark (DARIMONT et al. 1971, SVENSSON 1972). Although cytokinins may regulate meristem activity in roots (see Sect. 4) their

---

[1] As to the role of ABA (and auxin) in the gravitropism of roots, the reader is referred to Vol. 7, Chap 5.2.3, p. 615ff., this series

effect on growth may be a more direct effect on the elongation zone. In isolated corn roots kinetin retards elongation and division. Surprisingly, inhibition of wheat root elongation by cytokinin can be overcome by antiauxin (SVENSSON 1972).

Kinetin promotes the formation of ethylene in the elongation zone of corn roots. Although enhancement of ethylene production by kinetin is greater than by auxin, SVENSSON (1972) concluded on the basis of other data that ethylene does not mediate the inhibitory effect of kinetin on root elongation. A similar conclusion was reached by DUBUCQ (1976).

GASPAR and XHAUFFLAIRE (1967) noted that kinetin-treated lentil roots have more IAA oxidase activity than control roots. They suggested that kinetin inhibition of root elongation may be due to a reduction in the auxin content of the root. This was supported by DARIMONT et al. (1971) who found that kinetin reduces the IAA content of lentil roots.

2.7.3 Leaf Cells

2.7.3.1 Effects of Auxin

There are reports of both stimulatory and inhibitory effects of auxin on leaf cell expansion. Temporary immersion of intact *Avena* seedlings in IAA, for example, is reported to suppress the growth of the primary leaf (BALL and DYKE 1956). In leaves of *Xanthium*, auxin transport from the blade, and presumably synthesis within the blade, is highest during the period of maximal cell enlargement (DEGROOTE and MUIR 1977). There is also a correlation between increased auxin transport from the leaves and decreased cell division activity. This contrasts with a study by HUMPHRIES and WHEELER (1964), who found that the level of IAA was high in young primary and trifoliate leaves of *Phaseolus* but dropped nearly to zero in leaves in which cell division had ceased. They suggested that free IAA may be necessary for cell division in these leaves.

Auxin causes the lateral margins to curve upward and in toward the midrib in young leaves of *Phaseolus*. This is referred to as leaf hyponasty and appears to arise from a direct effect of auxin on cell expansion since neither ethylene nor kinetin causes blade curvature (LIPPINCOTT and LIPPINCOTT 1971). Auxin-induced leaf hyponasty suggests that endogenous auxin may normally play a role in regulating planar blade growth (HAYES and LIPPINCOTT 1976). Leaf blades which are submerged in auxin also exhibit hyponasty, indicating that the cells of the lower surface of the blade show a stronger growth response than cells of the upper surface.

There is evidence that the growth of petioles is dependent on auxin from the blade. Blade removal in *Helianthus* inhibits petiole elongation about 90%, an effect which is partially overcome if auxin is supplied to the cut surface of the petiole (PALMER 1972). Application of auxin to either the upper or lower surfaces of debladed petioles causes an increase in growth rate. DAYANADAN et al. (1976) have found that auxin, but not GA or kinetin, promotes cell elongation in parenchyma, collenchyma and vascular elements of the sheath pulvinus of *Avena* leaves.

2.7.3.2 Effects of Gibberellin

GA promotes cell expansion in cotyledons of certain light-grown seedlings (BANERJI and LALORAYA 1966, RIJVEN and PARKASH 1970, LETHAM 1971). GA also stimulates expansion of etiolated bean leaves, an effect which is synergistically enhanced by IAA (SARMA 1979). GA also enhances leaf growth in developing orchid protocorms (FONNESBECH 1972) and partially counteracts the inhibitory effect of ethylene on abaxial petiole cell elongation (PALMER 1972). In young developing leaf blades, however, GA is usually inhibitory. GA also inhibits leaf expansion in dwarf Japanese holly (WRIGHT and AUNG 1975), pea (BRIAN and HEMMING 1955) and *Coleus* (TRONCHET 1968). GA can also alter leaf shape if applied during the period of active leaf expansion. In trifoliate leaves of *Phaseolus*, for example, GA causes a shift in the length/breadth ratio with no effect on cell number (FELIPPE and DALE 1968).

2.7.3.3 Effects of Ethylene

Ethylene elicits a differential growth in the petioles of various leaves. In general ethylene promotes growth in cells of the upper half of the petiole relative to those in the lower half resulting in a downward curvature or epinasty of the leaf (PALMER 1972). Ethylene causes a transient enhancement of the growth rate of the upper half of the petiole of *Helianthus* and a transient inhibition of the elongation of the lower half of the petiole. The inhibition of elongation in the lower half can be reduced by GA_3 pretreatment (PALMER 1972). LYON (1970) has suggested that ethylene-induced petiole curvature arises from diversion of lamina auxin toward the upper half of the petiole or inhibition of auxin transport away from the upper half of the petiole. This is not supported by the finding (PALMER 1972) that ethylene induces epinastic curvature in debladed petioles exhibiting little or no extension growth and that ethylene-induced epinasty in *Coleus* leaves occurs with no effect on the distribution of IAA in the petiole (PALMER 1976).

2.7.3.4 Effects of Abscisic Acid

Abscisic acid inhibits frond cell expansion in duckweed (NEWTON 1977), reducing both the increase in width and length of the developing frond. ABA has similar though somewhat weaker effects on *Spirodela* (VAN STADEN and BORNMAN 1970), and in both cases the inhibitory effects of ABA can be partially overcome by cytokinin (VAN OVERBEEK et al. 1968, VAN STADEN and BORNMAN 1970).

2.7.3.5 Effects of Cytokinin

Cytokinin stimulates cell expansion in cotyledons of a variety of dicotyledonous plants, whether grown in light or darkness (NARAIN and LALORAYA 1974). Cytokinins also enhance expansion in young bean leaves (LEOPOLD and KAWASE 1964) and in other leaf types (LETHAM 1969), and the increased expansion occurs with little enhancement of cell division (LETHAM 1969, TSUI et al. 1980). The effect of 6-benzylaminopurine (BAP) on leaf cell expansion is concentration-dependent. In discs from etiolated bean leaves, 4 g l^{-1} BAP stimulates expansion, while a lower concentration (0.001 mg l^{-1}) is inhibitory (KULAEVA and TSIBULYA 1974).

2.7.4 Other Cell Types

2.7.4.1 Pollen Tubes

MALIK and MEHAN (1975a, b) studied the effects of phytohormones on pollen tube elongation in *Calotropis procera*. IAA was found to inhibit pollen tube elongation, perhaps by inducing ethylene formation. Inhibition of GA biosynthesis was observed to inhibit pollen tube elongation, while treatment with either GA, cytokinin, or ABA resulted in enhancement of growth. MATSUBARA (1973) found that cytokinin would not enhance the growth of pollen tubes of lily in vitro but did act in vivo to allow continued growth in cultivars in which pollen tube growth is normally prematurely terminated. The data indicate that the growth-promoting effect of cytokinin in vivo arises from its capacity to mobilize materials toward the developing pollen tube.

2.7.4.2 Epidermal Hairs

TING and coworkers (BEASLEY and TING 1973, 1974, DHINDSA et al. 1976) examined the effects of plant hormones on the growth of cotton fibers in vitro. In fertilized ovules, GA causes a large increase in fiber development while IAA causes a smaller increase. Both kinetin and ABA are inhibitory to fiber development but do not prevent the promotive effect of GA. The authors suggest that the larger promotive effect of GA relative to that of IAA may be due in part to differences in the content of these hormones in the isolated ovules.

2.7.5 Lower Plants

The process of cell expansion in lower plants is undoubtedly similar mechanistically to cell expansion in higher plants. Turgor, wall loosening, and wall biosynthesis must be indispensible to continued elongation in both. But is the chemical regulation of cell expansion similar in higher and lower plants? This question will be addressed by examining the occurrence of higher plant hormones in lower plants and the effect of applied hormones on cell expansion in the algae and fungi. For information on hormone effects on the ferns, mosses, and liverworts the reader may see BRIAN (1963), MILLER et al. (1970), BÄHRE (1976), and Bopp (1980, 1983).

2.7.5.1 Algae

The effects of phytohormones on growth and development in the algae were summarized by LANG (1965). Since then a great deal has been published on the occurrence of phytohormones and their effects on cell expansion in the algae. Reviews of the hormonal regulation of cell expansion and division in the algae can be found in PROVASOLI and CARLUCCI (1974) and in a series of seven articles by AUGIER (1978 and references cited there). There is ample evidence for the occurrence of substances with auxin-like (BUGGELN and CRAIG 1971, AUGIER 1976a), gibberellin-like (ABE et al. 1974, AUGIER 1976b, TAYLOR and WILKINSON 1977), and cytokinin-like (BENTLEY-MOWAT and REID 1968, VAN STADEN and BREEN 1973, MELKONIAN and WEBER 1975, AUGIER 1976b) activity in algae. There is also at least one report of ethylene production by algae (OSBORNE 1975), and a report of small enhancement of growth in *Scenedesmus* by low levels of ethylene (CHAPMAN and VANCE 1973). Ethylene is also reported to stimulate division in the diatom *Nitzschia putrida* (WAGNER 1970).

There appears to be no published evidence for the occurrence of ABA in algae, although ABA has been found to inhibit morphogenesis in *Petroglossum* and growth in the gametophyte of *Ecklonia* (cited in AUGIER 1976c). The dearth of reports on ethylene and ABA in algae may be due in part to the concentration of research activity on the role of auxins, gibberellins, and cytokinins in these organisms. However, many algae contain a protein capable of interfering with ethylene biosynthesis in higher plants (WATANABE and KONDO 1976) and it is possible that this inhibitor may also interfere with ethylene production in the algae.

There are numerous reports of stimulation of cell expansion and cell division by auxin, GA, and cytokinin in a variety of multicellular and in some unicellular algae. Enhancement of growth by these higher plant hormones has been reported in the Chlorophyceae, Charophyceae, Xanthophyceae, Phaeophyceae, Rhodophyceae, and Cyanophyceae (see discussion by AUGIER 1976c).

Since there are several thorough reviews of the effects of higher plant hormones on algae, only a few studies will be described here as examples.

Effects of Auxin. In a study of the effect of auxin on the unicellular green alga, *Acetabularia*, THIMANN and BETH (1959) found that IAA promotes stalk elongation and cap formation but inhibits rhizoid elongation. Although some of the effects were slow to develop, the results are interesting in view of the analogous stimulation of stem elongation and inhibition of root elongation by auxin in higher plants.

IAA is also reported to enhance growth in other genera of green algae and in certain brown algae (MOHSEN et al. 1974, PROVASOLI and CARLUCCI 1974) but the effects are often small. PROVASOLI and CARLUCCI (1974) point out that many reports of auxin promotion of algal growth may be due to the ethanol in the solutions used or to the nutrient benefit of the excessively high concentrations used. In the brown alga, *Alaria esculenta*, BUGGELN (1976) was unable to demonstrate a promotive effect of auxin on blade growth.

In the blue-green algae auxin exerts a concentration-dependent inhibition (in *Plectonema*) or promotion (in *Lyngbya*) of growth and induces the formation of akinete-like structures as well as heterocysts and hormogonia (DHALIWAL et al. 1973). Auxin stimulates growth and swarmer production in the green alga, *Ulva fasciata* (MOHSEN et al. 1974), and promotes cell division in the desmid, *Micrasterias thomasiana* (WOOD and BERLINER 1979).

Effects of Gibberellin. In a great many studies, GA's were found to inhibit algal growth (see Table 8 in AUGIER 1976c). In contrast, GA enhances cell division in the green alga, *Ulothrix* (CONRAD et al. 1959) and promotes growth in cultured *Ulva lactuca* and to a lesser extent in various species of *Chlorella* (PROVASOLI and CARLUCCI 1974). A substantial promotion of elongation by GA was obtained in *Enteromorpha prolifera* and in gametophytes of *Ecklonia* by JENNINGS (1968). GA also accelerates growth and RNA synthesis in *Euglena*, but the effect appears not to be GA-specific (BRALCZYK et al. 1978).

Effects of Cytokinin. Cytokinin activity has been reported in various species of green, brown and red algae (KENTZER et al. 1980). Reports of enhancement of algal cell growth by cytokinins include studies of the green alga, *Ulva* (PROVASOLI 1958), the brown alga, *Ectocarpus* (PEDERSEN 1968), and the red alga, *Hypnea* (JENNINGS 1969). By contrast, kinetin at concentrations up to 10^{-5} g ml^{-1} has no growth-promoting effect in cultures of *Fritschiella*, while higher concentrations are inhibitory (MELKONIAN and WEBER 1975). MELKONIAN and WEBER point out that reports of cytokinin stimulation of algal growth usually stem from work done with multicellular algae while the single-celled algae, with a few exceptions (BENTLEY-MOWAT and REID 1967), do not respond. Cytokinin-like substances have been extracted from the thallus of *Fucus vesiculosus*, and it is reported that cytokinins enhance adventitious branch formation on the thallus, especially in specimens collected in the fall (KENTZER et al. 1980). In a similar study, kinetin was found to enhance leaf growth in fronds of the *Conchocelis* phase of *Porphyra tenera* but to inhibit stem elongation (SHIMO and NAKATANI 1964 — cited by PROVASOLI and CARLUCCI 1974). These studies are of interest in view of the parallel nature of these effects with the effects of cytokinins in certain higher plants.

2.7.5.2 Fungi

Effects of Auxin. Although IAA is a common metabolite in fungi, there may be no major role for auxin in fungal growth regulation (GRUEN 1959). Auxin effects on fungal growth are usually inhibitory, and BRIAN (1963) has suggested that the few reports of promotion by auxin may be due to the nutrient contribution of the nitrogen-containing auxins.

In certain yeasts auxin has been reported to stimulate cell elongation, the effect occurring under culture conditions in which cell division is absent (YANAGISHIMA 1963). Yeast sexual hormones which stimulate cell expansion in opposite mating types have also been reported to stimulate both wall loosening and elongation in *Avena* coleoptile segments (MASUDA et al. 1970). The growth-promoting effect of auxin in yeast is obtainable with IAA and NAA but not with inactive auxin analogues. The response appears to be a true auxin response but it is probably not of wide significance since it can be obtained only in certain strains of yeast and only under specific culture conditions (SHIMODA and YANAGISHIMA 1969). In at least one case, a strain unresponsive to auxin became auxin-sensitive after exposure to GA (SHIMODA and YANAGISHIMA 1969). In *Saccharomyces cerevisiae*, BAREA et al. (1974) found that cell division and expansion were promoted by IAA, but that even larger effects could be obtained with kinetin or GA_3, so the response is not auxin-specific in all strains.

Auxin stimulates the growth of *Agaricus campestris* (GUHA and BANERJEE 1974) though the effect is small. GENTILE and KLEIN (1955) showed that IAA could overcome growth inhibition by an auxin antagonist in the fungus *Diplodia natalensis*. They interpreted this as evidence of a role for auxin in the growth of the fungus, even though very low levels of auxin were inhibitory in the absence of an auxin antagonist.

Effects of Gibberellin. The gibberellins were first discovered in the fungus *Gibberella fujikuroi* and this organism has since been shown to produce at least 18 different GA's. However, this should not be taken as evidence that the gibberellins are widely occurring in fungi or that they play a role in fungal growth and development, even in *Gibberella fujikuroi*. In a survey of 1500 culture filtrates from 100 different fungal and actinomycete genera, CURTIS (1957) was unable to detect gibberellin-like substances. PEGG (1973) reported that sporophores of *Agaricus* contain a GA-like substance but, in contrast to an earlier report by ALEXSANDROV (cited by PEGG 1973), PEGG found that neither this nor authentic GA's stimulated growth in *Agaricus* sporangiophores or vegetative mycelia.

In general, GA's have no effect on cell expansion in the fungi except for inhibition at high concentration. In view of the dramatic effects of GA on cell expansion in higher plants, it is unlikely that any comparable action in the fungi would continue to go undetected, and it seems logical to conclude that the GA's play no major role in growth regulation in the fungi.

Effects of Ethylene. The production of ethylene is a common but by no means ubiquitous activity among the fungi (ILAG and CURTIS 1968, ABELES 1973, CONSIDINE and PATCHING 1975). The ethylene-producing systems of fungi and higher plants may be similar, at least under some culture conditions (CHALUTZ and LIEBERMAN 1977). Fungi growing in soil produce ethylene (LYNCH and HARPER 1974), and some have suggested that this, along with ethylene produced by higher plants, may be important in restricting the growth of soil fungi (SMITH 1973). However, ethylene appears to be relatively ineffective as an inhibitor of fungal growth in culture (ARCHER 1976), and it seems unlikely that the low levels of ethylene in the soil exert a general inhibitory effect on fungal growth and development. In *Fusarium oxysporum* (SWART and KAMERBEEK 1977) and *Penicillium digitatum* (SPALDING and LIEBERMAN 1965) rates of ethylene production and growth show no correlation. The cultivated mushroom, *Agaricus bisporus*, exhibits a peak in ethylene production at a certain stage in development in the fruiting body (WOOD and HAMMOND 1977). The ethylene is produced by the mycelium, not the fruiting body, and there is no direct evidence that the ethylene plays a role in the life cycle of this fungus.

The sporangiosphores of *Phycomyces* produce ethylene, and ethylene enhances elongation of the sporangiophore and interferes with its normal avoidance or autochemotropic response (RUSSO et al. 1977). RUSSO (1977) postulated that class I mutants of *Phycomyces* lack autochemotropism because they fail to produce ethylene even though they are capable of responding to exogenously applied ethylene. He showed that class I and III mutants lack autochemotropism even though they do produce ethylene. In this case, however, the mutants have an abnormal growth response to ethylene. The data are indicative of a specific growth-regulating role for ethylene in this particular fungus. In more recent experiments, however, the role of ethylene in the avoidance response in *Phycomyces* has been brought into question. KENDE et al. (1982) noted that sporangiophores bend away from a barrier even when the barrier is coated with an ethylene-absorbing material, indicating that the avoidance response does not depend on asymmetric accumulation of ethylene.

Effects of Cytokinin. As mentioned earlier, kinetin promotes cell division and expansion in certain strains of yeast but the response is not cytokinin-specific. In *Agaricus campestris* cytokinins promote growth slightly at low concentrations but inhibit growth at concentrations above 1 mg l^{-1}. Cytokinins also promote growth in the fungus, *Verticillium albo-atrum*. However, adenine and other nitrogen bases have similar effects at the same concentration (ATMAR et al. 1976). Kinetin has a strong stimulatory effect on perithecia production in *Neurospora crassa* (LEE 1961).

2.7.6 Summary

The dramatic effects of auxins, gibberellins, and cytokinins on growth and development in higher plants are, for the most part, lacking in the fungi and the algae. There is evidence that these hormones or substances similar to them occur in many algae and fungi but in most instances there has been no rigorous chemical identification of endogenous "hormones." We therefore continue to speak of the presence of "auxin-like", "gibberellin-like", or "cytokinin-like" substances in lower plants.

BRIAN (1963) pointed out that the responses to auxin of coenocytic and multicellular algae are more widespread than those of unicellular algae. He suggested that hormonal control of growth has evolved along with structural complexity so that the multicellular green algae might be expected to be more similar to higher plants in their hormonal responsiveness.

Although there are many reports of higher plant hormone-like substances in the algae and fungi and numerous studies demonstrating effects of exogenously applied hormones on growth, especially in the algae, this does not in itself constitute evidence that hormones similar or identical to those in higher plants play a role in the regulation of growth and development in these organisms. Evidence for this must stem from more detailed studies of correlations between hormone levels and development in lower plants. Studies of mutants of algae or fungi incapable of producing or responding to hormones should be useful in this regard (see RUSSO 1977).

References

Abe H, Uchiyama RM, Sato R, Muto S (1974) Plant growth regulators occurring in marine algae. In: Plant growth substances 1973. Hirokawa, Tokyo, pp 201–206

Abeles FB (1966) Auxin stimulation of ethylene evolution. Plant Physiol 41:585–588

Abeles FB (1973) Ethylene in plant biology. Academic Press, London New York

Abeles FB, Lonski J (1969) Stimulation of lettuce seed germination by ethylene. Plant Physiol 44:277–280

Abeles FB, Rubinstein B (1964) Regulation of ethylene evolution and leaf abscission by auxin. Plant Physiol 39:963–969

Acton GJ, Murray PB (1974) The roles of auxin and gibberellin in reversing radiation inhibition of hypocotyl lengthening. Planta 117:219–226

Adams DO, Yang SF (1979) Ethylene biosynthesis: identification of 1-amino-cyclopropane-1-carboxylic acid as an intermediate in the conversion of methionine to ethylene. Proc Natl Acad Sci USA 76:170–174

Adams PA, Montague MH, Tepfer M, Rayle DL, Ikuma H, Kaufman PB (1975) Effect of gibberellic acid on the plasticity and elasticity of *Avena* stem segments. Plant Physiol 56:757–760

Addicott FT, Lyon JL (1969) Physiology of abscisic acid and related substances. Annu Rev Plant Physiol 20:139–146

Aharoni N (1977) Enhancement of ethylene production in water-stressed leaves. Plant Physiol 59:Suppl, p 39

Aitchison PA, Yeoman MM (1974) Control of periodic enzyme synthesis in dividing plant cells. In: Padilla GM, Cameron IL, Zimmerman A (eds) Cell cycle controls. Academic Press, London New York

Altman A, Goren R (1971) Promotion of callus formation by abscisic acid in citrus bud cultures. Plant Physiol 47:844–846

Andreae WA, Venis MA, Jursic F, Dumas T (1968) Does ethylene mediate root growth inhibition by indole-3-acetic acid? Plant Physiol 43:1375–1379

Anker L (1973) The auxin production of the physiological tip of the *Avena* coleoptile and the repression of tip regeneration by indoleacetic acid (not by naphthylacetic acid and 2,4-dichlorophenoxyacetic acid). Acta Bot Neerl 22:221–227

Anker L (1975) Auxin-synthesis inhibition by abscisic acid, and its reversal by gibberellic acid. Acta Bot Neerl 24:339–347

Anker L (1977) An analysis of kinetin effects on coleoptile growth. Acta Bot Neerl 26:369–375

Anonymous (1978) Growth regulators identified in plants. Chem Eng News Sept 18, 19–20

Apelbaum A, Burg SP (1972a) Effect of ethylene on cell division and deoxyribonucleic acid synthesis in *Pisum sativum*. Plant Physiol 50:117–124

Apelbaum A, Burg SP (1972b) Effects of ethylene and 2,4-dichlorophenoxyacetic acid on cellular expansion in *Pisum sativum*. Plant Physiol 50:125–131

Archer SA (1976) Ethylene and fungal growth. Trans Brit Mycol Soc 67:325–326

Arias I, Williams PM, Bradbeer JW (1976) Studies in seed dormancy. IX. The role of gibberellin biosynthesis and the release of bound gibberellin in the post-chilling accumulation of gibberellin in seeds of *Corylus avellana* L. Planta 131:135–139

Atkin RK, Barton GE, Robinson DK (1973) Effect of root-growing temperature on growth substances in xylem exudate of *Zea mays*. J Exp Bot 24:475–487

Atmar VT, Throneberry GO, Kuehn GD (1976) Effects of adenine and cytokinins on growth and protein kinase activity of *Verticillium albo-atrum*. Mycopathologia 59:171–174

Audus LJ (1972) Plant growth substances Vol I. Chemistry and physiology, 3rd edn. Hill, London

Augier H (1976a) Les hormones des algues. État actuel des connaissances. I. Recherche et tentatives d'identification des auxins. Bot Mar 19:127–143

Augier H (1976b) Les hormones des algues. État actuel des connaissances. II. Recherche et tentatives d'identification des gibberellins, des cytokinines et de diverses autres substances de nature hormonale. Bot Mar 19:245–254

Augier H (1976c) Les hormones des algues. État actuel des connaissances. III. Rôle des hormones dans les modalités de croissance et de développement des thalles. Bot Mar 19:351–377

Augier H (1978) Les hormones des algues. État actuel des connaissances. VII. Applications, conclusion, bibliographie. Bot Mar 21:175–197

Avanzi MG (1951) Ricerche sulla poliploidia somatica nei tessuti differenziati della radico di alcune graminaceae. Caryologia 3:351–369

Back A, Richmond AE (1971) Interrelations between gibberellic acid, cytokinins and abscisic acid in retarding leaf senescence. Physiol Plant 24:76–79

Bähre R (1976) Regulation des Protonemawachstums von *Athyrium filix femina* (L.) Roth III. Wirkung von Auxin (IAA) und Antiauxin (PCIB) unter verschiedenen Lichtbedingungen. Z Pflanzenphysiol 77:323–335

Balasimha D, Ram G, Tewari MN (1977) Cytokinin–coumarin interaction in relation to growth, sulfhydryl, chlorophylls, and peroxidase-activity in *Phaseolus radiatus* L. seedlings. Biochem Physiol Pflanz 171:49–54

Ball NG, Dyke IJ (1956) The effects of indoie-3-acetic acid and 2:4-dichlorophenoxyacetic acid on the growth rate and endogenous rhythm of intact *Avena* coleoptiles. J Exp Bot 7:25–41

Bandurski RS, Schulze A (1977) A light-induced change in indole-3-acetic acid concentration. Plant Physiol 59:Suppl, p 10

Banerji D, Laloraya MM (1966) Comparative effects of indole-3-acetic acid, kinetin and gibberellic acid on the growth of isolated *Cucurbita pepo* cotyledons. Curr Sci 35:601–602

Barea JM, Navarro E, Palomares A, Montoya E (1974) A rapid microbiological assay method for auxins, gibberellic acid and kinetin using yeast. J Appl Bacteriol 37:171–174

Barkley GM, Evans ML (1970) Timing of the auxin response in etiolated pea stem sections. Plant Physiol 45:143–147

Barlow PW (1969) Cell growth in the absence of division in a root meristem. Planta 88:215–223

Barlow PW (1973) Mitotic cycles in root meristems. In: Balls M, Billet FS (eds) The cell cycle in development and differentiation. Cambridge Univ Press, Cambridge, pp 133–165

Barlow PW (1976a) Towards an understanding of the behavior of root meristems. J Theor Biol 57:433–451

Barlow PW (1976b) The effect of ethylene on root meristems of *Pisum sativum* and *Zea mays*. Planta 131:235–243

Barlow PW, Macdonald PDM (1973) An analysis of the mitotic cell cycle in the root meristem of *Zea mays*. Proc Roy Soc Lond B 183:385–398

Basrur SV, Fletcher RA, Basrur PK (1976) In vitro effects of 2,4-dichlorophenoxy acetic acid (2,4-D) on bovine cells. Can J Comp Med 40:408–415

Batra MW, Edwards KL, Scott TK (1975) Auxin transport in roots: Its characteristics and relationships to growth. In: Torrey JG, Clarkson DT (eds) The development and function of roots. Academic Press, London New York, pp 299–325

Beasley CA, Ting IP (1973) The effects of plant growth substances on in vitro fiber development from fertilized cotton ovules. Am J Bot 60:130–139

Beasley CA, Ting IP (1974) Phytohormone effects on in vitro cotton seed development. In: Plant growth substances 1973. Hirokawa, Tokyo, pp 917–924

Bentley-Mowat JA, Reid SM (1967) Effect of gibberellins, kinetin and other factors on the growth of unicellular marine algae in culture. Bot Mar 12:185–199

Bentley-Mowat JA, Reid SM (1968) Investigation of the radish leaf bioassay for kinetins, and demonstration of kinetin-like substances in algae. Ann Bot 32:23–32

Bergmann L (1965) The effect of kinetin on the metabolism of plant tissue cultures. In: White PR (ed). Plant tissue culture. McCutchan, Berkeley, pp 171–181

Berrie AMM, Robertson J (1976) Abscisic acid as an endogenous component in lettuce fruits, *Lactuca sativa* L. cv. Grand Rapids. Does it control thermodormancy? Planta 131:211–215

Beyer EM, Morgan PW (1970) Effect of ethylene on the uptake, distribution, and metabolism of indoleacetic acid-1-C^{14} and -2-C^{14} and naphthaleneacetic acid-1-C^{14}. Plant Physiol 46:157–162

Biro RL, Hunt ER Jr, Erner Y, Jaffe MJ (1980) Thigmomorphogenesis: Changes in cell division and elongation in the internodes of mechanically perturbed or ethrel-treated bean plants. Ann Bot 45:655–664

Black M, Vlitos AJ (1972) Possible interrelationships of phytochrome and plant hormones. In: Mitrakos K, Shropshire W (eds) Phytochrome. Academic Press, London New York, pp 517–550

Blakely LM, Rodaway SJ, Hollen LB, Croker SG (1972) Control and kinetics of branch root formation in cultured root segments of *Haplopappus ravenii*. Plant Physiol 50:35–42

Blumenfeld A, Gazit S (1970) Interaction of kinetin and abscisic acid in the growth of soybean callus. Plant Physiol 45:535–536

Boeken G, Van Oostveldt P (1977) Gibberellic-acid-induced cell elongation in pea epicotyls: Effect on polyploidy and DNA content. Planta 135:89–91

Bopp M (1980) The hormonal regulation of morphogenesis in mosses. In: Skoog F (ed) Plant growth substances 1979. Springer Verlag, Berlin, Heidelberg, pp 351–361

Bopp M (1983) Developmental physiology of bryophytes. In: Schuster RM (ed) New manual of bryology. The Hattori Botanical Laboratory, Nichinan, Miyazaki, Japan, Vol. 1, pp 276–324

Boyle WS, Evans JO (1974) Effects of glyphosate and ethephon on meiotic chromosomes of *Secale cereale* L. J Hered 65:250

Bradford KJ, Dilley DR (1977) Ethylene production, growth, and epinasty of tomato plants as influenced by oxygen supply to the root. Plant Physiol 59:Suppl 37

Bralczyk J, Wielgat B, Wasilewska-Dabrowska LD, Kleczdowski K (1978) Growth-accelerating response of *Euglena gracilis* Z. to gibberellic acid. Plant Sci Lett 12:265–271

Brian PW (1963) Growth-regulating hormones in lower plants. Sci Prog 51:226–238

Brian PW, Hemming HG (1955) The effect of gibberellic acid on shoot growth of pea seedlings. Physiol Plant 8:669–681

Briggs WR (1963) Red light, auxin relationships and the phototropic responses of corn and oat coleoptiles. Am J Bot 50:196–207

Broughton WJ (1969) Relations between DNA, RNA and protein synthesis, and the cellular basis of the growth responses in gibberellic acid treated pea internodes. Ann Bot 33:227–243

Bryant JA (1976) The cell cycle. In: Bryant JA (ed) Molecular aspects of gene expression in plants. Academic Press, London New York, pp 177–216

Buggeln RG (1976) Auxin, an endogenous regulator of growth in algae? J Phycol 12:355–358

Buggeln RG, Craig JS (1971) Evaluation of evidence for the presence of indole-3-acetic acid in marine algae. Planta 97:173–178

Burg SP (1973) Ethylene in plant growth. Proc Natl Acad Sci USA 70:591–597

Burg SP, Burg EA (1966) The interaction between auxin and ethylene and its role in plant growth. Proc Natl Acad Sci USA 55:262–269

Burg SP, Burg EA (1967) Inhibition of polar auxin transport by ethylene. Plant Physiol 42:1224–1228

Burg SP, Dijkman MJ (1967) Ethylene and auxin participation in pollen induced fading of *Vanda* orchid blossoms. Plant Physiol 42:1648–1650

Burström H (1950) Studies on growth and metabolism of roots. IV. Positive and negative auxin effects on cell elongation. Physiol Plant 3:277–292

Burström H (1960) Influence of iron and gibberellic acid on the light sensitivity of roots. Physiol Plant 13:597–615

Burström H (1969) Influence of the tonic effect of gravitation and auxin on cell elongation and polarity in roots. Am J Bot 56:679–684

Butcher DN, Street HE (1960a) Effects of kinetin on the growth of excised tomato roots. Physiol Plant 13:46–55

Butcher DN, Street HE (1960b) The effects of gibberellins on the growth of excised tomato roots. J Exp Bot 11:206–216

Chadwick AV, Burg SP (1967) An explanation of the inhibition of root growth caused by indole-3-acetic acid. Plant Physiol 42:415–420

Chadwick AV, Burg SP (1970) Regulation of root growth by auxin-ethylene interaction. Plant Physiol 45:192–200

Chalutz E, Lieberman M (1977) Ethylene production by *Penicillium digitatum*. Plant Physiol 59:Suppl, p 38

Chapman RW, Vance BD (1973) The effects of ethrel on the growth of *Scenedesmus quadricauda*. J Phycol 9:Suppl 13

Chin T-Y, Beevers L (1970) Changes in endogenous growth regulators in nasturtium leaves during senescence. Planta 92:178–188

Chin T-Y, Meyer MM Jr, Beevers L (1969) Abscisic-acid-stimulated rooting of stem cuttings. Planta 88:192–196

Cleland R, Thompson M, Rayle DL, Purves WK (1968) Difference in effects of gibberellins and auxins on wall extensibility of cucumber hypocotyls. Nature 219:510–511

Clowes FAL (1954) The promeristem and the minimal construction centre in grass root apices. New Phytol 53:108–116

Clowes FAL (1969) Anatomical aspects of structure and development. In: Whittington WJ (ed) Root growth. Butterworth, London, pp 3–19

Clowes FAL (1972) Regulation of mitosis in roots by their caps. Nature New Biol 235:143–144

Clowes FAL (1975) The cessation of mitosis at the margins of a root meristem. New Phytol 74:263–271

Clowes FAL (1976) The root apex. In: Yeoman MM (ed) Division in higher plants. Academic Press, London New York, pp 253–284

Colby SR (1967) Calculating synergistic and antagonistic responses of herbicide combinations. Weeds 15:20–22

Conrad H, Saltman P, Eppley R (1959) Effects of auxin and gibberellic acid on growth of *Ulothrix*. Nature 184:556–557

Considine PJ, Patching JW (1975) Ethylene production by micro-organisms grown on phenolic acids. Ann Appl Biol 81:115–119

Constabel F, Kurz WGW, Chatson KB, Kirkpatrick JW (1977) Partial synchrony in soybean cell suspension cultures induced by ethylene. Exp Cell Res 105:263–268

Cooke RJ, Kendrick RE (1976) Phytochrome-controlled gibberellin metabolism in etioplast envelopes. Planta 131:303–307

Cooke RJ, Saunders PF (1975) Phytochrome-mediated changes in extractable gibberellin activity in a cell-free system from etiolated wheat leaves. Planta 123:299–302

Cooke RJ, Saunders PF, Kendrick RE (1975) Red-light-induced production of gibberellin-like substances in homogenates of etiolated wheat leaves and in suspensions of intact etioplasts. Planta 124:319–328

Corgan JN, Peyton C (1970) Abscisic acid levels in dormant peach flower buds. J Am Hortic Sci 95:770–774

Craker LE, Abeles FB, Shropshire W (1973) Light-induced ethylene production in Sorghum. Plant Physiol 51:1082–1083

Croes AF, Dodemont HJ, Stumm C (1976) Induction of méiosis in *Saccharomyces cerevisiae* at different points in the mitotic cycle. Planta 130:131–136

Curtis RW (1957) Survey of fungi and actinomycetes for compounds possessing gibberellin-like activity. Science 125:646

Darimont E, Gaspar Th, Hofinger M (1971) Auxin-kinetin interaction on the lentil root growth in relation to indoleacrylic acid metabolism. Z Pflanzenphysiol 64:232–240

Davidson D, MacLeod RD (1966) Changes in mitotic indices in roots of *Vicia faba*. I. Antagonistic effects of colchicine and IAA. Chromosoma 18:421–437

Davies PJ, Doro JA, Tarbox AW (1976) The movement and physiological effect of indoleacetic acid following point applications to root tips of *Zea mays*. Physiol Plant 36:333–337

Dayanadan P, Hebard FV, Kaufman PB (1976) Cell elongation in the grass pulvinus in response to geotropic stimulation and auxin application. Planta 131:245–252

Dayanadan P, Hebard FV, Kaufman PB (1977) Role of indole-3-acetic acid and gibberellin in the control of internodal elongation in *Avena* stem segments. Plant Physiol 59: Suppl, p 72

DeGroote D, Muir RM (1977) Physiologic auxin in relation to leaf development. Plant Physiol 59: Suppl, p 11

Dhaliwal AS, Dhaliwal GK, Moloney RJ (1973) Effect of indoleacetic acid on the growth patterns of blue-green algae. J Phycol 9: Suppl 14

Dhindsa RS, Beasley CA, Ting IP (1976) Effects of abscisic acid on in vitro growth of cotton fiber. Planta 130: 197–201

Digby J, Wareing PF (1966) The effect of growth hormones on cell division and expansion in liquid suspension cultures of *Acer pseudoplatanus*. J Exp Bot 17: 718–725

Dimalla GG, van Staden J (1977) Effect of ethylene on the endogeneous cytokinin and gibberellin levels in tuberizing potatoes. Plant Physiol 60: 218–221

Dravnieks DE, Skoog F, Burris RH (1969) Cytokinin activation of de novo thiamine biosynthesis in tobacco callus cultures. Plant Physiol 44: 866–870

Dubucq M (1976) Action comparée de l'auxine, de la kinétine et de l'éthrel sur la croissance et les isoperoxydases de *Lens culinaris* Med. Bull Soc. R Bot Belg 109: 93–108

Durley RC, Bewley JD, Railton ID, Pharis RP (1976) Effects of light, abscisic acid, and ^{6}N-benzyladenine on the metabolism of [^{3}H] gibberellin A_4 in seeds and seedlings of lettuce, cv. Grand Rapids. Plant Physiol 57: 699–703

Durnam DJ, Stuart DA, Jones RL (1977) Cell elongation and cell division in elongating lettuce hypocotyls. Plant Physiol 59: Suppl, p 72

Dyer AF (1976) The visible events of mitotic cell division. In: Yeoman MM (ed) Cell division in higher plants. Academic Press, London New York, pp 49–110

Edwards KL, Scott TK (1974) Rapid-growth responses of corn root segments: Effect of pH on elongation. Planta 119: 27–37

Edwards KL, Scott TK (1976) Rapid-growth responses of corn root segments: Effect of citrate-phosphate buffer on elongation. Planta 129: 229–233

Edwards KL, Scott TK (1977) Rapid-growth responses of corn root segments: Effect of auxin on elongation. Planta 135: 1–5

Edwards ME, Miller JH (1972) Growth regulation by ethylene in fern gametophytes. II. Inhibition of cell division. Am J Bot 59: 450–457

El-Antably HMM, Larsen P (1974) Distribution of gibberellins and abscisic acid in geotropically stimulated *Vicia faba* roots. Physiol Plant 32: 322–329

Eliasson L (1969) Growth regulators of *Populus tremula*. I. Distribution of auxin and growth inhibitors. Physiol Plant 22: 1288–1301

Eliasson L (1975) Effect of indoleacetic acid on the abscisic acid level in stem tissue. Physiol Plant 34: 117–120

Elliott MC (1977) Auxins and the regulation of root growth. In: Pilet PE (ed) Plant growth regulation. Springer, Berlin Heidelberg New York, pp 100–108

Elliott MC, Hall JF, Weston GD (1977) A role for gibberellins in the regulation of cell division. Plant Physiol 59: Suppl, p 74

Enu-Kwesi L, Dumbroff EB (1977) Changes in abscisic acid and other inhibitors in seeds of *Acer saccharum* during stratification. Plant Physiol 59: Suppl, p 76

Ernest LC, Valdovinos JG (1971) Regulation of auxin levels in *Coleus blumei* by ethylene. Plant Physiol 48: 402–406

Evans LS, Van't Hof J (1973) Cell arrest in G_2 in root meristems: A control factor from the cotyledons. Exp Cell Res 82: 471–473

Evans LS, Van't Hof J (1975) Is polyploidy necessary for tissue differentiation in higher plants? Am J Bot 62: 1060–1064

Evans ML (1974) Rapid responses to plant hormones. Annu Rev Plant Physiol 25: 195–223

Evans ML (1976a) Effect of cytokinins on the auxin response in coleoptiles. Plant Physiol 57: Suppl, p 54

Evans ML (1976b) A new sensitive root auxanometer: Preliminary studies of the interaction of auxin and acid pH in the regulation of intact root elongation. Plant Physiol 58: 599–601

Evans ML (1976c) Comparative effects of indole-3-acetic acid and 3-methyleneoxindole on intact lentil root elongation. Plant Sci Lett 7: 447–453

Evans ML, Schmitt MR (1975) The nature of spontaneous changes in growth rate in isolated coleoptile segments. Plant Physiol 55:757–762

Evans ML, Simon M, Vesper MJ (1977) Further characterization of the spontaneous growth response in *Zea* coleoptile segments. Plant Cell Physiol 18:441–452

Feldman LJ (1975) Cytokinins and quiescent center activity in roots of *Zea*. In: Torrey JG, Clarkson DT (eds) The development and function of roots. Academic Press, London New York, pp 55–72

Feldman LJ (1979) Cytokinin biosynthesis in corn roots. Planta 145:315–321

Feldman LJ (1981) Effect of auxin on acropetal auxin transport in roots of corn. Plant Physiol 67:278–281

Felippe GM, Dale JE (1968) Effects of a growth retardant, CCC, on leaf growth in *Phaseolus vulgaris*. Planta 80:328–343

Fletcher RA, Zalik S (1964) Effect of light quality on growth and free indoleacetic acid content in *Phaseolus vulgaris*. Plant Physiol 39:328–333

Fonnesbech M (1972) Growth hormones and propagation of *Cymbidium* in vitro. Physiol Plant 27:310–316

Fosket DE, Short KC (1973) The role of cytokinin in the regulation of growth, DNA synthesis and cell proliferation in cultured soybean tissues (*Glycine max* var. Biloxi). Physiol Plant 28:14–23

Frankland B, Wareing PF (1960) Effect of gibberellic acid on hypocotyl growth of lettuce seedlings. Nature 185:255–256

Friedberg SH, Davidson D (1970) Duration of S phase and cell cycles in diploid and tetraploid cells of mixoploid meristems. Exp Cell Res 61:216–218

Fuchs Y, Lieberman M (1968) Effects of kinetin, IAA, and gibberellin on ethylene production, and their interactions in growth of seedlings. Plant Physiol 43:2029–2036

Gaither DH, Lutz DH, Forrence LE (1975) Abscisic acid stimulates elongation of excised pea root tips. Plant Physiol 55:948–949

Gamburg KZ, Osharova LM (1973) The action of auxin on the mitotic cycle of tobacco cells growing in suspension culture. Tsitologiia 15:681–689

Gaspar Th, Xhaufflaire A (1967) Effect of kinetin on growth, auxin catabolism, peroxidase and catalase activities. Planta 72:252–257

Gaspar Th, Xhaufflaire A, Bouiellenne-Walrand M (1965) Acide gibberellique et activité auxines-oxydasique. Bull Soc R Sci Liege 34:149–175

Gaspar Th, Bouchet M, Fries D (1972) The localization of inhibitory oxidation products of indole-3-acetic acid near abscisic acid in the inhibitor-β complex of lentil root. Z Pflanzenphysiol 67:78–85

Gentile AC, Klein RM (1955) The apparent necessity of indoleacetic acid for the growth of *Diplodia* (Fungi imperfecti). Physiol Plant 8:291–299

Gepstein S, Thimann KV (1980) The effect of light on the production of ethylene from 1-aminocyclopropane-1-carboxylic acid by leaves. Planta 149:196–199

Gertman E, Fuchs Y (1972) Effect of abscisic acid and its interactions with other plant hormones on ethylene production in two plant systems. Plant Physiol 50:194–195

Gibbons GSB, Wilkins MB (1970) Growth-inhibitor production by root caps in relation to geotropic responses. Nature 226:558–559

Godlewski M (1980) Auxin as a factor regulating initiation of mitosis in antheridial filaments of *Chara vulgaris* L. Biochem Physiol Pflanz 175:314–321

Goeschl JD, Pratt HK (1968) Regulatory roles of ethylene in the etiolated growth habit of *Pisum sativum*. In: Wightman F, Setterfield G (eds) Biochemistry and physiology of plant growth substances. Runge, Ottawa, pp 1229–1242

Goldschmidt EE, Goren R, Even-Chen Z, Bittner S (1973) Increase in free and bound abscisic acid during natural and ethylene-induced senescence of citrus fruit peel. Plant Physiol 51:879–882

Goldschmidt EE, Aharoni Y, Eilati SK, Riov JW, Monselise SP (1977) Differential counteraction of ethylene effects by gibberellin A_3 and N_6-benzyladenine in senescing citrus peel. Plant Physiol 59:193–195

Gonzalez-Fernandez A, Diez JL, Gimenez-Martin G, DeLa Torre C (1972) Direct measurement of interphase shortening produced by kinetin plus indoleacetic acid in meristematic cells of *Allium cepa* L. Experientia 28:247–248

Goren R, Bukovac MJ (1973) Mechanism of naphthaleneacetic acid conjugation. No effect of ethylene. Plant Physiol 51:907–913

Goto N, Esashi Y (1974a) Differential hormone responses in different growing zones of the bean hypocotyl. Planta 116:225–241

Goto N, Esashi Y (1974b) Regulation of hypocotyl growth by ethylene. In: Plant growth substances 1973. Hirokawa, Tokyo, pp 853–863

Gougler J, Evans ML (1979) Short term effects of abscisic acid and 3,5-diiodo-4-hydroxybenzoic acid (DIHB) on intact root elongation. Plant Physiol 63, Suppl 36

Gougler J, Evans ML (1981) Adaptation of corn roots to exogenously applied auxin. Physiol Plant 51:394–398

Gould AR, Bayliss MW, Street HE (1974) Studies on the growth in culture of plant cells XVII. Analysis of the cell cycle of asynchronously dividing *Acer pseudoplatanus* L. cells in suspension culture. J Exp Bot 25:468–478

Green PB (1976) Growth and cell pattern formation on an axis: Critique of concepts, terminology and modes of study. Bot Gaz 137:187–202

Grotha R (1976) Der Einfluß des Antiauxins p-Chlorphenoxyisobuttersäure auf die Bildung meristematischer Zentren bei der Regeneration isolierter Gewebefragmente von *Riella helicophylla* (Bory et Mont.) Mont. Planta 129:235–238

Gruen HE (1959) Auxins and fungi. Annu Rev Plant Physiol 10:405–440

Guha AK, Banerjee AB (1974) Effect of indole-3-acetic acid and kinetin on submerged growth of *Agaricus campestris*. Acta Microbiol Pol Ser B 6:133–134

Guttman R (1956) Effects of kinetin on cell division, with special reference to initiation and duration of mitosis. Chromosoma 8:341–350

Guttman R, Back A (1958) Effects of kinetin on cell division in *Paramecium caudatum*. Nature 181:852

Haber AH (1962) Nonessentiality of concurrent cell divisions for degree of polarization of leaf growth. I. Studies with radiation-induced mitotic inhibition. Am J Bot 49:583–589

Hagen GL, Marcus A (1975) Cytokinin effects on growth of quiescent tobacco pith cells. Plant Physiol 55:90–93

Hayes AB, Lippincott JA (1976) Growth and gravitational response in the development of leaf blade hyponasty. Am J Bot 63:383–387

Hemberg T (1972a) Interaction of kinetin and abscisic acid in the *Avena* straight growth test. Physiol Plant 26:108–109

Hemberg T (1972b) The effect of kinetin on the occurrence of acid auxin in *Coleus blumei*. Physiol Plant 26:98–103

Hemberg T, Larsson U (1972) Interaction of kinetin and indoleacetic acid in the *Avena* straight-growth test. Physiol Plant 26:104–107

Henson IE, Wareing PF (1974) Cytokinins in *Xanthium strumarium*: A rapid response to short-day treatment. Physiol Plant 32:185–187

Hewett EW, Wareing PF (1973) Cytokinins in *Populus* x *robusta* (Schneid): Light effects on endogenous levels. Planta 114:119–129

Hiraki Y, Ota Y (1975) Relationship between growth-inhibition and ethylene production by mechanical stimulation in *Lilium longiflorum*. Plant Cell Physiol 16:185–189

Hiron RWP, Wright STC (1973) The role of endogenous abscisic acid in the response of plants to stress. J Exp Bot 24:769–781

Howard A, Pelc SR (1953) Synthesis of desoxyribonucleic acid in normal and irradiated cells and its relation to chromosome breakage. Heredity 6 Suppl:261–273

Hughes EWD, Street HE (1960) Effects of inhibitory concentrations of 3-indolylacetic acid and 3-indolylacetonitrile on cell division and tissue differentiation in excised tomato roots. J Exp Bot 11:198–205

Huisinga B (1967) Influence of irradiation on the distribution of growth in dark-grown *Avena* seedlings. Acta Bot Neerl 16:197–201

Huisinga B (1976) The export of auxin from tips and from sections of *Avena* coleoptiles as influenced by red light. Acta Bot Neerl 25:313–320

Humphries EC, Wheller AW (1964) Cell division and growth substances in leaves. In: Nitsch JP (ed) Regulateurs naturels de la croissance végétale. Ed Cent Natl Rech Sci, Paris, pp 505–515

Huxter TJ, Reid DM, Thorpe TA (1977) Role of ethylene in the growth of tobacco callus. Plant Physiol 59: Suppl, p 17

Ilag L, Curtis RW (1968) Production of ethylene by fungi. Science 159:1357–1358

Imaseki H, Pjon C-J (1970) The effect of ethylene on auxin-induced growth of excised rice coleoptile segments. Plant Cell Physiol 11:827–829

Imaseki H, Pjon C-J, Furuya M (1971) Phytochrome action in *Oryza sativa* L. IV. Red and far-red reversible effect on the production of ethylene in excised coleoptiles. Plant Physiol 49:631–633

Imaseki H, Kondo K, Watanabe A (1975) Mechanism of cytokinin action on auxin-induced ethylene production. Plant Cell Physiol 16:777–787

Itai C, Vaadia Y (1965) Kinetin-like activity in root exudate of water-stressed sunflower plants. Physiol Plant 18:941–944

Itai C, Vaadia Y (1971) Cytokinin activity in water-stressed shoots. Plant Physiol 47:87–90

Jacobson A, Corcoran MR (1977) Tannins as gibberellin antagonists in the synthesis of α-amylase and acid phosphatase by barley seeds. Plant Physiol 59:129–133

Jacqmard A (1968) Early effects of gibberellic acid on mitotic activity and DNA synthesis in the apical bud of *Rudbeckia bicolor*. New Phytol 69:269–271

Jaffe M (1980) Morphogenetic responses of plants to mechanical stimuli or stress. Bioscience 30:239–243

Janes HW, Loercher L, Frenkel C (1976) Effects of red light and ethylene on growth of etiolated lettuce seedlings. Plant Physiol 57:420–423

Jennings RC (1968) Gibberellins as endogenous growth regulators in green and brown algae. Planta 80:34–42

Jennings RC (1969) Cytokinins as endogenous growth regulators in the algae *Ecklonia* (Phaeophyta) and *Hypnea* (Rhodophyta). Aust J Biol Sci 22:621–627

Jindal KK, Hemberg T (1976) Influence of gibberellic acid on growth and endogenous auxin level in epicotyl and hypocotyl tissues of normal and dwarf bean plants. Physiol Plant 38:78–82

Jones R (1973) Gibberellins: Their physiological role. Annu Rev Plant Physiol 24:571–598

Jones RA, Kaufman PB (1971) Regulation of growth and invertase activity by kinetin and gibberellic acid in developing *Avena* internodes. Physiol Plant 24:491–497

Jones RL, Lang A (1968) Extractable and diffusible gibberellins from light- and dark-grown pea seedlings. Plant Physiol 43:629–634

Jordan WR, Skoog F (1971) Effects of cytokinins on growth and auxin in coleoptiles of derooted *Avena* seedlings. Plant Physiol 48:97–99

Jouanneau JP (1970) Renouvellement des protéines et effet spécifique de la kinétine sur des cultures de cellules de tabac. Physiol Plant 23:232–244

Jouanneau JP (1971) Control par les cytokinines de la synchronisation des mitoses dans les cellules de tabac. Exp Cell Res 67:329–337

Jouanneau JP (1975) Protein synthesis requirement for the cytokinin effect upon tobacco cell division. Exp Cell Res 91:184–190

Jouanneau JP, Tandeau De Marsac N (1973) Stepwise effects of cytokinin activity and DNA synthesis upon mitotic cycle events in partially synchronized tobacco cells. Exp Cell Res 77:167–174

Kamisaka S (1973) Requirement of cotyledons for gibberellic acid-induced hypocotyl elongation in lettuce seedlings. Isolation of the cotyledon factor active in enhancing the effect of gibberellic acid. Plant Cell Physiol 14:747–755

Kamisaka S, Sano H, Katsumi M, Masuda Y (1972) Effects of cyclic AMP and gibberellic acid on lettuce hypocotyl elongation and mechanical properties of its cell wall. Plant Cell Physiol 13:167–173

Kang BG, Burg SP (1973) Influence of ethylene on nucleic acid synthesis in etiolated *Pisum sativum*. Plant Cell Physiol 14:981–988

Kang BG, Ray PM (1969) Role of growth regulators in the bean hypocotyl hook opening response. Planta 87:193–205

Kang BG, Newcomb W, Burg SP (1971) Mechanism of auxin-induced ethylene production. Plant Physiol 47:504–509

Katsumi M, Kazama H (1974) Interrelationship between auxin and gibberellin in the elongation of cucumber hypocotyl sections. In: Plant growth substances 1973. Hirokawa, Tokyo, pp 845–852

Katsumi M, Purves WK, Phinney BO, Kato J (1965) The role of the cotyledons in gibberellin- and auxin-induced elongation of the cucumber hypocotyl. Physiol Plant 18:550–556

Kaufman PB, Jones RA (1974) Regulation of growth in *Avena* (oat) stem segments by gibberellic acid and abscisic acid. Physiol Plant 31:39–43

Kaufman PB, Petering LB, Adams PA (1969) Regulation of growth and cellular differentiation in developing *Avena* internodes by gibberellic acid and indole-3-acetic acid. Am J Botany 56:918–927

Kazama H, Katsumi M (1973) Auxin-gibberellin relationships in their effects on hypocotyl elongation of light-grown cucumber seedlings. Responses of sections to auxin, gibberellin, and sucrose. Plant Cell Physiol 14:449–458

Kazama H, Katsumi M (1974) Auxin-gibberellin relationships in their effects on hypocotyl elongation of light-grown cucumber seedlings II. Effect of GA_3-pretreatment on IAA-induced elongation. Plant Cell Physiol 15:307–314

Kende H, Acaster MA, Jones JF, Metraux J-P (1982) On the mode of action of ethylene. In: Plant growth substances 1982. Proc 11th Int Conf Plant Growth Substances, Aberystwyth, pp 269–277

Kentzer T, Synak R, Burkiewicz K, Banas A (1980) Cytokinin-like activity in sea water and *Fucus ventriculosus* L. Biol Plant 22:218–225

Khan AA, Downing RD (1968) Cytokinin reversal of abscisic acid inhibition of growth and α-amylase synthesis in barley seed. Physiol Plant 21:1301–1307

Kögl F, Elema J (1960) Wirkungsbeziehungen zwischen Indol-3-Essigsäure und Gibberellinsäure. Naturwissenschaften 47:90

Köhler D (1977) Zellteilung und Zellstreckung unter dem Einfluß von Gibberellinsäure und Rotlicht bei Zwerg- und Normalerbsen (*Pisum sativum*). Z Pflanzenphysiol 82:125–136

Köhler K-H, Dörfler M, Göring H (1980) The influence of light on the cytokinin content of *Amaranthus* seedlings. Biol Plant 22:128–134

Kondo K, Watanable A, Imaseki H (1975) Relationships in actions of indoleacetic acid, benzyladenine and abscisic acid in ethylene production. Plant Cell Physiol 16:1001–1007

Kondo N, Fujii T, Yamaki T (1969) Effect of light on auxin transport and elongation of *Avena* mesocotyl. Dev Growth Differ 11:46–61

Kopcewicz J, Khan AA (1977) Evidence for the involvement of auxin in seed germination. Plant Physiol 59:Suppl, p 33

Kriedemann PE, Loveys BR, Fuller GL, Leopold AC (1972) Abscisic acid and stomatal regulation. Plant Physiol 49:842–847

Ku HS, Suge H, Rappaport L, Pratt HK (1970) Stimulation of rice coleoptile growth by ethylene. Planta 90:333–339

Kulaeva OM, Tsibulya LV (1974) Cytokinin effect on the growth of discs from etiolated bean leaves. Fiziol Rast 21:709–713

Kundu KK, Audus LJ (1974) Root-growth inhibitors from root tips of *Zea mays* L. Planta 117:183–186

Lagerstedt HB, Langston RG (1967) The mobilizing force of kinetin. Life Sci 6:145–149

Lang A (1965) Physiology of growth and development in algae. A synopsis. In: Ruhland W (ed) Encyclopedia of plant physiology Vol XV. Springer, Berlin Göttingen Heidelberg, pp 680–715

Lau O-L, Yang SF (1973) Mechanism of a synergistic effect of kinetin on auxin-induced ethylene production. Suppression of auxin conjugation. Plant Physiol 51:1011–1014

Lau O-L, Yang SF (1976) Stimulation of ethylene production in the mung bean hypocotyls by cupric ion, calcium ion, and kinetin. Plant Physiol 57:88–92

Lau O-L, John WW, Yang SF (1977a) Effect of different cytokinins on ethylene production by mung bean hypocotyls in the presence of indole-3-acetic acid or calcium ion. Physiol Plant 39:1–3

Lau O-L, John WW, Yang SF (1977b) Inactivity of oxidation products of IAA to stimulate ethylene production in mungbean hypocotyls. Plant Physiol 59: Suppl, p 11

Lawson VR, Weintraub RL (1975) Interactions of microtubule disorganizers, plant hormones, and red light in wheat coleoptile segment growth. Plant Physiol 55: 1062–1066

Lee BO(1961) Effect of kinetin on the fertility of some strains of *Neurospora crassa*. Nature 192: 288

Lee TT (1971) Cytokinin-controlled indoleacetic acid oxidase isoenzymes in tobacco callus cultures. Plant Physiol 47: 181–185

Leguay JJ, Guern J (1977) Quantitative effects of 2,4-dichlorophenoxyacetic acid on growth of suspension-cultured *Acer pseudoplatanus* cells. II. Influence of 2,4-D metabolism and intracellular pH on the control of cell division by intracellular 2,4-D concentration. Plant Physiol 60: 265–270

Leopold AC, Kawase M (1964) Benzyladenine effects on bean leaf growth and senescence. Am J Bot 51: 294–298

Le Page-Degivry MT (1973) Intervention d'un inhibiteur lié dans la dormance embryonnaire de *Taxus baccata* L. CR Acad Sci Paris Ser D 277: 177–180

Letham DS (1969) Cytokinins and their relation to other phytohormones. Bioscience 19: 309–316

Letham DS (1971) Regulators of cell division in plant tissues. XII. A cytokinin bioassay using excised radish cotyledons. Physiol Plant 25: 391–396

Libbenga KR, Torrey JG (1973) Hormone-induced endoreduplication prior to mitosis in cultured pea root cortex cells. Am J Bot 60: 293–299

Lieberman M, Kunishi AT (1971) Abscisic acid and ethylene production. Plant Physiol 47, Suppl, p 22

Lintilhac PM (1974) Differentiation, organogenesis, and the tectonics of cell wall orientation. III. Theoretical considerations of wall mechanics. Am J Bot 61: 230–237

Lintilhac PM (1977) The effects of localized stress on secondary growth patterns in *Coleus*. An example of mechanical control of cell wall disposition. Plant Physiol 59: Suppl, p 26

Lippincott BB, Lippincott JA (1971) Auxin-induced hyponasty of the leaf blade of *Phaseolus vulgaris*. Am J Bot 58: 817–826

List A Jr (1969) Transient growth responses of the primary roots of *Zea mays*. Planta 87: 1–19

Livne A, Vaadia Y (1972) Water deficits and hormone relations. In: Kozlowski TT (ed) Water deficits and plant growth Vol III. Academic Press, London New York, pp 255–275

Liu PBW, Loy JB (1976) Action of gibberellic acid on cell proliferation in the subapical shoot meristem of watermelon seedlings. Am J Bot 63: 700–704

Lockhart JA (1965) The analysis of interactions of physical and chemical factors on plant growth. Annu Rev Plant Physiol 16: 37–52

Loveys BR (1979) The influence of light quality on levels of abscisic acid in tomato plants, and evidence for a novel abscisic acid metabolite. Physiol Plant 46: 79–84

Loveys BR, Wareing PF (1971) The hormonal control of wheat leaf unrolling. Planta 98: 117–127

Loveys BR, Leopold AC, Kriedemann PE (1974) Abscisic acid metabolism and stomatal physiology in *Betula lutea* following alteration in photo-period. Ann Bot 38: 85–92

Low VHK (1975) Role of gibberellins in root and shoot growth. In: Krishnamoorthy HN (ed) Gibberellins and plant growth. Wiley and Sons, New York, pp 101–114

Loy JB (1977) Hormonal regulation of cell division in the primary elongating meristems of shoots. In: Rost TL, Gifford EM Jr (eds) Mechanisms and control of cell division. Dowden, Hutchinson and Ross, Stroudsburg, pp 92–110

Loy JB, Pollard JE (1977) Interaction of a cytokinin and ethylene in promoting hypocotyl elongation in dwarf watermelon. Plant Physiol 59: Suppl, p 37

Lynch JM, Harper SHT (1974) Fungal growth rate and the formation of ethylene in soil. J Gen Microbiol 85: 91–96

Lyndon RF (1976) The shoot apex. In: Yeoman MM (ed) Cell division in higher plants. Academic Press, London New York, pp 285–314

Lyon CJ (1970) Ethylene inhibition of auxin transport by gravity in leaves. Plant Physiol 45:644–646

MacKenzie IA, Konar A, Street HE (1972) Cytokinins and the growth of cultured sycamore cells. New Phytol 71:633–638

MacLeod RD (1968) Changes in the mitotic cycle in lateral root meristems of *Vicia faba* following kinetin treatment. Chromosoma 24:177–187

MacLeod RD (1971) The response of apical meristems of primary roots of *Vicia faba* L. to colchicine treatments. Chromosoma 35:217–232

MacLeod RD (1976) Cap formation during the elongation of lateral roots of *Vicia faba* L. Ann Bot 40:877–885

MacLeod RD, Davidson O (1966) Changes in mitotic indices in roots of *Vicia faba*. II. Long term effects of colchicine and IAA. New Phytol 65:532–546

Malik CP, Mehan M (1975a) Correlative effects of auxin, gibberellic acid and kinetin on the elongation of pollen tubes in *Calotropis procera*. Biochem Physiol Pflanz 167:295–300

Malik CP, Mehan M (1975b) Interaction between cycocel and gibberellic-acid in pollen-tube elongation of *Calotropis procera*. Curr Sci 44:785–786

Manos GE (1961) The effects of growth substances on attached and detached root tips of *Pisum sativum* L. Physiol Plant 14:697–711

Manos PJ, Goldthwaite J (1976) An improved cytokinin bioassay using cultured soybean hypocotyl sections. Plant Physiol 57:894–897

Marinos NG (1960) Some responses of *Avena* coleoptiles to ethylene. J Exp Bot 11:227–235

Masuda Y, Yamamoto R (1972) Control of auxin-induced stem elongation by the epidermis. Physiol Plant 27:109–115

Masuda Y, Shimoda C, Takao N, Yanagishima N (1970) Effect of yeast sexual hormones on elongation of *Avena* coleoptile. Dev Growth Differ 12:207–215

Matsubara S (1973) Overcoming self-incompatability by cytokinins treatment on *Lilium longiflorum*. Bot Mag Tokyo 86:43–46

Mayak S, Dilley DR (1976) Regulation of senescence in carnation (*Dianthus caryophyllus*): Effect of abscisic acid and carbon dioxide on ethylene production. Plant Physiol 58:663–665

Mayak S, Halevy AH (1972) Interrelationships of ethylene and abscisic acid in the control of rose petal senescence. Plant Physiol 50:341–346

McBride R, Evans ML (1977) Auxin inhibition of acid- and fusicoccin-induced elongation in lentil roots. Planta 136:97–102

Melkonian M, Weber A (1975) Der Einfluß von Kinetin auf das Wachstum von *Fritschiella tuberosa* Iyeng. (Chaetophorineae, Chlorophyceae) in axenischer Massenkultur. Z Pflanzenphysiol 76:120–129

Mertz D (1966) Hormonal control of root growth. Plant Cell Physiol 7:125–135

Metzger JD, Zeevaart JAD (1980) Comparison of the levels of six endogenous gibberellins in roots and shoots of spinach in relation to photoperiod. Plant Physiol 66:679–683

Meyer RF, Boyer JS (1972) Sensitivity of cell division and cell elongation to low water potentials in soybean hypocotyls. Planta 108:77–87

Milborrow BV (1966) The effects of synthetic DL-dormin (abscisin II) on the growth of the oat mesocotyl. Planta 70:155–171

Milborrow BV (1970) Interaction of plant hormones: Abscisic and gibberellic acids. Science 168:875–877

Milborrow BV (1972) The biosynthesis and degradation of abscisic acid. In: Carr DJ (ed) Plant growth substances 1970. Springer, Berlin Heidelberg New York, pp 281–290

Milborrow BV (1974) The chemistry and physiology of abscisic acid. Annu Rev Plant Physiol 25:259–307

Milborrow BV, Robinson DR (1973) Factors affecting the biosynthesis of abscisic acid. J Exp Bot 24:537–548

Miller PM, Sweet HC, Miller JH (1970) Growth regulation by ethylene in fern gametophytes. I. Effects on protonemal and rhizoidal growth and interaction with auxin. Am J Bot 57:212–217

Milthorpe FL, Newton P (1963) Studies on the expansion of the leaf surface. III. The influence of radiation on cell division and leaf expansion. J Exp Bot 14:483–495

Minocha SC (1979) Abscisic acid promotion of cell division and DNA synthesis in Jerusalem artichoke tuber tissue cultured in vitro. Z Pflanzenphysiol 92:327–339

Mitchison JM, Creanor J (1969) Linear synthesis of sucrose and phosphatases during the cell cycle of *Schizosaccharomyces pombe*. J Cell Sci 5:373–391

Mohr H, Noblé A (1960) Die Steuerung der Schließung und Öffnung des Plumulahakens bei Keimlingen von *Lactuca sativa* durch sichtbare Strahlung. Planta 55:327–342

Mohsen AF, Khaleafa AF, Hashem MA, Metwalli A (1974) Effect of some auxins on growth, reproduction, amino acid, fat and sugar contents in *Ulva fascita* Delile. Bot Mar 17:213–217

Mondal MH (1975) Effects of gibberellic acid, calcium, kinetin, and ethylene on growth and cell wall composition of pea epicotyls. Plant Physiol 56:622–625

Morgan PW, Gausman HW (1966) Effects of ethylene on auxin transport. Plant Physiol 41:45–52

Morgan PW, Beyer E, Gausman HW (1968) Ethylene effects on auxin physiology. In: Wightman F, Setterfield G (eds) Biochemistry and physiology of plant growth substances. Runge, Ottawa, pp 1255–1273

Morris DA, Briant RE, Thomson PG (1969) The transport and metabolism of ^{14}C-labelled indoleacetic acid in intact pea seedlings. Planta 89:178–197

Muir RM, Chang KD (1974) Effect of red light on coleoptile growth. Plant Physiol 54:286–288

Mulkey TJ, Evans ML, Kuzmanoff KM (1982a) Promotion of growth and shift in the auxin dose/response relationship in maize roots treated with the ethylene biosynthesis inhibitors, aminoethoxyvinylglycine and cobalt. Plant Sci Lett 25:43–48

Mulkey TJ, Kuzmanoff KM, Evans ML (1982b) Promotion of growth and hydrogen ion efflux by auxin in roots of maize pretreated with ethylene biosynthesis inhibitors. Plant Physiol 70:186–188

Musgrave A, Kende H (1970) Radioactive gibberellin A_5 and its metabolism in dwarf peas. Plant Physiol 45:56–61

Musgrave A, Jackson MB, Ling E (1972) *Callitriche* stem elongation is controlled by ethylene and gibberellin. Nature New Biol 238:93–96

Nadeau R, Rappaport L, Stolp CF (1972) Uptake and metabolism of ^{3}H-gibberellin A_1 by barley aleurone layers: response to abscisic acid. Planta 107:315–324

Nagl W (1971) Gibberellinsäure-stimulierte Genaktivität im Endosperm von *Phaseolus*. Planta 96:145–151

Nagl W (1972) Selective inhibition of cell cycle stages in the *Allium* root meristem by colchicine and growth regulators. Am J Bot 59:346–351

Nagl W, Rücker W (1974) Shift of DNA replication from diploid to polyploid cells in cytokinin-controlled differentiation. Cytobios 10:137–144

Naqvi SM, Gordon SA (1967) Auxin transport in *Zea mays* coleoptiles II. Influence of light on the transport of indoleacetic acid-2-^{14}C. Plant Physiol 42:138–143

Narain A, Laloraya MM (1974) Cucumber cotyledon expansion as a bioassay for cytokinins. Z Pflanzenphysiol 71:313–322

Negi SS, Olmo HP (1972) Certain embryological and biochemical aspects of cytokinin SD 8339 in converting sex of a male *Vitis vinifera* (sylvestris). Am J Bot 59:851–857

Newton RJ (1977) Abscisic acid effects on fronds and roots of *Lemna minor* L. Am J Bot 64:45–49

Nishi A, Sugano N (1970) Growth and division of carrot cells in suspension culture. Plant Cell Physiol 11:757–765

Nishi A, Kato K, Takahashi M, Yoshida R (1977) Partial synchronization of carrot cell culture by auxin deprivation. Physiol Plant 39:9–12

Nishinari N, Yamaki T (1976) Relationship between cell division and endogenous auxin in synchronously-cultured tobacco cells. Bot Mag (Tokyo) 89:73–81

Nitsch JP, Lance-Nougarède A (1967) L'action conjuguée des auxines et des cytokinines sur les cellules de moelle de tabac: Étude physiologique et microscopie electronique. Bull Soc Fr Physiol Veg 13:81–118

Okerse R, Galston AW (1967) Gibberellin-auxin interaction in pea stem elongation. Plant Physiol 42:47–54

Osborne DJ (1975) Ethylene as a natural regulator of the growth of plants. Ann Appl Biol 81:95–98

Paleg LG (1965) Physiological effects of gibberellins. Annu Rev Plant Physiol 18:291–322

Palmer JH (1972) Roles of ethylene and indolyl-3-acetic acid in petiole epinasty in *Helianthus annuus* and the modifying influence of gibberellic acid. J Exp Bot 23:733–743

Palmer JH (1976) Failure of ethylene to change the distribution of indoleacetic acid in the petiole of *Coleus blumei frederici* during epinasty. Plant Physiol 58:513–515

Peaud-Lenoël C (1977) The hormonal regulation of the cell division cycle. In: Pilet PE (ed) Plant growth regulation. Springer, Berlin Heidelberg New York, pp 240–248

Peaud-Lenoël C, Jouanneau JP (1971) Controle du cycle mitotique dans les suspensions de cellules de tabac cultivées en milieu liquide. In: Les Cultures de Tissus de Plantes. Coll Int CNRS 193. CNRS, Paris, pp 95–102

Pedersen M (1968) *Ectocarpus fasciculatus*: marine brown alga requiring kinetin. Nature 218:776

Pegg GF (1973) Gibberellin-like substances in the sporangiophores of *Agaricus bisporus* (Lange) Imbach. J Exp Bot 24:675–688

Pernet JJ, Pilet PE (1976) Indoleacetic acid movement in the root cap. Planta 128:183–184

Philipson JJ, Hillman JR, Wilkins MB (1973) Studies on the action of abscisic acid on IAA-induced rapid growth of *Avena* coleoptile segments. Planta 114:87–93

Phillips DA (1971) Abscisic acid inhibition of root nodule initiation in *Pisum sativum*. Planta 100:181–190

Phillips HL Jr, Torrey JG (1972) Duration of cell cycles in cultured roots of *Convolvulus*. Am J Bot 59:183–188

Phillips IDJ (1964) Root-shoot hormone relations. II. Changes in endogenous auxin concentrations produced by flooding the root system in *Helianthus annuus*. Ann Bot 28:36–45

Phillips IDJ, Miners J, Roddick JG (1980) Effects of light and photoperiodic conditions on ABA in leaves and roots of *Acer pseudoplatanus* L. Planta 149:118–122

Phinney BO (1956) Growth response of single-gene dwarf mutants in maize to gibberellic acid. Proc Natl Acad Sci USA 42:185–189

Phinney BO, West CA (1960) Gibberellins as native plant growth regulators. Annu Rev Plant Physiol 11:411–436

Phinney BO, West CA (1961) Gibberellins and plant growth. In: Ruhland W (ed) Encyclopedia of plant physiology Vol XIV. Springer, Berlin Göttingen Heidelberg, pp 1185–1227

Pieniazek J (1971) Regulatory role of abscisic acid in the petiole abscission of apple explants. Bull Acad Pol Sci 19:125–129

Pilet PE (1970) The effect of auxin and abscisic acid on the catabolism of RNA. J Exp Bot 21:446–451

Pilet PE (1974) Control by the root cap on growth and georeaction of roots. In: Plant growth substances 1973. Hirokawa, Tokyo, pp 1104–1110

Pilet PE (1975a) Abscisic acid as a root growth inhibitor: Physiological analyses. Planta 122:299–302

Pilet PE (1975b) Action of fusicoccin and abscisic acid on root growth. Plant Sci Lett 5:137–140

Pratt HK, Goeschl JD (1969) Physiological roles of ethylene in plants. Annu Rev Plant Physiol 20:541–584

Provasoli L (1958) Effect of plant hormones on *Ulva*. Biol Bull 114:375–384

Provasoli L, Carlucci AF (1974) Vitamins and growth regulators. In: Stewart WDP (ed) Algal physiology and biochemistry. Blackwell, Oxford, pp 741–787

Radin JW, Loomis RS (1969) Ethylene and carbon dioxide in the growth and development of cultured radish roots. Plant Physiol 44:1584–1589

Railton ID, Wareing PF (1973b) Effects of daylength on endogenous gibberellins in *Solanum andigena* III. Gibberellin production by the leaves. Physiol Plant 29:430–433

Railton ID, Wareing PF (1973a) Effects of abscisic acid on the levels of endogenous gibberellin-like substances in *Solanum andigena*. Planta 112:65–69

Rappaport L (1979) Does GA_3 regulate cell division? Plant Physiol 63:Suppl, p 32
Rauser WE, Horton RF (1975) Rapid effects of indoleacetic acid and ethylene on the growth of intact pea roots. Plant Physiol 55:443–447
Rehm MM, Cline MG (1973) Rapid growth inhibition of *Avena* coleoptile segments by abscisic acid. Plant Physiol 51:93–96
Reid DM, Crozier A (1971) Effects of waterlogging on the gibberellin content and growth of tomato plants. J Exp Bot 22:39–48
Reid DM, Railton ID (1974) The influence of benzyladenine on the growth and gibberellin content of shoots of waterlogged tomato plants. Plant Sci Lett 2:151–156
Reid DM, Crozier A, Harvey BMR (1969) The effect of flooding on the export of gibberellins from the root to the shoot. Planta 89:376–379
Rijven AHGC, Parkash V (1970) Cytokinin-induced growth responses by fenugreek cotyledons. Plant Physiol 46:638–640
Rivier L, Pilet PE (1974) Indolyl-3-acetic acid in cap and apex of maize roots: Identification and quantification by mass fragmentography. Planta 120:107–112
Roberts K, Northcote DH (1970) The structure of sycamore callus cells during division in a partially synchronized suspension culture. J Cell Sci 6:299–321
Roberts LW (1976) Cytodifferentiation in plants. Xylogenesis as a model system. Cambridge Univ Press, London
Romanova LV, Ivanova OA, Tkachenko VN (1975) Effect of light and temperature conditions of potato tuber germination on growth, yield and hormonal activity of the plants. Fiziol Biokhim Kul't Rast 7:625–631
Rosenberry TL, Bernhard SA (1972) Studies of catalysis by acetylcholinesterase. Synergistic effects of inhibitors during the hydrolysis of acetic acid esters. Biochemistry 11:4308–4321
Ross JD, Bradbeer JW (1971) Studies in seed dormancy. VI. The effects of growth retardants on the gibberellin content and germination of chilled seeds of *Corylus avellana* L. Planta 100:303–308
Rummens FHA (1975) An improved definition of synergistic and antagonistic effects. Weed Sci 23:4–6
Russo VEA (1977) Ethylene-induced growth in *Phycomyces* mutants abnormal for autochemotropism. J Bacteriol 130:548–551
Russo VEA, Halloran B, Gallori E (1977) Ethylene is involved in the autochemotropism of *Phycomyces*. Planta 134:61–67
Sakai S, Imaseki H (1971) Auxin-induced ethylene production by mung bean hypocotyl segments. Plant Cell Physiol 12:349–359
Salminen SO, Gagianas AA, Berg AR (1972) The effect of NAA (1-naphthalene-acetic acid) and morphactin (methyl 2-chloro-9-hydroxyfluorene-9-carboxylate) on RNA and protein synthesis in senescing bean endocarp. Physiol Plant 26:186–190
Sargent JA, Atack AV, Osborne DJ (1973) Orientation of cell growth in the etiolated pea stem. Effect of ethylene and auxin on cell wall deposition. Planta 109:185–192
Sarma CM (1979) Interactions between indole-3-acetic acid and gibberellic acid on the expansion growth of leaf disks of bean. Indian J Plant Physiol 22:242–246
Sastry KSK, Muir RM (1965) Effects of gibberellic acid on utilization of auxin precursors by apical segments of the *Avena* coleoptile. Plant Physiol 40:294–298
Schurzmann M, Hild V (1980) Effect of indoleacetic acid, abscisic acid, root tips and coleoptile tips on growth and curvature of maize roots. Planta 150:32–36
Scott TK (1972) Auxins and roots. Annu Rev Plant Physiol 23:235–258
Scott TK, Briggs WR (1960) Auxin relationships in the Alaska pea (*Pisum sativum* L.). Am J Bot 47:492–499
Seth AK, Davies CR, Wareing PF (1966) Auxin effects on the mobility of kinetin in the plant. Science 151:587–588
Setterfield G (1963) Growth regulation in excised slices of Jerusalem artichoke tuber tissue. Symp Soc Exp Biol 17:98–126
Sharma PK, Baruah P (1975) Effect of auxins on yeast. Nucleus 18:124–128
Shen-Miller J, McNitt RE, Wojciechowski M (1978) Regions of differential cell elongation and mitosis, and root meristem morphology in different tissues of geotropically stimulated maize root apices. Plant Physiol 61:7–12

Shimo S, Nakatani S (1964) The effect of kinetin on the growth of *Porphyra*. J Agricult Lab No 5. Central Res Lab of Electricity 55–60
Shimoda C, Yanagishima N (1969) Relation of special RNA to sensitivity of yeast strains to auxin. Plant Cell Physiol 10:61–67
Shininger TL (1975) The morphological, anatomical and cytological effects of gibberellins. In: Krishnamoorthy HN (ed) Gibberellins and plant growth. Wiley and Sons, New York, pp 203–224
Short KC, Torrey JG (1972) Cytokinins in seedling roots of pea. Plant Physiol 49:155–160
Short KC, Tepfer DA, Fosket DE (1974) Regulation of polyribosome formation and cell division in cultured soybean cells by cytokinin. J Cell Sci 15:75–87
Simard A (1971) Initiation of DNA synthesis by kinetin and experimental factors in tobacco pith tissue in vitro. Can J Bot 49:1541–1549
Simpson SF (1977) The effect of cytokinin pulses on DNA synthesis in the first cell cycle in pea root cortical explants. Plant Physiol 59:Suppl, p 74
Skene KGM (1970) The relationship between the effects of CCC on root growth and cytokinin levels in the bleeding sap of *Vitis vinifera* L. J Exp Bot 21:418–431
Skene KGM (1975) Cytokinin production by roots as a factor in the control of plant growth. In: Torrey JG, Clarkson DT (eds) The development and function of roots. Academic Press, London New York, pp 365–396
Skene KGM, Kerridge GH (1967) Effect of root temperature on cytokinin activity in root exudate of *Vitis vinifera* L. Plant Physiol 42:1131–1139
Skoog F, Hamzi HQ, Sweykowska AM, Leonard NJ, Carraway KL, Fujii T, Helgeson JP, Loeppky RN (1967) Cytokinins: structure/activity relationships. Phytochemistry 6:1169–1192
Smith AM (1973) Ethylene as a cause of soil fungistasis. Nature 246:311–313
Soni SL, Kaufman PB (1972) Regulation of invertase activity and growth in rice (*Oryza*) coleoptile sections by gibberellic acid and sucrose. Can J Bot 50:1185–1190
Spalding DH, Lieberman M (1965) Factors affecting the production of ethylene by *Penicillium digitatum*. Plant Physiol 40:645–648
Stebbins GL (1965) Some relationships between mitotic rhythm, nucleic acid synthesis, and morphogenesis in higher plants. Brookhaven Symp Biol 18:204–220
Stebbins GL, Suryakant SS, Jamin D, Jura P (1967) Changed orientation of the mitotic spindle of stomatal guard cell divisions in *Hordeum vulgare*. Am J Bot 54:71–80
Stewart JM, Hsu CL (1977) Influence of phytohormones on the response of cultured cotton ovules to (2-chloroethyl)phosphonic acid. Physiol Plant 39:79–85
Stolp CF, Nadeau R, Rappaport L (1973) Effect of abscisic acid on uptake and metabolism of [^{3}H]gibberellin A_1 and [^{3}H]pseudogibberellin A_1 by barley half-seeds. Plant Physiol 52:546–548
Street HE (1966) The physiology of root growth. Annu Rev Plant Physiol 17:315–344
Stuart DA, Jones RL (1977) Roles of extensibility and turgor in gibberellin- and dark-stimulated growth. Plant Physiol 59:61–68
Suge H (1971a) 2-Chloroethylphosphonic acid as ethylene releasing agent for the stimulation of rice and the inhibition of wheat, barley, oat and rye coleoptile growth. Proc Crop Sci Soc Jpn 40:127–131
Suge H (1971b) Stimulation of oat and rice mesocotyl growth by ethylene. Plant Cell Physiol 12:831–837
Suzuki T, Kondo N, Fujii T (1979) Distribution of growth regulators in relation to the light-induced geotropic responsiveness in *Zea* roots. Planta 145:323–329
Svarinskaya RA, Gavrilova NS (1976) Effect of gibberellin on the duration of mitotic cycle and on the intensity of DNA synthesis. Genetika 12:30–36
Svensson SB (1972) A comparative study of the changes in root growth induced by coumarin, auxin, ethylene, kinetin and gibberellic acid. Physiol Plant 26:115–135
Swart A, Kamerbeek GA (1977) Ethylene production and mycelium growth of the tulip strain of *Fusarium oxysporum* as influenced by shaking of and oxygen supply to the culture medium. Physiol Plant 39:38–44
Syono K, Newcomb W, Torrey JG (1976) Cytokinin production in relation to the development of pea root nodules. Can J Bot 54:2155–2162

Takahashi K (1972) Abscisic acid as a stimulator for rice mesocotyl growth. Nature New Biol 238:92–93

Takahashi K (1973) Interaction between ethylene, abscisic acid and gibberellic acid in elongation of rice mesocotyl. Planta 109:363–364

Tanada T (1972) Antagonism between indoleacetic acid and abscisic acid on a rapid phytochrome-mediated process. Nature 236:460–461

Tanada T (1973) Indoleacetic acid and abscisic acid antagonism. I. On the phytochrome-mediated attachment of mung bean root tips on glass. Plant Physiol 51:150–153

Taylor IEP, Wilkinson AJ (1977) The occurrence of gibberellins and gibberellin-like substances in algae. Phycologia 16:37–42

Teltscherova L (1970) Changes in the level of endogenous cytokinins in apical buds of *Chenopodium rubrum* L. Biol Plant 12:134–138

Thimann KV (1936) Auxins and the growth of roots. Am J Bot 23:561–599

Thimann KV (1937) On the nature of inhibitions caused by auxin. Am J Bot 24:407–412

Thimann KV, Beth K (1959) Action of auxins on *Acetabularia* and the effect of enucleation. Nature 183:946–948

Thomas TH, Wareing PF, Robinson PM (1965) Chemistry and physiology of "dormins" in sycamore. Action of the sycamore "dormin" as a gibberellin antagonist. Nature 205:1270–1272

Thomas TH, Wurr DCE (1976) Gibberellin and growth inhibitor changes in potato tuber buds in response to cold treatment. Ann Appl Biol 83:317–320

Tietz A (1974) Der Einfluß von Licht auf Wachstum und Abscisinsäuregehalt der Erbsenwurzel. Biochem Physiol Pflanz 165:387–392

Tillberg E (1974) Levels of indol-3yl-acetic acid and acid inhibitors in green and etiolated bean seedlings (*Phaseolus vulgaris*). Physiol Plant 31:106–111

Torrey JG (1961) Kinetin as a trigger for mitosis in mature endomitotic plant cells. Exp Cell Res 23:281–299

Torrey JG (1972) On the initiation of organization in the root apex. In: Miller MW, Kuehnert CC (eds) Dynamics of meristem cell populations. Plenum, New York, pp 1–13

Trewavas A (1981) How do plant growth substances work? Plant Cell Environ 4:204–228

Tronchet A (1968) Quelques effets de la kinétine et de l'acide gibbérellique sur la forme et la structure anatomique des feuilles de *Coleus blumei*. Ann Sci Univ Besancon Bot 5:3–8

Tsuboi M, Yanagishima N (1974) Changes in sporulation ability during the vegetative cell cycle in *Saccharomyces cerevisiae*. Bot Mag (Tokyo) 87:183–193

Tsui C, Guo-qing T, Hui-ying C, Yan-ru S, Han-ping L, Zhe T, Shu-huan L, Xian-zhang L (1980) Effect of cytokinins on the expansion and metabolism of excised cucumber cotyledons. Aust J Plant Physiol 7:227–236

Van Oostveldt P, Van Goethem G, Van Parijs R (1976) Effect of light on cell elongation, nucleic acid and protein synthesis in hypocotyls of *Lupinus angustifolius*. Planta 129:259–263

Van Overbeek J, Loeffler JE, Mason MIR (1968) Mode of action of abscisic acid. In: Wightman P, Setterfield J (eds) Biochemistry and physiology of plant growth substances. Runge, Ottawa, pp 1593–1607

Van Staden J, Bornman CH (1970) Cytokinin and gibberellin effects on abscisic acid induced inhibition of growth in *Spirodela*. J S Afr Bot 36:207–214

Van Staden J, Breen CM (1973) Cytokinins in freshwater algal cultures. Plant Sci Lett 1:325–330

Van Staden J, Wareing PF (1972) The effect of light on endogenous cytokinin levels in seeds of *Rumex obtusifolius*. Planta 104:126–133

Van Staden J, Olatoye ST, Hall MA (1973) Effect of light and ethylene upon cytokinin levels in seed of *Spergula arvensis*. J Exp Bot 24:662–666

Van't Hof J (1965) Relationships between mitotic cycle duration, S period duration and the average rate of DNA synthesis in the root meristem cells of several plants. Exp Cell Res 39:48–58

Van't Hof J (1967) The site of action of kinetin and IAA in the mitotic cycle of cultured pea roots. Am J Bot 54:639 (Abst)

Van't Hof J (1968) The action of IAA and kinetin on the mitotic cycle of proliferative and stationary phase excised root meristems. Exp Cell Res 51:167–176

Van't Hof J (1973) Two principal points of control in the mitotic cycle of pea meristem cells: energy considerations, characterization, and radiosensitivity. In: Duplan JF, Chapiro A (eds) Advances in radiation research. Biology and medicine Vol 2. Gordon and Breach, London, pp 881–894

Van't Hof J (1974) Control of the cell cycle in higher plants. In: Padilla GM, Cameron IL, Zimmerman AM (eds) Cell cycle controls. Academic Press, London New York, pp 77–85

Van't Hof J, Bjerknes CA (1977) 18 μm replication units of chromosomal DNA fibers of differentiated cells of Pea (*Pisum sativum*). Chromosoma 64:287–294

Vanderhoef LN, Briggs WR (1977) Red light-inhibited mesocotyl elongation in maize seedlings. The auxin hypothesis. Plant Physiol 59:Suppl, p 49

Vanderhoef LN, Stahl CA (1975) Separation of two responses to auxin by means of cytokinin inhibition. Proc Natl Acad Sci USA 72:1822–1825

Vanderhoef LN, Stahl CA, Seigel N, Zeigler R (1973) The inhibition by cytokinin of auxin-promoted elongation in excised soybean hypocotyl. Physiol Plant 29:22–27

Wada M, Furuya M (1970) Photocontrol of the orientation of cell division in *Adiantum*. I. Effects of the dark and red periods in the apical cell of gametophytes. Dev Growth Differ 12:109–118

Wada M, Furuya M (1971) Photocontrol of the orientation of cell division in *Adiantum*. II. Effects of the direction of white light in the apical cell of gametophytes. Planta 98:177–185

Wagner J (1970) Über die Wirkung des Äthylens auf *Nitzschia putrida* Benecke. Protoplasma 70:225–231

Walker GWR, Dietrich J (1961) Kinetin-induced meiotic-prophase acceleration and stasis in *Tradescantia* anthers cultured on media deficient in sugar. Nature 192:889–890

Walton DC, Harrison MA, Cote P (1976) The effects of water stress on abscisic-acid levels and metabolism in roots of *Phaseolus vulgaris* L. and other plants. Planta 131:141–144

Wardlaw CW (1965) The organization of the shoot apex. In: Ruhland W (ed) Encyclopedia of plant physiology Vol XV 1. Springer, Berlin Göttingen Heidelberg, pp 966–1076

Wareing PF, Good JEG, Manuel J (1968) Some possible physiological roles of abscisic acid. In: Wightman F, Setterfield G (eds) Biochemistry and physiology of plant growth substances. Runge, Ottawa, pp 1561–1579

Wareing PF, Horgan R, Henson IE, Davis W (1977) Cytokinin relations in the whole plant. In: Pilet PE (ed) Plant growth regulation. Springer, Berlin Heidelberg New York, pp 147–153

Watanabe T, Kondo N (1976) Isolation of a proteinaceous inhibitor of ethylene biosynthesis from marine algae. Agric Biol Chem 40:1877–1878

Webster PL (1967) Cell cycle kinetics in meristems: Effects of colchicine and IAA. PhD Thesis, Case-Western Res Univ, Cleveland, Ohio

Webster PL, Langenauer HD (1973) Experimental control of the activity of the quiescent centre in excised root tips of *Zea mays*. Planta 112:91–100

White KL, Hill AC, Bennett JH (1974) Synergistic inhibition of apparent photosynthesis rate of alfalfa by combinations of sulfur dioxide and nitrogen dioxide. Environ Sci Technol 8:574–576

Whittington WJ (Ed) (1969) Root growth. Proc 15th Easter Sch Agric Sci, Univ Nottingham. Butterworth, London

Wilkins H, Wain RL (1974) The root cap and control of root elongation in *Zea mays* L. seedlings exposed to white light. Planta 121:1–8

Wood DA, Hammond JBW (1977) Ethylene production by axenic fruiting cultures of *Agaricus bisporus*. Appl Environ Microbiol 34:228–229

Wood NL, Berliner MD (1979) Effects of indoleacetic acid on the desmid *Micrasterias thomasiana*. Plant Sci Lett 16:285–289

Woolley DJ Wareing PF (1972b) The role of roots, cytokinins and apical dominance in the control of lateral shoot form in *Solanum andigena*. Planta 105:33–42

Woolley DJ, Wareing PF (1972a) The interaction between growth promoters in apical dominance II. Environmental effects on endogenous cytokinin and gibberellin levels in *Solanum andigena*. New Phytol 71:1015–1025

Wright RD, Aung LH (1975) Effects of applied gibberellins on the growth of Japanese holly. Hortscience 10:181–182

Wright STC, Hiron RWP (1972) The accumulation of abscisic acid in plants during wilting and other stress conditions. In: Carr DJ (ed) Plant growth substances 1970. Springer, Berlin Heidelberg New York, pp 291–298

Yamaki T, Shimojo E, Yoshizawa K, Namekawa K (1974) Auxin in cancer tissue. In: Plant growth substances 1973. Hirokawa, Tokyo, pp 44–51

Yanagishima N (1963) Effect of auxin and antiauxin on cell elongation in yeast. Plant Cell Physiol 4:257–264

Yang SF (1980) Regulation of ethylene biosynthesis. Hortscience 15:238–243

Yeoman MM, Brown R (1971) Effects of mechanical stress on the plane of cell division in developing callus cultures. Ann Bot 35:1101–1112

Yu Y-B, Adams DO, Yang SF (1979) Regulation of auxin-induced ethylene production in mung bean hypocotyls: role of 1-aminocyclopropane-1-carboxylic acid. Plant Physiol 63:589–590

Zeevaart JAD (1971) (+)-Abscisic acid content of spinach in relation to photoperiod and water stress. Plant Physiol 48:86–90

Zeevaart JAD (1980) Changes in the levels of abscisic acid and its metabolites in excised leaf blades of *Xanthium strumarium* during and after water stress. Plant Physiol 66:672–678

Zobel RW, Roberts LW (1978) Effects of low concentrations of ethylene on cell division and cytodifferentiation in lettuce pith explants. Can J Bot 56:987–990

3 Transport and Other Modes of Movement of Hormones (Mainly Auxins)

H. KALDEWEY

3.1 Introduction

In the course of writing this article, it became more and more evident that it would be scarcely possible to review the immense number of studies of translocation of plant hormones since the appearance of the previous Encyclopedia (LEOPOLD 1961). Therefore, instead of aiming at covering all papers and considering all aspects of hormone movement, this review—still numbering more than 500 references—is restricted to three aspects of the topic: first, a brief historical sketch (Sect. 3.2); next, a consideration of the different methods used to estimate translocation and a critical evaluation of each. The section on General Remarks on Hormone Movement (Sect. 3.3), is the main theme of the review and, although there is some disadvantage in scattering results in this section, it is hoped that this is compensated far by illustrating pros and cons as well as advantages and limitations of experimental designs, and overcome by cross-referencing. The third and last section (Evidence of Hormone Translocation, Sect. 3.4) is a brief resume of the results thus far obtained in examining different types of hormone movement and hormone movement systems. Space allows neither a review of the manifold interactions between external and internal factors on hormone translocation, nor an in-depth consideration of the role of hormone distribution and its influence on patterns of growth and development, although they are occasionally referred to in different sections of this survey.

3.2 Historical Sketch

3.2.1 Indirect Evidence for Hormone Translocation

Well before the discovery of auxins and gibberellins during the 1920's hormone movement had been suggested as occurring in plants. As early as in 1879 Julius SACHS, considering observations on correlations between main axes and side shoots, proposed substances other than nutrients as involved in these phenom-

Abbreviations: ABA = abscisic acid; 2,4-D = 2,4-dichlorophenoxyacetic acid; DNP = 2,4-dinitrophenol; GA_3 = gibberellic acid; IAA = indole-3-acetic acid; IBA = indole-3-butyric acid; NAA = naphthalene-1-acetic acid; NPA = naphthylphthalamic acid; PEP = phosphoenolpyruvate; 2,3,5-T = 2,3,5-trichlorophenoxyacetic acid; TIBA = 2,3,5-triiodobenzoic acid.

ena. In later papers he characterized hormones as regulators of developmental processes and described their translocation within the plant. Almost at the same time, DARWIN (1880) discovered the failure of decapitated coleoptiles to react phototropically and deduced that in the intact organ, some "influence" was translocated from the tip toward the bending region. FITTING (1907) detected that after gravitropically stimulating solely the tip of a coleoptile, the stimulus moved basipetally and could cross a transverse cut. Thorough experiments led BOYSEN JENSEN (1910, 1913) and PAÁL (1914, 1919) — both students of WILHELM PFEFFER—to the conclusion that a "growth-promoting substance" (BOYSEN JENSEN) or "growth regulators" (PAÁL), arising in the tip, were translocated basipetally to the growing and reacting regions of coleoptiles. A "hormone theory" of root formation and gravitropic curvature was put forward by LOEB (1917). He observed an increase in georeaction and rooting of debladed *Bryophyllum* cuttings if only one young leaf remained at its apical end, indicating that "geotropic and root-forming hormones" were assumed to be "associated or identical", and they were formed in the leaf and moved along the stem.

3.2.2 Direct Evidence for Hormone Translocation

During the early 1920's an increasing number of papers appeared indicating the dependence of growth processes upon hormone-like substances originating from the plant's apical region, thus calling for a basipetal movement of such growth hormones (Wuchshormone) (SÖDING 1923, 1925). Basically, the results and deductions from experiments with coleoptiles (BRAUNER 1923, CHOLODNY 1924, NIELSEN 1924, SEUBERT 1925) were confirmed in seedlings of dicots (BEYER 1925, STARK 1925) as well as in adult plants (SÖDING 1926). WENT (1928) collected and isolated a "growth substance" in agar blocks at the basal cut surface of *Avena* coleoptile tips. This agar block, when placed on top of a coleoptile stump, caused elongation of the stump, or induced curvature of the stump if the block was asymmetrically applied. This was the crucial experiment that led WENT to postulate: "Ohne Wuchsstoff kein Wachstum" (no growth without growth substances). Moreover, WENT'S *Avena* curvature test provided a tool for quantitative estimation, thus initiating a new era of "growth substance" or auxin research.

3.2.3 Hormone Transport Experiments

3.2.3.1 The Early Period

Very soon the classic papers by DOLK (1930, 1936) appeared on auxin transport in gravitropically stimulated coleoptiles and by VAN DER WEIJ (1932, 1934) on the characteristics of auxin transport in short sections of etiolated *Avena* coleoptiles.

VAN DER WEIJ introduced the intercept method (see Sect. 3.3.2) and calculated the velocity (Transportgeschwindigkeit) of the auxin stream and its intensity (Transportintensität), i.e., the amount of auxin passing the cross-section of a coleoptile section in unit time. He demonstrated the transport to be strictly basipetal, independent of the physical orientation of the tissue. The auxin stream was able to proceed against a strong concentration gradient. Identical transport characteristics appeared in coleoptile sections of different length, although the basal sections transported less auxin than apical ones. Increasing the auxin concentration in the agar donor blocks did not alter the transport velocity. The basipetal auxin stream did continue after removal of the donor block, although part of the auxin had been "used up" in growth by the sections. With rising temperatures, up to about 30 °C, the intensity of transport increased at unchanged transport velocity (see however Sect. 3.2.3.2). At 0 °C, auxin translocation resembled physical diffusion and in the 1934 paper VAN DER WEIJ reported that the auxin transport system could be completely paralyzed by ether narcosis, auxin translocation then becoming apolar, resembling physical diffusion.

3.2.3.2 The Period of Confirmation, Extension, and Criticism

VAN DER WEIJ's experiments could almost pass for a catalog of auxin transport research of the subsequent three decades. In fact, this period, which has been reviewed in the first *Encyclopedia* by LEOPOLD (1961, for literature see this article), was characterized by mainly confirming and extending VAN DER WEIJ's observations and to include plant parts other than coleoptiles. Basipetal auxin translocation which "forms the basis... of apical dominance, tropistic movements of plant parts, and the basal orientation of root formation" was described "for many types of stems including hypocotyls, epicotyls, and rhizomes, and for leaves" (LEOPOLD 1961, p 671). For different plants, polarity was shown to decrease in the stem with increasing distance from the apical growing point. Some acropetal auxin movement, different from xylem transfer, was observed, especially in older tissues and flowering plants (JACOBS 1952, 1954, LEOPOLD and GUERNSEY 1953). Experiments with roots led to contradictory results concerning the polarity of auxin movement (GORTER 1932, FABER 1936, YEOMANS and AUDUS 1964, PILET 1964, BONNETT and TORREY 1965a, b, 1966). Narcotics other than ether were found to inhibit auxin transport, such as various respiratory poisons and metabolic inhibitors, especially sulfhydryl inhibitors, such as TIBA which, among others, blocks the transport system without detectable effect upon respiration (KUSE 1953, NIEDERGANG-KAMIEN and SKOOG 1956, NIEDERGANG-KAMIEN and LEOPOLD 1957, CHRISTIE and LEOPOLD 1965a). On the other hand, an adequate supply of oxygen and sugar, possibly as respiratory substrate, was determined to be necessary for maintaining the auxin transport system (v. GUTTENBERG and ZETSCHE 1956, REIFF and v. GUTTENBERG 1961, NAQVI and GORDON 1964, SMITH and JACOBS 1968). In contrast to VAN DER WEIJ's (1932) conclusion, the correlation between metabolic activity and auxin transport was underlined by demonstration that the transport velocity did depend upon temperature, being optimal at about 30 °C, typical for many metabolic processes (PILET 1951, GREGORY and HANCOCK 1955, JACOBS 1961, HERTEL 1962, HERTEL and LEOPOLD 1963, NEWMAN 1963, VARDAR 1964, KALDEWEY 1965a).

Estimations of the velocity of auxin transport ranged from 10 to 15 mm h^{-1} in coleoptiles; the velocity appeared to be lower in other tissues (e.g., JACOBS 1951, 1961, SCOTT and BRIGGS 1962, 1963, KALDEWEY 1963, 1965a, 1966, 1967a).

With the availability of synthetic growth regulators, including IAA instead of the urine auxin used by VAN DER WEIJ, and with the observation that several of these synthetic chemicals (i.e., phenoxy acids) inhibited IAA transport, speculations grew about the molecular requirements of substances having auxin activity and/or of being transportable. An early bioelectric theory, assuming such substances to move along an electric gradient in the plant, appeared as a promising approach to explain polar auxin transport (WENT 1932, CLARK 1937a, LUND 1947, SCHRANK 1951). However, electrical potential changes observed in connection with auxin movement, it now seems, are probably a consequence rather than the cause of auxin transport (CLARK 1937b, 1938, NEWMAN 1959, 1963, SCOTT 1967, see also Sects. 3.3.2.5, and 3.3.4.2).

VAN DER WEIJ's initial observation of "a consumption" of part of the auxin applied to the transport sections was given little attention in subsequent auxin transport experiments. Rather, information on this subject came from biochemical studies on auxin action and occurrence. These studies revealed that auxin was enzymatically destroyed by a so-called IAA-oxidase which was, accordingly, considered a regulator of IAA levels in the living plant. However, there was also evidence that such IAA destruction was restricted to damaged cells and would thus not account for natural auxin consumption. This field of auxinology—which has not really been recognized as being intimately linked to the physiology of auxin transport during the period here described—has been reviewed in the *Encyclopedia* by GALSTON and HILLMAN (1961; see also RAY 1958, HARE 1964). Toward the end of this period, the period of auxin studies with nonradioactive IAA and other growth regulators, the detection by ANDREAE and GOOD (1955) of a conjugation of IAA to aspartic acid reopened the issue of interrelations between transport and consumption of auxin.

In characterizing the studies on auxin transport of the 30 years since VAN DER WEIJ's fundamental observations, LEOPOLD (1961, p 671) stated: "Thus far the research has succeeded in describing the changes in polar transport through plants, it has described some of the quantitative properties of transport, it has led to some concepts of the possible mechanisms, and it has evoked some implications of this knowledge in understanding of polarity itself".

3.2.3.3 The "Modern" Period

The above phrases of LEOPOLD might also be taken as an introduction to the following survey of the past two decades of research on the movement of plant hormones. There is some progress, but none of the problems indicated has so far been solved. One may say that today's state of knowledge is even more puzzling than before. With the detection of whole new classes of plant hormones, cytokinins and abscisins, the re-detection of gibberellins and, in a sense, of ethylene, evidence has increased that there are interactions between the different natural growth substances, not only with regard to their growth-regulating function, but also with regard to the levels of hormones in the tissue and the interdependence of translocation.

The most striking improvement in the techniques involved with transport studies, since the end of the 1950's, came from the availability of radioactive growth regulators. The possibility of easy detection and measurement of minute amounts of labeled growth regulators led to reinvestigations and critical reconsiderations of existing results and concepts of hormone transport phenomena. Many of the earlier findings were confirmed—others were not (GOLDSMITH and THIMANN 1962).

Despite the great advantages, the interpretations of results of the first published experiments with ^{14}C-IAA concerning the redistribution of radioactivity after phototropic (BÜNNING et al. 1956, GORDON and EIB 1956, REISENER 1958, PILET 1960) or gravitropic (REISENER 1957, CHING and FANG 1958, REISENER and SIMON 1960) stimulation proved to have been misleading. This was largely due to the use of IAA of low specific activity at nonphysiologically high concentration, together with the unawareness of the fact that a large portion of IAA is immobilized after application of such high amounts or concentrations of the auxin. The doctoral theses of GOLDSMITH (1959), HERTEL (1962, HERTEL and LEOPOLD 1963), and NAQVI (1963) and the tropism studies of GILLESPIE and THIMANN (1961, 1963, PICKARD and THIMANN 1964) and GOLDSMITH and WILKINS (1964) were early attempts to indicate, and to successfully overcome these "teething troubles".

Estimations with different methods of the so-called transport velocity (see Sect. 3.3.3.3) revealed that this transport parameter may vary between 3 to 4 mm h^{-1} up to almost 20 mm h^{-1} at laboratory temperature, depending on the species and the plant part; velocities of the same order of magnitude were observed in intact plants if the auxin was applied to young leaves at the apex. However, after application to mature leaves, the auxin moved at a tenfold increase in speed; the movement obviously taking place in the sieve tubes (see Sect. 3.3.2.4). However, this finding does not indicate that sieve tubes are the natural channels for auxin transport, rather autoradiographic, aphid-feeding, and surgical studies point to phloem cells other than sieve tubes to be involved (WANGERMANN 1968, 1970, 1974, BONNEMAIN 1971, MORRIS and KADIR 1972, SHELDRAKE 1973a, BOURBOULOUX et al. 1973, BOURBOULOUX and BONNEMAIN 1973, 1979, MORRIS and THOMAS 1978, WANGERMANN and WITHERS 1978, ELIEZER and MORRIS 1980). Nevertheless, a shoot-to-root transport was demonstrated for radioactive labeled auxins applied to the shoot apex and auxin was shown to be transported acropetally in the root (toward the root tip) mainly in the stele, while smaller amounts moved apolarly in the cortex (see Sects. 3.3.6.2 and 3.4.3).

Studies using other compounds, many closely related chemically yet differing in auxin activity, showed a pronounced correlation between auxin activity and the ability of polar transport (e.g., HERTEL et al. 1969). However, this ability appeared not to be correlated with promotion of cell elongation, since polar transport was shown to persist or even to increase when growth was suppressed by the use of osmotica. Experiments of this kind clearly indicated that the often observed and confirmed decrease in auxin transport and/or polarity with increasing age of a tissue is not due to a corresponding alteration of cell elongation, as suggested by MCCREADY and JACOBS (1963b, 1967).

The use of radioactive hormones greatly facilitated the study of the fate of the compounds applied to plant tissues. For example, it allowed precise qualitative and quantitative comparisons between the processes of auxin transport and immobilization, utilization, and for metabolism of auxin. Studies of this kind not only extended the knowledge of conjugation and breakdown products, but also pointed to an inverse interrelationship between the ability of a tissue to immobilize and to transport auxin. In addition, both processes are influenced by internal (e.g., interactions between endogenous growth regulators) and external physical (e.g., gravity, light) and chemical (e.g., oxygen, ethylene) factors (see GOLDSMITH 1977, LETHAM et al. 1978, JACOBS 1979).

The search for the site(s) of auxin transport has followed three main lines: (1) transport experiments using tissues of differing anatomical-histological characteristics; (2) macro- and microautoradiographic studies of tissue transporting auxin; and (3) analyses of membrane fractions with specific binding properties.

No final conclusion can so far be drawn in spite of an immense and still increasing number of papers on this subject. The results of experiments of type (1) and (2) do, however, indicate that phloem cells in the vascular bundles might serve as the predominant transport paths, in spite of the fact that stem pith cylinders, which did not contain vascular tissue, are capable of basipetal auxin transport (see Sect. 3.3.6). Conclusions which might be drawn from studies on auxin binding to membrane fractions are similarly ambiguous. Studies initiated by HERTEL and co-workers at the beginning of the 1970's (LEMBI et al. 1971) indicated that several membrane fractions specifically bind auxins. Indeed, there are indications that binding sites on the plasma membrane (plasmalemma) might be involved in auxin transport. However, it has been shown that specific auxin binding occurs also to some sites localized in the cytosol (see Sect. 3.3.5). Auxin binding to the plasma membrane would fit the demands of the secretion theory of polar auxin transport put forward about 20 years ago (HERTEL and LEOPOLD 1963, CHRISTIE and LEOPOLD 1965a, b). According to this concept, auxin moves in the cytoplasm of a cell with a basipetal polarity of transport resulting from an energy-dependent, carrier-mediated secretion of the auxin molecules preferentially through the basal membrane of each cell (see Sect. 3.3.4.3, HERTEL 1983 and JACOBS and GILBERT 1983).

A re-detection and a renewed appreciation of the dependence on pH of IAA uptake and transport led to the development of the so-called chemiosmotic theory of polar auxin diffusion (RUBERY and SHELDRAKE 1973, 1974, RAVEN 1975, see also GOLDSMITH 1977 and Sect. 3.3.4.4). This theory suggests that the lipophilic weak acid IAA (IAA-H) permeates the cell membrane more easily than the anion (IAA^-); and deduces from experimental data that the membrane permeability for the former is 2 to 3 orders of magnitude greater than for the latter. IAA will then accumulate in the cytoplasm, the pH of which is presumed to be about 7, with an external pH of less than 7 by a thermodynamically downhill migration of both IAA-H and IAA^-. The acid (IAA-H), entering the cell by diffusion, will dissociate "and this maintains a high concentration gradient for its further influx" (GOLDSMITH 1977, p 441). The anion (IAA^-), on the other hand, will diffuse out of the cell along the outwardly directed anion concentration gradient. However, because of the limited anion permeability of cell membranes, the influx of IAA-H will be greater than IAA^--efflux, thus resulting in a net accumulation of IAA^- in the cytoplasm. The degree of equilibrium is inversely related to the anion permeability of the membrane. In a series of cells, then, basipetal auxin translocation will result if the anion permeability is smaller in the apical than in the basal cell membrane.

A principal difference between the chemiosmotic and the secretion theory is the involvement of metabolic energy in the process of the polar transport. According to the secretion theory, energy is directly coupled "to carriers, in those cases, where auxin transport appears to be uphill" whereas "the new theory postulates that auxin transport is thermodynamically downhill, but with metabolic energy expended to maintain the pH and electrical gradients" in addition to maintaining "an ordered membrane with asymmetric permeability to IAA^-" (GOLDSMITH 1977, pp 446 and 491).

Although auxin dominates the research in transport of plant hormones, an increasing number of papers are now appearing dealing with the movement of other plant growth regulators. Again, the availability of radioactive compounds has facilitated such studies. Nevertheless there again remain many uncertainties as to how and where these substances are moved and how they influence each other in their translocation. "Gibberellic acid and thiamine... have been shown to move with strong basipetal polarity through petiole sections; still other hormones, such as cytokinins and probably abscisic acid, move but with no polarity. In roots gibberellic acid moves with basipetal polarity, but thiamine, kinetin and probably zeatin move without polarity.... Each hormone affects the movement of another, judging by combinations tested so far. For instance, indoleacetic acid increases the basipetal movement of kinetin (and makes it polar); gibberellic acid increases the basipetal movement of indoleacetic acid in stems and decreases it in petioles, whereas abscisic acid decreases the basipetal movement of indoleacetic acid in petioles". This picture drawn by JACOBS (1978b, p 361) may serve to outline some aspects and possible interpretations of studies on this subject. It should be kept in mind, however, that it is designed from an, as yet, rather limited basis of experimental data.

Readers may refer to the following reviews and summarizing articles: (a) for auxin transport by LEOPOLD 1963, GOLDSMITH 1968b, 1977, KALDEWEY 1968a, VARDAR 1968b and in roots by SCOTT 1972, BATRA et al. 1975; (b) for translocation of other growth regulators by MCCREADY 1966, GOLDSMITH 1969, ZIEGLER 1973, JACOBS 1978b; and (c) for general phenomena of transport by THIMANN 1972, LETHAM et al. 1978, JACOBS 1979, and Proceedings of the Izmir Conference on *Transport of Plant Hormones,* edited by VARDAR (1968a), which consider transport phenomena generally.

3.3 General Remarks on Hormone Movement

3.3.1 Terminology and the Transport Equation

Were it the case that auxin transport takes place within a plant part—a stem, a petiole, a root—by travelling in a solution of constant concentration (density) in a continuous stream of constant velocity (speed), then, the efficiency of the transport system may be described by the mass of auxin passing the cross section of the plant part in unit time (see CANNY 1960, 1971, 1973, KALDEWEY 1967b, 1968a, NEWMAN 1974). This mass transfer or flux per cross-sectional area was defined as transport intensity by VAN DER WEIJ (1932) depending on the concentration and the velocity of the stream.

Thus,

$$\text{mass transfer (flux)} = \text{area} \times \text{concentration} \times \text{velocity} \tag{1}$$

the appropriate units would be:

$$\text{ng h}^{-1} = \text{mm}^2 \times \text{ng mm}^{-3} \times \text{mm h}^{-1}.$$

Concentration in this equation means the mass of auxin per volume of the transporting system just as area means the cross-sectional area of the system. The area would be identical with the cross section of the plant part examined if auxin were moved across the total area of all cells of the tissue. It is quite certain that this is not the case; it is, however, uncertain what part of the tissue is involved in transport (see Sect. 3.3.6). Thus the quantities *area* and *concentration* are as yet not accessible to experimental determination. However, in the product "*area* × *concentration*" the area of the unknown transport channels is eliminated. It remains a quantity with the dimension "mass per length" which means the amount of mobile auxin present in unit length of the transport system, identical with unit length of the plant part. This quantity, which can be estimated experimentally, was termed *capacity* by VAN DER WEIJ (1932).

VAN DER WEIJ compared the transport system with a conveyor-belt and used the term *capacity* (*Kapazität*) to describe the "load of a unit length of this belt" ("die Kapazität..., d.h. die Stoffmenge, welche jedes cm des Transportbandes enthalten kann," VAN DER WEIJ 1932, p 482). This expression has led to some confusion in the literature. The term capacity has been used, misleadingly, in the sense of transport intensity or maximum intensity. Since it means "length density", the term *transport density* has been proposed, and this term will be used in this article (KALDEWEY 1968a).

Thus, the transport process may be described by the following equation:

$$\text{transport intensity} = \text{transport density} \times \text{transport velocity}$$

$$I\,(\text{ng h}^{-1}) = d\,(\text{ng mm}^{-1}) \times v\,(\text{mm h}^{-1}) \tag{2}$$

The term *velocity* instead of rate of movement is used intentionally in the transport equation. "The expression 'rate of movement of auxin' has been used by some authors to mean the distance moved in unit time..., and by others to mean the amount passing through the tissue in unit time.... The term 'rate' has therefore become incurably ambiguous, and ought no longer to be employed..." (MCCREADY 1966, p 284). Unfortunately, this suggestion has not been always followed and the term is still being used. Similarly, the term *transport* which has been proposed to be restricted to active polar translocation processes is used in "most recent literature... synonymously with movement, polar or nonpolar, without implying anything further regarding the mechanism involved" (GOLDSMITH 1977, p 440).

From Eq. (2) it is evident that the cross-sectional area of the plant part examined may be taken as unity. Transport intensity therefore means mass transfer across the total cross-sectional area. This now is probably the only appropriate unit which auxin transfer of a plant part may be referred to, although evidence is increasing from surgical and autoradiographic studies that the different cell types contribute very differently to the total amount of auxin transported (see Sect. 3.3.6). Thus, the calculation of a specific mass transfer (CANNY 1960, p 509, KALDEWEY 1967b, p 487), i.e., mass transfer per unit area (e.g., flux mm^{-2} of the plant part's cross section), may well be misleading since the specific area is not known. However, comparisons of specific fluxes in similar plant parts of different size may give information of their efficiency.

A majority of the studies on transport of auxin (and other plant hormones) published during the years of research since the first papers of the "Utrecht School" (WENT 1928, VAN DER WEIJ 1932, 1934, DOLK 1930, 1936) have been designed to estimate the parameters in the transport equation. Using varying experimental conditions and many kinds of plant parts, and presupposing a uniform auxin stream, usually only two parameters, intensity and velocity, were determined in order to calculate the third, density, of the equation. Whether such calculations are justified, will be critically discussed in Section 3.3.3.

3.3.2 Methods to Estimate Hormone Translocation Characteristics

3.3.2.1 Intercept Method of van der Weij

The classic technique for studying hormone transport in plants is the intercept method, introduced by VAN DER WEIJ (1932) to estimate characteristics of basipetal auxin transport in *Avena* coleoptiles. He applied agar blocks containing auxin, prepared from urine or collected from *Zea* coleoptile tips, the so-called donors, to the apical cut end of short (1–5 mm) sections of etiolated coleoptiles, previously depleted of endogenous auxin, and he placed plain agar blocks, the receivers, on their basal cut surface (Fig. 3.1). The amount of auxin found in the receivers after different transport periods was then determined by the *Avena* curvature test (WENT 1928) and plotted against time, beginning with donor application. The reasonably linear time course of auxin accumulation was taken as evidence that there was a uniform auxin stream with a sharp front moving through the coleoptile section. On that assumption, the intercept on the time axis of the least squared line gave an estimation of the time required for the front of the auxin stream to reach the base of the transport tissue section. Hence, transport velocity could be calculated. VAN DER WEIJ estimated the velocity, for coleoptiles, to be 10 mm h^{-1}.

In early auxin transport studies, the straight line was drawn by eye and the time intercept determined from graphical extrapolation. This subjective method has occasionally led to misinterpretations. For instance, from data thus treated, VAN DER WEIJ (1932) concluded that the transport velocity remained constant in varying temperatures, and that the observed increase in transport intensity was due to an increase in transport density. However, statistical treat-

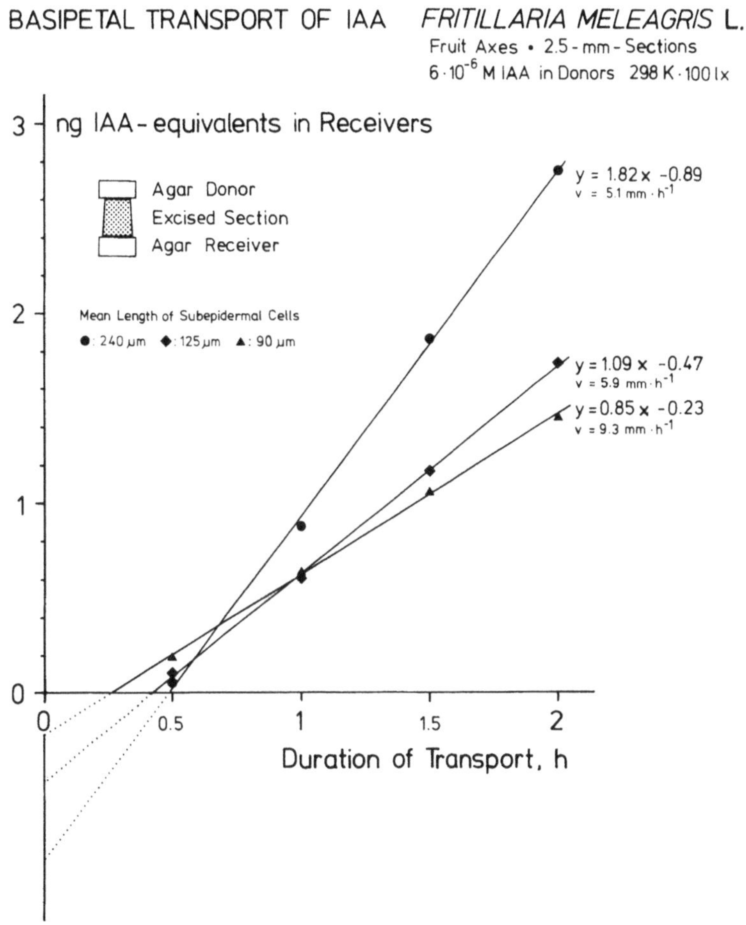

Fig. 3.1. Time course of basipetal IAA transport through sections excised from different regions of fruit axes of *Fritillaria meleagris* L. as an example to demonstrate the intercept method of VAN DER WEIJ (1932). Sets of six sections, previously depleted of their endogenous auxin, were placed upright on large agar blocks (receivers) which were subdivided into 12 small blocks suitable for assay with the *Avena* daylight test according to SÖDING (1952); similar blocks containing IAA at a concentration of 3–6 µM (donors) were applied to the apical cut surface of the sets. Receivers and donors were replaced every half hour. The accumulated auxin in the receivers, expressed as ng IAA equivalents transported per section, are plotted against time. The linear regression equations and the lines of closest fit were estimated by the least-squares method. The intercepts of these lines with the time axis served to determine the transport velocity (v). The mean lengths of subepidermal cells, measured on epidermal strips taken from all sections after transport, were used to characterize section "age" (see KALDEWEY 1957). Mean values of three to four replicates. (Data from KALDEWEY 1967a)

ment of VAN DER WEIJ's data by linear regression analysis — as was done by GREGORY and HANCOCK (1955) — indicates that there is, in fact, both an increase in velocity as well as in intensity with temperature; both having a Q_{10} of approximately 2.

The linear equation of best fit to transport data

$$y = bx + a, \tag{3}$$

where y = amount of auxin in receivers, x = duration of transport, b = slope and a = y-intercept of the straight line, gives all the information on transport characteristics, namely the intensity (I), the velocity (v), and the density (d).

1. The transport intensity corresponds to the slope of the straight line, thus

$$I = b. \tag{4}$$

2. The intercept of the straight line with the x-axis corresponds to the time intercept, called "time delay" (t_d) by KALDEWEY (1968a), and allows to calculate the transport velocity by

$$v = l/t_d,$$

where l = length of the transport section.
Since the x-intercept results from setting Eq. (3) equal to zero

$$t_d = -a/b,$$

thus

$$v = -l \times b/a. \tag{5}$$

3. Rearranging the transport Eq. (2) leads to

$$d = I/v,$$

and when substituting I by Eq. (4) and v by Eq. (5) the transport density may be calculated by

$$d = b/(-l \times b/a),$$

thus

$$d = -a/l.$$

Calculations of this kind are valid only if the hormone moves in a uniform stream, at constant velocity (for further discussion and criticism see KALDEWEY 1968a, p 12 and Sect. 3.3.3).

If auxin determinations are done by bioassay, the transport sections must be depleted of their endogenous auxin before the donors are applied. Usually, batches of ten or more sections are combined in order to collect enough auxin in a receiver for quantitative determinations. VAN DER WEIJ used three or more batches of coleoptile sections, and assayed for auxin in the receiver blocks at three or more intervals. Another method is to replace receiver blocks periodically with fresh (blank) agar blocks. The amounts of auxin in the respective receivers were then used for plotting and/or computing data (see Fig. 3.1). In order to guarantee a constant auxin concentration in the donors, they, too, were replaced periodically in longer experiments (see Sects. 3.3.3.4, 3.3.3.5).

The advent of labeled hormones, especially those of high specific activity, facilitated precise measurements of minute quantities, allowed studies with individual transport sections, and shortened collecting intervals to a few minutes. Further, transport sections, not depleted of their endogenous hormones, could be used. There have been detected by this means (1) fluctuations of IAA-movement after blue light irradiation (THORNTON and THIMANN 1967), (2) oscillations of auxin export to basal receivers (HERTEL and FLORY 1968), (3) the existence of a faster and a slower component of auxin transport (NEWMAN 1965, 1970, DE LA FUENTE and LEOPOLD 1972), and (4) a nonlinearity of the initial IAA accumulation in basal receivers (see Figs. 3.3 and 3.4 and Sect. 3.3.3).

When a constant delivery of auxin to the receivers is reached, the export "curve" becomes linear, which is a prerequisite for the equations and calculations outlined above. Linearity of the export curve usually lasts only for a few hours after donor application. The onset of the decline in rate varies with the type and length of the transport section, with the concentration and nature of the hormone applied, and with experimental conditions.

Basipetal IAA export, in donor-supplied sections, about 5 mm in length, cut from coleoptiles, stems, and petioles has the following characteristics: delivery begins to decline 2 to 6 h after auxin application (e.g., GOLDSMITH and THIMANN 1962, NAQVI 1963, PILET 1965, GREENWOOD and GOLDSMITH 1970, KALDEWEY et al. 1977, DRAKE and CARR 1978); its linearity is prolonged at lowered temperature (McCREADY 1968a, b); it varies with tissue age (KALDEWEY 1967a, KALDEWEY et al. 1973) and anatomical structure (JACOBS and McCREADY 1967, SHELDRAKE 1973a). A constant auxin delivery to apical receivers (typically weak) may last for 24 h with shoot segments (e.g., McCREADY and JACOBS 1963a, McCREADY 1968a, b) and for 6 to about 12 h with root segments (e.g., YEOMANS and AUDUS 1964, PILET 1964, KIRK and JACOBS 1968, SCOTT and WILKINS 1968, 1969, HILLMAN and PHILLIPS 1970, WILKINS and CANE 1970, SHAW and WILKINS 1974). The duration of constant export rate appears to increase with increasing length of the sections (GOLDSMITH and THIMANN 1962) and to decrease with increasing hormone concentration in the donors (GOLDSMITH and THIMANN 1962, McCREADY 1963, THORNTON and THIMANN 1967, KALDEWEY and GINKEL 1974). Hormones, such as NAA, IBA (LEOPOLD and LAM 1961), and also GA_3 (JACOBS and PRUETT 1972, 1973) seem to behave similar to IAA with respect to the decline in the export rate from short sections. In contrast, almost no decline was observed up to 24 h for the slower (1–2 mm h^{-1}) moving 2,4-D (NAQVI 1963, McCREADY 1963, McCREADY and JACOBS 1963a, McCREADY 1968a, b), for ABA (INGERSOLL and SMITH 1970, DÖRFFLING et al. 1973), and the apolar-moving kinetin (VEEN and JACOBS 1969b).

A cause of the decline of auxin export rate has been traced to the limited availability of energy sources, especially in short sections of etiolated plants. This view is supported by the enhancement of auxin transport caused by ATP (REIFF and v. GUTTENBERG 1961), or by addition of sugars (v. GUTTENBERG and ZETSCHE 1956, SMITH and JACOBS 1968, NAQVI 1974, KRUL and COLCLASURE 1977). In fact, NAQVI and GORDON (1964) observed a prolonged period of constant IAA delivery to basal receivers from *Zea* coleoptile sections with supplementation of sucrose through the donor block.

On the other hand, DE LA FUENTE and LEOPOLD (1970a) demonstrated, for hypocotyl sections of light-grown *Helianthus* seedlings, that a fresh cut at the receiver end of sections, "aged" for 24 h, made before donor application, greatly eliminated the otherwise drastic reduction of auxin delivery to basal receivers. Thus, deteriorative changes at tissue cut surfaces may cause a decline or even act as a barrier to transport (e.g., BRIGGS et al. 1955; STEEVES and BRIGGS 1960, SCOTT and BRIGGS 1960, 1963, THIMANN and WARDLAW 1963, WANGERMANN 1974), and cause a pile-up of mobile auxin within the tissue adjacent to the receivers. Such accumulation has been reported to occur after about

2 h of basipetal IAA transport through 5-mm sections of green sunflower (LEOPOLD and LAM 1962) and pea stems (KALDEWEY et al. 1974, Fig. 4). However, the recovery from the "aging effect" by a recutting appears not to be a general phenomenon. In segments of *Phaseolus* petioles, aging periods between 5 and 12 h caused a drastic reduction in export to basal receivers of both auxins IAA and 2,3,5-T and cytokinins even when receivers were applied to freshly cut aged sections (OSBORNE et al. 1968, OSBORNE and MULLINS 1969). Further, apical application of auxin or kinetin during the aging perioa apparently maintained the transport capability of excised sections (see also LEOPOLD and LAM 1962, HERTEL and FLORY 1968, RAYLE et al. 1969, VEEN 1969) similarly to storage of the tissue at low temperature during this period (WANGERMANN 1974). Thus results of this kind indicate that there is a disturbance of the transport system itself, possibly through an inactivation or loss of transport sites (LEOPOLD and LAM 1962, OSBORNE 1968, OSBORNE and MULLINS 1969, DE LA FUENTE and LEOPOLD 1970a) during auxin depletion or aging, rather than a deterioration of cut surfaces. The occurrence of both conditions is, indeed, likely. Therefore, transport experiments with freshly cut, nondepleted tissue sections, using hormone concentrations resembling those of the natural state, and employing shorter time periods, may be the best way to duplicate properties of the natural transport system.

A pronounced hindrance of export of gibberellins and ABA through cut surfaces has been observed in segments of coleoptiles, stems, and roots (PHILLIPS and HARTUNG 1974, 1976, HARTUNG and PHILLIPS 1974, HARTUNG and BEHL 1974, WILKINS and NASH 1974). In these cases the estimation of transport characteristics by the hormone content of agar receivers is misrepresentative. Therefore, instead of using agar blocks, the extraction of parts of long tissue sections, cut at a definite distance from the donor end after different transport periods has been used as a substitute or as "tissue receivers" (e.g., BONNETT and TORREY 1965a, WANGERMANN 1974, HARTUNG and BEHL 1974, HARTUNG 1975, KRUL 1977). This necessitates, of course, that the transport data must be taken from different individual (or sets of) segments.

In spite of the uncertainties, difficulties, and complications mentioned so far and to be considered further below (see Sect. 3.3.3), the results obtained are generally believed to reflect and describe the properties of the hormone transport systems in intact plants. This belief is supported by results obtained with other methods which, at least partly, confirm the estimations derived from the intercept method.

3.3.2.2 Short-Term Collecting Method

Although the most accurate way to estimate the characteristics of natural hormone transport is in a system as intact as possible, the method outlined below has not been adopted to any great extent.

In studies of the basipetal auxin transport in flower stalks of *Fritillaria,* a method has been used which may be likely to estimate the in situ situation (KALDEWEY 1968b, 1971). The basic idea of this method was to replace the stream of endogenous auxin, delivered to the stalks by the bud or fruit (KAL-

DEWEY 1957, 1964), with radioactive IAA and to follow its behavior in the plant, left as intact as possible. The natural auxin source was replaced by a lanolin paste containing IAA (in this case, ^{14}C-IAA). The new auxin source sustained natural elongation and growth movements of the stalk (KALDEWEY 1957, 1962, 1965b). Seven hours later, the axis of the plant was cut several centimeters below the apex and supplied at the cut end of its base with agar receivers, twice each for 15-min periods. Then, a 5-mm section was excised and stood on fresh agar receivers to allow all the mobile auxin to move out, and the new basal cut surface of the remaining axis was successively supplied with another pair of receivers for 15-min periods. The radioactivity collected in the 15-min receivers and the section receivers, respectively, allowed an estimation of the transport intensity and transport density. It was possible, therefore, to calculate the transport velocity from these quantities [see Eq. (2)]. Further, auxin immobilized within the 5-mm sections could be determined by extraction (see Fig. 3.6).

A similar procedure to the one described above was employed by NEWMAN (1970) to determine the flux of applied ^{14}C-IAA, and the concentration of mobile auxin in different regions of *Avena* coleoptiles. In this case, the auxin-depleted (see legend to Fig. 3.2B) coleoptiles were fed with ^{14}C-IAA for a period of 41 min. Thereafter, they were subdivided into successive 1-mm sections which were then placed on a series of agar receivers for the collection of mobile IAA between 2 and 10 min up to a period of about half an hour. Since ^{14}C from fractions not belonging to mobile IAA "emerging almost immediately from damaged tissue" (NEWMAN 1970, p 268) appeared to contaminate the first 2-min sample, the flux of the second 2-min period was taken as measure of the transport intensity. The concentration of moving IAA [i.e., density $\times$ area^{-1}, see Equs. (1) and (2)] was calculated using the assumption that all the IAA moved as a stream at a velocity of 9.4 mm h^{-1}, a figure which was determined in separate experiments (see p 99). A comparison of this value with the concentration of total mobile auxin present in the 1-mm pieces along the coleoptile, as determined by exhaustive basipetal drainage of the sections, revealed mobile auxin to be approximately 40% to 60% below the corresponding values of the total mobile auxin. Further, it took >30 min to drain the 1-mm pieces, which was much longer than expected if the auxin were moving in a uniform stream at 9.4 mm h^{-1} (see similar results of DE LA FUENTE and LEOPOLD 1970a, 1972 for *Helianthus* hypocotyls and of DAVIES (1974) for *Pisum* stem sections). Thus, these results mean "that at least half the IAA that can emerge from the base of a 1-mm section... has been in a mobile form in the tissue but not actually moving in a well-defined transport stream at uniform speed" (NEWMAN 1970, p 269).

Similar conclusions have been drawn from studies of ^{14}C-IAA transport in *Pisum* internodes (KALDEWEY et al. 1974) and *Citrullus* hypocotyls (KALDEWEY et al. 1977), as affected by water stress and transport inhibitors, respectively. Water stress, as well as the inhibitor lycoricidinol, drastically decreased the transport density, as estimated by the intercept method, although the total amount of mobile auxin emerging from the basal cut surfaces of subsections after transport periods of 4 to 5 h increased or was only slightly decreased. Further, both treatments caused a higher portion of total mobile auxin to remain near the apical donor blocks and, consequently, less reached the base and less emerged in the receiver blocks than in the controls. It was speculated that both treatments might weaken the basipetal polarity of auxin transport and lead to "an 'extra-transport-systemic' fraction of radiocarbon which is not immobolized, obviously not transported basipetally, and nevertheless diffuses out of the basal cut surface into agar blocks when the subsections are quantitatively drained" (KALDEWEY et al. 1974, p 572). However, a basipetal decrease of mobile auxin concentration within the sections, observed also in controls (though less steep), as well as in the above-mentioned short-term experiments of NEWMAN (1970), points to the possibility of an acropetal component in the transport systems. In fact, NEWMAN found that about one third the amount of IAA that emerged from the base of the 1-mm sections, was delivered to agar receivers placed on the top of the sections instead of at the basal cut surface (see also GOLDSMITH 1966b). Similar results are reported for sections cut from the hypocotyl and the primary root of *Gossypium* seedlings which had been fed from the apex with ^{14}C-IAA. Such sections for transport periods up to 14 h delivered radiocarbon to both acropetal and basipetal agar receivers; the ratio between acropetal and basipetal export depended on the duration of transport and the region from which the sections were cut (KALDEWEY and KRAUS 1972).

These results suggest that in the short-term collection method basal and apical receivers should be applied simultaneously. By this means, the acropetal component of the transport system is accounted for and the auxin then delivered to basal receivers probably will give a more accurate measure for the density of the basipetal auxin stream. Moreover, the calculation of the acropetal/basipetal ratio may reflect the polarity of hormone translocation.

3.3.2.3 Pulse Experiments

Short-term application of auxin to the apical cut surface of coleoptile sections, combined with an estimation of auxin accumulation with time in basal receivers which were replaced at brief intervals, was demonstrated by VAN DER WEIJ (1932, p 442ff) to be a means of calculating transport velocity. He observed that the auxin export rate (i.e., the transport intensity) increased initially to a maximum and then decreased. He assumed that the arrival time of the peak of transport intensity was the period of time needed by the auxin stream to traverse the segment. The velocity thus estimated (8 mm h^{-1}) was similar to the values of about 10 mm h^{-1} obtained with the intercept method. When labeled hormones became available, such pulse experiments were refined and modified. The duration of the pulse application could be reduced to 60 s (SHEN-MILLER 1973a, b) and the receivers could be changed with great frequency to improve the estimation of the peak.

Currently, the pulse method is run as a pulse-chase experiment; that is, the pulse donors containing radioactive labeled hormone are replaced by agar donor blocks containing the unlabeled hormone at similar concentration during the transport periods. Normally there is no detectable effect of the chase by unlabeled auxin on the radioactive pulse migration in *Avena* coleoptiles (GOLDSMITH and THIMANN 1962) and *Helianthus* hypocotyls (DE LA FUENTE and LEOPOLD 1970a) although IAA appeared to be "pushing" a ^{14}C-IAA pulse in *Lens* stem sections (PILET 1965).

HERTEL (1962) and HERTEL and LEOPOLD (1963) were the first to use the pulse-chase method with ^{14}C-IAA and they observed that, in corn coleoptiles, the speed of the "auxin wave" was equal to the transport velocity as measured by the intercept method (Fig. 3.2A). The wave of NAA was shown to move

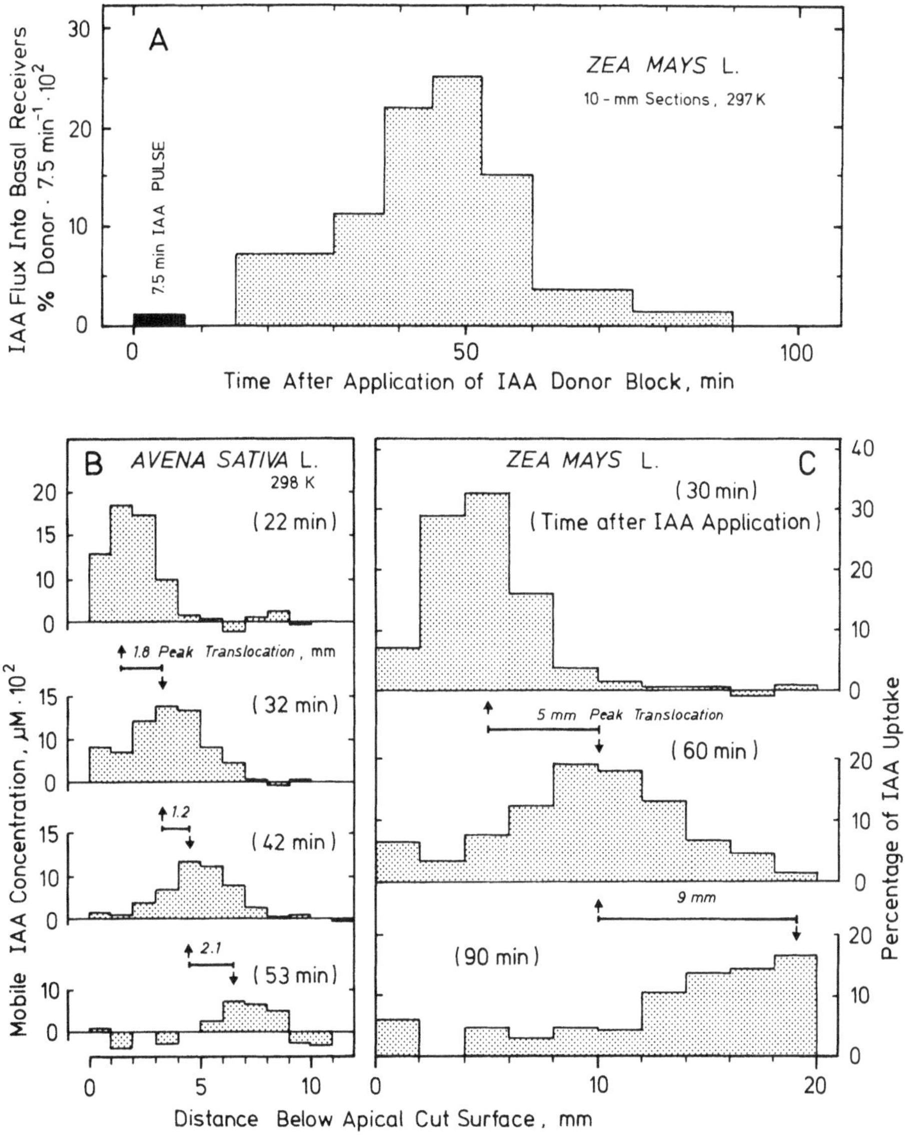

Fig. 3.2. **A** IAA flux into basal agar receivers from 10-mm sections of corn coleoptiles apically supplied with a 5-min pulse of ^{14}C-IAA (5 μM in agar). At the time indicated by each column, the sections (30 per transport block) were transferred to new receivers. (Data from HERTEL and FLORY 1968). **B** Distribution of mobile auxin in oat coleoptiles apically supplied with a 10-min pulse of ^{14}C-IAA (0.7 μM in agar). The mobile auxin concentration was estimated from the radioactivity in basal agar receivers taken from samples of nine coleoptiles subdivided serially into 1-mm sections at the times measured from the start of donor application. Note the differences in translocation of the peak, indicated by *arrows*, in different regions of the coleoptile. (Data from NEWMAN 1970). **C** Distribution of radioactivity within 20-mm sections of corn coleoptiles apically supplied with a 15-min pulse of ^{14}C-IAA (10 μM in agar). The pulse donors were replaced by blocks containing unlabeled IAA at an identical concentration. The sections were cut into successive 2-mm pieces at the times indicated, and the level of tissue radioactivity was measured. Note the differences in translocation of the peak, indicated by *arrows*, during the two 30-min transport periods. (Data from GOLDSMITH 1967b)

at about 9 mm h^{-1}, which is slower than that of IAA (about 13 mm h^{-1}); a peak of a very flat wave of 2,4-D moved at a speed of 1.2 mm h^{-1} (HERTEL and FLORY 1968). This latter result was similar to that determined in bean petioles by the use of the intercept method (McCREADY and JACOBS 1963a). Pretreatment of tissue sections with auxin at low concentrations which has been shown to enhance subsequent auxin transport (e.g., LEOPOLD and LAM 1962, HERTEL and LEOPOLD 1963, HERTEL and FLORY 1968, DE LA FUENTE and LEOPOLD 1970a), was also shown to enhance the speed and to increase the height of an IAA-wave in corn coleoptiles; a pretreatment period as short as 20 min was sufficient to stimulate IAA transport (RAYLE et al. 1969).

The availability of labeled hormones allowed a further modification of pulse experiments as was first demonstrated by GOLDSMITH (1967a, b). Instead of collecting the pulse by means of agar receivers, its migration was followed in segments of 20-mm transport sections (Fig. 3.2 C). After the pulse application, the sections were placed on agar receivers for different periods of time, each of which was short enough to insure that the peak of the label remained within the tissue. The sections were then subdivided into successive 2-mm pieces and the radioactivity of the tissue determined. A disadvantage of this method is that the data are obtained from different individuals or different sets of individuals, though with enough replications a representative average may be obtained.

This was the first method to allow the separation of the actual physiological process of auxin transit from the artificial, yet unavoidable, uptake and/or exit phases which take place across injured cells in the other methods. Indeed, with this method, GOLDSMITH demonstrated in corn coleoptiles that auxin transit itself is oxygen-dependent. In air, a distinct though somewhat dispersing peak of radioactivity moved at 12 to 15 mm h^{-1}, which is consistent with the transport velocity measured with other methods. However, the velocity of basipetal peak migration was reduced to 1 to 2 mm h^{-1} about 10 min after the tissue was transferred to a nitrogen atmosphere. The observed shape of the pulse, its basipolar migration at nearly constant velocity, and its dispersion with time of movement can be described by mathematical models based upon assumptions of the hypotheses of chemiosmotic polar diffusion (GOLDSMITH et al. 1981, see also Sect. 3.3.4.4) or of a diffusion-superposed drift movement (HASENSTEIN 1982, HASENSTEIN and KALDEWEY 1983).

Using essentially the same method, GOLDSMITH (1982) demonstrated that there is a saturable site responsible for polar transport of IAA in maize coleoptiles. The advantage in this method of not having to handle the sections during the transport periods was extended by GOLDSMITH and RAY (1973). They showed that centrifugation of corn coleoptiles slightly enhanced the basipetal migration of the pulse if the centrifugal force worked toward the base of the sections. This treatment also had the effect of displacing the cytoplasm toward the base of the cells. On the contrary, the pulse migration was drastically reduced if the centrifugal force was in the direction toward the apex. A useful conclusion gained from these results of this study was "that the metabolic component of the transport is a polar secretion of auxin localized in the basal plasma membrane of each cell" (GOLDSMITH and RAY 1973, p. 297; see Sects. 3.3.4.3, 3.3.4.4).

A variation from applying the hormone pulse to excised sections, has been to apply labeled IAA to decapitated but otherwise intact plants, which were either depleted of their endogenous auxin (NEWMAN 1970) or not depleted (SHEN-MILLER 1973a, b). NEWMAN followed the migration of the peak of mobile auxin by collecting radiocarbon in basal receiver blocks from 1-mm pieces of *Avena* coleoptiles which were subdivided following different transport periods. He showed that the peak of mobile auxin migrated at a velocity resembling that determined by the other methods (Fig. 3.2 B). However, the results indicate that the speed of migration of the peak varied in different regions of the coleoptiles. This too, was the case in GOLDSMITH's experiments with corn coleoptiles (Fig. 3.2 C and GOLDSMITH 1982) as was also shown by SHEN-MILLER (1973a, b) for the migration of the front of IAA pulses in corn and oat coleoptiles (see Sect. 3.3.3.4).

Pulse application combined with the calculation of the transport velocity by pulse migration and the so-called transport capacity and transport index has been used in studies on the translocation of auxin(s), ABA, GA, and kinetin in stems and petioles of *Gossypium* (BEYER and MORGAN 1969 and papers cited below). In these experiments, hormone pulses of 20 to 30 min were allowed to migrate at two time periods, through 20- to 60-mm-long tissue sections. These were then subdivided into 2–3 mm subsections, the radioactivity of which was determined by extraction and expressed as the percentage of total activity found in the section. The so-called transport capacity was calculated by subtracting the percent activity of each subsection of the shorter transport period from the values of corresponding subsections of the longer transport period and summing the positive values. Thus, this quantity represents the percentage of radioactivity translocated between the shorter and longer transport periods which is independent of the distance through which the label has moved (BEYER and MORGAN 1969, DAVENPORT et al. 1977b). To account for the distance component, the positive differences in subsection activity between the short- and long-term experiments were multiplied by the median distance of each of the subsections from the hormone source. The sum of these values provided the value for the "transport index" (DAVENPORT et al. 1980). The authors showed that aging of stem sections decreased the velocity of ^{14}C-IAA-pulse migration, as had been observed following pretreatment of stem sections with ethylene (BEYER and MORGAN 1969) and following water stressing cotton seedlings before excision of the cotyledonary petiole segments (DAVENPORT et al. 1977b). The latter treatments also reduced the "transport capacity". Finally, the "transport capacity", "transport index", and the velocity of ^{14}C-IAA were demonstrated to decrease with increasing maturity of petioles excised from different regions of 90-day-old plants, and they could be further reduced by water stress at any age of tissue.

ABA, GA_3, and kinetin appeared to move slowly in *Gossypium* (1 to 2 mm h^{-1}) and without polarity. The pulse migrations were unaffected by water stress or anaerobic conditions (DAVENPORT et al. 1977a, 1979).

DA CRUZ and AUDUS (1978) used short-term pulses (45 s) of ^{3}H-IAA which were applied by micro-injection into the cortex near the apex of intact stolons of *Saxifraga sarmentosa* to follow basipetal auxin movement in a subsequent transport period of 15–18 h. Separate analyses of cortex and stele revealed two distinct types of translocation. The main flow was in the stele traveling with a distinct peak at about 7 mm h^{-1}. The profiles of radioactivity in the cortex, however, pointed to a diffusion type of flow. On the basis of fitting theoretical diffusion profiles to the observed cortex profiles, auxin movement in the cortex was referred to as an "accelerated" diffusion with a diffusion coefficient 20 to 30 times greater than that estimated for IAA in aqueous agar (LARSEN 1955). TIBA, NPA, a morphactin, and the herbicide Endothal blocked both types of movement, as did cycloheximide, DNP, and KCN. However, actinomcycin D and puromycin had no effect, which suggests that the inhibitor action was associated with an uncoupling of oxidative phosphorylation rather than with an inhibition of synthesis of a protein carrier of auxin. Pulse application made 15 cm below the tip of the stolon revealed a "pulsed" basipetal IAA-movement at about 10 mm h^{-1}, which was blocked by TIBA. In addition, there was an acropetal peak dislocation at a velocity in excess of 25 mm h^{-1}, which probably took place in the stele and was not blocked by TIBA. Thus, there are obviously three distinct channels of translocation of IAA in the stolons, each having its own mechanism.

Extremely short pulses of a few seconds of extremely small amounts of labeled hormones, injected by means of special microcapillaries, have been used to follow the translocation of IAA during gravitropic stimulation in intact coleoptiles of *Avena* and *Zea* (SHAW et al. 1973); movement against gravity of GA_3 injected into the lower side of horizontally placed roots and coleoptiles of *Zea* (WEBSTER and WILKINS 1974); downward and upward dislocation of ABA, microinjected near to the root tip at the upper and lower sides, respectively, in intact horizontal roots of *Phaseolus* (HARTUNG 1976a).

Long-term pulses of 6 h, without chase, applied to young leaves have been used to follow the migration of labeled IAA along the shoot and root of *Scrophularia arguta* (MIGINIAC et al. 1978) and *Vicia faba* (BONNEMAIN and BOURBOULOUX 1973, BOURBOULOUX et al. 1973, BOURBOULOUX and BONNEMAIN 1979, EVERAT-BOURBOULOUX and BONNEMAIN 1980) for periods up to 2 and 8 days, respectively. The peak of radioactivity was shown to move down the stem at about 7 mm h^{-1} and to reach the root system, where a portion of radioactive material was probably recycled to the shoot.

3.3.2.4 Penetration-Depletion Method

The Principle. A rather simple method to determine the velocity and the intensity of endogenous auxin transport was introduced in estimating auxin transport in woody shoots for the first time (GREGORY and HANCOCK 1955). Excised stem sections of crab apple (*Malus* sp.), 50 mm in length, and known to contain endogenous auxin, were stored for definite time periods in a humid atmosphere after excision, having agar receivers in contact with the basal cut surfaces. The sections were then subdivided into short subsections from which diffusible auxin was collected by means of basal agar receivers and the auxin content of each was determined by bioassay. It was shown that as the storage period increased, the apical portions of the sections became depleted of endogenous auxin and that the increasing number of subsections evacuated and the amount of auxin collected during the storage periods allowed an estimation of both the velocity and the intensity of endogenous auxin transport. A temperature dependence of both parameters was found with the optimum around 30 °C and with a Q_{10} of approximately 2. The maximum velocity ranged between 6 to 10 mm h^{-1}. Further, a reduction of both parameters was observed at decreasing oxygen concentrations in the atmosphere.

Modifications. This method was modified and extended by SCOTT and BRIGGS (1960, 1962, 1963) and KALDEWEY (1963, 1965a) by following the time of the progress of both the depletion of endogenous auxin in decapitated but otherwise intact plants as well as the penetration of IAA applied to the apical cut surface of previously auxin-depleted plants. By the use of bioassay of short-term ether extracts of *Pisum* seedlings, subdivided into 1-cm stem sections after treatment, SCOTT and BRIGGS estimated transport velocities of 10 to 12 mm h^{-1} and 6 to 7 mm h^{-1} for both endogenous and applied auxins in light-grown and dark-grown seedlings, respectively. They observed a decrease in both velocity and in the amount of auxin transported in the more basal cut sections, or those with an increasing age of tissue. KALDEWEY bioassayed diffusible auxin collected in basal agar receivers of successive 5-mm sections of decapitated flower stalks of *Fritillaria*. Collections were made 1–4 h after either drainage or the transport of IAA which had been applied to previously fully or partly depleted stems. The transport velocity was shown to vary between 2.5 and 10.5 mm h^{-1} depending on the experimental conditions. The velocity was higher for genuine auxin than for applied IAA in more apical (younger) than in more basal stem regions, and higher in the upper rather than in the lower sides of stems kept horizontally for >12 h before IAA application. Further, a temperature dependence in the transport velocity of both natural and applied auxin was demonstrated which was similar to that reported for *Malus* shoots (GREGORY and HANCOCK 1955) with a Q_{10} of about 1.5.

NEWMAN (1965, 1970) modified the penetration method by determining the profiles of mobile ^{14}C within decapitated, auxin depleted *Avena* coleoptiles supplied with agar donors containing ^{14}C-IAA. He plotted the amounts of radiocarbon collected in basal receivers of 1-mm sections against distance from the top. At the lowest concentration of auxin applied (0.3 μM), there was a linear decrease in the profile, while at higher donor concentrations the profiles were exponential. By determining the intersections of the profile plots with the distance axis, a transport velocity of about 10 mm h^{-1} was calculated. This agrees closely with estimations from other methods.

Criticism. The depletion method has been criticized, since the time needed to evacuate a limited amount of endogenous auxin is a function both of the velocity and the density of the auxin stream (BROWN and WETMORE 1959) as well as of auxin metabolism (JACOBS 1979, p 206) and it is not possible to discriminate between the processes. This criticism holds true also for the penetration method and the same rule must also apply in reverse to "filling" a section with auxin with both processes preceding the onset of auxin delivery to receivers. The intercept method must therefore also be similarly criticized (see KALDEWEY 1965a, p 139f.). In both methods the calculation of a velocity of the hormone stream, when assuming it is uniform, becomes even more questionable, since a "thinning" of the front is observed when the auxin is penetrating and a "tailing" observed when it is leaving the tissue (SCOTT and BRIGGS 1960, KALDEWEY 1963, 1965a). As these observations have since been confirmed in numerous studies (see Sect. 3.3.3.1), there appears to be movement of auxin molecules at differing velocities. As a consequence, by following penetration or depletion respectively, the velocity of the fastest or the slowest molecules would be determined. The estimation is, moreover, strongly dependent upon the sensitivity of detection (for further criticism see Sect. 3.3.3).

Modes of Hormone Application. After labeled hormones became available, the penetration method was applied with a wide range of modifications to follow hormone movement, frequently using intact plants. The hormones have been applied by means of droplets (with or without wetting agents), agar blocks, or lanolin pastes either to intact or injured surfaces of plants, depleted of endogenous auxin or not. They have been injected either with microsyringes or with feeding capillaries into young fruits (e.g., BONNEMAIN 1970, BONNEMAIN and BOURBOULOUX 1973, BOURBOULOUX and BONNEMAIN 1973), young seeds (GROCHOWSKA 1968), receptacles (ANTOSZEWSKI and LIS 1968), and apical buds (DÖRFFLING et al. 1973), all of which are loci of natural auxin production. Hormones have been injected: into the bark and xylem of woody shoots, showing a two-way interchange between xylem and extra cambial tissue for GA_3 and kinetin (BOWEN and WAREING 1969) and auxin (FIELD and PEEL 1971, LEPP and PEEL 1971, ZAMSKI and WAREING 1974) as well as showing basipetal translocation of the latter; into the cortex (DA CRUZ and AUDUS 1978) or pith (BASLER 1974, BASLER and MCBRIDE 1977) of herbaceous stems; into mid or side veins of leaves (e.g., ESCHRICH 1968, GOLDSMITH et al. 1974, HARTUNG 1976b); into cotyledons, endosperm (WHITEHOUSE and ZALIK 1968); scutellum (BROSSARD and TEPPER 1969); and into the hollow of coleoptiles (MARTIN et al. 1978). The hormones have also been allowed to be taken up from larger volumes of solutions via tubes of various materials sealed to stem or petiole stumps (e.g., FULFORD et al. 1968, SCOTT and PRITCHARD 1968), by dipping leaves into the solution (HOAD et al. 1971), or from nutrient solutions supplied to the roots. Labeled IAA has been applied to root tips by agar blocks adhering to, yet remaining movable, on Plexiglas supports so that the growing roots neither lost contact nor were damaged (KONINGS 1968), or as pinpoint application by means of anion exchange resin beads (shown to be suitable for carriers of growth substances by GEE and GREYSON 1969) previously soaked in the auxin solution (DAVIES et al. 1976), or in 1-μl droplets which are readily absorbed (HARTUNG 1977).

Results. In general, by use of these various techniques, auxin applied to loci of natural auxin production of shoots was shown to have these characteristics: it moves basipetally at around 10 mm h^{-1} along the axis; its transport may be inhibited by light in the shoot-root transient zone (YAMAKI and FUJII 1968, see also KALDEWEY and KRAUS 1972); it continues into and, at reduced velocity, moves down the root to its tip (MORRIS et al. 1969, MORRIS 1970, 1977, 1979, MORRIS and THOMAS 1978, HOLLIS and TEPPER 1971, BONNEMAIN 1971, BONNEMAIN and BOURBOULOUX 1973, ELIASSON 1972, BOURBOULOUX et al. 1973, BOURBOULOUX and BONNEMAIN 1973, 1974, 1979, BATRA et al. 1975, CHENOU et al. 1978, MARTIN et al. 1978, MIGINIAC et al. 1978, MORRIS and THOMAS 1978, ROWNTREE and MORRIS 1979, EVERAT-BOURBOULOUX and BONNEMAIN 1980). Furthermore, the movement is independent of a sink activity for the transported auxin (ELIEZER and MORRIS 1979), it is specific for auxins (MORRIS and WINFIELD 1972, MORRIS and THOMAS 1974), it is blocked by TIBA (KALDEWEY and KRAUS 1972, MORRIS et at al. 1973), it is not influenced by the length of the putative auxin-transporting cells, namely the young, differentiating vascular elements, and it is enhanced by light in both green and etiolated plants (ELIEZER and

MORRIS 1980). Auxin is apparently not available for sieve-tube-feeding aphids although LEPP and PEEL (1971) have reported auxin activity in aphid honeydew.

By contrast, auxins when applied to mature leaves, at non-physiologically high concentrations, are exported along with sugars, amino acids, and along with indoleacetylaspartic acid formed from the applied IAA, at 100 to 250 mm h^{-1} which is typical for the movement of assimilates in sieve elements (e.g., DAY 1952, FANG and BUTTS 1957, LITTLE and BLACKMAN 1963, FLETCHER and ZALIK 1965, BOLLAG and GALUN 1966, CORDES 1966, WHITEHOUSE and ZALIK 1967, FULFORD et al. 1968, BONNEMAIN 1971, 1972, MORRIS and THOMAS 1974, 1978, MORRIS 1977). The applied auxin is picked up by the sieve-tube-feeding aphids (ESCHRICH 1968, HOAD et al. 1971, MORRIS and KADIR 1972), and its movement is not blocked by TIBA (BARRY 1971, MORRIS et al. 1973, GOLDSMITH et al. 1974). No information is yet available as to the physiological importance of such—possibly artificial—hormone translocation.

There is evidence, however, that auxin applied to unfolded leaves at physiological concentrations, does move down petioles in a typical auxin transport system (KALDEWEY and JACOBS 1974, 1975), and such transport is inhibited by TIBA (own unpublished data). In these experiments with intact *Coleus* plants, a dramatic increase of radioactivity was observed in the region of the abscission layer of petioles, the base of the leaf blades of which had been supplied with droplets of ^{14}C-IAA solutions. This result may indicate that auxin translocation observed in these experiments reflects the pathway of natural auxin which is produced in the leaf blade (MYERS 1940, WETMORE and JACOBS 1953, SCOTT and JACOBS 1964, BÖTTGER 1970) is moved down the petiole and accumulates in the abscission zone (RUBINSTEIN and LEOPOLD 1963, JACOBS 1964, JACOBS et al. 1966, RASMUSSEN and BUKOVAC 1966, HALLIDAY and WANGERMANN 1972a, b), and regulates abscission (LAIBACH 1933, MAI 1934, ADDICOTT 1970, BÖTTGER 1970, KOZLOWSKI 1973, OSBORNE 1974, DAVENPORT et al. 1977b, 1980, NOODÉN and LEOPOLD 1978, JACOBS 1979, p 174ff.).

When injected into storage tissues (endosperm of *Zea*, cotyledon of *Phaseolus*) IAA (and tryptophan) is translocated upward and downward in the young seedling (FLETCHER and ZALIK 1965, WHITEHOUSE and ZALIK 1967, 1968, BATRA et al. 1975, TSURUMI and WADA 1980). In later studies on *Zea*, it has been shown that IAA-myo-inositol, which is one of the major genuine indolylic endosperm constituents (UEDA and BANDURSKI 1969, BANDURSKI 1978, EPSTEIN et al. 1980), is translocated from the endosperm of germinating corn kernels to the coleoptile at a rate 400 times that of free IAA and 40 times that of tryptophan-derived IAA (HALL and BANDURSKI 1978, NOWACKI and BANDURSKI 1980). The high rate of translocation and the rate of hydrolysis of the ester in the shoot appear adequate both to maintain a constant concentration of ester IAA in the growing shoot and to supply precursor for the free IAA moving down from the coleoptile tip (NOWACKI and BANDURSKI 1980).

3.3.2.5 Electric Potential Wave Method

Changes of electric potential have been observed in *Avena* coleoptiles following phototropic and gravitropic stimuli (e.g., BOSE 1907, BRAUNER 1927, 1959, BACKUS and SCHRANK 1952, GRAHM and HERTZ 1962) and initially were considered to cause the asymmetry in auxin distribution occurring after tropistic stimulation (see reviews of SCOTT 1967, HIGINBOTHAM 1973). However, extensive studies with refined methods (GRAHM and HERTZ 1964, GRAHM 1964, 1971, JOHNSSON 1965, 1967, HERTZ 1971) on the appearance of electrical potentials across the organ following either tropistic stimuli or unilateral auxin application indicate that electrical responses follow rather than precede auxin asymmetry (e.g., WILKINS and WOODCOCK 1965, FILNER et al. 1970, WOODCOCK and WILKINS 1970, 1971, MORATH 1972, WOODCOCK and HERTZ 1972). This is also true in the case of trans-membrane potentials which are increased by hormone application (ETHERTON 1970, HIGINBOTHAM 1973, MARRÈ et al. 1974, CLELAND et al. 1977).

NEWMAN (1959, 1963) detected surface potential changes in the form of a wave moving down phototropically stimulated oat coleoptiles and coleoptiles to which auxin had been supplied at freshly cut tips. These results strongly support the view that the electric responses are the consequence of an increase in the auxin concentration of the cells. NEWMAN (1963, p 643) suggested "that the advancing IAA stream triggers off a series of processes in the tissue as it passes and these produce the electric oscillations." One such process is surely the well-established hormone-induced secretion of H^+ ions (e.g., CLELAND 1975, 1976, RAVEN 1975, SMITH and RAVEN 1976, JACOBS and RAY 1976). If "cells with an increased auxin concentration secrete more H^+ ions", then (1) unilateral illumination of the coleoptiles, followed by "a step-up in the auxin concentration on the shaded side" will lead to an asymmetry in H^+ secretion, and (2) H^+ secretion will be stimulated in auxin-depleted cells as soon as they are resupplied by the downward-migrating IAA molecules. Thus, one can conclude that "the oscillations in surface potential... probably reflect underlying oscillations in H^+ secretion in response to an increment in the concentration of auxin" (GOLDSMITH 1977, p 448f).

NEWMAN's observation of the downward migration of an electric wave at a speed just characteristic for the velocity of auxin transport is further support for an interpretation of this kind. In addition, he points out "the measurement of the wave provides a method of detecting translocated IAA... while it is actually moving in the plant. The effect of various environmental conditions and changes in them may therefore be observed while the wave is travelling and a clear distinction made between processes of uptake and of translocation" (NEWMAN 1963, p 643). He also demonstrated a temperature dependence of the speed of wave migration which was similar to that reported for auxin transport velocity in shoots (GREGORY and HANCOCK 1955, KALDEWEY 1963, 1965a) and roots (WILKINS and CANE 1970) and that the speed of migration of the wave increased with increased IAA concentrations in the donor blocks applied to the auxin depleted coleoptiles (see Sect. 3.3.3.2). Further, the wave moved more slowly in more basal, older regions of the coleoptile, and it was not observed after IAA application to the basal cut surface of auxin depleted coleoptile sections nor following an apical application of 2,4-D. Finally, he observed that there is "just one cycle of oscillations, without subsequent oscillations of decreasing magnitude" after a 10-min pulse application of IAA (NEWMAN 1963, p 638).

NEWMAN's observations, and the reports on oscillations in the transport of auxin in coleoptiles (HERTEL and FLORY 1968, SHEN-MILLER 1973a, b) point to "the operation of feedback loops... in the establishment of the stable pH and electical gradients that are associated with the new cellular auxin concentration and an increased cellular flux of auxin" (GOLDSMITH 1977, p 449; see also SCOTT 1967). Clearly, these processes are oxygen-dependent since no potentials develop in cases even where there is an asymmetric auxin distribution in corn coleoptiles which have been geostimulated under anaerobic conditions (DEDOLPH et al. 1965, NAQVI et al. 1965).

3.3.3 Evaluation of Estimation and Interpretation of Hormone Movement

3.3.3.1 Does Auxin Move in a Stream?

Measurements and calculations of velocities, densities, and intensities of transport to characterize hormone translocation usually imply that these quantities be constant and that the hormone moves in a stream. They do not, however, allow for degradation and/or immobilization, i.e., leakage of molecules out of the stream, to take place. Yet such phenomena do take place, and do vary with time and distance from the hormone source. Variations in the density of mobile auxin have been demonstrated even within short transport sections (e.g., KALDEWEY 1963 in *Fritillaria* axes; NEWMAN 1965, 1970 in *Avena* coleoptiles; KALDEWEY and KRAUS 1972 in *Gossypium* seedlings; KALDEWEY et al. 1974 in *Pisum* internodes; KALDEWEY 1976 in *Tulipa* axes). The commonly observed decline of mobile auxin as a function of distance from the auxin source indicates that not all auxin molecules move with the same velocity. The same conclusion may be drawn from the tpyical initial gradual increase of hormone flux into basal receivers which occurs before linearity of the time course is reached (e.g., HERTEL 1962, HERTEL and LEOPOLD 1963, DE LA FUENTE and LEOPOLD 1973 in *Helianthus* hypocotyls; MCCREADY and JACOBS 1963a, b, in petioles and SMITH and JACOBS 1968 in hypocotyls of *Phaseolus;* DE LA FUENTE and LEOPOLD 1966 in *Coleus* internodes; THORNTON and THIMANN 1967 in *Avena* coleoptiles; GREENWOOD and GOLDSMITH 1970 in *Pinus* embryonic hypocotyls; WILKINS and CANE 1970, WILKINS et al. 1972 and SHAW and WILKINS 1974 in *Zea* roots; KALDEWEY et al. 1974 in *Pisum* internodes; TSURUMI and OHWAKI 1978 in *Vicia* roots).

A further complication is evident in that there exist different transport channels with different properties. The auxin stream seems to contain a fast fraction of low density which is separate from a main and slower fraction (e.g., VARDAR 1964, NEWMAN 1965, RAYLE et al. 1969, DE LA FUENTE and LEOPOLD 1972, KALDEWEY and KRAUS 1972, PATRICK and WOODLEY 1973, KRUL 1977, SHELDRAKE 1979, see also GOLDSMITH 1977, p 452ff.). The fast and slow transport fractions may be associated with different compartments of the cells, possibly the cytoplasm and the vacuole, respectively. This possibility is based on the multiphasic efflux and elution profiles of plant sections supplied with labeled auxin (DE LA FUENTE and LEOPOLD 1970b, 1972, DAVIES 1974; see also GOLDSMITH 1977, p 453f.).

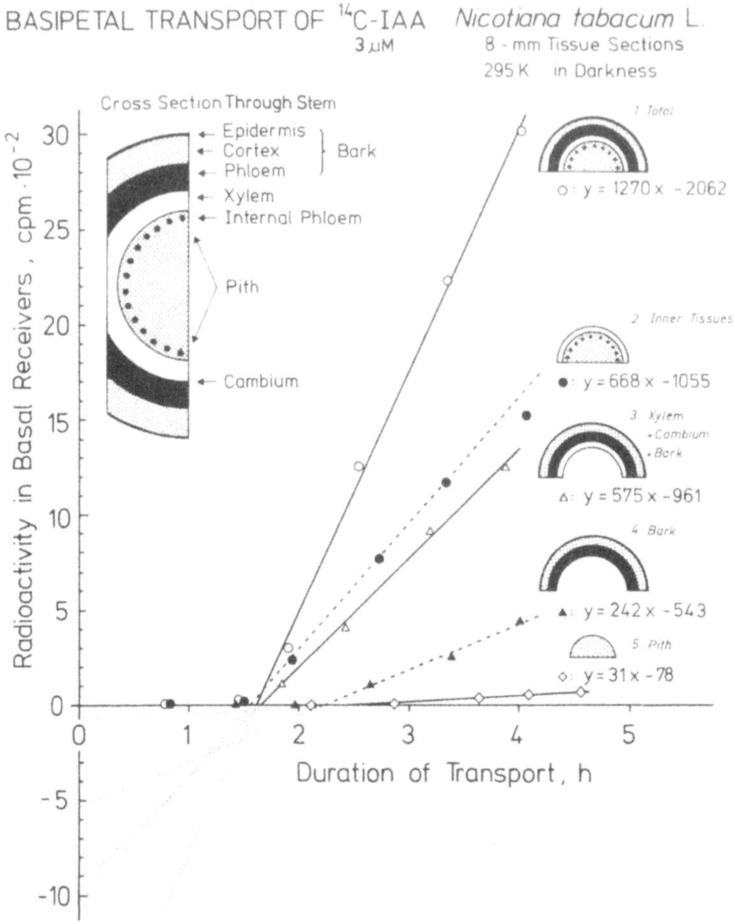

Fig. 3.3. Time course of radiocarbon accumulation in basal agar receivers of 8-mm sections of tobacco stem tissues, apically supplied with agar donor blocks containing 1-^{14}C-IAA (3 μM). The linear regression equations and the lines of closest fit were estimated by the least-squares method from the data beyond 1.5 h for tissue types 1, 2 and 3, and 2.5 h for tissue types 4 and 5 (the last time value was ignored for the inner tissues since it showed a decline in the ^{14}C export rate after a transport period of about 3.5 h in other experiments). Note that small amounts of radioactivity were found in the receivers from all tissue types before the intersections of the straight lines. (Data from SHELDRAKE 1973a)

However, such differing profiles would also be expected if cell types of the system were differently suited for transport. Such a case may be deduced from transport experiments using autoradiography and/or separated tissues in roots (e.g., BOWEN et al. 1972, HARTUNG and PHILLIPS 1974, SHAW and WILKINS 1974, MITCHELL and DAVIES 1975, TSURUMI and OHWAKI 1978; see also Sect. 3.3.6.2) as well as in shoots (e.g., SHELDRAKE 1973a, IWAMI and MASUDA 1976, DA CRUZ and AUDUS 1978, MORRIS and THOMAS 1978, WANGERMANN and WITHERS 1978, BOURBOULOUX and BONNEMAIN 1979; see also Sect. 3.3.6.1).

SHELDRAKE (1973a) demonstrated by separation of the stem tissues of *Nicotiana* internodes that the great majority of the strongly basipetal auxin transport took place in cells of the internal phloem and in cells close to the cambium. Very small amounts were transported in bark and pith preparations and none in xylem tissues. Using the intercept method of VAN DER WEIJ (1932), he estimated velocities of about 5 mm h^{-1} for the transport of 1–^{14}C-IAA in complete stem sections, in inner tissue segments containing the internal phloem, and in xylem + cambium + bark section. The transport densities were similar to each other in the two latter preparations. Lower transport densities were found in the bark and in exclusively pith sections having transport velocities of 3.8 and 3.1 mm h^{-1}, respectively. In all cases, however, a small amount of radioactivity was found in basal receivers considerably in advance of the time intercepts calculated from the linear parts of the auxin arrival curves (Fig. 3.3). Thus, even in homogenous tissues, auxin molecules appear to move at different velocities.

3.3.3.2 Loading of the Transport System

As mentioned above, differing velocities of hormone molecules will lead to an initial gradual increase in the slope in the arrival curves. This would also occur if there were a gradual generation of transport sites, as has been suggested by DE LA FUENTE and LEOPOLD (1970a). In both cases, the time required for linearity of the arrival curves should be independent of the hormone concentra-

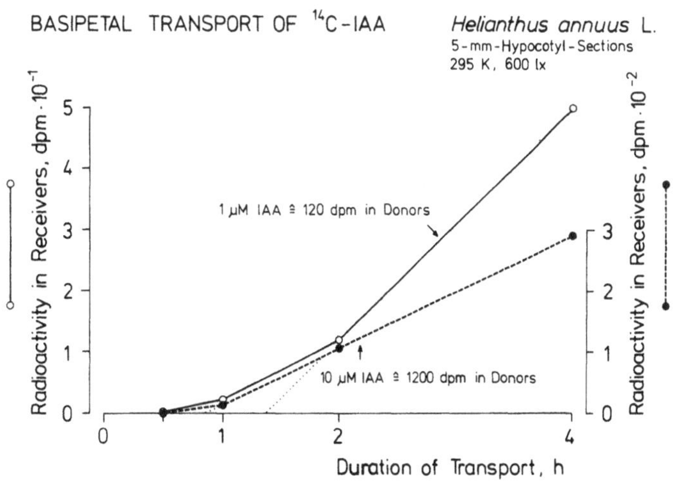

Fig. 3.4. Time course of basipetal IAA transport through hypocotyl sections of 6-day-old light-grown seedlings of *Helianthus annuus* L. demonstrating the initial nonlinearity of ^{14}C-accumulation in basal receivers. Individual sections, supplied with apical donors immediately after excision, were placed for different time periods on basal receivers. In the 4-h experiment, donors and receivers were replaced after 2 h. *Points* represent average values of four replicates, each with three individual sections. (Data from an unpublished Diploma Thesis of ANNELIE PFANGER, Faculty of Nat. Sci., Univ. of the Saarland, Saarbrücken, 1980)

tion in the donor blocks. However, a gradual increase in hormone export may, as well, reflect a gradual "loading" of the transport system up to saturation by the hormone molecules which diffuse physically into the tissue from the donor blocks and which are in part subjected to degradation and/or immobilization. In this case, the time required for saturation which leads to linearity of the arrival curve should be inversely related to the concentration of hormone molecules available for the transport system. The results of transport studies in our laboratory with hypocotyl sections of *Helianthus* appear to meet this requirement (Fig. 3.4): at a concentration of 10 μM ^{14}C-IAA in apical donor blocks strictly linear accumulation of radioactivity in basal receivers occurs after 1 h of transport. In contrast, the slope of the arrival curve increases up to 4 h with a one-tenth donor concentration. This result supports the "loading-hypothesis".

3.3.3.3 Validity of Velocity Estimations

Whatever characteristics of the transport system might cause the initial nonlinearity of the arrival curve, its existence calls into question the validity of calculating transport velocities from intersections of the extrapolated linear regression lines with the time axis. If the hormone molecules move at different speeds, whether in a homogenous tissue or in different specialized parts of a transport section lacking homogeneity, such a calculation will yield something approaching an "average velocity". If, however, the initial nonlinearity is the consequence of a gradual generation of transport sites or a gradual loading of the transport system (see Sect. 3.3.3.1), the absolute onset of linearity— not an extrapolation— would be the estimate of the time elapsed between application of the donor block and appearance of the full hormone stream in the receivers. Since this period would include both the time needed for maximum generation of transport sites or for maximum loading of the system, and the time required for the saturated system to traverse the transport section, the calculation of a transport velocity from this period of time (as well as from the time intercept of the extrapolated linear regression) is misleading (see KALDEWEY 1965a, p 138ff.). Therefore, reports on transport velocities and, especially, on differences in the transport velocity at different hormone concentrations may be wrong.

By use of the intercept method of VAN DER WEIJ (1932) the transport velocities in sunflower hypocotyl sections (Fig. 3.4), as estimated from the intersections with the time axis of the extrapolated linear regression lines, amount to 5.9 mm h^{-1} and 3.7 mm h^{-1} at the higher and lower donor concentrations, respectively. Similar estimated relationships between donor concentration and transport velocity estimations have been found for basipetal IAA transport in *Coleus* stems (NAQVI 1963), *Zea* coleoptiles (NAQVI and GORDON 1964, NAQVI 1976), and *Avena* coleoptiles (NEWMAN 1963, 1970) and may be deduced from the shape of the arrival curves derived mathematically by NEWMAN (1974) from his experimental data. On the other hand, velocities estimated by the intercept method have been reported not to be significantly different at different auxin concentrations in the donor, though the calculated values tended to be higher at increased donor concentration (e.g., VAN DER WEIJ 1932 in *Avena* coleoptiles;

HERTEL 1962 in *Zea* coleoptiles; MCCREADY and JACOBS 1963a in *Phaseolus* petioles; PILET 1965 in *Lens* shoots). The contradictory reports of estimated transport velocities may result from the choice of the auxin concentration in relation to maximum loading capacity of the transport system. Since, if the "loading hypothesis" is correct, the time required for a saturated loading of the transport system will be a function of both the concentration of the available hormone molecules as well as the maximum loading capacity of the system, then at a given donor concentration, as the time needed for saturation increases the calculated transport velocities should accordingly decrease with an increasing loading capacity of the system. This has been observed in fruit axes of *Fritillaria* (KALDEWEY 1967a) for IAA transport in short sections, which were excised at different distances from the fruit and depleted of their endogenous auxin. Calculated transport velocities decreased with increasing intensity and density of the auxin stream (see Fig. 3.1).

A consequence drawn from these considerations is that the onset of hormone delivery to receivers had to be determined to estimate the velocity of hormone movement. Velocity estimations of this kind will give higher values than calculations by the intercept method. In the 5-mm hypocotyl sections of sunflower (Fig. 3.4, see also DE LA FUENTE and LEOPOLD 1973 for similar results), small but significant amounts of radioactivity appeared in basal receivers 30 min after application of both high and low concentrations of ^{14}C-IAA. This reveals a velocity of at least 10 mm h^{-1}, which is in contrast to and is considerably higher than the 5.9 or 3.7 mm h^{-1}, as calculated from the time axis intersections mentioned above. Essentially the same conclusion was drawn by THORNTON and THIMANN (1967) from their ^{14}C-IAA transport studies with *Avena* coleoptile sections in which they replaced the receiver blocks every 5 min. While a velocity of approximately 10 mm h^{-1} was calculated by the intercept method, radioactivity appeared in receivers of 4-mm sections 10 and 15 min after donor application which indicates a transport velocity between 15 and 26 mm h^{-1}, or perhaps higher. This velocity corresponds with the migration speed of the phototropic response in intact coleoptiles. Clearly, such determinations of auxin movement depend strongly upon the sensitivity of the method used.

Similarly, high velocities can be determined from pulse transport experiments with labeled auxin (HERTEL 1962, HERTEL and LEOPOLD 1963, GOLDSMITH 1967a, b, HERTEL and FLORY 1968, RAYLE et al. 1969, BEYER and MORGAN 1969, SHEN-MILLER 1973a, b, GOLDSMITH et al. 1981, GOLDSMITH 1982, HASENSTEIN 1982, HASENSTEIN and KALDEWEY 1983). In such experiments, after removal of the auxin source, a pulse of label migrates basipetally through the tissue, the peak of which moves at a velocity similar to that determined by the intercept method. Nevertheless, small amounts of radioactivity are detectable very considerably ahead of the peak, either by scanning the distribution of radioactivity within the transport section, or by following the time course of appearance of the pulse in basal receivers (see Fig. 3.2). Inferring from the distance covered by these "fast" molecules in *Zea* coleoptiles, the velocities of more than 20 mm h^{-1} for IAA and NAA, and about 6 mm h^{-1} for 2,4-D are in sharp contrast to 10 to 15 mm h^{-1} and 1.2 mm h^{-1} calculated by the migration of the peak of a pulse, or by the intercept method (e.g., HERTEL and FLORY 1968).

3.3.3.4 Fluctuations of Transport Velocity

SHEN-MILLER (1973a) applied 1-min pulses of ^{14}C-IAA to individual oat and corn coleoptiles after removal of an apical 1-mm tip. She followed the migration of radioactivity for as long as 1.5 h, by determining the radiocarbon content of successive 2-mm sections which were subdivided at 6-min intervals. Corn coleoptiles appeared to have transport velocities of 41, 13, and 34 mm h^{-1}, in the apical 4-mm region, the middle, and basal parts, respectively. Oat coleoptiles had values of 28 mm h^{-1} for the apical 4-mm region and 14 mm h^{-1} for the remainder.

The pattern of transport velocities of this kind, in a streaming system, would be expected to lead to an increase in the density of the hormone molecules within the region of low velocity; as though molecules were moving against a barrier. This has been suggested to be the cause of variations in the concentration of mobile auxin along the axes of geostimulated fruit stalks of *Fritillaria* (KALDEWEY 1963). The variation with time of the radiocarbon profiles along the ^{14}C-IAA-fed *Zea* coleoptiles seems also to confirm this hypothesis (Fig. 3.5). There is an initial and fast increase in the content of auxin from the apical pulse which moves down to a distance of about 6 mm. After some 30 min a plateau is reached in this region. Though this plateau remains for about another 30 min, some auxin moves slowly down the coleoptile towards the

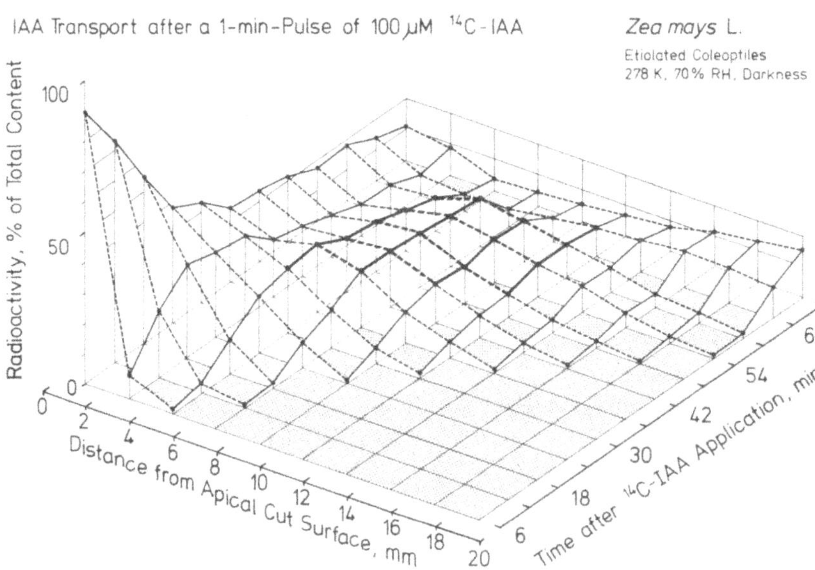

Fig. 3.5. Distribution of radioactivity within decapitated but otherwise intact etiolated coleoptiles of *Zea mays* supplied with a 1-min pulse of 100 µM ^{14}C-IAA, incorporated in an agar donor block. Individual coleoptiles were subdivided at 6-min intervals into 2-mm pieces, and the radioactivity in each counted. Note the generation of a transient plateau between about 6 and 12 mm below the tip (made plain by *heavily drawn lines*). This is the region of reduced auxin transport velocity as compared with the apical and the more basal parts of the coleoptiles. (Each *point* is based on the means of eight experiments with three coleoptiles each; data from SHEN-MILLER 1973a)

basal region of high transport velocity. Thereafter, however, the auxin quickly disappears from the barrier region and piles up at the very base of the coleoptile.

Though it was not discussed, differences in the velocity of auxin transport along coleoptiles may also be deduced from the displacement of the peak of radioactivity with time in the pulse experiments of GOLDSMITH (1967a, b) and NEWMAN (1970). In 20-mm segments of corn coleoptiles, cut 3 mm below the tip and supplied with a 15-min pulse of ^{14}C-IAA, the peak appeared about 5 mm below the apical cut surface after 30 min, it moved basipetally another 5 mm during the following 30 min, but then shifted nearly 10 mm downward to the base of the section in a subsequent half-hour period (Fig. 3.2C and GOLDSMITH 1982). In oat coleoptiles, depleted of their endogenous auxin before application of a 10 min pulse of ^{14}C-IAA, the peak of mobile auxin moved downward 1.8, 1.2, and 2.1 mm, respectively, in three successive 10-min periods occurring 20 to 50 min after donor application (Fig. 3.2B). Thus, this pattern of velocity alterations resembled that observed along "nondepleted" corn coleoptiles by SHEN-MILLER (1973a, b).

3.3.3.5 Flexibility of the Transport System

The examples described above which show variations of transport characteristics occurring along the axes of the plant part, or to take place as the result of variations in the duration of the experiments, basically alter the picture of an essentially uniform auxin stream moving basipetally at a constant velocity from an apically located auxin source. That the transport system is flexible may also be concluded from earlier estimations of the movement of diffusible native auxin along plant axes. SCOTT and BRIGGS (1960), using light-grown pea seedlings, collected the endogenous auxin delivered to basal agar receivers from cross sections of the axes cut at different levels of the stem. The amount of diffusible auxin decreased basipetally within the strongly elongating subapical internode, while it remained constant in the next lower but scarcely elongating internode and then fell to almost zero further down at the more mature base. A basipetal decrease in the amount of diffusible auxin collected from 10-mm sections was likewise observed in prefloral flower stalks of *Fritillaria;* the highest amounts diffused from the sections just below the bud and fruit, these being the sole auxin sources for the stalks (KALDEWEY 1957).

This kind of profile in diffusible auxin may be explained by the use or "consumption" of auxin within the elongating region of the stem and may not necessarily indicate differences in the properties of the auxin transport system along that portion of the axis. In *Fritillaria,* however, in a later stage of development when stem elongation had entirely ceased, twice as much diffusible auxin was collected from sections excised at more basal regions of the stem than from either the fruit or the subapical segment. These observations do point to variations along the axes in the properties of the transport system, though it cannot be determined whether higher amounts of diffusible auxin in the basal stem region were the consequence of a decrease of the transport velocity resulting in an increase in the density of mobile auxin molecules (barrier effect, see Sect. 3.3.3.4). However, an interpretation of this kind may be derived from prefloral flower stalks of *Fritillaria* (KALDEWEY 1968b, 1971) in which transport characteristics were studied in detail in the apical geoepinasticly bent region. Elongation and growth movements of the stalks are known to be regu-

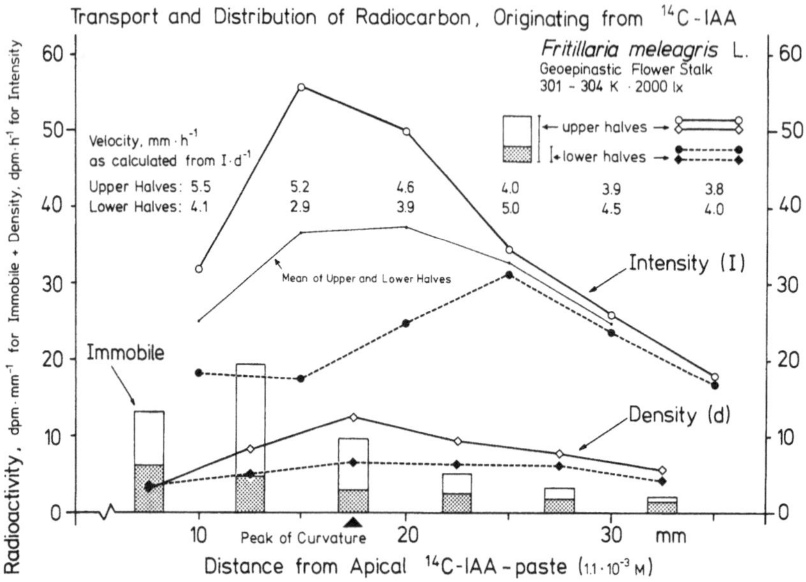

Fig. 3.6. Transport and distribution of radioactivity originating from 2-^{14}C-IAA, applied to decapitated apices of geoepinastically bent flower stalks of *Fritillaria meleagris* L. The transport characteristics have been evaluated by the "short-term collecting method" (see Sect. 3.3.2.2). The transport intensity was calculated from the radioactivity collected in agar receivers applied for two 15-min periods to the basal cut surface of a 5-cm explant, continuously supplied with an auxin paste source. The cut was separated by a mica plate into upper and lower halves or the convex and concave sides of the bent axis. After the second collection period, a 5-mm section was excised and radioactivity similarly collected from the new cut surface, and so forth up to the apex. The excised 5-mm sections were separated in upper and lower halves and placed individually on agar receivers for 1 h, to determine the transport densities. They were then exhaustively extracted with ethanol to give the immobilized fractions. (Mean values of four plants, data from KALDEWEY 1968b)

lated by the flower bud. The bud is the only natural auxin source and it can be effectively replaced by a lanolin paste containing IAA (KALDEWEY 1957). Using the method described in Section 3.3.2.2 transport intensities and densities as well as immobilized fractions were determined along the axis following replacement of the flower bud with ^{14}C-IAA in lanolin paste. These experiments revealed a basipetal increase in both the transport density and the intensity to or just beyond the peak of the curvature followed by a decrease in both parameters (Fig. 3.6). The transport velocity (v), as calculated from $v = I\,d^{-1}$, did gradually slow down basipetally from 5.5 to 3.8 mm h^{-1} in the upper (convex) halves of the curved region of the axis, which may account for part of the basipetal increase in the density. However, while the density quadrupled toward the peak of this region, the velocity slowed by only about 10%, thus it does not account for all of the increase in the density. In the lower (concave) halves of the axis, irregular alterations of the calculated velocities appeared not to correspond to the less pronounced variations of the transport density. Perhaps the most surprising result of this investigation is the initial basipetal increase in the transport intensity which appears to mean there is an augmentation of the basipetal auxin flux with an increasing distance from the apical auxin source; this is something of a paradox particularly in respect to a continuous stream, since an efflux would not be expected to exceed the supply.

Lateral redistribution of auxin in epinastic main and side shoots of *Coleus* and petioles of different species as postulated by LYON (1963a, b, 1965a, b) would be the equivalent of a supplementary auxin source for the receiving half of the axis and it may cause an increase in the transport intensity in that half. However, this is not a suitable explanation in the case of *Fritillaria*, since regions of increasing transport intensity do not coincide in upper and lower halves, in spite of the fact that overall intensity does increase basipetally initially. Thus an alternative explanation for the intensity alterations might be that an exchange may exist between the mobile and immobile auxin fractions, and that this may take place between the cytoplasm and the vacuoles.

The examples mentioned so far point to considerable flexibility of the auxin transport system. This view is also supported by the observation of oscillations of electric potential moving down *Avena* coleoptiles after illumination or after the supply of auxin (NEWMAN 1959, 1963), by the report of fluctuations of IAA movement in segments of oat coleoptiles after blue light illumination (THORNTON and THIMANN 1967), and in individual plant parts, by the demonstration of even more pronounced oscillations of the export rate of radiocarbon from auxin-depleted segments of oat and corn coleoptiles, supplied with labeled IAA (HERTEL and FLORY 1968). It is further supported by previously mentioned (see Sect. 3.3.3.4) experiments of SHEN-MILLER (1973a), where rhythmic fluctuations of the IAA transport intensity in intact coleoptiles of oat and corn were observed, moreover the rhythmicity was "out of phase between the upper and lower halves of geostimulated coleoptiles" (SHEN-MILLER 1973b, p 169).

In summary, then, the concept of a continuous stream of auxin molecules moving at a uniform and a constant velocity and density has been based mainly on, and derived, from studies of the time course of auxin transport through sections which were similar to the classic experiments of VAN DER WEIJ (1932, 1934). It is now clear that this concept has not only "constrained the thinking about auxin transport for the best part of half a century" (GOLDSMITH 1977, p 454), but that it has no basis in reality. It is unfortunate that there is as yet no generally accepted substitute for this too inflexible picture of the auxin transport system.

3.3.4 Models for Transport Mechanisms

Any model of the mode of action for a transport system of a polarly transported hormone must account for the preferential direction of movement. For auxins, the direction of movement is predominantly basipetally in shoots and acropetally, i.e., toward the tip, in roots. Basipetal auxin transport has been shown to be established, to a certain degree, in hepatics (MARAVOLO 1976) and in embryonic axes of vascular plants (GREENWOOD and GOLDSMITH 1970, FRY and WANGERMANN 1976). This polarity develops gradually up to a maximum with increasing age of the plant (e.g., JACOBS 1950, SMITH and JACOBS 1968, 1969), and then declines, either because of a decrease in basipetal transport (e.g., WERBLIN and JACOBS 1967, OSBORNE 1968, SMITH and JACOBS 1968, 1969, VEEN and JACOBS 1969a, SHELDRAKE 1973a), or because of an increase in acropetal movement coupled with the decrease in basipetal transport (for examples and further references see p 122 and p 123). The basipetal polarity of auxin transport has been shown to persist even in stems of inverted cuttings which had been induced to form roots at their apical end and shoots at the base (WENT 1941, SHELDRAKE 1974).

3.3.4.1 Conveyor Belt Model

VAN DER WEIJ's (1932) comparison of the transport system with a unidirectionally moving conveyor belt loaded with different amounts of auxin molecules was a first attempt to conceptualize the basipetal auxin transport mechanism. It is noteworthy that this rather mechanistic picture, which has found little experimental basis, has had something of a revival by assuming plasmalemmasomes to be possible "vehicles" of auxin transport (WANGERMANN and WITHERS 1978; see Fig. 3.7).

3.3.4.2 Electrical Polarity Hypothesis

Like the conveyor model, the electrical polarity hypothesis has not been accepted as an explanation for the mechanism. Here, it was assumed that negatively charged auxin ions move in an electrical field toward the positively charged base of the plant (WENT 1932, CLARK 1937a, described in detail by LUND 1947 and SCHRANK 1951), or as the case may be along an electrical potential gradient across a geostimulated organ (see Sect. 3.3.2.5) and in this way the polarity is established. Although electrical phenomena are involved in auxin transport (NEWMAN 1959, 1963) it has turned out that surface potential gradients are the result of rather than the cause of polar transport (CLARK 1937b, 1938, SCOTT 1967).

3.3.4.3 Polar Secretion Theory

Studies using inhibitors specific to auxin transport, such as TIBA, first used by KUSE (1953), NIEDERGANG-KAMIEN and SKOOG (1956), NIEDERGANG-KAMIEN and LEOPOLD (1957), NPA, first demonstrated to reduce auxin transport by MORGAN and SÖDING (1958) and MORGAN (1964), and more recently the wide range of synthetic compounds with inhibitory qualities (see, e.g., ZWAR and RIJVEN 1956, KEITT and BAKER 1966, KRELLE and LIBBERT 1968, BEYER 1972, BROWN et al. 1972, 1973, KATEKAR and GEISSLER 1975, 1977a, b, KATEKAR 1976, BEYER et al. 1976, MARIGO and BOUDET 1977) were responsible for the development of the theory of polar secretion of auxin. Experiments of this kind revealed that inhibitors were more effective at blocking polar transport when applied to basal receiver blocks than to the apical donor blocks (HERTEL and LEOPOLD 1963, LEOPOLD 1963, CHRISTIE and LEOPOLD 1965a, b, HERTEL and FLORY 1968, WINTER 1967, 1968, GAGIANAS and BERG 1977). Such inhibitors when added to donors in fact had little or no effect on auxin uptake (e.g.,

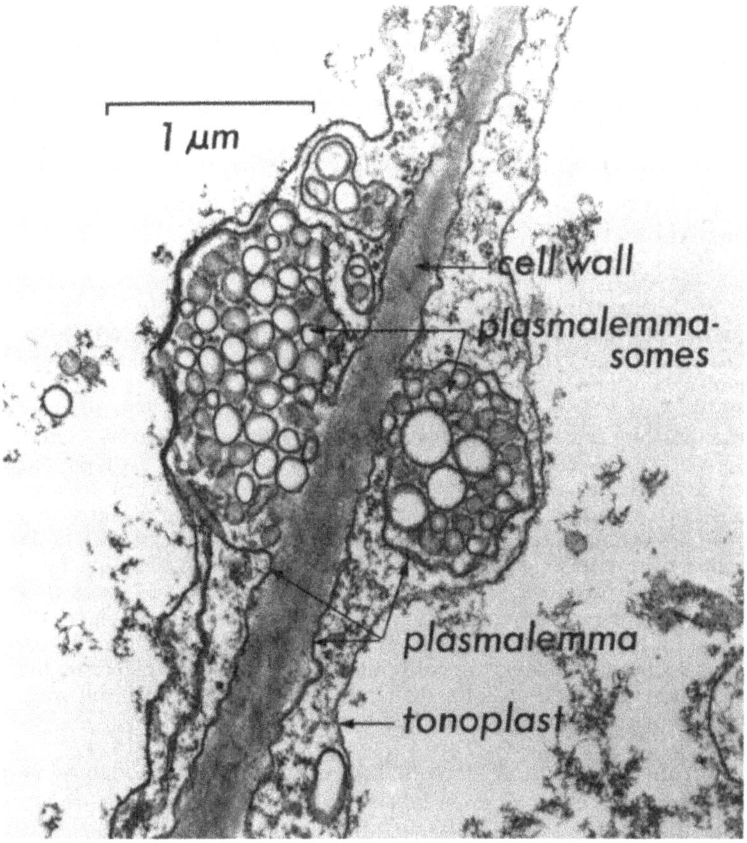

Fig. 3.7. An electrophotomicrograph of a transverse section through two adjacent parenchyma cells of the primary phloem region of a young internode of *Coleus blumei*. The plasmalemmasomes are located between plasmalemma and cell walls. (By courtesy of late Dr. ELISABETH WANGERMANN, Univ. of Leicester, Department of Botany, Leicester, U.K.)

LEOPOLD 1963, CHRISTIE and LEOPOLD 1965a, b, WINTER 1968, VARDAR 1970, KALDEWEY and KRAUS 1972, MORRIS et al. 1973, GAGIANAS and BERG 1977, KALDEWEY et al. 1977, WANGERMANN and WITHERS 1978) which thus lead to an intracellular accumulation of auxin (HERTEL and LEOPOLD 1963, CHRISTIE and LEOPOLD 1965a, b, HERTEL and FLORY 1968, WINTER 1968, BEYER 1972, BOPP 1972, DAVIES 1973, 1974, CANDE and RAY 1976, see also HERTEL 1983).

The results of such studies are consistent with the concept that the polarity of auxin transport is imposed by the active secretion of auxin molecules out of the cytoplasm through the membrane at the base of the cell and that the molecules then diffuse passively into the adjacent cell below. This model, first developed by HERTEL (1962), HERTEL and LEOPOLD (1962, 1963), LEOPOLD (1963), assumed that secretion is mediated by metabolically driven carriers which is a feature consistent with the facts that auxin transport depends upon an adequate oxygen supply (e.g., GREGORY and HANCOCK 1955, BROWN and WET-

MORE 1959, NAQVI et al. 1965, GOLDSMITH 1966a, b, 1967a, b; 1968a, WILKINS and MARTIN 1967, WILKINS and SHAW 1967, WILKINS and SCOTT 1968, WILKINS and WHYTE 1968) and is inhibited by metabolic poisons (e.g., VAN DER WEIJ 1934, DU BUY and OLSON 1940, NIEDERGANG-KAMIEN and LEOPOLD 1957, REIFF and V. GUTTENBERG 1961, HERTEL 1962, HERTEL and LEOPOLD 1963, CHRISTIE and LEOPOLD 1965b, WILKINS and MARTIN 1967, MCCREADY 1968a, b, WILKINS and WHYTE 1968, GOLDSMITH and RAY 1973, CANDE and RAY 1976, DA CRUZ and AUDUS 1978).

The theory of polar secretion has been followed and extended by many workers (e.g., GOLDSMITH 1966a, DE LA FUENTE and LEOPOLD 1966, 1970a, b, LEOPOLD and DE LA FUENTE 1968, MCCREADY 1968a, b, OSBORNE 1968, OSBORNE and MULLINS 1969, CANDE and RAY 1976; and literature there cited). Cytoplasmic streaming has been shown not to be a prerequisite for polar transport (CANDE et al. 1973, GOLDSMITH and RAY 1973). Also, plasmodesmata, which had been emphasized in the concept of symplast transport are not absolutely essential (ARISZ 1969, ZIEGLER 1973, see also ELIEZER and MORRIS 1980). Disruption of almost all the plasmodesmata by plasmolysis strongly reduced but did not completely block basipetal auxin transport in *Avena* coleoptile sections, and polar transport properties were fully restored after deplasmolysis (CANDE and RAY 1976, DRAKE and CARR 1978, see also WOODCOCK and WILKINS 1971). These observations support the view that polar auxin transport is due to secretion by the plasmalemma into the apoplast, where auxin molecules will diffuse and enter the neighboring cell where they are polarly secreted again. Further, it has been shown that auxin can be taken up into and can be lost from the auxin stream via the apoplast by feeding IAA to and collecting it from corn coleoptiles using agar blocks placed on the slightly abraded epidermis. Thus, this passive exchange occurs at the outer cell wall which does not contain plasmodesmata (CANDE and RAY 1976).

Probably the most convincing evidence for the location in the basal cell membrane of energy-dependent processes in polar auxin transport came from centrifugation studies initiated by OUITRAKUL and HERTEL (1969). Results are similar and in line with the inhibition of auxin transport in inverted sections, as was earlier suggested by VAN DER WEIJ (1932) and confirmed in more recent studies (HERTEL and LEOPOLD 1962, NAQVI and GORDON 1966, LITTLE and GOLDSMITH 1967, WHITEHOUSE and ZALIK 1968). GOLDSMITH and RAY (1973), centrifuged coleoptile sections, treated with cytochalasin B to suppress protoplasmic streaming, and they showed that displacement of the cytoplasm to the apical end of the cells strongly inhibited basipetal auxin transport. On the other hand, transport in similar sections was promoted by centrifugation toward the base. Inhibition of transport by acropetal displacement of the cytoplasm was not due to the absence of amyloplasts, since a short reverse centrifugation which restored amyloplasts to the base of the cells did not restore transport. However, the inhibitory effect of acropetal displacement of the cytoplasm could be partially overcome by feeding ATP or phosphoenole pyruvate. Thus, GOLDSMITH and RAY suggested that a removal of mitochondria from the basal cell membrane was a cause of the inhibitory effect of centrifugation. Acropetal auxin translocation was not significantly affected by centrifugation and transport remained basipetal after recovery from acropetal displacement of the cytoplasm. These results support the evidence that acropetal movement of auxin takes place predominantly by diffusion and does not depend on metabolism, as expressed by MCCREADY (1963, 1968a, b) and GOLDSMITH (1966a, b, 1968a) (see also review of GOLDSMITH 1968b, p 353ff.). The findings are in contrast to the model of active secretion at both directions in the cell (DE LA FUENTE and LEOPOLD 1966) and the achievement of polarity by a preference of secretion at the basal end of each cell. By use of a mathematical model, LEOPOLD and HALL (1966) demonstrated that very slight differences in the secretion at opposite ends of individual cells lead to an increasing overall polarity with an increasing number of cells arranged in a linear file, thus explaining the increase in polarity with increasing tissue section length (WENT and WHITE 1939, DE LA FUENTE and LEOPOLD 1966) or increasing cell number within transport sections (JACOBS 1950, 1961; see however ELIEZER and MORRIS 1980). However, since GOLDSMITH and RAY (1973) convincingly showed that uptake of auxin was only slightly affected by centrifugation they concluded that "the model, metabolic secretion from the base and passive uptake at the apical end (HERTEL and LEOPOLD 1963, CHRISTIE and LEOPOLD 1965a, b), predicts correctly the observed effects of centrifugation on transport. Centrifugation experiments therefore provide evidence that transport of auxin is due to polar metabolic secretion of auxin localized at the basal plasma membrane of each cell" (GOLDSMITH and RAY 1973, p 312).

Recently, JACOBS and GILBERT (1983) succeeded in visualizing in pea stem tissue by means of an indirect immunofluorescence technique the presumptive auxin transport carrier in the plasma membranes at the basal ends of parenchyma cells sheathing the vascular bundles.

3.3.4.4 Theory of Chemiosmotic Polar Diffusion

There remain basically two phenomena regarding auxin transport not readily explainable by the polar secretion theory, which led RUBERY and SHELDRAKE (1973, 1974) and RAVEN (1975) to suggest chemiosmotic processes as the foundation of the polarity of transport. First, auxin uptake increases with a decrease of pH of the medium or agar donor blocks which had already been observed in earlier reports (ALBAUM et al. 1937, REINHOLD 1953, VAN DE WESTRINGH and VELDSTRA 1958, POOLE and THIMANN 1964, SMITH and JACOBS 1968, 1969) and second, the concentration in the tissue may exceed that of the source (GOLDSMITH 1968a). The studies which initiated the development of the chemiosmotic theory were done with suspension cultures of crown gall cells of *Parthenocissus* (RUBERY and SHELDRAKE 1973, 1974, RUBERY 1977, 1978, 1979) and giant cells of the algae *Hydrodictyon* (RAVEN 1975). Results which support the theory have now been extended to sections of *Pisum* stems (DAVIES and RUBERY 1978, JACOBS and GILBERT 1983) and *Zea* coleoptiles (EDWARDS and GOLDSMITH 1980).

This new theory has been extensively discussed in the review of GOLDSMITH in 1977 (see also RUBERY 1980), and her introductory sentences (p 441) outline the basic concept. "Cells, being more permeable to undissociated auxin molecules than to auxin anions, can accumulate auxin when the pH of the cytoplasm is above that of the walls... Carriers may be but are not necessarily involved in passage across the cell membrane... Unlike the conventional theory of polar secretion, which presumes active transport of auxin with direct coupling of energy to carriers in those cases when auxin transport appears to be uphill, the new theory postulates that auxin transport is thermodynamically downhill, but with metabolic energy expended to maintain the pH and electrical gradients and the polar permeability... This theory... holds the promise of accounting for the known properties of polar auxin transport", and it "predicts that polarity depends on pH and electrical gradients as well as the permeability ratios between the two ends of the cell. For this reason, the new theory" ... may be called *"proton and emf*[1] *dependent polar diffusion,* or more simply just *chemiosmotic polar diffusion."*

Thus, according to this theory, polarity of transport is produced by an asymmetry in cellular permeability to the auxin anions (see also Sect. 3.2.3.3), "but polar transport is basically an asymmetric diffusion" (GOLDSMITH 1977, p 457). Therefore, the theory requires that profiles of the distribution of radioactivity in plant parts which had been supplied with a constant concentration of labeled auxin should fit to a solution of FICK's second law of diffusion, which relates the concentration to both the time of transport and the distance from the source, namely to error function curves. Such curves are constructed from

$$C_x/C_0 = 1\text{-erf}\, x/2\sqrt{D \times t},$$

where C_x is the radioactivity (concentration) in the tissue at distance x from the point of application or from the peak of radioactivity; C_0 is the concentration of the donor or the concentration at the peak of radioactivity; D is the diffusion coefficient, t is time, and 1-erf is the error function complement, available from tables.

Profiles of radioactivity from labeled IAA entering petiole sections of *Coleus* (VEEN and FRISSEL 1975), applied to intact stolons of *Saxifraga sarmentosa* (DA CRUZ and AUDUS 1978), and applied to the apical bud of intact *Pisum* seedlings (ELIEZER and MORRIS 1979, 1980) have been shown to fit closely to such error function curves. However, the values of D, which were used to fit the curves to the experimental data, were up to more than one order of magnitude greater than expected for diffusion of IAA in water. The "asymmetric diffusion", as postulated by the theory of chemiosmotic polar diffusion, calling for an "accelerated diffusion", accounts for this discrepancy. Mathematical models have meanwhile been presented which allow to describe the shape and the migration of auxin pulses as well as auxin profiles after IAA application to plant tissue (MITCHINSON 1980, GOLDSMITH et al. 1981, GOLDSMITH and GOLDSMITH 1981, HASENSTEIN 1982, HASENSTEIN and KALDEWEY 1983).

3.3.5 Binding Sites

The concept of an involvement of membrane-associated carriers in hormone transport initiated the search for membrane fractions with specific binding properties. In such studies, started by HERTEL and coworkers (e.g., LEMBI et al. 1971, HERTEL et al. 1972, THOMSON et al. 1973, THOMSON and LEOPOLD 1974, DOHRMANN et al. 1978) and extended, e.g., by BATT and VENIS (1976), BATT

[1] emf = electromotive force

et al. (1976), VENIS (1977), RAY et al. (1977), MURPHY (1980), and SUSSMAN and GOLDSMITH (1980), different membrane fractions in preparations of *Zea* coleoptiles were isolated which appeared to have possible specific binding properties for auxins as well as different inhibitors (for a brief review see LAMB 1976; for extensive reviews KENDE and GARDNER 1976, RUBERY 1981). Perhaps the most appropriate candidate for such an auxin transport carrier was found by JACOBS and HERTEL (1978) in subcellular fractions from *Cucurbita* hypocotyls.

WANGERMANN and WITHERS (1978) have recently put forward the interesting hypothesis that polar auxin transport is linked to the presence of plasmalemmasomes (Fig. 3.7). These structures were shown to be present in parenchyma cells in sections known to exhibit a functional auxin transport system but were absent in the absence of transport. The authors speculated that transport channels might be initiated by the production of plasmalemmasomes in the cells.

3.3.6 Pathways of Hormone Movement

Basically two lines have been followed in the search for the pathways of hormone translocation: (1) surgical methods and; (2) autoradiographic methods which have come with the advent of radioactive hormones. Phloem-feeding aphids have been used as a speciality to detect translocation in sieve elements (for references see p 100f.).

There are obstacles in both methods. Surgery may lead to wounding effects which can mask the transport properties of intact tissues while autoradiography, without special precaution, cannot distinguish between mobile hormones and the products of immobilization or metabolism. A major difficulty in getting reliable microautoradiographs is the degree of solubility of hormones in both the polar and nonpolar solvents which are used in the conventional methods to prepare specimens for microscope investigation. For macroautoradiographs, however, quickly frozen and lyophylized preparations are used (e.g., ESCHRICH 1968, MORRIS et al. 1969, IWAMI and MASUDA 1976, TSURUMI and OHWAKI 1978, CHENOU et al. 1978, ROWNTREE and MORRIS 1979). Unless insoluble material is to be looked for (e.g., LIAO and HAMILTON 1966, VEEN 1966, ROBINSON et al. 1968, SABINS et al. 1969, WANGERMANN 1968, 1970, GEE 1972, DAVIES 1973), extreme precautions must be exercised to preserve the original distribution of the radiotracers in the preparation of microautoradiographs. The tissue to be sectioned must be quickly frozen in freezing mixtures (e.g., WANGERMANN 1970, 1974, MORRIS and THOMAS 1978) or liquid nitrogen, sliced with precooled knives in cryostat chambers, and the sections then mounted on dry emulsions of stripping film without thawing (e.g., WANGERMANN 1968, 1970, 1974, BOWEN et al. 1972, ZAMSKI and WAREING 1974, MORRIS and THOMAS 1978, BOURBOULOUX and BONNEMAIN 1979). The tissues may also be freeze-dried (BONNEMAIN 1971, BOURBOULOUX and BONNEMAIN 1973, 1974, BOURBOULOUX et al. 1973) or freeze-dried and paraffin infiltrated (GOLDSMITH et al. 1974, MIGINIAC et al. 1978) before sectioning and mounting on stripping film. It is important, yet not always done, to run control sections of identical plant material which was not treated with the radioactive hormone, since a blackening of the emulsion is sometimes produced also by nonradioactive substances in the preparations.

3.3.6.1 Shoots

By use of these various techniques it has been shown, especially for auxins, that hormones applied to the unfolded leaves of intact plants are generally translocated in sieve tubes at a velocity typical for the transport of assimilates. The physiological significance of such movement is unknown. On the other hand, radioactive auxin, when applied to young leaves, the apex or the cut top, moves toward the roots at about 10 mm h^{-1} (see p 100). In this case, although auxin was confined to the phloem and cambial zones, which include the differentiating secondary vascular elements (TEPPER and BROSSARD 1969, BONNEMAIN 1970, 1971, BOURBOULOUX and BONNEMAIN 1973, 1979, BOURBOULOUX et al. 1973, ZAMSKI and TSIVION 1977, MIGINIAC et al. 1978), the fully differentiated phloem and xylem elements were shown to be free of radiotracers (MORRIS and THOMAS 1978) as was the honeydew from the phloem-feeding aphids (MORRIS and KADIR 1972). Similarly, in intact stolons of *Saxifraga sarmentosa,* which were separated into cortex and stele after ^{3}H-IAA transport, basipetal movement of radioactivity with the characteristics of auxin transport was observed to occur in the stele, while translocation in the cortex resembled accelerated diffusion (DA CRUZ and AUDUS 1978).

There is increasing evidence, both from surgical techniques and autoradiographs, that stem sections behave similarly to intact stems with respect to the pathway of auxin transport. In stem sections of *Coleus, Helianthus, Vicia* (WANGERMANN 1968, 1970, 1974, 1977, WANGERMANN and WITHERS 1978), and of *Nicotiana* (SHELDRAKE 1973a) the transport of IAA was mainly associated with the phloem part of vascular tissue (see Fig. 3.3). On the other hand, IAA also moved basipetally in explants of *Coleus* free of vascular tissue (MAI 1934, JACOBS and MCCREADY 1967) while in coleoptiles of *Avena, Triticum,* and *Zea* it has been shown that both parenchyma and vascular tissues take part in active basipetal auxin transport (VAN DER WEIJ 1932, HERTEL and LEOPOLD 1963, WANGERMANN and WITHERS 1978).

3.3.6.2 Roots

In roots, like shoots, the active transport of auxin, though acropetal (BOWEN et al. 1972, DAVIES and MITCHELL 1972, BRIDGES et al. 1973, GREENWOOD et al. 1973, HARTUNG and PHILLIPS 1974, SHAW and WILKINS 1974, TSURUMI and OHWAKI 1978, TSURUMI and WADA 1980; see also review of BATRA et al. 1975), and the translocation of ABA (HARTUNG and BEHL 1975a, b) appear to be associated with the phloem of vascular tissues or at least with the stele. This is true also for gibberellins (HARTUNG and PHILLIPS 1974), which have been shown to move basipetally in roots (JACOBS and PRUETT 1973). In the cortical tissue, both auxin and gibberellins move apolarly (BOWEN et al. 1972, HARTUNG and PHILLIPS 1974, OHWAKI et al. 1974, SHAW and WILKINS 1974), as does ABA (HARTUNG and BEHL 1975a, b). However, in the root tip region, which has been demonstrated to contain high amounts of endogenous inhibitors and auxins (RIVIER and PILET 1974, PILET 1977), the cortex (MITCHELL and DAVIES 1975) is probably the channel of a basipetal transport component (KONINGS 1968, 1969, KONINGS and GAYADIN 1971, DAVIES et al. 1976, PERNET and PILET 1976). In gravistimulated roots, asymmetric redistribution of both genuine and applied hormones has been observed in this region (e.g., KONINGS 1968, 1969, PILET 1977 and literature there cited; see also Vol. 7, this Series).

3.4 Evidence of Hormone Translocation

3.4.1 Hormone Movement in Vascular Systems

Endogenous growth regulators may be translocated in the plant's vascular systems. This would seem to hold especially true for gibberellins and cytokinins, since activity has been detected in the xylem sap of many different herbaceous as well as woody species. Obviously these hormones are also exported from photosynthesizing leaves, as indicated by their presence in sieve tube sap. Although with less frequency, also auxin and ABA-like activity have been detected in both xylem and phloem sap (Table 3.1).

When applied to intact photosynthesizing mature leaves, auxins (see p 101), gibberellins (ZWEIG et al. 1961, MCCOMB 1964, CHIN and LOCKHART 1965), and ABA (HOCKING et al. 1972, SHINDY et al. 1973, DÖRFFLING et al. 1973, BELLANDI and DÖRFFLING 1974) are translocated like assimilates. Also, when applied to roots of intact plants or cuttings, to basal leaves, or injected into stems, cytokinins (LAGERSTEDT and LANGSTON 1967, GUERN and SADORGE 1967, GUERN et al. 1968, CHVOJKA et al. 1971, MORRIS and WINFIELD 1972, WOOLEY and WAREING 1972, GORDON et al. 1974a, b, MOZES and ALTMAN 1977), gibberellins (ZWEIG et al. 1961, ASAKAWA et al. 1974, BARENDSE 1974, CHAILAKHYAN et al. 1974, DAVIES and RAPPAPORT 1975, ALTMAN and MOZES 1977), and ABA (MITTELHEUSER and VAN STEVENINCK 1971, HOCKING et al. 1972) have been reported to be readily distributed within the plants, which points to a translocation within the transpiration stream. There is also evidence for an exchange of hormones between phloem and xylem which has been shown to occur in both directions of stems of *Salix*, for GA_3 (CHAILAKHYAN et al. 1974) and kinetin (BOWEN and WAREING 1969), and auxins (FIELD and PEEL 1971, LEPP and PEEL 1971, ZAMSKI and WAREING 1974).

After feeding the tips of intact primary roots of *Phaseolus* seedlings with aqueous solutions of labeled hormones by means of 1 µl-droplets (which were quickly absorbed), percentages of 30, 10, 4, and 1 of the label from kinetin, ABA, GA_1, and IAA, respectively, were translocated into the shoot system during the subsequent 24 h. Most of the shoot's kinetin label was detected in the apical bud and primary leaves, although in all cases radioactivity was found in lateral and adventitious roots, about 10% for the ABA and 3% for the other hormones (HARTUNG 1977).

Although these observations point to a circulation of plant hormones within the vascular system, there is little certainty concerning the physiological significance of such movement. Some growth response was observed in which phloem translocation of gibberellin was known to occur (CHIN and LOCKHART 1965); ABA, fed via the transpiration stream, led to rapid stomatal closure (CUMMINS et al. 1971, MITTELHEUSER and VAN STEVENINCK 1971). CROZIER and REID (1971) speculated from their observations on *Phaseolus* seedlings that inactive GA_{19} is produced in the shoot, translocated to the root, where it is converted to the active GA_1 and is redistributed within the plant. On the other hand, gibberellin-like activity of bleeding sap has been found to occur in butanol-extractable conjugates (SEMBDNER et al. 1968), which may be reserve or *inactivated forms*

Table 3.1. Evidence for genuine phytohormones in root exudates (*Xylem sap*) and phloem sap (partly aphid honey dew) in herbaceous and woody plants

	Herbaceous plants		Woody plants	
Xylem sap				
Auxins	*Avena, Hordeum, Triticum, Zea*	Sheldrake (1973c)	*Malus*	Luckwill and Whyte (1968)
	Ricinus	Hall and Medlow (1974)		
Abscisins	*Zea*	Atkin et al. (1973)	*Prunus*	Davison and Young (1973, 1974)
			Salix	Davison (1965), Bowen and Hoad (1968), Lenton et al. (1968), Alvim et al. (1976)
	Helianthus	Hoad (1975)		
	Nicotiana	Itai et al. (1973), Itai and Ben Zioni (1974)		
	Phaseolus	Itai et al. (1973), Itai and Ben Zioni (1974)		
Gibberellins	*Hordeum, Triticum*	Couillerot and Bonnemain (1975)	*Pseudotsuga*	Lavender et al. (1973) Sweet et al. (1974)
	Oryza	Murakami (1968)		
	Zea	Atkin et al. (1973)	*Acer*	Carr and Reid (1968), Reid and Burrows (1968), Sembdner et al. (1968)
	Cucumis	Atsmon et al. (1968)		
	Daucus	Carr and Reid (1968)		
	Helianthus	Phillips and Jones (1964), Kende and Sitton (1967), Sitton et al. (1967)	*Betula*	Reid and Burrows (1968)
			Coffea	Browning (1971, 1973a)
	Impatiens	Carr et al. (1964), Reid and Carr (1967)	*Malus*	Jones and Lacey (1968), Luckwill and Whyte (1968)
	Ipomoea, Pharbitis	Murakami (1968)	*Pyrus*	Jones and Lacey (1968)
	Lupinus	Carr et al. (1964)		
	Pisum	Carr et al. (1964), Carr and Reid (1968)	*Ulmus*	Sembdner et al. (1968)
	Lycopersicon	Reid et al. (1969), Reid and Crozier (1971), Menhenett and Wareing (1975)	*Vitis*	Skene (1967)
Cytokinins	*Oryza, Zea*	Yoshida et al. (1971), Yoshida and Oritani (1974)	*Pseudotsuga*	Morris et al. (1976)
	Coleus	Banko and Boe (1975), van Staden and Davey (1976a)	*Alnus*	Henson and Wheeler (1977)
	Perilla	Beever and Woolhouse (1973, 1974)	*Acer*	Nitsch and Nitsch (1965), Carr and Reid (1968), Horgan et al. (1973), Hewett and Wareing (1974), Purse et al. (1976)
	Helianthus	Kende (1964, 1965), Itai and Vaadia (1965); Kende and Sitton (1967), Carr and Reid (1968), Itai et al. (1968),		

		Klämbt (1968), Burrows and Carr (1969), Wagner and Michael (1971), Sitton et al. (1972)	Betula	Carr and Reid (1968), Šmid and Vardjan (1970)
			Coffea	Browning (1971, 1973b)
	Xanthium	Carr and Burrows (1966), Purse (1975)	Malus	Luckwill and Whyte (1968), Jones (1973), Greene (1975)
	Impatiens	Carr and Burrows (1966), van Staden and Davey (1976a)	Populus	Hewett (1973), Hewett and Wareing (1973a, b, 1974)
	Lycopersicon	Tal et al. (1970), Menhenett and Wareing (1975), Davey and van Staden (1976) Menary and van Staden (1976), van Staden and Davey (1976a) van Staden and Menary (1976), van Staden and Dimalla (1977)	Quercus	Smith and Schwabe (1980)
			Salix	Hewett and Wareing (1974), Alvim et al. (1976)
			Vitis	Loeffler and van Overbeek (1964), Nitsch and Nitsch (1965), Skene and Kerridge (1967), Nitsch (1968), Skene and Osmond (1968), Skene (1972), Skene and Antcliff (1972)
	Nicotiana	Kulaeva (1962), Itai and Vaadia (1971), Itai et al. (1973), Itai and Ben Zioni (1974)		
	Phaseolus	Itai et al. (1973), Itai and Ben Zioni (1974), van Staden and Davey (1976a)		
	Pisum, Lupinus	Carr and Burrows (1966)		
	Protea	van Staden and Davey (1976b)		
Phloem sap				
Auxins	Ricinus	Hall and Baker (1972), Hoad (1973a), Hall and Medlow (1974)	Fagus	Huber et al. (1937)
			Malus	Hoad (1973a)
			Salix	Hoad et al. (1971), Hoad (1973a)
Abscisins	Ricinus	Hoad (1973a, b), Zeevaart (1977)	Abies	Milborrow (1969)
			Malus	Hoad (1973a)
			Salix	Hoad (1967, 1973a)
Gibberellins	Ricinus	Hall and Baker (1972), Hoad (1973a)	Fagus, Quercus	Kluge et al. (1964)
	Taraxacum	Hoad and Bowen (1968), Hoad (1973a)	Malus	Hoad (1973a)
	Vicia	Hoad and Bowen (1968), Hoad (1973a)	Robinia	Kluge et al. (1964)
			Salix	Hoad and Bowen (1968), Hoad (1973a)
			Tilia	Kluge et al. (1964)
Cytokinins	Ricinus	Hall and Baker (1972), Hoad (1973a)	Salix	van Staden (1976)
	Xanthium	Phillips and Cleland (1972)	Yucca	Vonk (1974, 1978, 1979)

which are no longer required. Also, cytokinins have been shown to be translocated in vascular systems in the form of conjugation products (e.g., PHILLIPS and CLELAND 1972, HORGAN et al. 1973, JONES 1973, HEWETT and WAREING 1974, VONK 1974, 1978, 1979, ALVIM et al. 1976, DAVEY and VAN STADEN 1976, VAN STADEN 1976). Nevertheless, the coincidence of cytokinin activity in root exudates and cytokinin-regulated developmental events in the shoot indicates that "root-produced cytokinins play a vital role in the growth and senescence of the aerial parts of the flowering plant" (LETHAM et al. 1978, Vol I, p 252). Similarly, in grass seedlings genuine IAA-myo-inositol appears to be translocated in the vascular system from the endosperm to the coleoptile tip and to serve as the precursor for IAA which is then transported basipetally and regulates coleoptile growth (UEDA and BANDURSKI 1969, HALL and BANDURSKI 1978, EPSTEIN et al. 1980, NOWACKI and BANDURSKI 1980; see p 101).

3.4.2 Hormone Translocation in Shoot Sections

3.4.2.1 Auxins

Occurrence of Basipetal Transport. The basipetal polarity of auxin transport, first detected in etiolated *Avena* coleoptiles (c.f. WENT 1928, VAN DER WEIJ 1932), appears to be a common feature of the shoot system. It has been demonstrated in many kinds of shoot tissues, including coleoptiles of *Zea* (HERTEL and LEOPOLD 1963, NAQVI and GORDON 1964, 1966, GOLDSMITH 1967a, b, HERTEL and FLORY 1968, NAQVI 1975), hypocotyls of *Phaseolus* (JACOBS 1950, SMITH and JACOBS 1968) and *Helianthus* (ABROL and AUDUS 1973), epicotyls of *Helianthus* (LEOPOLD and LAM 1961, 1962, REIFF and v. GUTTENBERG 1961), embryonic axes of *Pinus* (GREENWOOD and GOLDSMITH 1970), *Acer*, and *Phaseolus* (FRY and WANGERMANN 1976), seedlings of *Pinus* (ZAERR 1968), stems of *Malus* (GREGORY and HANCOCK 1955), *Lens* (PILET 1965), *Pisum* (SCOTT and BRIGGS 1960, MORRIS 1977), *Saccharum* (SCOTT and MOST 1972) and *Selaginella* (WOCHOK and SUSSEX 1973), internode sections cut from different regions of *Coleus* (JACOBS 1952, 1954, 1961, 1977, NAQVI 1963, SCOTT and JACOBS 1963, NAQVI and GORDON 1965, JACOBS and MCCREADY 1967, LEOPOLD and DE LA FUENTE 1967, WANGERMANN 1968, KOEVENIG and JACOBS 1972) and *Nicotiana* (BAYER 1972, SHELDRAKE 1973a), petioles of *Phaseolus* (MCCREADY 1963, MCCREADY and JACOBS 1963a, b, BLACK and OSBORNE 1965, JACOBS 1967, OSBORNE 1968, OSBORNE et al. 1968, HAREL 1969), *Coleus* (WERBLIN and JACOBS 1967, VEEN and JACOBS 1969a), and *Xanthium* (JACOBS 1978a), leaves of monocotyledons (SHELDRAKE 1972), fruit stalks of *Fritillaria* (KALDEWEY 1966), gynophores of *Arachis* (JACOBS 1951), stamen of *Digitalis* (WEIS 1969) and *Cleome* (KOEVENIG and SILLIX 1973), stamen and styles of *Narcissus* (unpublished data), and flower stalks, stamen, styles and ovary sections of *Lilium* (unpublished data); basipetal auxin transport has been shown not necessarily to depend on the presence of vascular tissues (MAI 1934, JACOBS and MCCREADY 1967, SHELDRAKE 1973a, WANGERMANN 1974, 1977, WANGERMANN and WITHERS 1978) and to be even rather pronounced in parenchyma of coleoptiles (VAN DER WEIJ 1932, HERTEL and LEOPOLD 1963, WANGERMANN and WITHERS 1978).

Degree of Polarity. Most early studies, using bioassays in which auxin activity was detected solely in basal and none in apical receivers, pointed to an absolute basipolarity of auxin transport. However, JACOBS (1952, 1954) found that in sections cut from young internodes of light-grown *Coleus*, about one-third of the amount of auxin which is delivered to basal receivers during basipetal transport can be collected in apical receivers following application of IAA to the basal cut surface of the sections (see also LEOPOLD and GUERNSEY 1953). Use of radioactive auxins not only confirmed this result (NAQVI and GORDON 1965, LEOPOLD and DE LA FUENTE 1967), but showed that some acropetal auxin translocation was also a general phenomenon. The degree of polarity, expressed as the ratio between activity in basal and apical receivers after uniform periods of basipetal and acropetal transport, greatly varies [e.g., 300:1 in *Saccharum* stems (SCOTT and MOST 1972) vs. 3:1 ratio in young *Coleus* internodes (JACOBS 1952, 1954)]. The ratio increases with an increase in the length of section (e.g., WENT and WHITE 1939, LEOPOLD and HALL 1966, DE LA FUENTE and LEOPOLD 1966; see also Sect. 3.3.4.3), and decreases with age. This decrease may be because of a decline with age in basipetal transport [e.g., petioles of *Coleus* (WERBLIN and JACOBS 1967, VEEN and JACOBS 1969a) and *Phaseolus* (OSBORNE 1968, OSBORNE et al. 1968), stems of *Coleus* (WANGERMANN 1968) and *Nicotiana* (SHELDRAKE 1973a)], and an increase with age in acropetal movement [e.g., petioles of *Phaseolus* (MCCREADY and JACOBS 1963b, LEOPOLD and DE LA FUENTE 1968) and *Xanthium* (JACOBS 1978a), hypocotyls of *Phaseolus* (SMITH and JACOBS 1968, 1969), stamen filaments of *Cleome* (KOEVENIG and SILLIX 1973)].

Specificity of the Auxin Transport System. Comparative studies with chemically closely related compounds reveal a positive correlation between auxin activity and degree of basipetal polarity in the transport of these substances (e.g., WENT and WHITE 1939, LEOPOLD and LAM 1961, JACOBS 1967, HERTEL et al. 1969, VEEN 1972). Further, the basipetal transport of auxins is specifically inhibited by substances such as TIBA, naphthylphthalamic acid (NPA), 3,3a-dihydro-2-(p-methoxyphenyl)-8H-pyrazolo[5,1a] isoindol-8-one (DPX 1840) (see Sect. 3.3.4.3) not only in treated sections but also when they are applied to intact plants at the loci of natural auxin production (see p 100). In summary, the numerous transport studies make certain the presence of a unique transport system, which is specific for auxin molecules, and which moves the hormone basipetally from the natural auxin sources in the shoot and then on to the root to regulate growth and other developmental processes. It is clear that this system is fully functional in tissue sections which is the material which has been used most in transport studies.

3.4.2.2 Hormones Other than Auxins

In contrast to the convincing demonstration of a specific basipetal auxin transport system, mostly conflicting results have so far been obtained concerning the translocation of other hormones.

Cytokinins. The first experiments with cytokinins which showed that kinetin, when applied exogenously to the surface of leaves (MOTHES and ENGELBRECHT 1961, MOTHES 1964) or stems (SACHS and THIMANN 1964) appeared immobile and acted instead as an "attraction center" especially for amino acids. However, when benzyladenine was applied to the cut surface of *Phaseolus* petiole segments (OSBORNE and BLACK 1964, OSBORNE and MCCREADY 1965, BLACK and OSBORNE 1965, LAGERSTEDT and LANGSTON 1966), or kinetin to stumps of decapitated *Phaseolus* seedlings (SETH et al. 1966) or sections from petioles and hypocotyls of very young *Raphanus* seedlings (RADIN and LOOMIS 1974), predominantly basipetal movement was observed with an enhancement in presence of IAA. However, in contrast, an absence of polarity of cytokinin translocation was reported for petiole segments of *Coleus* (FOX and WEIS 1965, VEEN and JACOBS 1969b, JACOBS 1972, 1976, KOEVENIG 1973) and *Xanthium* (OSBORNE and MCCREADY 1965, LAGERSTEDT and LANGSTON 1966), for stem sections of *Pisum* (FOX and WEIS 1965) and *Lens* (PILET et al. 1967, PILET 1968), for sections from hypocotyls and cotyledonary petioles of 3-day-old *Raphanus* seedlings (RADIN and LOOMIS 1974), for *Avena* coleoptiles (FOX and WEIS 1965, KOEVENIG 1973) and *Zea* mesocotyls and coleoptiles (EL-SAIDI 1972), and for segments from different floral organs of *Cleome* (KOEVENIG 1973). Moreover, FOX and WEIS (1965) were not able to confirm the results of OSBORNE and coworkers who, in all likelihood, had used unphysiologically high hormone concentrations. Thus, a clear picture of the mechanism and polarity of cytokinin translocation is lacking.

Gibberellins. The results of gibberellin translocation studies in shoots are similarly conflicting. That gibberellin can move basipetally through stem sections has been shown in studies with *Helianthus* (KENTZER and LIBBERT 1961, LIBBERT and GERDES 1964) and *Pisum* (COHEN et al. 1966, JONES and LANG 1968) and, in addition, LIBBERT and coworkers even observed an inhibiting effect on this translocation by TIBA. The first studies on the polarity of gibberellin movement, although unphysiologically high concentrations were applied in the donor blocks, revealed GA_3 to move very slowly (approximately 1 mm h^{-1}) and without polarity (KATO 1958, CLOR 1967). Similar results were obtained using ^3H-GA_1 with segments of *Zea* coleoptiles (HERTEL et al. 1969, WILKINS and NASH 1974), short sections of *Phaseolus* stems (PHILLIPS and HARTUNG 1974), and *Cucumis* hypocotyl segments (RUDICH et al. 1976). However, basipetal polarity of GA_3 movement has been reported for sections of young *Coleus* petioles (JACOBS and KALDEWEY 1970, JACOBS 1972, JACOBS and PRUETT 1972) and of *Saccharum* internodes (SCOTT and MOST 1972). In segments of *Phaseolus* internodes, several cm long, PHILLIPS and HARTUNG (1974) observed GA_1 to move preferentially toward the apical elongating region and ŠEBÁNEK and HINK (1967) showed in *Pisum* epicotyls a possible polarization of gibberellin movement towards applied IAA.

Abscisins and Thiamine. The few studies on ABA movement in shoot sections has likewise led to contradictory results. DÖRFFLING and BÖTTGER (1968) and MILBORROW (1968) found fast (20–30 mm h^{-1}) ABA movement through petiole segments with weak basipetal polarity, especially in young tissues. In *Gossypium* cotyledonary petioles ^{14}C-ABA moved similarly fast, but without polarity (INGERSOLL and SMITH 1970, 1971). No polarity was detected for a slow (2–3 mm h^{-1}) movement of ABA through stem segments of *Pisum* (DÖRFFLING et al. 1973). Finally, VEEN (1975) and VEEN and FRISSEL (1975) found no radioactivity at all in apical and basal receivers of *Coleus* petiole segments with ^{14}C-ABA applied oppositely to the basal and apical cut surface, respectively. The distribution of radiocarbon within the sections gave no sign of polar movement. Similar results have been obtained with hypocotyl sections of *Citrullus*, although in this case, the distribution of radioactivity within the sections pointed to a slight acropetal translocation of ^{14}C-ABA (own unpublished data). When ABA is added to the apical bud of intact *Pisum* seedlings, from which IAA is transported basipetally into the root (see p 100), it remains near the application point and does not reach the root, even after a period of 24 h (DÖRFFLING et al. 1973, BELLANDI and DÖRFFLING 1974).

Thiamine, which probably serves as a root growth hormone, and which is produced in the leaves of higher plants has been shown to move strongly basipetally in large amounts through petiole segments and acropetally through stem segments of *Lycopersicon* (KRUSZEWSKI and JACOBS 1974).

3.4.3 Hormone Translocation in Root Sections

3.4.3.1 Auxins

Auxin transport in roots has long been believed to be basipetal, i.e., from the root tip toward the shoot (see LEOPOLD 1961, 1963). This view was stressed by the CHOLODNY-WENT theory of auxin and root gravitropism, which implied that auxin synthesized in the root tip is translocated basipetally, and accumulates in the lower side of a geostimulated root to a concentration which is growth-inhibiting (CHOLODNY 1924, 1926, 1934). However, there were indications from early studies which pointed to an acropetal auxin translocation (GORTER 1932, FABER 1936), derived indirectly from acropetal xylem differentiation and root primordia induction studies (see WIGHTMAN and THIMANN 1980, WIGHTMAN et al. 1980). In fact, the first studies on ^{14}C-IAA movement in roots of *Vicia* (YEOMANS and AUDUS 1964) and *Lens* (PILET 1964) indicated preferential acropetal translocation. The idea of acropetal auxin movement was further extended by BONNETT and TORREY (1965a, b; 1966), who observed lateral root formation predominantly at the cut surface near the tip of *Convolvulus* root sections together with a higher accumulation of radiocarbon at the apical than at the basal end of sections previously supplied with ^{14}C-IAA. Starting with the paper of KIRK and JACOBS (1968) and followed by the studies of WILKINS and SCOTT (1968) and SCOTT and WILKINS (1968) acropetal auxin transport through root sections into apical receivers was established for different species (see reviews of SCOTT 1972, BATRA et al. 1975, GOLDSMITH 1977, p 466ff.). The acropetal transport of auxin appeared to occur in the stele (BOWEN et al. 1972, HARTUNG and PHILLIPS 1974, OHWAKI et al. 1974, SHAW and WILKINS 1974) whereas non-polar movement combined with rapid metabolization (AASHEIM and IVERSEN 1971) was observed in the cortical tissues (BOWEN et al. 1972, GREENWOOD et al. 1973, HARTUNG and PHILLIPS 1974, OHWAKI et al. 1974, SHAW and WILKINS 1974). However, auxin applied to the root tip, as well as perhaps endogenous auxin translocated to or produced in the tip, appears to move basipetally toward the elongation zone (for references see Sect. 3.3.6.2).

3.4.3.2 Hormones Other than Auxins

Basipetal movement of GA_3 was found in segments of *Zea* roots (JACOBS and PRUETT 1973) and was shown to occur in the stele of *Phaseolus* roots while it moved apolarly in the cortex (HARTUNG and PHILLIPS 1974). Apolar movement of cytokinins was observed in root sections of *Raphanus* (RADIN and LOOMIS 1974) and *Pisum* (JACOBS 1976), whereas basipetal translocation was reported for *Zea* root segments (EL-SAIDI 1971, 1972). ABA movement was shown to be acropetal in root sections of *Phaseolus* (HARTUNG and BEHL 1974). This polar movement was restricted to the stele (HARTUNG and BEHL 1975a) and was increased by illumination (HARTUNG and BEHL 1975b). Thiamine which has been shown to move strongly basipetally in *Coleus* petioles is translocated apolarly in root sections of *Pisum* (JACOBS and DE MUTH 1977). (For hormone movement in the root tip region see Sect. 3.3.6.3).

Acknowledgements. I wish to thank Mrs. REGINA KARMANN for drawing the figures and for her steady engagement in typing the diverse versions of the manuscript. I wish further to thank Professor TOM K. SCOTT and Dr. JANE SHEN-MILLER for their suggestions and editorial assistance in preparing this chapter.

References

Aasheim T, Iversen T-H (1971) Decarboxylation and transport of auxin in segments of sunflower and cabbage roots. II. A chromatographic study using IAA-1-^{14}C and IAA-5-^{3}H. Physiol Plant 24:325–329

Abrol BK, Audus LJ (1973) The lateral transport of 2,4-dichlorophenoxyacetic acid in horizontal hypocotyl segments of *Helianthus annuus*. J Exp Bot 24:1209–1223

Addicott FT (1970) Plant hormones in the control of abscission. Biol Rev Camb Philos Soc 45:485–524

Albaum HG, Kaiser S, Nestler HA (1937) The relation of hydrogen-ion concentration to the penetration of 3-indoleacetic acid into *Nitella* cells. Am J Bot 24:513–518

Altman A, Mozes R (1977) Comparative basipetal transport of 6-benzylaminopurine-8-^{14}C, gibberellin A_3-^{3}H, IAA-2-^{14}C, and sucrose-^{14}C in the root of intact *Citrus aurantium* seedlings. Physiol Plant 39:233–235

Alvim R, Hewett EW, Saunders PF (1976) Seasonal variation in the hormone content of willow. I. Changes in abscisic acid content and cytokinin activity in the xylem sap. Plant Physiol 57:474–476

Andreae WA, Good NE (1955) The formation of indoleacetylaspartic acid in pea seedlings. Plant Physiol 30:380–382

Antoszewski R, Lis E (1968) Translocation of some radioactive compounds from the strawberry receptacle to the mother plant. Bull Acad Pol Sci 16:443–446

Arisz WH (1969) Intercellular polar transport and the role of plasmodesmata in coleoptiles and *Vallisneria* leaves. Acta Bot Neerl 18:14–38

Asakawa Y, Tamari K, Inoue K, Kaji J (1974) Studies on metabolism of gibberellin A_3 in plant. 1. Translocation and intracellular distribution of tritiated gibberellin A_3. Agric Biol Chem 38:713–717

Atkin RK, Barton GE, Robinson DK (1973) Effect of root-growing temperature on substances in xylem exudate of *Zea mays*. J Exp Bot 24:475–487

Atsmon D, Lang A, Light EN (1968) Contents and recovery of gibberellins in monoecious and gynoecious cucumber plants. Plant Physiol 43:806–810

Backus GE, Schrank AR (1952) Electrical and curvature responses of the *Avena* coleoptile to unilateral illumination. Plant Physiol 27:251–262

Bandurski RS (1978) Chemistry and physiology of myo-inositol esters of indole-3-acetic acid. In: Wells W, Eisenberg F (eds) Cyclitols and phosphoinositides. Academic Press, London New York, pp 35–54

Banko TJ, Boe AA (1975) Effects of pH, temperature, nutrition, ethephon, and chloromequat on endogenous cytokinin levels of *Coleus blumei* Benth. J Am Soc Hortic Sci 100:168–172

Barendse GWM (1974) Accumulation and metabolism of radioactive gibberellic acid in seedlings of *Pharbitis nil* Chois. In: Plant Growth Substances 1973. Tokyo: Hirokawa 1974, pp 332–341

Barry AJ (1971) The effect of 2,3,5-triiodobenzoic acid on the transport and metabolism of indoleacetic acid in intact pea seedlings. B Sc Diss, Univ Southampton

Basler E (1974) Abscisic acid and gibberellic acid as factors in the translocation of auxin. Plant Cell Physiol 15:351–361

Basler E, McBride R (1977) Interaction of coumarin, gibberellic acid and abscisic acid in the translocation of auxin in bean seedlings. Plant Cell Physiol 18:939–947

Batra MW, Edwards KL, Scott TK (1975) Auxin transport in roots: Its characteristics and relationship to growth. In: Torrey JG, Clarkson DT (eds) The development and function of roots. Academic Press, London New York, pp 299–325

Batt S, Venis MA (1976) Separation and localization of two classes of auxin-binding sites in corn coleoptile membranes. Planta 130:15–21

Batt S, Wilkins MB, Venis MA (1976) Auxin binding to corn coleoptile membranes: kinetics and specificity. Planta 130:7–13

Bayer MH (1972) Transport and accumulation of IAA-^{14}C in tumor-forming *Nicotiana* hybrids. J Exp Bot 23:801–812

Beever JE, Woolhouse HW (1973) Increased cytokinin from root system of *Perilla* and flower and fruit development. Nature 246:31–32

Beever JE, Woolhouse HW (1974) Increased cytokinin export from the roots of *Perilla frutescens* following disbudding or floral induction. In: Bieleski RL, Ferguson AR, Cresswell MM (eds) Mechanism of regulation of plant growth. Roy Soc N Z, Wellington, pp 681–686

Bellandi DM, Dörffling K (1974) Transport of abscisic acid-2-C-14 in intact pea seedlings. Physiol Plant 32:365–368

Beyer A (1925) Untersuchungen über den Traumatotropismus der Pflanzen. Biol Zentralbl 45:683–702, 746–768

Beyer EM (1972) Auxin transport, a new synthetic inhibitor. Plant Physiol 50:322–327

Beyer EM Jr, Morgan PW (1969) Ethylene modification of an auxin pulse in cotton stem sections. Plant Physiol 44:1690–1694

Beyer EM, Johnson AL, Sweetser PB (1976) A new class of synthetic auxin transport inhibitors. Plant Physiol 57:839–841

Black MK, Osborne DJ (1965) Polarity of transport of benzyladenine, adenine and indole-3-acetic acid in petiole segments of *Phaseolus vulgaris*. Plant Physiol 40:676–680

Bollag JM, Galun E (1966) Distribution of labelled indolyl-3-acetic acid in intact cucumber plants. Nature 211:647–648

Bonnemain JL (1970) Transport de l'AIA marqué et de ses derivés à partir de jeunes fruits. CR Acad Sci Paris Ser D 270:1326–1329

Bonnemain JL (1971) Transport et distribution des traceurs après application de AIA-2-^{14}C sur les feuilles de *Vicia faba*. CR Acad Sci Paris Ser D 273:1699–1702

Bonnemain JL (1972) La plante supérieure, siège d'une circulation complexe. Physiol Vég 10:697–729

Bonnemain JL, Bourbouloux A (1973) The transport and metabolism of ^{14}C-indoleacetic acid in intact plants. Proc Res Inst Pom Skierniewice Pol Ser E Conf Symp 3:207–214

Bonnett HT Jr, Torrey JG (1965a) Auxin transport in *Convolvulus* roots cultured in vitro. Plant Physiol 40:813–818

Bonnett HT Jr, Torrey JG (1965b) Chemical control of organ formation in root segments of *Convolvulus* cultured in vitro. Plant Physiol 40:1228–1236

Bonnett HT Jr, Torrey JG (1966) Comparative anatomy of endogenous bud and lateral root formation in *Convolvulus arvensis* roots cultured in vitro. Am J Bot 53:496–507

Bopp M (1972) On the effect of morphactin. In: Kaldewey H, Vardar Y (eds) Hormonal regulation of plant growth and development. Verlag Chemie, Weinheim, pp 333–348

Bose JC (1907) Comparative electrophysiology. Longmans, London

Böttger M (1970) Die hormonale Regulation des Blattfalls bei *Coleus rehneltianus* Berger. I. Die Wechselwirkung von Indol-3-essigsäure, Gibberellin- und Abscisinsäure auf Explantate. II. Die natürliche Rolle von Abscisinsäure im Blattfallprozeß. Planta 93:190–213

Bourbouloux A, Bonnemain JL (1973) Transport de l'auxine-^{14}C en provenance de jeunes gousses de *Vicia faba* L. Planta 115:161–172

Bourbouloux A, Bonnemain JL (1974) Transport, distribution et metabolisme de l'auxine dans la racine de *Vicia faba* L. après application de (^{14}C)AIA ou de (^{3}H)AIA sur le bourgeon. Planta 119:169–182

Bourbouloux A, Bonnemain JL (1979) The different components of the movement and the areas of retention of labelled molecules after application of [^{3}H]-indolyl-acetic acid to the apical bud of *Vicia faba*. Physiol Plant 47:260–268

Bourbouloux A, Bonnemain JL, Vasseur J (1973) Caractéristiques du transport dans la tige de l'auxine-^{14}C en provenance de jeunes feuilles de *Vicia faba* L. CR Acad Sci Paris Ser D 276:501–504

Bowen MR, Hoad GV (1968) Inhibitor content of phloem and xylem sap obtained from willow (*Salix viminalis*) entering dormancy. Planta 81:64–70

Bowen MR, Wareing PF (1969) The interchange of ^{14}C-kinetin and ^{14}C-gibberellic acid between the bark and xylem of willow. Planta 89:108–125

Bowen MR, Wilkins MB, Cane AR, McCorquodale I (1972) Auxin transport in roots. VIII. The distribution of radioactivity in the tissues of *Zea* root segments. Planta 105:273–292

Boysen Jensen P (1910) Über die Leitung des phototropischen Reizes in *Avena*-Keimpflanzen. Ber Dtsch Bot Ges 28:118–120

Boysen Jensen P (1913) Über die Leitung des phototropischen Reizes in der Avenakoleoptile. Ber Dtsch Bot Ges 31:559–566

Brauner L (1923) Über den Einfluß der Koleoptilspitze auf die geotropische Reaktion der Avenakeimlinge. Ber Dtsch Bot Ges 41:208–211

Brauner L (1927) Untersuchungen über das geoelektrische Phänomen. Jahrb Wiss Bot 66:381–428

Brauner L (1959) Neue Versuche zur Analyse des geoelektrischen Effekts. Planta 53:449–483

Bridges IG, Hillman JR, Wilkins MB (1973) Identification and localisation of auxin in primary roots of *Zea mays* by mass spectroscopy. Planta 115:189–192

Briggs WR, Steeves TA, Sussex IM, Wetmore RH (1955) A comparison of auxin destruction by tissue extracts and intact tissues of the fern *Osmunda cinnamomea* L. Plant Physiol 30:148–155

Brossard D, Tepper H (1969) Vitesse du transport de l'auxine exogène dans les racines du *Zea mays*. CR Acad Sci Paris Ser D 276:1301–1303

Brown BT, Johansen O, Sasse WHF (1972) New inhibitor of auxin transport. Experientia 28:1290–1291

Brown BT, Johansen O, Katekar GF, Sasse WH (1973) The effect on root geotropism of certain ortho-carboxyphenylpropanones. Pestic Sci 4:473–484

Brown CL, Wetmore RH (1959) Auxin transport in the long shoots of pine. Am J Bot 46:586–590

Browning G (1971) The hormonal regulation of flowering and cropping in *Coffea arabica* L. Ph D Diss, Univ Bristol

Browning G (1973a) Flower bud dormancy in *Coffea arabica* L. 1. Studies of gibberellin in flower buds and xylem sap and of abscisic acid in flower buds in relation to dormancy release. J Hortic Sci 48:29–41

Browning G (1973b) Flower bud dormancy in *Coffea arabica* L. 2. Relation of cytokinins in xylem sap and flower buds to dormancy release. J Hortic Sci 48:297–310

Bünning E, Reisener HJ, Weygand F, Simon H, Klebe JF (1956) Versuche mit radioaktiver Indolessigsäure zur Prüfung der sogenannten Ablenkung des Wuchsstoffstromes durch Licht. Z Naturforsch 11b:363–364

Burrows WJ, Carr DJ (1969) Effects of flooding the root system of sunflower plants on the cytokinin content in the xylem sap. Physiol Plant 22:1105–1112

Cande WZ, Ray PM (1976) Nature of cell-to-cell transfer of auxin in polar transport. Planta 129:43–52

Cande WZ, Goldsmith MHM, Ray PM (1973) Polar auxin transport and auxin-induced elongation in the absence of cytoplasmic streaming. Planta 111:279–296

Canny MJ (1960) The rate of translocation. Biol Rev 35:507–532

Canny MJ (1971) Translocation: mechanism and kinetics. Annu Rev Plant Physiol 22:237–260

Canny MJ (1973) Phloem translocation. Univ Press, Cambridge

Carr DJ, Burrows WJ (1966) Evidence for the presence in xylem sap of substances with kinetin-like activity. Life Sci 5:2061–2077

Carr DJ, Reid DM (1968) The physiological significance of the synthesis of hormones in roots and of their export to the shoot system. In: Wightman F, Setterfield G (eds) Biochemistry and physiology of plant growth substances. Runge, Ottawa, pp 1169–1185

Carr DJ, Reid DM, Skene KGM (1964) The supply of gibberellins from the root to the shoot. Planta 63:382–392

Chailakhyan MKh, Khlopenkova LP, Khazhakyan KhK (1974) Movement of gibberellins and their effect on shoot growth and stem thickening in entire plants. Dokl Bot Sci 215:484–487

Chenou E, Sossountzov L, Lefebvre M-F (1978) Distribution de l'AIA-^{14}C en relation avec l'inhibition de croissance des bourgeons: étude d'une fougère, le *Marsilea drummondii* A. Br. Physiol Vég 16:137–156

Chin TY, Lockhart JA (1965) Translocation of applied gibberellin in bean seedlings. Am J Bot 52:828–833

Ching TM, Fang SC (1958) The redistribution of radioactivity in geotropically stimulated plants pretreated with radioactive indoleacetic acid. Physiol Plant 11:722–727

Cholodny N (1924) Über die hormonale Wirkung der Organspitze bei der geotropischen Krümmung. Ber Dtsch Bot Ges 42:536–562

Cholodny N (1926) Beiträge zur Analyse der geotropischen Reaktion. Jahrb Wiss Bot 65:447–459

Cholodny N (1934) Über die Bildung und Leitung des Wuchshormons bei den Wurzeln. Planta 21:517–530

Christie AE, Leopold AC (1965a) On the manner of triiodobenzoic acid inhibition of auxin transport. Plant Cell Physiol 6:337–345

Christie AE, Leopold AC (1965b) Entry and exit of indoleacetic acid in corn coleoptiles. Plant Cell Physiol 6:453–465

Chvojka L, Stohr J, Hejmova L, Benes J (1971) Transport and localization of 6-(dimethylallylamino)purine-8-^{14}C cytokinin in apple trees. Biol Plant 13:65–68

Clark WG (1937a) Electrical polarity and auxin transport. Plant Physiol 12:409–440

Clark WG (1937b) Polar transport and electrical polarity in coleoptile of *Avena*. Plant Physiol 12:737–757

Clark WG (1938) Electrical polarity and auxin transport. Plant Physiol 13:529–552

Cleland RE (1975) Auxin-induced hydrogen ion excretion: correlation with growth, and control by external pH and water stress. Planta 127:233–242
Cleland RE (1976) Kinetics of hormone-induced H^+ excretion. Plant Physiol 58:210–213
Cleland RE, Prins H, Harper R, Higinbotham N (1977) Rapid hormone-induced hyperpolarization of the oat coleoptile transmembrane potential. Plant Physiol 59:395–397
Clor MA (1967) Translocation of tritium-labelled gibberellic acid in pea stem segments and potato tuber cylinders. Nature 214:1263–1264
Cohen D, Robinson JB, Paleg LG (1966) Decapitated peas and diffusible gibberellins. Aust J Biol Sci 19:535–543
Cordes W (1966) Translocation of ^{14}C-labelled indoleacetic acid in *Coleus* stems, as influenced by humidity. Plant Cell Physiol 7:183–184
Couillerot J-P, Bonnemain J-L (1975) Transport et devenir des molécules marquées après l'application d'acide gibbérellique-^{14}C sur les jeunes feuilles de tomate. CR Acad Sci Paris Ser D 280:1453–1454
Crozier A, Reid DM (1971) Do roots synthesize gibberellins? Can J Bot 49:967–975
Cummins WR, Kende H, Raschke K (1971) Specificity and reversibility of the rapid stomatal response to abscisic acid. Planta 99:347–351
Da Cruz GS, Audus LJ (1978) Studies on the transport of tritiated indol-3-ylacetic acid in stolones of *Saxifraga sarmentosa*. Ann Bot 42:577–594
Darwin C (1880) The power of movement in plants. Murray, London
Davenport TL, Jordan WR, Morgan PW (1977a) Movement and endogenous levels of abscisic acid during water stress-induced abscission in cotton seedlings. Plant Physiol 59:1165–1168
Davenport TL, Morgan PW, Jordan WR (1977b) Auxin transport as related to leaf abscission during water stress in cotton. Plant Physiol 59:554–557
Davenport TL, Jordan WR, Morgan PW (1979) Movement of kinetin and gibberellic acid in leaf petioles during water stress-induced abscission in cotton. Plant Physiol 63:152–155
Davenport TL, Morgan PW, Jordan WR (1980) Reduction of auxin transport capacity with age and internal water deficits in cotton petioles. Plant Physiol 65:1023–1025
Davey JE, van Staden J (1976) Cytokinin translocation: Changes in zeatin and zeatin-riboside levels in the root exudate of tomato plants during their development. Planta 130:69–72
Davies LJ, Rappaport L (1975) Metabolism of tritiated gibberellins in d-5 dwarf maize. I. In excised tissues and intact dwarf and normal plants. Plant Physiol 55:620–625
Davies PJ (1973) The uptake and fractional distribution of differentially labeled indoleacetic acid in light grown stems. Physiol Plant 28:95–100
Davies PJ (1974) The uptake and elution of indoleacetic acid by pea stem sections in relation to auxin induced growth. In: Plant growth substances 1973. Hirokawa, Tokyo, pp 767–779
Davies PJ, Mitchell EK (1972) Transport of indoleacetic acid in intact roots of *Phaseolus coccineus*. Planta 105:139–154
Davies PJ, Rubery PH (1978) Components of auxin transport in stem segments of *Pisum sativum* L. Planta 142:211–219
Davies PJ, Doro JA, Tarbox AW (1976) The movement and physiological effect of indoleacetic acid following point applications to root tips of *Zea mays*. Physiol Plant 36:333–337
Davison RM (1965) Some properties of a plant growth inhibitor present in xylem sap of woody plants. Austr J Biol Sci 18:475–486
Davison RM, Young H (1973) Abscisic acid content of xylem sap. Planta 109:95–98
Davison RM, Young H (1974) Seasonal changes in the level of abscisic acid in xylem sap of peach. Plant Sci Lett 2:79–82
Day BE (1952) Absorption and translocation of 2,4-dichlorophenoxyacetic acid by bean plants. Plant Physiol 27:143–152
Dedolph RR, Breen JJ, Gordon SA (1965) Geoelectric effect and geotropic curvature. Science 148:1100–1101
de la Fuente RK, Leopold AC (1966) Kinetics of polar auxin transport. Plant Physiol 41:1481–1484

de la Fuente RK, Leopold AC (1970a) The transportable auxin pool. Plant Physiol 45:19–24

de la Fuente RK, Leopold AC (1970b) Turnover in the transportable auxin pool. Plant Physiol 45:642–643

de la Fuente RK, Leopold AC (1972) Two components of auxin transport. Plant Physiol 50:491–495

de la Fuente RK, Leopold AC (1973) A role for calcium in auxin transport. Plant Physiol 51:845–847

Dohrmann U, Hertel R, Kowalik H (1978) Properties of auxin binding sites in different subcellular fractions from maize coleoptiles. Planta 140:97–106

Dolk HE (1936) Geotropie en groeistof. Diss Utrecht 1930. English translation: Geotropism and the growth substance. Rec Trav Bot Néerl 33:509–585

Dörffling K, Böttger M (1968) Transport von Abscisinsäure in Explantaten, Blattstiel- und Internodialsegmenten von *Coleus rehneltianus*. Planta 80:299–308

Dörffling K, Bellandi DM, Böttger M, Lückel H, Menzer U (1973) Abscisic acid: Properties and transport and effect on distribution of potassium and phosphorus. Proc Res Inst Pomol Skierniewice Pol Ser E Conf Symp 3:259–272

Drake G, Carr DJ (1978) Plasmodesmata, tropisms, and auxin transport. J Exp Bot 29:1309–1318

Du Buy HG, Olson RA (1940) The relation between respiration, protoplasmic streaming, and auxin transport in the *Avena* coleoptile. Am J Bot 27:401–414

Edwards KL, Goldsmith MHM (1980) pH-dependent accumulation of indoleacetic acid by corn coleoptile sections. Planta 147:457–466

Eliasson L (1972) Translocation of shoot-applied indolylacetic acid into the roots of *Populus tremula*. Physiol Plant 27:412–416

Eliezer J, Morris DA (1979) Effects of temperature and sink activity on the transport of ^{14}C-labelled indol-3yl-acetic acid in the intact pea plant (*Pisum sativum* L.) Planta 147:216–224

Eliezer J, Morris DA (1980) Cell length, light and ^{14}C-labelled indol-3yl-acetic acid transport in *Pisum sativum* L. and *Phaseolus vulgaris* L. Planta 149:327–331

El-Saidi MT (1971) Transport and metabolism of kinetin-8-^{14}C in *Zea mays* L. roots. Ann Bot 35:1073–1083

El-Saidi MT (1972) Transport and metabolism of kinetin-8-^{14}C in the mesocotyl and coleoptile of *Zea mays* L. Ann Bot 36:99–107

Epstein E, Cohen ID, Bandurski RS (1980) Concentration and metabolic turnover of indoles in germinating kernels of *Zea mays* L. Plant Physiol 65:415–421

Eschrich W (1968) Translokation radioaktiv markierter Indolyl-3-essigsäure in Siebröhren von *Vicia faba*. Planta 78:144–157

Etherton B (1970) Effect of indole-3-acetic acid on membrane potentials of oat coleoptile cells. Plant Physiol 45:527–528

Everat-Bourbouloux A, Bonnemain J-L (1980) Distribution of labelled auxin and derivatives in stem tissues of intact and decapitated broad-bean plants in relation to apical dominance. Physiol Plant 50:145–152

Faber E-R (1936) Wuchsstoffversuche an Keimwurzeln. Jahrb Wiss Bot 83:439–469

Fang SC, Butts JS (1957) Studies on carboxyl-^{14}C-labelled 3-indoleacetic acid in plants. Plant Physiol 32:253–259

Field RJ, Peel AJ (1971) The movement of growth regulators and herbicides into the sieve elements of willow. New Phytol 70:997–1003

Filner B, Hertel R, Steele Ch, Fan V (1970) Some aspects of geotropism in coleoptiles. Planta 94:333–354

Fitting H (1907) Die Leitung tropistischer Reize in parallelotropen Pflanzenteilen. Jahrb Wiss Bot 44:177–253

Fletcher RA, Zalik S (1965) Effect of light of several spectral bands on the metabolism of radioactive IAA in bean seedlings. Plant Physiol 40:549–552

Fox JE, Weis JS (1965) Transport of the kinin, N_6-benzyladenine: non-polar or polar? Nature 206:678–679

Fry SC, Wangermann E (1976) Polar transport of auxin through embryos. New Phytol 77:313–317

Fulford RM, Quinlan JD, Lacey HJ, Barlow HWB (1968) The acropetal movement of growth substances from young leaves on woody shoots. In: Wightman F, Setterfield G (eds) Biochemistry and physiology of plant growth substances. Runge, Ottawa, pp 1187–1203

Gagianas AA, Berg AR (1977) The effect of morphactin (methyl 2-chloro-9-hydroxyfluorene-9-carboxylate) on basipetal transport of indol-3-ylacetic acid in hypocotyl sections of *Phaseolus vulgaris* L. Ann Bot 41:1135–1148

Galston AW, Hillman WS (1961) The degradation of auxin. In: Ruhland W (ed) Encyclopedia of plant physiology vol 14. Springer, Berlin Göttingen Heidelberg, pp 647–670

Gee H (1972) Localization and uptake of ^{14}C-IAA in relation to xylem regeneration in *Coleus* internodes. Planta 108:1–9

Gee H, Greyson RI (1969) Ion exchange resin beads as a carrier for growth substances. Can J Bot 47:1505–1506

Gillespie B, Thimann KV (1961) The lateral transport of indoleacetic acid-C^{14} in geotropism. Experientia 17:126–129

Gillespie B, Thimann KV (1963) Transport and distribution of auxin during tropistic response. I. The lateral migration of auxin in geotropism. Plant Physiol 38:214–225

Goldsmith MHM (1959) Characteristics of the translocation of indoleacetic acid in the coleoptile of *Avena*. PhD Thesis, Radcliffe Coll, Cambridge, Mass

Goldsmith MHM (1966a) Movement of indoleacetic acid in coleoptiles of *Avena sativa* L. II. Suspension of polarity by total inhibition of the basipetal transport. Plant Physiol 41:15–27

Goldsmith MHM (1966b) Maintenance of polarity of auxin movement by basipetal transport. Plant Physiol 41:749–754

Goldsmith MHM (1967a) Separation of transit of auxin from uptake: average velocity and reversible inhibition by anaerobic conditions. Science 156:661–663

Goldsmith MHM (1967b) Movement of pulses of labeled auxin in corn coleoptiles. Plant Physiol 42:258–263

Goldsmith MHM (1968a) Comparison of aerobic and anaerobic movement of 3-indoleacetic acid in coleoptiles of oats and corn. In: Wightman F, Setterfield G (eds) Biochemistry and physiology of plant growth substances. Runge, Ottawa, pp 1037–1050

Goldsmith MHM (1968b) The transport of auxin. Annu Rev Plant Physiol 19:347–360

Goldsmith MHM (1969) Transport of plant growth regulators. In: Wilkins MB (ed) The physiology of plant growth and development. McGraw-Hill, London, pp 127–162

Goldsmith MHM (1977) The polar transport of auxin. Annu Rev Plant Physiol 28:439–478

Goldsmith MHM (1982) A saturable site responsible for polar transport of indole-3-acetic acid in sections of maize coleoptiles. Planta 155:68–75

Goldsmith MHM, Goldsmith TH (1981) Quantitative predictions for the chemiosmotic uptake of auxin. Planta 153:25–33

Goldsmith MHM, Ray PM (1973) Intracellular localization of the active process in polar transport of auxin. Planta 111:297–314

Goldsmith MHM, Thimann KV (1962) Some characteristics of movement of indoleacetic acid in coleoptiles of *Avena*. I. Uptake, destruction, immobilization, and distribution of IAA during basipetal translocation. Plant Physiol 37:492–505

Goldsmith MHM, Wilkins MB (1964) Movement of auxin in coleoptiles of *Zea mays* L. during geotropic stimulation. Plant Physiol 39:151–162

Goldsmith MHM, Cataldo DA, Karn J, Brenneman T, Trip P (1974) The rapid non-polar transport of auxin in the phloem of intact *Coleus* plants. Planta 116:301–317

Goldsmith MHM, Goldsmith TH, Martin MH (1981) Mathematical analysis of the chemosmotic polar diffusion of auxin through plant tissues. Proc Natl Acad Sci USA 78:976–980

Gordon ME, Letham DS, Parker CW (1974a) Regulators of cell division in plant tissues XVIII. The metabolism and translocation of zeatin in intact radish seedlings. Ann Bot 38:809–825

Gordon ME, Wilson MM, Parker CW, Letham DS (1974b) The metabolism of cytokinins by radish seedlings. In: Bieleski RL, Ferguson AR, Cresswell MM (eds) Mechanism of regulation of plant growth. Roy Soc N Z, Wellington, pp 773–780

Gordon SA, Eib M (1956) Auxin transport in the phototropic response (abstract). Plant Physiol Suppl 31:XIV
Gorter CJ (1932) Groeistofproblemen bij Wortels. Diss Univ Utrecht
Grahm L (1964) Measurements of geoelectric and auxin-induced potentials in coleoptiles with refined vibrating electrode technique. Physiol Plant 17:231–261
Grahm L (1971) A vibrating-reed electrometer for the measurement of transverse and longitudinal potential differences in plants. In: Gordon SA, Cohen MJ (eds) Gravity and the organism. Univ Chicago Press, Chicago London, pp 159–162
Grahm L, Hertz CH (1962) Measurements of the geoelectric effect in coleoptiles by a new technique. Physiol Plant 15:96–114
Grahm L, Hertz CH (1964) Measurements of the geoelectric effect in coleoptiles. Physiol Plant 17:186–201
Greene DW (1975) Cytokinin activity in xylem sap and extracts of MM 106 apple rootstocks. Hort Science 10:73–74
Greenwood MS, Goldsmith MHM (1970) Polar transport and accumulation of indole-3-acetic acid during root regeneration by *Pinus lambertiana* embryos. Planta 95:297–313
Greenwood MS, Hillman JR, Shaw S, Wilkins MB (1973) Localisation and identification of auxin in roots of *Zea mays*. Planta 109:369–374
Gregory FG, Hancock CR (1955) The rate of transport of natural auxin in woody shoots. Ann Bot 19:451–465
Grochowska MJ (1968) Translocation of indole-3-acetic acid-2-^{14}C injected into seeds of five-week-old apple fruits. Bull Acad Pol Sci 16:577–580
Guern J, Sadorge P (1967) Polarité du transport de la 6-benzylaminopurine dans les jeunes plantes étiolées de *Cicer arietinum* L. CR Acad Sci Paris Ser D 264:2106–2109
Guern J, Dorée M, Sadorge P (1968) Transport, metabolism and biological activity of some cytokinins. In: Wightman F, Setterfield G (eds) Biochemistry and physiology of plant growth substances. Runge, Ottawa, pp 1155–1167
Guttenberg H v., Zetsche K (1956) Der Einfluß des Lichtes auf die Auxinbildung und den Auxintransport. Planta 40:99–134
Hall PL, Bandurski RS (1978) Movement of indole-3-acetic acid and tryptophan-derived indole-3-acetic acid from the endosperm to the shoot of *Zea mays* L. Plant Physiol 61:425–429
Hall SM, Baker DA (1972) The chemical composition of *Ricinus* phloem exudate. Planta 106:131–140
Hall SM, Medlow GC (1974) Identification of IAA in phloem and root pressure saps of *Ricinus communis* L. by mass spectrometry. Planta 119:257–261
Halliday MBW, Wangermann E (1972a) Leaf abscission in *Coleus*. I. Abscission zone formation and the effect of auxin on abscission. New Phytol 71:649–663
Halliday MBW, Wangermann E (1972b) Leaf abscission in *Coleus*. II. The distribution and fate of [^{14}C]-indolylacetic acid in the petiole. New Phytol 71:665–670
Hare RC (1964) Indoleacetic acid oxidase. Bot Rev 30:129–165
Harel S (1969) Modification of 2,4-dichlorophenoxyacetic acid movement in bean petioles by light. Plant Physiol 44:615–617
Hartung W (1975) Die Wirkung von niedriger Temperatur, Stickstoffatmosphäre und Entkopplern auf die longitudinale Radioaktivitätsverteilung nach akropetaler und basipetaler Applikation von [2-^{14}C]-ABA in langen Wurzelsegmenten von *Phaseolus coccineus* L. Planta 126:173–179
Hartung W (1976a) Der basipetale [2-^{14}C]Abscisinsäuretransport in Wurzeln intakter Bohnenkeimlinge und seine Bedeutung für den Wurzelgeotropismus. Planta 128:59–62
Hartung W (1976b) Effect of water stress on transport of [2-^{14}C]abscisic acid in intact plants of *Phaseolus coccineus* L. Oecologia 26:177–183
Hartung W (1977) Der Transport von [2-^{14}C]Abscisinsäure aus dem Wurzelsystem intakter Bohnenkeimlinge in die oberirdischen Teile der Pflanze. Z Pflanzenphysiol 83:81–84
Hartung W, Behl R (1974) Transport und Stoffwechsel von 2-[^{14}C]-Abscisinsäure in Wurzelsegmenten von *Phaseolus coccineus* L. Planta 120:299–305
Hartung W, Behl R (1975a) Lokalisation des akropetalen Transports von 2-[^{14}C]Abscisinsäure und Hinweis für einen Radialtransport von ABA zwischen Zentralzylinder und Rindenzylinder in Wurzeln von *Phaseolus coccineus* L. Planta 122:53–59

Hartung W, Behl R (1975b) Die Wirkung von Licht auf den Transport von 2-[^{14}C]Abscisinsäure in Bohnenwurzelsegmenten. Planta 122:61–65

Hartung W, Phillips IDJ (1974) Basipetally polarised transport of [^{3}H]gibberellin A_1 and [^{14}C]gibberellin A_3, and acropetal polarity of [^{14}C]indole-3-acetic acid transport in stelar tissues of *Phaseolus coccineus* roots. Planta 118:311–322

Hasenstein K-H (1982) Transport und Immobilisierung von Auxin im Hypokotyl von *Helianthus annuus* L. Eine experimentelle und mathematische Analyse. Diss Univ Saarbrücken

Hasenstein K-H, Kaldewey H (1983) Movement of auxin pulses through hypocotyls of *Helianthus annuus;* a mathematical analysis and deduced conclusions. Proc 3rd Int Symp Plant Growth Regulators, Oct 1981, Varna, Bulgaria, pp. 65–73

Henson IE, Wheeler CT (1977) Hormones in plants bearing nitrogen-fixing root nodules: Distribution and seasonal changes in levels of cytokinins in *Alnus glutinosa* (L.) Gaertn. J Exp Bot 28:205–214

Hertel R (1962) Der Auxintransport in der Koleoptile von *Zea mays* L. Diss Ludwig Maximilians-Univ, München

Hertel R (1983) The mechanism of auxin transport as a model for auxin action. Z. Pflanzenphysiol 112:53–67

Hertel R, Flory R (1968) Auxin movement in corn coleoptiles. Planta 82:123–144

Hertel R, Leopold AC (1962) Auxintransport und Schwerkraft. Naturwissenschaften 49:377–378

Hertel R, Leopold AC (1963) Versuche zur Analyse des Auxintransports in der Koleoptile von *Zea mays* L. Planta 59:535–562

Hertel R, Evans ML, Leopold AC, Sell HM (1969) The specificity of the auxin transport system. Planta 85:238–249

Hertel R, Thomson K-St, Russo VEA (1972) In vitro auxin binding to particulate cell fractions from corn coleoptiles. Planta 107:325–340

Hertz CH (1971) Bioelectric phenomena in graviperception. In: Gordon SA, Cohen MJ (eds) Gravity and the organism. Univ Chicago Press, Chicago London, pp 151–158

Hewett EW (1973) Cytokinins in woody plants. Ph D Diss Univ Wales

Hewett EW, Wareing PF (1973a) Cytokinins in *Populus × robusta:* Changes during chilling and bud burst. Physiol Plant 28:393–399

Hewett EW, Wareing PF (1973b) Cytokinins in *Populus × robusta:* Qualitative changes during development. Physiol Plant 29:386–389

Hewett EW, Wareing PF (1974) Cytokinin changes during chilling and bud burst in woody plants. In: Bieleski RL, Ferguson AR, Cresswell MM (eds) Mechanism of regulation of plant growth. Roy Soc N Z, Wellington, pp 693–701

Higinbotham M (1973) Electropotentials of plant cells. Annu Rev Plant Physiol 24:25–46

Hillman SK, Phillips IDJ (1970) Transport and metabolism of indol-3yl-(acetic acid-2-^{14}C) in pea roots. J Exp Bot 21:959–967

Hoad GV (1967) (+)-Abscisin II in phloem exudate of willow. Life Sci 6:1113–1118

Hoad GV (1973a) Hormones in the phloem of higher plants. Proc Res Inst Pomology Skierniewice, Poland Ser E 3:17–30

Hoad GV (1973b) Effect of moisture stress on abscisic acid levels in *Ricinus communis* L. with particular reference to phloem exudate. Planta 113:367–372

Hoad GV (1975) Effect of osmotic stress on abscisic acid levels in xylem sap of sunflower (*Helianthus annuus* L.). Planta 124:25–29

Hoad GV, Bowen MR (1968) Evidence for gibberellin-like substances in the phloem exudate of higher plants. Planta 82:22–32

Hoad GV, Hillman SK, Wareing PF (1971) Studies on the movement of indole auxins in willow (*Salix viminalis* L.). Planta 99:73–88

Hocking TJ, Hillman JR, Wilkins MB (1972) Movement of abscisic acid in *Phaseolus vulgaris* plants. Nature New Biol 235:124–125

Hollis CA, Tepper HB (1971) Auxin transport within intact dormant and active white ash shoots. Plant Physiol 48:146–149

Horgan R, Hewett EW, Purse JG, Horgan JM, Wareing PF (1973) Identification of a cytokinin in sycamore sap by gas chromatography-mass spectrometry. Plant Sci Lett 1:321–324

Huber B, Schmidt E, Jahnel H (1937) Untersuchungen über den Assimilatstrom. I. Mitteilung aus der sächsischen fortstlichen Versuchsanstalt Tharandt, Abteilung Botanik. Tharandt Forstl Jahrb 88:1017–1050

Ingersoll RB, Smith OE (1970) Movement of (RS)-abscisic acid in the cotton explant. Plant Physiol 45:576–578

Ingersoll RB, Smith OE (1971) Transport of abscisic acid. Plant Cell Physiol 12:301–309

Itai C, Ben-Zioni A (1974) Regulation of plant responses to high temperature. In: Bieleski RL, Ferguson AR, Cresswell MM (eds) Mechanism of regulation of plant growth. Roy Soc N Z Wellington, pp 477–482

Itai C, Vaadia Y (1965) Kinetin-like activity in root exudate of water-stressed sunflower plants. Physiol Plant 18:941–944

Itai C, Vaadia Y (1971) Cytokinin activity in water-stressed shoots. Plant Physiol 47:87–90

Itai C, Richmond A, Vaadia Y (1968) The role of root cytokinins during water and salinity stress. Isr J Bot 17:187–193

Itai C, Ben-Zioni A, Ordin L (1973) Correlative changes in endogenous hormone levels and shoot growth induced by short heat treatments to the root. Physiol Plant 29:355–360

Iwami S, Masuda Y (1976) Distribution of labeled auxin in geotropically stimulated stems of cucumber and pea. Plant Cell Physiol 17:227–237

Jacobs M, Gilbert SF (1983) Basal localization of the presumptive auxin transport carrier in pea stem cells. Science 220:1297–1300

Jacobs M, Hertel R (1978) Auxin binding to subcellular fractions from *Cucurbita* hypocotyls: in vitro evidence for an auxin transport carrier. Planta 142:1–10

Jacobs M, Ray PM (1976) Rapid auxin-induced decrease in free space pH and its relationship to auxin-induced growth in maize and pea. Plant Physiol 58:203–209

Jacobs WP (1950) Auxin-transport in the hypocotyl of *Phaseolus vulgaris* L. Am J Bot 37:551–555

Jacobs WP (1951) Auxin relationships in an intercalary meristem: further studies on the gynophore of *Arachis hypogaea* L. Am J Bot 38:307–310

Jacobs WP (1952) The role of auxin in differentiation of xylem around a wound. Am J Bot 39:301–309

Jacobs WP (1954) Acropetal auxin transport and xylem regeneration — a quantitive study. Am Natur 88:327–337

Jacobs WP (1961) The polar movement of auxin in the shoots of higher plants: its occurrence and physiological significance. In: Klein RM (ed) Plant growth regulation. Iowa State Univ Press Ames, pp 397–409

Jacobs WP (1964) The role of native growth substances in controlling the shedding of organs and abscission. Proc 16th Int Hortic Congr 1962, 5:619–625

Jacobs WP (1967) Comparison of the movement and vascular differentiation effects of the endogenous auxin and of phenoxyacetic weed-killers in stems and petioles of *Coleus* and *Phaseolus*. Ann NY Acad Sci 144:102–117

Jacobs WP (1972) The movement of plant hormones: auxins, gibberellins, and cytokinins. In: Carr DJ (ed) Plant growth substances 1970. Springer, Berlin Heidelberg New York, pp 701–709

Jacobs WP (1976) Apolar movement of zeatin through *Coleus* petioles and *Pisum* roots as estimated by bioassay and radioactive labelling. Am J Bot 63:571–577

Jacobs WP (1977) Polarity of indoleacetic acid in young *Coleus* stems. Plant Physiol 60:95–97

Jacobs WP (1978a) Does the induction of flowering by photoperiodic change the polarity or other characteristics of IAA transport in petioles for the short-day plant *Xanthium?* Plant Physiol 61:307–310

Jacobs WP (1978b) Regulation of development by the differential polarity of various hormones as well as by effects of one hormone on the polarity of another. In: Schütte HR, Gross D (eds) Regulation of developmental processes in plants. Fischer, Jena, pp 361–380

Jacobs WP (1979) Plant hormones and plant development. Cambridge Univ Press, Cambridge London

Jacobs WP, De Muth PJ (1977) Non-polar movement of thiamine in stems of tomato and roots of pea. Bot Gaz 138:266–269

Jacobs WP, Kaldewey H (1970) Polar movement of gibberellic acid through young *Coleus* petioles. Plant Physiol 45:539–541

Jacobs WP, McCready CC (1967) Polar transport of growth regulators in pith and vascular tissues of *Coleus* stems. Am J Bot 54:1035–1040

Jacobs WP, Pruett P (1972) The polar movement of gibberellin through *Coleus* petioles. In: Kaldewey H, Vardar Y (eds) Hormonal regulation of plant growth and development. Verlag Chemie, Weinheim, pp 45–55

Jacobs WP, Pruett P (1973) The time-course of polar movement of gibberellin through *Zea* roots. Am J Bot 60:896–900

Jacobs WP, McCready CC, Osborne DJ (1966) Transport of the auxin 2,4-dichlorophenoxyacetic acid through abscission zones, pulvini, and petioles of *Phaseolus vulgaris*. Plant Physiol 41:725–730

Johnsson A (1965) Photoinduced lateral potentials in *Zea mays*. Physiol Plant 18:574–576

Johnsson A (1967) Relationships between photo-induced and gravity-induced electrical potentials in *Zea mays*. Physiol. Plant 20:562–579

Jones OP (1973) Effect of cytokinins in xylem sap from apple trees on apple shoot growth. J Hortic Sci 48:181–188

Jones OP, Lacey HJ (1968) Gibberellin-like substances in the transpiration stream of apple and pear trees. J Exp Bot 19:526–531

Jones RL, Lang A (1968) Extractable and diffusible gibberellins from light- and dark-grown pea seedlings. Plant Physiol 43:629–634

Kaldewey H (1957) Wachstumsverlauf, Wuchsstoffbildung und Nutationsbewegungen von *Fritillaria meleagris* L. im Laufe der Vegetationsperiode. Planta 49:300–344

Kaldewey H (1962) Plagio- und Diageotropismus der Sprosse und Blätter, einschließlich Epinastie, Hyponastie, Entfaltungsbewegungen. In: Ruhland W (ed) Encyclopedia of plant physiology Vol 17/2. Springer, Berlin Göttingen Heidelberg, pp 200–321

Kaldewey H (1963) Geschwindigkeit, Intensität und Kapazität des Wuchsstofftransports in geotropisch gereizten Fruchtstielen der Schachblume *Fritillaria meleagris* L. Planta 60:178–204

Kaldewey H (1964) Papier- und dünnschichtchromatographische Trennung, Farbreaktionen und biologischer Test einiger Wachstumsregulatoren aus Fruchtstielen von *Fritillaria meleagris* L. In: Nitsch JP (ed) Régulateurs naturels de la croissance végétale. CNRS, Paris, pp 421–443

Kaldewey H (1965a) Wuchsstofftransport, Temperatur und Pflanzenalter. Ber Dtsch Bot Ges 78:(128)–(143)

Kaldewey H (1965b) Thermonastische Bewegungen der Sproßachsen von *Fritillaria meleagris* L. in Abhängigkeit von Wachstum und Wuchsstoff. Planta 67:55–74

Kaldewey H (1966) Basipetal transport of auxins and their detection by thin-layer chromatography. Proc Int Symp Plant Stimulation. Sofia, pp 211–221

Kaldewey H (1967a) Weitere Beobachtungen zum basipetalen Auxintransport in Fruchtstielen von *Fritillaria meleagris* L. Intensität, Kapazität und Geschwindigkeit des Transports von Indol-3-essigsäure in Abhängigkeit vom Pflanzenalter. Ber Dtsch Bot Ges 80:238–251

Kaldewey H (1967b) Auxintransport in ungereizten Pflanzen. Wiss Z Univ Rostock, Math-Nat Reihe 16:487–494

Kaldewey H (1968a) Auxin transport: general remarks concerning the terminology and the methods. In: Vardar Y (ed) The transport of plant hormones. North-Holland, Amsterdam, pp 1–23

Kaldewey H (1968b) Transport und Verteilung von Indol-3-(essigsäure-2-^{14}C) in nickenden Sproßachsen. Vortr Gesamtgeb Bot 2:90–106

Kaldewey H (1971) Geoepinasty, an example of gravimorphism. In: Gordon SA, Cohen MJ (eds) Gravity and the organism. Univ Chicago Press, Chicago London, pp 333–339

Kaldewey H (1976) Considerations of geotropism in plants. In: Sneath PHA (ed) Life sciences and space research Vol 14. Akademie Verlag, Berlin, pp 21–36

Kaldewey H, Ginkel U (1974) Influence of lycoricidinol on uptake and transport of

IAA-2-^{14}C in sections cut from different regions of the hypocotyl of light-grown seedlings of *Citrullus edulis*. In: Plant growth substances 1973. Hirokawa, Tokyo, pp 1079–1089

Kaldewey H, Jacobs WP (1974) Movement and distribution of ^{14}C within petioles of intact *Coleus blumei* (Labiatae) after application of indole-3-(acetic acid-2-^{14}C) to the leaf surface. Am J Bot 61:699–706

Kaldewey H, Jacobs WP (1975) The influence of humidity and gibberellic acid on movement and immobilization of radiocarbon within petioles of *Coleus blumei* Benth. after application of indole-3-(acetic acid-2-^{14}C) to leaves of intact plants. Biochem Physiol Pflanz 168:401–410

Kaldewey H, Kraus L (1972) Translocation and immobilization of radiocarbon in the hypocotyl and the primary root of *Gossypium hirsutum* L. after application of IAA-2-^{14}C to intact light-grown seedlings. In: Kaldewey H, Vardar Y (eds) Hormonal regulation in plant growth and development. Verlag Chemie, Weinheim, pp 137–153

Kaldewey H, Ginkel U, Lehmann I, Seiwert R (1973) Transport and immobilization of indoleacetic acid as affected by morphactins. I. Time course of auxin transport in sections excised from different hypocotyl regions of light-grown seedlings of *Citrullus edulis*. Proc Res Inst Pomol Skierniewice Pol Ser E Conf Symp 3:215–226

Kaldewey H, Ginkel U, Wawczyniak G (1974) Auxin transport and water stress in pea (*Pisum sativum* L.). Ber Dtsch Bot Ges 87:563–576

Kaldewey H, Ginkel U, Karmann R, Paland I (1977) Interaction in translocation of growth regulators in shoots with special reference to auxin transport as affected by two inhibitors, a morphactin and lycoricidinol. In: Pilet PE (ed) Plant growth regulation. Springer, Berlin Heidelberg New York, pp 91–99

Katekar GF (1976) Inhibitors of the geotropic response in plants: a correlation of molecular structures. Phytochemistry 15:1421–1424

Katekar GF, Geissler AE (1975) Auxin transport inhibitors. Fluorescein and related compounds. Plant Physiol 56:645–646

Katekar GF, Geissler AE (1977a) Auxin transport inhibitors II. 2-(1-Pyrenoyl)benzoic acid, a potent inhibitor of polar auxin transport. Aust J Plant Physiol 4:321–325

Katekar GF, Geissler AF (1977b) Auxin transport inhibitors III. Chemical requirements of a class of auxin transport inhibitors. Plant Physiol 60:826–829

Kato J (1958) Nonpolar transport of gibberellin through pea stem and a method for its determination. Science 128:1008–1009

Keitt GW, Baker RA (1966) Auxin activity of substituted benzoic acids and their effect on polar auxin transport. Plant Physiol 41:1561–1569

Kende H (1964) Preservation of chlorophyll in leaf sections by substances obtained from root exudate. Science 145:1066–1067

Kende H (1965) Kinetin-like factors in the root exudate of sunflowers. Proc Natl Acad Sci USA 53:1302–1307

Kende H, Gardner GM (1976) Hormone binding in plants. Annu Rev Plant Physiol 27:267–290

Kende H, Sitton D (1967) The physiological significance of kinetin- and gibberellin-like root hormones. Ann NY Acad Sci 144:235–243

Kentzer T, Libbert E (1961) Blockade des Gibberellinsäuretransports in Hypocotylsegmenten durch Trijodbenzoesäure. Zugleich ein neuer Agarblocktest auf Gibberellin. Planta 56:23–27

Kirk SC, Jacobs WP (1968) Polar movement of indole-3-acetic acid-^{14}C in the roots of *Lens* and *Phaseolus*. Plant Physiol 43:675–682

Klämbt HD (1968) Cytokinine aus *Helianthus annuus*. Planta 82:170–178

Kluge M, Reinhard E, Ziegler H (1964) Gibberellinaktivität von Siebröhrensäften. Naturwissenschaften 51:145–146

Koevenig JL (1973) Nonpolar movement of N_6-benzyladenine-^{14}C in coleoptile, stem, petiole and floral organ sections. Can J Bot 51:2079–2083

Koevenig JL, Jacobs WP (1972) Effect of light on basipetal movement of indoleacetic acid in green stem sections of *Coleus*. Plant Physiol 49:866–867

Koevenig JL, Sillix D (1973) Movement of IAA in sections from spider flower (*Cleome hassleriana*) stamen filaments. Am J Bot 60:231–235

Konings H (1968) Auxin in root geotropism. In: Vardar Y (ed) The transport of plant hormones. North-Holland, Amsterdam, pp 237–250

Konings H (1969) The influence of acropetally transported indoleacetic acid on the geotropism of intact pea roots and its modification by 2,3,5-triiodobenzoic acid. Acta Bot Neerl 18:528–537

Konings H, Gayadin AP (1971) Transport, binding, and decarboxylation of carboxyl-labeled IAA-^{14}C in intact pea roots. Acta Bot Neerl 20:646–654

Kozlowski TT (ed) (1973) Shedding of plant parts. Academic Press, London New York

Krelle E, Libbert E (1968) Inhibition of the polar auxin transport by a morphactin. Planta 80:317–320

Krul WR (1977) Differences in velocity of auxin movement obtained by intercept and peak arrival methods. Physiol Plant 41:259–264

Krul WR, Colclasure GC (1977) Effect of galactose and other monosaccharides on IAA movement in bean hypocotyl segments. Physiol Plant 41:249–253

Kruszewski SP, Jacobs WP (1974) Polarity of thiamine movement through tomato petioles. Plant Physiol 54:310–311

Kulaeva ON (1962) The effects of roots on leaf metabolism in relation to the action of kinetin on leaves. Sov Plant Physiol,9:181–189 (Engl transl)

Kuse G (1953) Effect of triiodobenzoic acid on the growth of lateral bud and on tropism of petiole. Mem Coll Sci Kyoto B 20:207–215

Lagerstedt HB, Langston RG (1966) Transport of kinetin-8-^{14}C in petioles. Physiol Plant 19:734–740

Lagerstedt HB, Langston RG (1967) Translocation of radioactive kinetin. Plant Physiol 42:611–622

Laibach F (1933) Wuchsstoffversuche mit lebenden Orchideenpollinien. Ber Dtsch Bot Ges 51:336–340

Lamb CJ (1976) Hormone binding in plants. Nature 274:312–314

Larsen P (1955) Growth substances in higher plants. In: Paech K, Tracey MV (eds) Modern methods of plant analysis Vol 3. Springer, Berlin Göttingen Heidelberg, pp 565–625

Lavender DP, Sweet GB, Zaerr JB, Hermann RK (1973) Spring shoot growth in Douglas-fir may be initiated by gibberellins exported from the roots. Science 182:838–839

Lembi CA, Morré DJ, Thomson KSt, Hertel R (1971) 1-N-naphthylphthalamic-acid (NPA)-binding activity of a plasma membrane-rich fraction from maize coleoptiles. Planta 99:37–45

Lenton JR, Bowen MR, Saunders PF (1968) Detection of abscisic acid in the xylem sap of willow by gas chromatography. Nature 220:86–87

Leopold AC (1961) The transport of auxin. In: Ruhland W (ed) Encyclopedia of plant physiology Vol 14. Springer, Berlin Göttingen Heidelberg, pp 671–682

Leopold AC (1963) The polarity of auxin transport. Brookhaven Symp Biol 16:218–234

Leopold AC, de la Fuente RK (1967) The polarity of auxin transport. Ann NY Acad Sci 144:94–101

Leopold AC, de la Fuente RK (1968) A view of polar auxin transport. In: Vardar Y (ed) The transport of plant hormones. North-Holland, Amsterdam, pp 24–47

Leopold AC, Guernsey FS (1953) Auxin polarity in the *Coleus* plant. Bot Gaz 115:147–154

Leopold AC, Hall OF (1966) A mathematical model of polar auxin transport. Plant Physiol 41:1476–1480

Leopold AC, Lam SL (1961) Polar transport of three auxins. In: Klein RM (ed) Plant growth regulation. Iowa State Univ Press, Ames, pp 411–418

Leopold AC, Lam SL (1962) The auxin transport gradient. Physiol Plant 15:631–638

Lepp LW, Peel AJ (1971) Patterns of translocation and metabolism of ^{14}C-labelled IAA in the phloem of willow. Planta 96:62–73

Letham DS, Goodwin PB, Higgins TJV (eds) (1978) Phytohormones and related compounds — a comprehensive treatise Vol I. The biochemistry of phytohormones and related compounds. Vol II. Phytohormones and the development of higher plants. Elsevier/North-Holland Biomedical Press, Amsterdam New York

Liao SH, Hamilton RH (1966) Intracellular localization of growth hormones in plants. Science 151:822–824

Libbert E, Gerdes I (1964) Können Gibberelline an tropistischen Krümmungen beteiligt sein? (Untersuchungen über den Einfluß von Gibberellinsäure auf Krümmungsreaktionen und über den Quertransport von Gibberellinsäure in der Pflanze). Planta 61:245–258

Little CHA, Goldsmith MHM (1967) Effect of inversion on growth and movement of indole-3-acetic acid in coleoptiles. Plant Physiol 42:1239–1245

Little ECS, Blackman GE (1963) The movement of growth regulators in plants. III. Comparative studies of transport in *Phaseolus vulgaris*. New Phytol 62:173–197

Loeb J (1917) Influence of leaf upon root formation and geotropic curvature in the stem of *Bryophyllum calycinum* and the possibility of a hormone theory of these processes. Bot Gaz 63:25–50

Loeffler JE, van Overbeek J (1964) Kinin activity in coconut milk. In: Nitsch JP (ed) Régulateurs naturels de la croissance végétale. CNRS, Paris, pp 77–82

Luckwill LC, Whyte P (1968) Hormones in the xylem sap of apple trees. Soc Chem Ind (Lond) Monogr 31:87–101

Lund EJ (1947) Bioelectric fields and growth. Univ Texas Press, Austin

Lyon CJ (1963a) Auxin factor in branch epinasty. Plant Physiol 38:145–152

Lyon CJ (1963b) Auxin transport in leaf epinasty. Plant Physiol 38:567–574

Lyon CJ (1965a) Auxin transport in geotropic curvatures of a branched plant. Plant Physiol 40:18–24

Lyon CJ (1965b) Action of gravity on basipetal transport of auxin. Plant Physiol 40:953–961

Mai G (1934) Korrelationsuntersuchungen an entspreiteten Blattstielen mit lebenden Orchideenpollinien als Wuchsstoffquelle. Jahrb Wiss Bot 79:681–713

Maravolo NC (1976) Polarity and localization of auxin movement in the hepatic *Marchantia polymorpha*. Am J Bot 63:526–531

Marigo G, Boudet AM (1977) Relations polyphénols-croissance: Mise en évidence d'un effect inhibiteur des composés phénoliques sur le transport polarisé de l'auxine. Physiol Plant 41:197–202

Marré E, Lado P, Ferroni A, Ballarin Denti A (1974) Transmembrane potential increase induced by auxin, benzyladenine and fusicoccin. Correlation with proton extrusion and cell enlargement. Plant Sci Lett 2:257–265

Martin HV, Elliott MC, Wangermann E, Pilet PE (1978) Auxin gradient along the root of the maize seedling. Planta 141:179–181

McComb AJ (1964) The stability and movement of gibberellic acid in pea seedlings. Ann Bot 28:669–687

McCready CC (1963) Movement of growth regulators in plants. I. Polar transport of 2,4-dichlorophenoxyacetic acid in segments from petioles of *Phaseolus vulgaris*. New Phytol 62:3–18

McCready CC (1966) Translocation of growth regulators. Annu Rev Plant Physiol 17:283–294

McCready CC (1968a) The polarity of auxin movement in segments excised from petioles of *Phaseolus vulgaris* L. In: Wightman F, Setterfield G (eds) Biochemistry and physiology of plant growth substances. Runge, Ottawa, pp 1005–1023

McCready CC (1968b) The acropetal movement of auxin through segments excised from petioles of *Phaseolus vulgaris* L. In: Vardar Y (ed) The transport of plant hormones. North-Holland, Amsterdam, pp 108–129

McCready CC, Jacobs WP (1963a) Movement of growth regulators in plants. II. Polar transport of radioactivity from indoleacetic acid-^{14}C and 2,4-dichlorophenoxyacetic acid-^{14}C in petioles of *Phaseolus vulgaris*. New Phytol 62:19–34

McCready CC, Jacobs WP (1963b) Movement of growth regulators in plants. IV. Relationship between age, growth, and polar transport in petioles of *Phaseolus vulgaris*. New Phytol 62:360–366

McCready CC, Jacobs WP (1967) Movement of growth regulators in plants. V. A further note on the relationship between polar transport and growth. New Phytol 66:485–488

Menary RC, van Staden J (1976) Effect of phosphorus nutrition and cytokinin on flowering in the tomato *Lycopersicon esculentum* Mill. Aust J Plant Physiol 3:201–205
Menhenett R, Wareing PF (1975) Possible involvement of growth substances in response of tomato plants (*Lycopersicon esculentum* Mill.) to different soil temperatures. J Hortic Sci 50:381–397
Miginiac E, Sossountzov L, Dugue N (1978) Modalités du transport et du métabolisme de l'AIA ^{14}C ou ^{3}H en relation avec la morphogenèse des bourgeons axillaires chez le *Scrophularia arguta*. Physiol Plant 44:335–344
Milborrow BV (1968) Identification and measurement of (+)-abscisic acid in plants. In: Wightman F, Setterfield G (eds) Biochemistry and physiology of plant growth substances. Runge, Ottawa, pp 1531–1545
Milborrow BV (1969) The occurrence and function of abscisic acid in plants. Sci Progr 57:533–558
Mitchell EK, Davies PJ (1975) Evidence for three different systems of movement of indoleacetic acid in intact roots of *Phaseolus coccineus*. Physiol Plant 33:290–294
Mitchinson GJ (1980) The dynamics of auxin transport. Proc R Soc Lond B 209:489–511
Mittelheuser CJ, van Steveninck RFM (1971) Rapid action of abscisic acid on photosynthesis and stomatal resistance. Planta 97:83–86
Morath M (1972) Some early effects of auxin and of geotropic exposure on coleoptiles. In: Kaldewey H, Vardar Y (eds) Hormonal regulation in plant growth and development. Verlag Chemie, Weinheim, pp 377–381
Morgan DG (1964) Influence of α-naphthylphthalamic acid on the movement of indolyl-3-acetic acid in plants. Nature 201:476–477
Morgan DG, Söding H (1958) Über die Wirkungsweise von Phthalsäuremono-α-naphthylamid (PNA) auf das Wachstum der Haferkoleoptile. Planta 52:235–249
Morris DA (1970) Light and the transport and metabolism of indoleacetic acid in normal and albino dwarf pea seedlings. Planta 91:1–7
Morris DA (1977) Transport of endogenous auxin in two-branched dwarf pea seedlings (*Pisum sativum*). Some implications for polarity and apical dominance. Planta 136:91–96
Morris DA (1979) Effects of temperature on the velocity of exogenous auxin transport in intact chilling-sensitive and chilling-resistant plants. Planta 146:603–605
Morris DA, Kadir GO (1972) Pathway of auxin transport in the intact pea seedling (*Pisum sativum* L.). Planta 107:171–182
Morris DA, Thomas AG (1974) The specificity of auxin transport in intact pea seedlings (*Pisum sativum* L.). Planta 118:225–234
Morris DA, Thomas AG (1978) A microautoradiographic study of auxin transport in the stem of intact pea seedlings (*Pisum sativum* L.). J Exp Bot 29:147–157
Morris DA, Winfield PJ (1972) Kinetin transport to axillary buds of dwarf pea (*Pisum sativum* L.). J Exp Bot 23:346–355
Morris DA, Briant RE, Thomson PG (1969) The transport and metabolism of ^{14}C-labelled indoleacetic acid in intact pea seedlings. Planta 89:178–197
Morris DA, Kadir GO, Barry AJ (1973) Auxin transport in intact pea seedlings (*Pisum sativum* L.): The inhibition of transport by 2,3,5-triiodobenzoic acid. Planta 110:173–182
Morris RO, Zaerr JB, Chapman RW (1976) Trace enrichment of cytokinins from Douglas-fir xylem exudate. Planta 131:271–274
Mothes K (1964) The role of kinetin in plant regulation. In: Nitsch JP (ed) Régulateurs naturels de la croissance végétale. CNRS, Paris, pp 131–140
Mothes K, Engelbrecht L (1961) Kinetin-induced directed transport of substances in excised leaves in the dark. Phytochemistry 1:58–62
Mozes R, Altman A (1977) Characteristics of root-to-shoot transport of cytokinin 6-benzylaminopurine in intact seedlings of *Citrus aurantium*. Physiol Plant 39:225–232
Murakami Y (1968) Gibberellin-like substances in roots of *Oryza sativa*, *Pharbitis nil* and *Ipomoea batatas* and the site of their synthesis in the plant. Bot Mag (Tokyo) 81:334–343
Murphy GJP (1980) A reassessment of the binding of naphthaleneacetic acid by membrane preparations from maize. Planta 149:417–426

Myers RM (1940) Effect of growth substances on the abcission layer in leaves of *Coleus*. Bot Gaz 102:323–338

Naqvi SM (1963) Transport studies with C^{14}-indoleacetic acid and C^{14}-2,4-dichlorophenoxyacetic acid in *Coleus* stems. Ph D Thesis, Princeton Univ, Princeton

Naqvi SM (1974) The effect of sugar supplement on the kinetics of indoleacetic acid-2-^{14}C transport in *Zea mays* L. coleoptile. Z Pflanzenphysiol 71:1–5

Naqvi SM (1975) Kinetics of auxin transport in light- and in dark-grown *Zea mays* L. coleoptile segments. Z Pflanzenphysiol 76:379–383

Naqvi SM (1976) Auxin in *Zea mays* L. coleoptile segments: saturation of absorption and transport at various concentrations. Pak J Bot 8:47–52

Naqvi SM, Gordon SA (1964) Auxin transport in corn coleoptile sections. I. The effect of carbohydrate supplement on transport polarity, velocity, and capacity. Argonne Nat Lab Report 6971:190–193

Naqvi SM, Gordon SA (1965) Auxin transport in flowering and vegetative shoots of *Coleus blumei* Benth. Plant Physiol 40:116–118

Naqvi SM, Gordon SA (1966) Auxin transport in *Zea mays* L. coleoptiles. I. Influence of gravity on the transport of indoleacetic acid-2-^{14}C. Plant Physiol 41:1113–1118

Naqvi SM, Dedolph RR, Gordon SA (1965) Auxin transport and geotropical potential in corn coleoptile sections. Plant Physiol 40:966–968

Newman IA (1959) Electrical determination of transport of 3-indole acetic acid in *Avena*. Nature 184:1728–1729

Newman IA (1963) Electric potentials and auxin translocation in *Avena*. Aust J Biol Sci 16:629–646

Newman IA (1965) Distribution of indolyl-3-acetic acid labelled with carbon-14 in *Avena*. Nature 205:1336–1337

Newman IA (1970) Auxintransport in *Avena*. I. Indoleacetic acid-^{14}C distribution and speeds. Plant Physiol 46:263–272

Newman IA (1974) Kinetics of auxin transport. In: Bieleski RL, Ferguson AR, Cresswell MM (eds) Mechanisms of regulation of plant growth. Roy Soc N Z, Wellington, pp 665–672

Niedergang-Kamien E, Leopold AC (1957) Inhibitors of polar auxin transport. Physiol Plant 10:29–38

Niedergang-Kamien E, Skoog F (1956) Studies on polarity and auxin transport in plants I. Physiol Plant 9:60–73

Nielsen N (1924) Studies on the transmission of stimuli in the coleoptile of *Avena*. Dansk Bot Ark 4:1–45

Nitsch JP (1968) Natural cytokinins. Soc Chem Ind (Lond) Monogr 31:111–123

Nitsch JP, Nitsch C (1965) Présence de phytokinines et autre substances de croissance dans la sève d'*Acer saccharum* et de *Vitis vinifera*. Bull Soc Bot Fr 112:11–18

Noodén LD, Leopold AC (1978) Phytohormones and the endogenous regulation of senescence and abscission. In: Letham DS, Goodwin PB, Higgins TJV (eds) Phytohormones and related compounds: a comprehensive treatise Vol 2. Elsevier, Amsterdam New York, pp 329–369

Nowacki J, Bandurski RS (1980) *Myo*-inositol esters of indole-3-acetic acid as seed auxin precursors of *Zea mays* L. Plant Physiol 65:422–427

Ohwaki Y, Tsurumi S, Nagao M (1974) Auxin transport in *Vicia* roots. In: Plant growth substances 1973. Hirokawa, Tokyo, pp 1071–1078

Osborne DJ (1968) A theoretical model for polar auxin transport. In: Vardar Y (ed) The transport of plant hormones. North-Holland, Amsterdam, pp 97–107

Osborne DJ (1974) Hormones and the shedding of leaves and bolls. Cotton Grow Rev 51:256–365

Osborne DJ, Black MK (1964) Polar transport of a kinin, benzyladenine. Nature 201:97

Osborne DJ, McCready CC (1965) Transport of the kinin, N_6-benzyladenine: Non-polar or polar? Nature 206:679–680

Osborne DJ, Mullins MG (1969) Auxin, ethylene and kinetin in a carrier-protein model system for the polar transport of auxins in petiole segments of *Phaseolus vulgaris*. New Phytol 68:977–991

Osborne DJ, Horton RF, Black MK (1968) Senescence in excised petiole segments:

The relevance to auxin and kinin transport. In: Vardar Y (ed) The transport of plant hormones. North-Holland, Amsterdam, pp 79–96

Ouitrakul R, Hertel R (1969) Effect of gravity and centrifugal acceleration on auxin transport in corn coleoptiles. Planta 88:233–243

Paál A (1914) Über phototropische Reizleitungen. Ber Dtsch Bot Ges 32:499–503

Paál A (1919) Über phototropische Reizleitung. Jahrb Wiss Bot 58:406–458

Patrick JW, Woodley DJ (1973) Auxin physiology of decapitated stems of *Phaseolus vulgaris* L. treated with indol-3yl-acetic acid. J Exp Bot 24:949–957

Pernet JJ, Pilet PE (1976) Indoleacetic acid movement in the root cap. Planta 128:183–184

Phillips DA, Cleland CF (1972) Cytokinin activity from the phloem sap of *Xanthium strumarium* L. Planta 102:173–178

Phillips IDJ, Hartung W (1974) Basipetal and acropetal transport of [3,4-^{3}H]gibberellin A_1 in short and long segments of *Phaseolus coccineus* second internode. Planta 116:109–121

Phillips IDJ, Hartung W (1976) Longitudinal and lateral transport of [3,4-^{3}H]gibberellin A_1 and 3-indolyl(acetic acid-2-^{14}C) in upright and geotropically responding green internode segments from *Helianthus annuus*. New Phytol 76:1–9

Phillips IDJ, Jones RL (1964) Gibberellin-like activity in bleeding sap of root systems of *Helianthus annuus* detected by a new dwarf pea epicotyl assay and other methods. Planta 63:269–278

Pickard BG, Thimann KV (1964) Transport and distribution of auxin during tropistic response. II. The lateral migration of auxin in phototropism of coleoptiles. Plant Physiol 39:341–350

Pilet PE (1951) Étude de la circulation des auxines dans la racine de *Lens culinaris* Medicus. Ber Schweiz Bot Ges 61:410–424

Pilet PE (1960) Action de la lumière sur le transport d'acide β-indolyl-acétique marqué par du C^{14}. Experientia 16:111

Pilet PE (1964) Auxin transport in roots of *Lens culinaris*. Nature 204:561–562

Pilet PE (1965) Polar transport of radioactivity from ^{14}C-labelled-β-indoleacetic acid in stems of *Lens culinaris*. Physiol Plant 18:687–702

Pilet PE (1968) In vitro and in vivo auxin and cytokinin transport. In: Wightman F, Setterfield G (eds) Biochemistry and physiology of plant growth substances. Runge, Ottawa, pp 993–1004

Pilet PE (1977) Growth inhibitors in growing and geostimulated maize roots. In: Pilet PE (ed) Plant growth regulation. Springer, Berlin Heidelberg New York, pp 115–128

Pilet PE, Guern J, Hugon E (1967) Sur le transport de la 6-benzylaminopurine-^{14}C. Physiol Vég 5:261–270

Poole RJ, Thimann KV (1964) Uptake of indole-3-acetic acid and indole-3-acetonitrile by *Avena* coleoptile sections. Plant Physiol 39:98–103

Purse JG (1975) Studies on endogenous cytokinins in plants. PhD Thesis Univ Wales

Purse JG, Horgan R, Horgan JM, Wareing PF (1976) Cytokinins of sycamore spring sap. Planta 132:1–8

Radin JW, Loomis RS (1974) Polar transport of kinetin in tissues of radish. Plant Physiol 53:348–351

Rasmussen HP, Bukovac MJ (1966) Naphthaleneacetic acid: localization in the abscission zone of the bean. Science 152:217–218

Raven JA (1975) Transport of indoleacetic acid in plant cells in relation to pH and electrical potential gradients, and its significance for polar IAA transport. New Phytol 74:163–172

Ray PM (1958) Destruction of auxin. Annu Rev Plant Physiol 9:81–118

Ray PM, Dohrmann U, Hertel R (1977) Characterization of naphthaleneacetic acid binding to receptor sites on cellular membranes of maize coleoptile tissue. Plant Physiol 59:357–364

Rayle DL, Ouitrakul R, Hertel R (1969) Effect of auxins on the auxin transport system in coleoptiles. Planta 87:49–53

Reid DM, Burrows WJ (1968) Cytokinin-like and gibberellin-like activity in the spring sap of trees. Experientia 24:189–190

Reid DM, Carr DJ (1967) Effects of a dwarfing compound, CCC, on the production and export of gibberellin-like substances in root systems. Planta 73:1–11
Reid DM, Crozier A (1971) Effects of waterlogging on the gibberellin content and growth of tomato plants. J Exp Bot 22:39–48
Reid DM, Crozier A, Harvey BMR (1969) The effects of flooding on the export of gibberellins from the root to the shoot. Planta 89:376–379
Reiff B, v. Guttenberg H (1961) Der polare Wuchsstofftransport von *Helianthus annuus* in seiner Abhängigkeit von Alter, Quellungszustand und Kohlenhydratversorgung des Gewebes. Flora 151:44–72
Reinhold L (1953) The uptake of indole-3-acetic acid by pea epicotyl segments and carrot disks. New Phytol 53:217–239
Reisener HJ (1957) Versuche zum Geotropismus mit radioaktiver β-Indolylessigsäure. Naturwissenschaften 44:120
Reisener HJ (1958) Untersuchungen über den Phototropismus der Haferkoleoptile. Z Bot 46:474–505
Reisener HJ, Simon H (1960) Weitere Geotropismus-Versuche mit radioaktiver β-Indolylessigsäure (IES-2-C^{14}). Z Bot 48:66–70
Rivier L, Pilet PE (1974) Indolyl-3-acetic acid in cap and apex of maize roots: identification and quantification by mass fragmentography. Planta 120:107–112
Robinson BJ, Forman M, Addicott FT (1968) Auxin transport and conjugation in cotton explants. Plant Physiol 43:1321–1323
Rowntree RA, Morris DA (1979) Accumulation of ^{14}C from exogenous labelled auxin in lateral root primordia of intact pea seedlings (*Pisum sativum* L.). Planta 144:463–466
Rubery PH (1977) The specificity of carrier-mediated auxin transport by suspension-cultured crown gall cells. Planta 135:275–283
Rubery PH (1978) Hydrogen ion dependence of carrier-mediated auxin uptake by suspension cultured crown gall cells. Planta 144:173–178
Rubery PH (1979) The effects of 2,4-dinitrophenol and chemical modifying reagents on auxin transport by suspension cultured crown gall cells. Planta 144:173–178
Rubery PH (1980) The mechanism of transmembrane auxin transport and its relation to the chemiosmotic hypothesis of the polar transport of auxin. In: Skoog F (ed) Plant growth substances 1979. Springer, Berlin Heidelberg New York, pp 50–60
Rubery PH (1981) Auxin receptors. Annu Rev Plant Physiol 32:569–596
Rubery PH, Sheldrake AR (1973) Effect of pH and surface charge on cell uptake of auxin. Nature 244:285–288
Rubery PH, Sheldrake AR (1974) Carrier-mediated auxin transport. Planta 118:101–121
Rubinstein B, Leopold AC (1963) Analysis of the auxin control of bean leaf abscission. Plant Physiol 38:262–267
Rudich J, Sell HM, Baker LR (1976) Transport and metabolism of ^{3}H-gibberellin A_3 in dioecious cucumber seedlings. Plant Physiol 57:734–737
Sabnis DD, Hirshberg G, Jacobs WP (1969) Radioautographic analysis of the distribution of label from ^{3}H-indoleacetic acid supplied to isolated *Coleus* internodes. Plant Physiol 44:27–36
Sachs J (1879) Über orthotrope und plagiotrope Pflanzenteile. Arb Bot Inst Würzburg 2:226–284
Sachs T, Thimann KV (1964) Release of lateral buds from apical dominance. Nature 201:939–940
Schrank AR (1951) Electrical polarity and auxins. In: Skoog F (ed) Plant growth substances. Univ Wisconsin Press, Madison, pp 123–140
Scott BIH (1967) Electric fields in plants. Annu Rev Plant Physiol 23:235–258
Scott TK (1972) Auxin and roots. Annu Rev Plant Physiol 23:235–258
Scott TK, Briggs WR (1960) Auxin relationship in the Alaska pea (*Pisum sativum*). Am J Bot 47:492–499
Scott TK, Briggs WR (1962) Recovery of native and applied auxin from the light-grown 'Alaska' pea seedlings. Am J Bot 49:1056–1063
Scott TK, Briggs WR (1963) Recovery of native and applied auxin from the dark-grown 'Alaska' pea seedlings. Am J Bot 50:652–657

Scott TK, Jacobs WP (1963) Auxin in *Coleus* stems: limitation of transport at higher concentration. Science 139:589–590
Scott TK, Jacobs WP (1964) Critical assessment of techniques for identifying the physiologically significant auxins in plants. In: Nitsch JP (ed) Régulateurs naturels de la croissance végétale. CNRS, Paris, pp 457–474
Scott TK, Most BH (1972) The movement of growth hormones in sugar cane. In: Kaldewey H, Vardar Y (eds) Hormonal regulation in plant growth and development. Verlag Chemie, Weinheim, pp 58–67
Scott TK, Pritchard JB (1968) The control of apical dominance in the Alaska pea. In: Vardar Y (ed) The transport of plant hormones. North-Holland, Amsterdam, pp 309–319
Scott TK, Wilkins MB (1968) Auxin transport in roots. II. Polar flux of IAA in *Zea* roots. Planta 83:323–334
Scott TK, Wilkins MB (1969) Auxin transport in roots. IV. Effects of light on IAA movement and geotropic responsiveness in *Zea* roots. Planta 87:249–258
Šebánek J, Hink J (1967) Die Bedeutung der Attraktion von Gibberellin durch Indolylessigsäure bei der apikalen Dominanz. Planta 76:124–128
Sembdner G, Weiland J, Aurich D, Schreiber K (1968) Isolation, structure and metabolism of a gibberellin glucoside. Soc Chem Ind Monogr 31:70–86
Seth AK, Davies CR, Wareing PF (1966) Auxin effects on the mobility of kinetin in the plant. Science 151:587–588
Seubert E (1925) Über Wachstumsregulatoren in der Koleoptile von *Avena*. Z Bot 17:49–88
Shaw S, Gardener G, Wilkins MB (1973) The lateral transport of IAA in intact coleoptiles of *Avena sativa* L. and *Zea mays* L. during geotropic stimulation. Planta 115:97–111
Shaw S, Wilkins MB (1974) Auxin transport in roots. X. Relative movement of radioactivity from IAA in the stele and cortex of *Zea* root segments. J Exp Bot 25:199–207
Sheldrake AR (1972) Polar auxin transport in leaves of monocotyledons. Nature 238:352–353
Sheldrake AR (1973a) Auxin transport in secondary tissues. J Exp Bot 24:87–96
Sheldrake AR (1973b) Do coleoptile tips produce auxin? New Phytol 72:433–447
Sheldrake AR (1973c) The production of hormones in higher plants. Biol Rev 48:509–559
Sheldrake AR (1974) The polarity of auxin transport in inverted cuttings. New Phytol 73:637–642
Sheldrake AR (1979) Effects of osmotic stress on polar auxin transport in *Avena* mesocotyl sections. Planta 145:113–117
Shen-Miller J (1973a) Rhythmicity in basipetal transport of indoleacetic acid through coleoptiles. Plant Physiol 51:615–619
Shen-Miller J (1973b) Rhythmic differences in the basipetal movement of indoleacetic acid between separated upper and lower halves of geotropically stimulated corn coleoptiles. Plant Physiol 52:166–170
Shindy WW, Asmundson CM, Smith OE, Kumamoto J (1973) Absorption and distribution of high specific radioactivity 2-^{14}C-abscisic acid in cotton seedlings. Plant Physiol 52:443–447
Sitton D, Richmond A, Vaadia Y (1967) On the synthesis of gibberellins in roots. Phytochemistry 6:1101–1105
Sitton D, Itai C, Kende H (1972) Decreased cytokinin production in the roots as a factor in shoot senescence. Planta 73:296–300
Skene KGM (1967) Gibberellin-like substances in root exudate of *Vitis vinifera*. Planta 74:250–265
Skene KGM (1972) Cytokinins in the xylem sap of grape vine canes. Changes in activity during coldstorage. Planta 104:89–92
Skene KGM, Antcliff AJ (1972) A comparative study of cytokinin levels in bleeding sap of *Vitis vinifera* (L.) and the two grapevine rootstocks, Salt Creek and 1613. J Exp Bot 23:283–293
Skene KGM, Kerridge GH (1967) Effect of root temperature on cytokinin activity in root exudate of *Vitis vinifera* L. Plant Physiol 42:1131–1139

Skene KGM, Osmond G (1968) Increases in the level of cytokinins in bleeding sap of *Vitis vinifera* L. after CCC treatment. Science 159:1477–1478

Šmid N, Vardjan M (1970) Les cytokinines dans la sève printanière du bouleau *Betula pendula* Roth. Biol Vestn 18:27–36

Smith CW, Jacobs WP (1968) The movement of IAA-^{14}C in the hypocotyl of *Phaseolus vulgaris*. In: Vardar Y (ed) The transport of plant hormones. North-Holland, Amsterdam, pp 48–64

Smith CW, Jacobs WP (1969) The movement of IAA-^{14}C in the hypocotyl of *Phaseolus vulgaris*. Am J Bot 56:492–497

Smith DJ, Schwabe WW (1980) Cytokinin activity in oak (*Quercus robur*) with particular reference to transplanting. Physiol Plant 48:27–32

Smith FA, Raven JA (1976) H^+ transport and regulation of cell pH. In: Lüttge U, Pitman MG (eds) Encyclopedia of plant physiology new series Vol 2A. Springer, Berlin Heidelberg New York, pp 317–346

Söding H (1923) Werden von der Spitze der Haferkoleoptile Wuchshormone gebildet? Ber Dtsch Bot Ges 41:396–400

Söding H (1925) Zur Kenntnis der Wuchshormone in der Haferkoleoptile. Jahrb Wiss Bot 64:587–603

Söding H (1926) Über den Einfluß der jungen Infloreszenz auf das Wachstum ihres Schaftes. Jahrb Wiss Bot 65:611–635

Söding H (1952) Die Wuchsstofflehre. Thieme, Stuttgart

Stark P (1925) Zur Methodik chemotropischer Reizung bei Dikotyledonenkeimstengeln I. Ber Dtsch Bot Ges 43:568–575

Steeves TA, Briggs WR (1960) Morphogenetic studies on *Osmunda cinnamomea* L. The auxin relationships of expanding fronds. J Exp Bot 11:45–67

Sussman MR, Goldsmith MHM (1980) Solubilization of the receptor for N-1-naphthylphthalamic acid. Plant Physiol 66:1074–1078

Sweet GB, Zaerr JB, Lavender DP (1974) The effect of some manipulations on the level of gibberellins in seedling Douglas fir. In: Bieleski RL, Ferguson AR, Cresswell MM (eds) Mechanism of regulation of plant growth. Roy Soc N Z, Wellington, pp 709–713

Tal M, Imber D, Itai C (1970) Abnormal stomatal behaviour and hormonal imbalance in flacca, a wilty mutant of tomato. Plant Physiol 46:367–373

Tepper HB, Brossard D (1969) Voie de transport de l'acide indolylacétique chez *Zea* et chez le *Coleus*. CR Acad Sci Paris Ser D 269:567–569

Thimann KV (1972) The natural plant hormones. In: Steward FC (ed) Plant physiology. A treatise Vol VIB: physiology of development: the hormones. Academic Press, London New York

Thimann KV, Wardlaw IF (1963) The effect of light on the uptake and transport of indoleacetic acid in the green stem of the pea. Physiol Plant 16:368–377

Thomson KSt, Leopold AC (1974) In vitro binding of morphactins and 1-N-naphthylphthalamic acid in corn coleoptiles and their effects on auxin transport. Planta 115:259–270

Thomson KSt, Hertel R, Müller S, Tavares JE (1973) 1-N-naphthylphthalamic acid and 2,3,5-triiodobenzoic acid: In vitro binding to particulate cell fractions and action on auxin transport in corn coleoptiles. Planta 109:337–352

Thornton RM, Thimann KV (1967) Transient effects of light on auxin transport in the *Avena* coleoptile. Plant Physiol 42:247–257

Tsurumi S, Ohwaki Y (1978) Transport of ^{14}C-labeled indoleacetic acid in *Vicia* root segments. Plant Cell Physiol 19:1195–1206

Tsurumi S, Wada S (1980) Transport of shoot- and cotyledon-applied indole-3-acetic acid to *Vicia faba* root. Plant Cell Physiol 21:803–816

Ueda M, Bandurski RS (1969) A quantitative estimation of alkalilabile indole-3-acetic acid compounds in dormant and germinating maize kernels. Plant Physiol 44:1175–1181

Van der Weij HG (1932) Der Mechanismus des Wuchsstofftransportes. Recl Trav Bot Néerl 29:379–496

Van der Weij HG (1934) Der Mechanismus des Wuchsstofftransportes II. Recl Trav Bot Néerl 31:810–857
Van de Westringh C, Veldstra H (1958) On the influence of the polar group on growth substance transport. Comparative investigations with ^{14}C-labelled indolylacetic acid and indolylmethanesulphonic acid. Recl Trav Chim Pays-Bas Belg 77:1114–1128
Van Staden J (1976) Occurrence of a cytokinin glucoside in the leaves and honeydew of *Salix babylonica*. Physiol Plant 36:225–228
Van Staden J, Davey JE (1976a) Cytokinin translocation in xylem sap of herbaceous plants. Z Pflanzenphysiol 77:377–382
Van Staden J, Davey JE (1976b) Note on cytokinins in the xylem sap of *Protea compacta*. J S Afr Bot 42:13–15
Van Staden J, Dimalla GGA (1977) A comparison of the endogenous cytokinins in the roots and xylem exudate of nematode-resistent and susceptible tomato cultivars. J Exp Bot 28:1351–1356
Van Staden J, Menary RC (1976) Identification of cytokinins in the xylem sap of tomato. Z Pflanzenphysiol 78:262–265
Vardar Y (1964) Experiments with *Helianthus annuus* hypocotyls on IAA-^{14}C transport in relation with temperature. Ber Schweiz Bot Ges 74:229–236
Vardar Y (ed) (1968a) The transport of plant hormones. North Holland, Amsterdam
Vardar Y (1968b) Agents modifying the longitudinal transport of auxin. In: Vardar Y (ed) The transport of plant hormones. North-Holland, Amsterdam, pp 156–191
Vardar Y (1970) Studies on the metabolic interaction of triiodobenzoic acid during the transport of externally applied indoleacetic acid. Ber Schweiz Bot Ges 80:438–440
Veen H (1966) Transport, immobilisation and location of naphthylacetic acid-2-^{14}C in *Coleus* explants. Acta Bot Neerl 15:419–433
Veen H (1969) Auxin transport, auxin metabolism and ageing. Acta Bot Neerl 18:447–454
Veen H (1972) Relationship between transport and metabolism of α-naphthaleneacetic acid and α-decalylacetic acid in segments of *Coleus*. Planta 103:35–44
Veen H (1975) Non-polar translocation of abscisic acid in petiole segments of *Coleus*. Acta Bot Neerl 24:54–62
Veen H, Frissel MJ (1975) Simulation of hormone transport in petiole segments of *Coleus*. Physiol Plant 34:208–215
Veen H, Jacobs WP (1969a) Transport and metabolism of indole-3-acetic acid in *Coleus* petiole segments of increasing age. Plant Physiol 44:1157–1162
Veen H, Jacobs WP (1969b) Movement and metabolism of kinetin-^{14}C and of adenine-^{14}C in *Coleus* petioles of increasing age. Plant Physiol 44:1277–1284
Venis MA (1977) Affinity labels for auxin binding sites in corn coleoptile membranes. Planta 134:145–149
Vonk CR (1974) Studies on phloem exudation from *Yucca flaccida* Haw. XIII. Evidence for the occurrence of a cytokinin nucleotide in the exudate. Acta Bot Neerl 23:541–548
Vonk CR (1978) Formation of cytokinin nucleotides in a detached inflorescence stalk and the occurrence of nucleotides in phloem exudate from attached *Yucca* plants. Physiol Plant 44:161–166
Vonk CR (1979) Origin of cytokinins transported in the phloem. Physiol Plant 46:235–240
Wagner H, Michael G (1971) Der Einfluß unterschiedlicher Stickstoffversorgung auf die Cytokininbildung in Wurzeln von Sonnenblumenpflanzen. Biochem Physiol Pflanz 162:147–158
Wangermann E (1968) The distribution of indolylacetic acid in *Coleus* stems. In: Vardar Y (ed) The transport of plant hormones. North-Holland, Amsterdam, pp 65–78
Wangermann E (1970) Autoradiographic localization of soluble and insoluble ^{14}C from [^{14}C]indolylacetic acid supplied to isolated *Coleus* internodes. New Phytol 69:919–927
Wangermann E (1974) The pathway of transport of applied indolyl-acetic acid through internode segments. New Phytol 73:623–636
Wangermann E (1977) Further localization of auxin transport through internode segments. New Phytol 79:501–504
Wangermann E, Withers LA (1978) Auxin transport characteristics and cellular ultrastructure of different types of parenchyma. New Phytol 81:1–17

Webster J, Wilkins MB (1974) Lateral movement of radioactivity from [^{14}C]gibberellic acid (GA_3) in roots and coleoptiles of *Zea mays* L. seedlings during geotropic stimulation. Planta 121:303–308

Weis A (1969) Untersuchungen über Wachstumsregulatoren in Blüten von *Narcissus pseudonarcissus* L. und zum Auxintransport in Griffeln von *Digitalis purpurea* L. Diss Univ d Saarlandes, Saarbrücken

Went FW (1928) Wuchsstoff und Wachstum. Rec Trav Bot Néerl 25:1–116

Went FW (1932) Eine botanische Polaritätstheorie. Jahrb Wiss Bot 76:528–557

Went FW (1941) Polarity of auxin transport in inverted *Tagetes* cuttings. Bot Gaz 103:386–390

Went FW, White R (1939) Experiments on the transport of auxin. Bot Gaz 100:465–484

Werblin TP, Jacobs WP (1967) Auxin transport and polarity in *Coleus* petioles of increasing age. Wiss Z Univ Rostock Math Naturwiss Reihe 16:495–497

Wetmore RH, Jacobs WP (1953) Studies on abscission: the inhibiting effect of auxin. Am J Bot 40:272–276

Whitehouse RL, Zalik S (1967) Transport of indole-3-acetic acid-1′-^{14}C and tryptophan-1-^{14}C in seedlings of *Phaseolus coccineus* L. and *Zea mays* L. Plant Physiol 42:1362–1372

Whitehouse RL, Zalik S (1968) Translocation of radioactively labelled 3-indoleacetic acid and tryptophan in seedlings of *Phaseolus coccineus* L. and *Zea mays* L. In: Wightman F, Setterfield G (eds) Biochemistry and physiology of plant growth substances. Runge, Ottawa, pp 977–991

Wightman F, Thimann KV (1980) Hormonal factors controlling the initiation and the development of lateral roots. I. Sources of primordia-inducing substances in the primary root of pea seedlings. Physiol Plant 49:13–20

Wightman F, Schneider EA, Thimann KV (1980) Hormonal factors controlling the initiation and development of lateral roots. II. Effects of exogenous growth factors on lateral root formation in pea roots. Physiol Plant 49:304–314

Wilkins MB, Cane AR (1970) Auxin transport in roots. V. Effects of temperature on the movement of IAA in *Zea* roots. J Exp Bot 21:195–211

Wilkins MB, Cane AR, McCorquodale I (1972) Auxin transport in roots. VII. Uptake and movement of radioactivity from IAA-^{14}C by *Zea* roots. Planta 105:93–113

Wilkins MB, Martin N (1967) The dependence of basipetal polar transport of auxin upon aerobic metabolism. Plant Physiol 42:831–839

Wilkins MB, Nash LJ (1974) Movement of radioactivity from [^{3}H]GA_3 in geotropically stimulated coleoptiles of *Zea mays*. Planta 115:245–251

Wilkins MB, Scott TK (1968) Auxin transport in roots. III. Dependence of polar flux of IAA in *Zea* roots upon metabolism. Planta 83:335–346

Wilkins MB, Shaw S (1967) Geotropic response of coleoptiles under anaerobic conditions. Plant Physiol 42:1111–1113

Wilkins MB, Whyte P (1968) Polar transport of auxin in *Zea* coleoptiles under anaerobic conditions. In: Wightman F, Setterfield G (eds) Biochemistry and physiology of plant growth substances. Runge, Ottawa, pp 1051–1062

Wilkins MB, Woodcock AER (1965) Origin of the geoelectric effect in plants. Nature 208:990–992

Winter A (1967) The promotion of the immobilization of auxin in *Avena* coleoptiles by triiodobenzoic acid. Physiol Plant 20:330–336

Winter A (1968) 2,3,5-Triiodobenzoic acid and the transport of 3-indoleacetic acid. In: Wightman F, Setterfield G (eds) Biochemistry and physiology of plant growth substances. Runge, Ottawa, pp 1063–1076

Wochok ZS, Sussex IM (1973) Morphogenesis in *Selaginella*. Auxin transport in the stem. Plant Physiol 51:646–650

Woodcock AER, Hertz CH (1972) The geoelectric effect in plant shoots. V. A discussion. J Exp Bot 23:953–957

Woodcock AER, Wilkins MB (1970) The geoelectric effect in plant shoots. III. Dependence upon auxin concentration gradients and aerobic metabolism. J Exp Bot 21:985–996

Woodcock AER, Wilkins MB (1971) The geoelectric effect in plant shoots. J Exp Bot 22:512–525

Wooley DJ, Wareing PF (1972) The interaction between growth promotors in apical dominance. I. Hormonal interaction, movement and metabolism of a cytokinin in rootless cuttings. New Phytol 71:781–793

Yamaki T, Fujii T (1968) Effect of light on auxin transport through the node tissue of *Avena* seedling. In: Wightman F, Setterfield G (eds) Biochemistry and physiology of plant growth substances. Runge, Ottawa, pp 1025–1036

Yeomans LM, Audus LJ (1964) Auxin transport in roots. *Vicia faba*. Nature 204:559–562

Yoshida R, Oritani T (1974) Effects of nitrogen topdressing on cytokinin content in the root exudate of rice plant. Proc Crop Sci Soc Jpn 43:47–51

Yoshida R, Oritani T, Nishi A (1971) Kinetin-like factors in the root exudate of rice plants. Plant Cell Physiol 12:89–94

Zaerr JB (1968) Transport gradient of indoleacetic acid in pine seedlings. Physiol Plant 21:1265–1269

Zamski E, Tsivion Y (1977) Translocation in plants possessing supernumerary phloem. I. ^{14}C-assimilates and auxin in the internal phloem of tobacco (*Nicotiana tabacum* L.). J Exp Bot 28:117–126

Zamski E, Wareing PF (1974) Vertical and radial movement of auxin in young sycamore plants. New Phytol 73:61–69

Zeevaart JAD (1977) Sites of abscisic acid synthesis and metabolism in *Ricinus communis*. Plant Physiol 59:788–791

Ziegler H (1973) Wasserumsatz und Stoffbewegungen. Bericht über die Arbeiten zum Wuchsstofftransport im Parenchym. Fortschr Bot 35:67–75

Zwar JA, Rijven AHGC (1956) Inhibition of transport of indole-3-acetic acid in the etiolated hypocotyl of *Phaseolus vulgaris* L. Aust J Biol Sci 9:528–538

Zweig G, Yamaguchi S, Mason GM (1961) Translocation of C^{14}-gibberellin in red kidney bean, normal corn and dwarf corn. Adv Chem Ser 28:122–134

4 Functions of Hormones at Tissue Level of Organization

W.P. JACOBS

4.1 Introduction

A major advance made since the 1965 edition of the volumes 14–16 on developmental physiology in the first series of this Encyclopedia was prepared has been the quantitative elucidation of hormonal control of sieve-tube differentiation during regeneration in stems. In addition, the earlier quantitative studies of tracheary differentiation have been extended and amplified. An unexpectedly close connection between sieve tubes and tracheary cells in their regeneration and normal differentiation has been revealed and a unifying picture of vascular differentiation in the shoot has resulted. A start has recently been made in unraveling the controlling factors for fiber differentiation. These advances will be described and discussed below.

For convenience we will discuss regeneration first, then normal differentiation. This distinction is somewhat artificial, because most experiments in which hormones have been added to plants have involved the wounding the plants (the hormone being substituted for excised organs). However, I shall restrict the term regeneration to studies in which the continuity of some tissue, such as a vascular strand, is severed and subsequently reforms from cells that would not normally differentiate into such tissue.

4.2 Role of Hormones in Differentiation of Vascular Tissues

4.2.1 Regeneration in Shoots

4.2.1.1 Xylem

Studies of hormonal control of tissue differentiation focused first and frequently on xylem. Xylem was often selected because of the relative ease of identifying the lignified secondary wall typical of tracheary cells. However, in the first quantitative study (JACOBS 1952), xylem was selected because it had been reported to regenerate in a basipetally polar direction (SIMON 1908, SINNOTT and BLOCH 1945), suggesting that xylem regeneration might be controlled by auxin, the only hormone then known to move with basipetal polarity. The other rationale for testing the potential xylogenic activity of auxin was that auxin had previously been shown to control so many other developmental activities of plants that it was a likely guess that auxin would be found to be controlling cell differentiation, also.

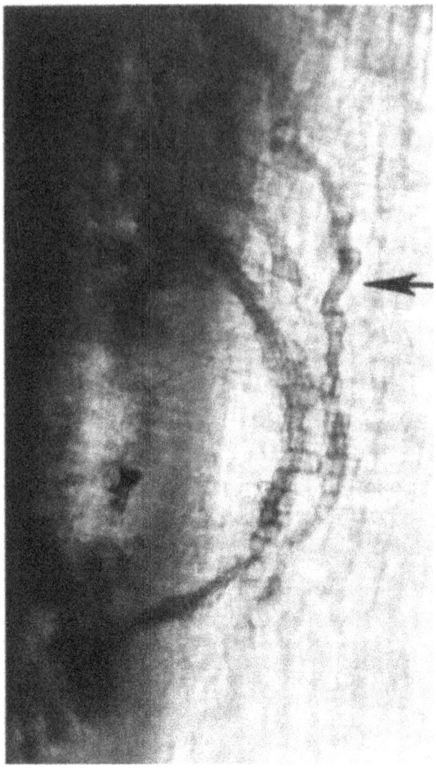

Fig. 4.1. A cleared and stained whole mount, showing strands of tracheary cells (*arrow*) that have regenerated around a wound (*left*) in a *Coleus* stem. (JACOBS 1952)

The main advance that JACOBS made was to use larger sample sizes (than the few plants characteristic of many earlier studies by developmental anatomists), to show reproducibility, and to analyze results with statistical tests of significance. Because he used whole-mount clearing methods (rather than serial sections of paraffin-embedded tissue), it was feasible to process the large samples. The stained tracheary cells could be seen through the cleared outer tissue (Fig. 4.1).

Coleus' elongating young internodes, a number of weeks before they would be old enough to develop cambium, were used as experimental material. Leaves were shown to be the major suppliers of diffusible auxin in the shoot, with the leaf pair just above the young internode studied producing the maximal amount of auxin judging by bioassay (JACOBS 1952). The pattern of auxin production in relation to leaf age was the same as earlier reported by GOODWIN (1937) and DELISLE (1938) in other genera. When a vascular strand in one corner of *Coleus'* four-cornered stem was severed, a substantial number of tracheary strands regenerated in a 5–7 day period, if the plants were otherwise intact. (JACOBS counted the number of multi-celled xylem strands regenerated around the wound as a measure of the number of xylem cells that had regenerated.) If the normal sources of auxin were removed by excising the leaves above the severed vascular strand in the young internode, there was a statistically significant decrease in the number of regenerating xylem strands. If synthetic

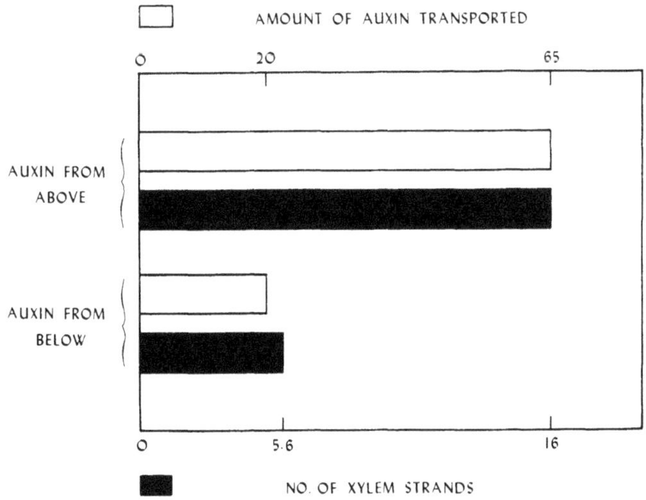

Fig. 4.2. Relation between the amount of auxin (IAA) transported basipetally and acropetally through excised stem sections (*open bars*) and the number of tracheary strands regenerated in plants when leaves above or below were excised (*solid bars*). (JACOBS 1954)

indole-3-acetic acid (IAA) was substituted for the leaves above the wound so that the normal amount of auxin was supplied, the normal number of regenerated xylem strands was restored in the subjacent internode (JACOBS 1952, 1956). The fact that exact replacement resulted from this exact substitution of IAA for the endogenous auxin supported the belief that *Coleus'* auxin was IAA. More specific evidence that the diffusible auxin of *Coleus* was indeed IAA and only IAA came from later studies that found only one zone of auxin activity on chromatograms of the diffusible auxin, with that zone containing IAA by Rf in two solvents, color tests, and activity in two bioassays (SCOTT and JACOBS 1964). Still more thorough purification of leaf extracts by HEMBERG (1972) confirmed that IAA was the endogenous auxin of *Coleus*.

Further quantitative evidence came from studies of the relative polarity of IAA movement in *Coleus* internodes of the same developmental stage used for the xylem regeneration studies. Unlike the etiolated seedling tissue used in most earlier studies of auxin transport, these elongating green stems showed a significant amount of IAA movement in the acropetal (shoot-base toward shoot-tip) direction. Judging by bioassay of receivers, one third as much IAA moved acropetally through excised internode sections as moved basipetally through matching sections to which the same amount of IAA was applied (JACOBS 1954). When *Coleus* plants had leaves excised so that the endogenous IAA would be available only from below or only from above the wound in the young internode, the ratio of the number of tracheary strands regenerated in the two treatments showed a close parallel with the ability of such internodes to transport IAA in the two directions (Fig. 4.2). For both IAA transport and tracheary regeneration the ratio was 3:1 (JACOBS 1954, 1956).

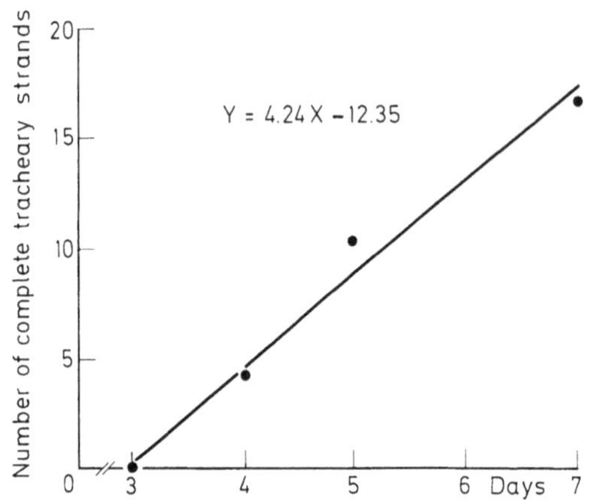

Fig. 4.3. Time course of regeneration of tracheary strands around a wound in young internodes of *Coleus*. (JACOBS 1979)

These first quantitative studies of cell differentiation under the influence of hormones are among the few cases in the literature where *exact* substitution of synthetic hormone for the normal hormone source has been demonstrated to give full and *exact* replacement of the hormone source in terms of a physiological effect. The prevention of abscission by IAA is another such case (Chap. 8 in JACOBS 1979). In both cases, the full effect of the leaves was replaceable by IAA alone. The results confirmed the hypothesis that, of all the chemicals undoubtedly required to differentiate a strand of regenerated tracheary cells, IAA from the leaves was the one that acted as a limiting factor in the process. Furthermore, the results with the polarity of the stem (Fig. 4.2) supported the view that the ability of the stem to transport IAA was a further check on how much IAA reached the regenerating area (JACOBS 1954).

Further confirmation of the closeness of the relation between the polarity of IAA movement and of tracheary regeneration came recently from studies of the same clone of *Coleus* growing under conditions that gave much faster growth. As expected from earlier studies showing that increased elongation was correlated with more polar auxin movement (e.g., MCCREADY and JACOBS 1963), the polarity of IAA movement was now almost exclusively basipetal (JACOBS 1977); and, correspondingly, the number of tracheary strands that regenerated when the leaves remained on only below the wound was now zero (ALONI and JACOBS 1977a). The time-course of tracheary strand regeneration was determined, also; no strands were yet complete around the wound after 3 days, but starting at 4 days there was a steady increase in number to the 17 completed after 7 days (Fig. 4.3).

IAA replaced the distal sources of endogenous auxin whether supplied as a 2 mg l^{-1} solution through the petiole of the debladed leaf immediately above the wound (JACOBS 1952, 1956) or mixed with lanolin in a 1.0% emulsion and added to the cut stem (JACOBS 1952; Figs. 5 and 6 of JACOBS et al. 1959).

More precise determinations were made by THOMPSON and JACOBS (1966), who made the wound in one of the smaller vascular strands in the flat sides of *Coleus* stems (rather than in the large strand in the corners), so that the regeneration would be almost all in one plane and would consist of fewer cells.

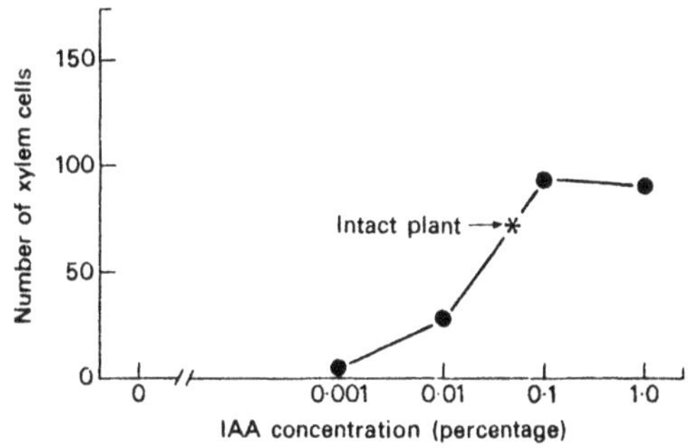

Fig. 4.4. The number of tracheary cells regenerating around a slit wound in excised young internodes of *Coleus* in relation to increasing doses of IAA in lanolin added to the apical cut surface. The *asterisk* marks the interpolated value for the number of tracheary cells regenerated around such a wound in otherwise intact control plants. (After JACOBS 1970, corrected data from THOMPSON and JACOBS 1966)

This allowed individual regenerated tracheary cells (rather than strands of cells) to be counted. They also further isolated the system by studying regeneration in a young internode from which not only all distal shoots had been excised but also all shoots and roots below had been excised. Various "concentrations" of IAA in lanolin were added to the apical cut of such explants and the number of regenerated tracheary cells counted (Fig. 4.4). By this more exact measurement, IAA alone could replace the whole rest of the shoot and all the root system in causing tracheary regeneration. Interpolation on the response curve of the regeneration counts from plants intact except for the wounded stem demonstrated that 0.05% IAA in lanolin exactly replaced all the rest of the plant in causing tracheary regeneration in this young internode (asterisk in Fig. 4.4). [The saturation effect in tracheary regeneration seen as the IAA was increased from 0.1% to 1.0% in its mix with lanolin was expected from the similar saturation of the basipetal movement of IAA through such excised young *Coleus* internodes (JACOBS 1961, SCOTT and JACOBS 1963)].

Because of the interest of JACOBS' laboratory in discovering the limiting factors in healthy green plants under the conditions approximating those of normal cultivation, the experiments described above were run in the full light of the greenhouse or in growth chambers at similarly high light intensity. If, by contrast, such experiments were run under lower light intensities and for 24 days (rather than 5–7), then sugar had to be added to obtain tracheary regeneration, although supplementing the sucrose with IAA gave still more tracheary cells (BESLOW and RIER 1969). Other experiments on the interactions of sugars and auxins are described below under "tissue cultures" (Sect. 4).

IAA, at the same "concentration" in lanolin that exactly restored tracheary regeneration in the young elongating internodes, also restored the normal amount of tracheary regeneration in *Coleus* internodes old enough to be starting cambium development (THOMPSON and JACOBS 1966). There was a similar relation to the polarity of IAA movement in these older internodes, too: IAA applied at the base of the excised internode neither moved acropetally through the tissue (JACOBS and MCCREADY 1967) nor restored tracheary regeneration (THOMPSON and JACOBS 1966). Added at the apical end it moved basipetally and replaced the whole rest of the plant in the latter's effect on tracheary regeneration. BENAYOUN et al. (1975) confirmed that 0.1% IAA in lanolin could replace the distal leaves in *Coleus*.

This effectiveness of IAA is not limited to *Coleus*. Tomato (THOMPSON and JACOBS 1966), peanut, and okra, *Hibiscus esculentus* (THOMPSON 1968, 1970) also showed stimulation of tracheary regeneration by IAA added at the apical end, but less or none by IAA added at the basal end of excised internodes. Similarly, IAA was said to replace the xylogenic action of the young shoot in pea seedlings (SACHS 1968a). The effectiveness of IAA derives apparently from its auxin properties, judging by limited tests on *Coleus* by THOMPSON (1965). Several other auxins (NAA, and the auxin herbicides 2,4-D and 2,4,5-T) were as effective as IAA in causing tracheary regeneration; in fact, 2,4-D was much more active (Fig. 15 of JACOBS 1970). By contrast, several nonauxin hormones were without significant effect (e.g., ascorbic acid, nicotinic acid, thiamine), although each was tested at only one concentration. *Gibberellic acid* (GA_3) was tested in several concentrations, added alone or with IAA (THOMPSON 1965). Alone it had no effect; added with IAA GA_3 increased the average number of tracheary cells that regenerated, but not to a statistically significant extent.

4.2.1.2 Sieve-Tube Elements

The same approach that had worked well with studies of tracheary regeneration (i.e., of using clearing techniques that allowed relatively large samples to be processed in a quantitative way) was applied to sieve-tube regeneration in *Coleus* (LAMOTTE and JACOBS 1962). They studied older internodes that were just starting to develop cambium because that stage allowed easier separation of the phloem regenerates from the central portion of the plant. Their "flap technique" is illustrated in Fig. 4.5. The flap of phloem tissue, once cleared, stained, and mounted with sieve tubes up, permitted easy counting of the regenerated sieve tubes (Fig. 4.6). A diagram of such a phloem flap collected after 5 days of regeneration is shown in Fig. 4.7, with the numbers illustrating the convention

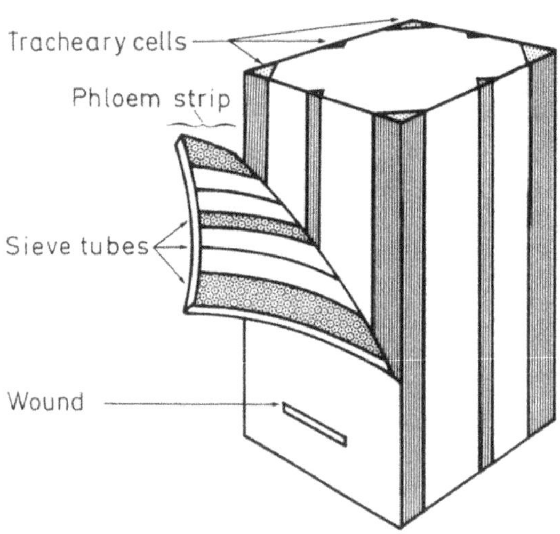

Fig. 4.5. Diagram of the technique used by LAMOTTE and JACOBS (1962) for removing a phloem strip from one side of a *Coleus* stem (JACOBS 1970). Note that the smallest strands of sieve tubes do not yet have tracheary counterparts differentiated on the other side of the cambium

used for counting regenerated sieve-tube strands. Each such strand contained an average of ten sieve-tube elements.

By excising different parts of the shoot, LAMOTTE and JACOBS (1963) found that the distal leaves and buds were more important than the proximal ones in supporting sieve-tube regeneration, excision of either distal leaves or distal buds causing a statistically significant drop in the number of sieve tubes regenerated. THOMPSON (1967) and HOUCK and LAMOTTE (1977) confirmed the effectiveness of the distal organs. The role of distal organs was also reflected in the pattern of regeneration around the wound in plants intact except for the transverse wound: early stages showed more regeneration above the wound (i.e., closer to the distal organs) (LAMOTTE and JACOBS 1963, ALONI and JACOBS 1977b).

This basipetal polarity suggested that auxins might be the controlling substance, because auxins were the only hormones known at that time to move

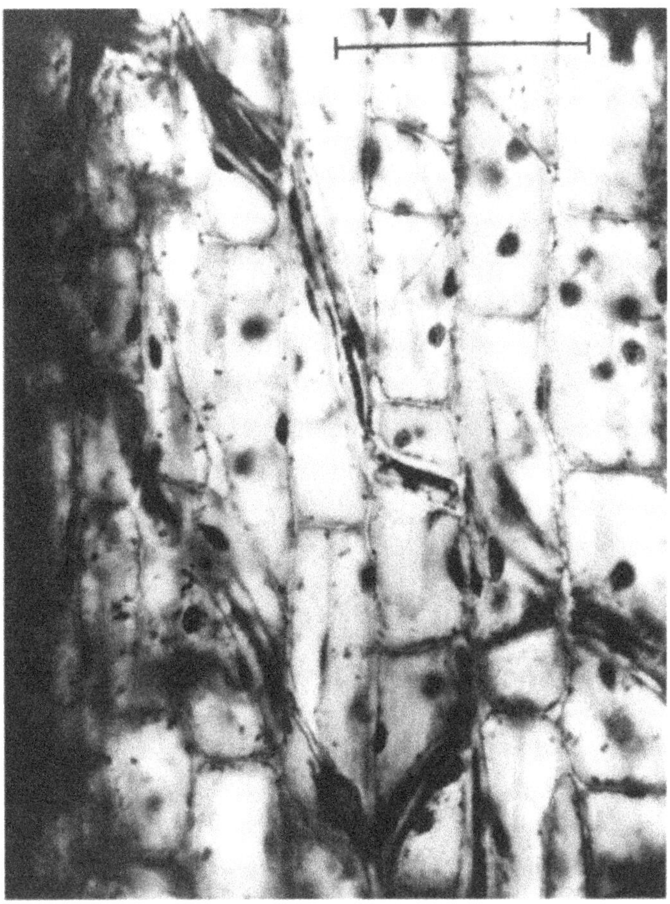

Fig. 4.6. A portion of a cleared and stained phloem strip, showing regenerated sieve tubes cutting diagonally across the interfascicular tissues. The scale represents 0.1 mm. (LAMOTTE and JACOBS 1962)

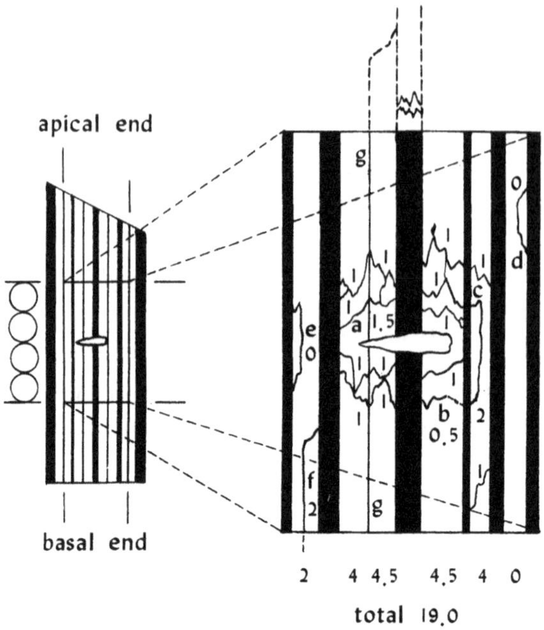

Fig. 4.7. Diagram of a phloem strip after 5 days of regeneration, illustrating the counting convention of LAMOTTE and JACOBS (1962)

Table 4.1. Effects on sieve tube regeneration of IAA and sucrose substituted for distal shoot organs of *Coleus blumei*. (Data from LAMOTTE and JACOBS 1963)

Treatment	Number of regenerated strands after 5 days (mean ± S.E., n = 4)
I. Intact control plants	23.2 ± 2.2
II. All shoot organs excised and the following substituted:	
Water	7.2 ± 3.4[a]
Sucrose (20 g l^{-1})	7.6 ± 2.5[a]
IAA (2 mg l^{-1})	23.9 ± 4.5
IAA + sucrose	20.9 ± 3.9

[a] Significantly different (at 5% level) from values from intact controls

in shoots with basipetal polarity – even though the older literature supported the view that different factors would be limiting sieve-tube and tracheary regeneration. Accordingly, IAA was substituted for the distal organs. At 2 mg l^{-1}, IAA completely replaced the distal organs in their effect on sieve-tube regeneration (Table 4.1) (LAMOTTE and JACOBS 1963). Sucrose added alone or with IAA had no effect. IAA at 0.1% in lanolin resulted in more sieve-tube strands than found in the "intact" plants. THOMPSON (1967) and HOUCK and LAMOTTE

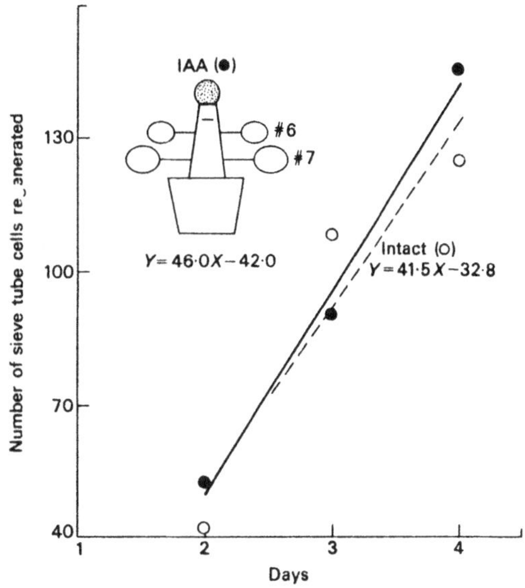

Fig. 4.8. The time-course of the regeneration of sieve tube cells around a slit wound in older internodes of *Coleus*, showing the effectiveness of IAA in replacing the leaves and shoots above the wound. (Graph and regression calculations from JACOBS 1979, other data from THOMPSON 1967)

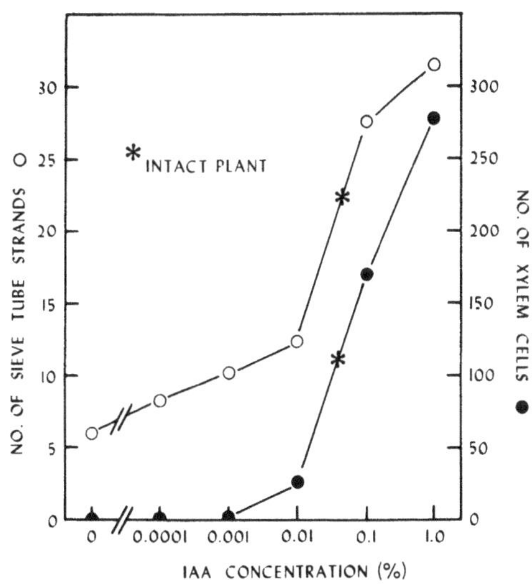

Fig. 4.9. The dose-response curve to IAA (added in lanolin to the apical end of excised older internodes of *Coleus*) of regenerating tracheary cells (*solid circles*) and sieve tube strands (*open circles*). *Asterisks* mark the interpolated values from "intact" controls (as in Fig. 4.4) (THOMPSON and JACOBS 1966). Note that each sieve tube strand contains an average of ten cells (LAMOTTE and JACOBS 1962); hence, the two curves directly reflect the number of *cells* that regenerated

(1977) confirmed that IAA could replace the distal organs, with THOMPSON adding such information for a 4-day time-course (Fig. 4.8).

In other experiments, internodes were cut free of the entire rest of the plant and regeneration tested in these isolated stem pieces. IAA added to the apical end of such an excised older internode increased sieve-tube regeneration over the controls (LAMOTTE and JACOBS 1963, THOMPSON and JACOBS 1966). The dose-response curve of apically applied IAA, as it affected both tracheary and sieve cells, is shown in Fig. 4.9: with no externally added IAA, a small number of sieve-tube cells but no tracheary cells regenerated; as the IAA level was increased, more vascular cells of both types regenerated (THOMPSON and JACOBS

1966). Plants intact except for the transverse wound regenerated numbers of vascular cells equivalent to those formed by about 0.05% IAA added to the isolated internode (asterisks in Fig. 4.9). Hence, 0.05% IAA in lanolin exactly replaced the root system and the rest of the shoot system in their effects on the regeneration of both sieve tubes and tracheary cells. At any one IAA level, more sieve cells than tracheary cells regenerated.

Like the effect of IAA on tracheary cells in the young, elongating internode, the effect of IAA on regeneration in this older internode was strictly polar (THOMPSON and JACOBS 1966), as was IAA movement through excised tissues of such internodes (JACOBS and MCCREADY 1967).

It was a surprise to have the same value of added IAA restore tracheary regeneration to "intact" levels in both the older internode No. 5 (Fig. 4.9) and the younger internode No. 2 (Fig. 4.4). To see if sieve-tube regeneration in internode No. 2 was restored as completely by IAA, the same internodes used for the tracheary regeneration studies of Fig. 4.4 were analyzed for sieve-tube regeneration (THOMPSON and JACOBS 1966). Although the number of sieve tubes regenerated did increase with added IAA, no IAA concentration brought the level close to that in the "intact" plants (Fig. 5 of THOMPSON and JACOBS 1966). Presumably, some factor in addition to IAA was limiting sieve-tube regeneration in the younger internode, although IAA was apparently the sole limiting factor for tracheary cells in those same internodes.

The additional factor required for sieve-tube regeneration in the younger internode is apt to be cytokinin, judging by the results of HOUCK and LAMOTTE (1977). They concentrated on regeneration of small vascular strands, at such an early stage of development that no tracheary cells, only sieve tubes, had differentiated from the procambium. (They used internodes at a developmental stage somewhat younger than internode No. 5, but substantially older than No. 2). Although IAA could replace the distal organs in their effect on sieve-tube regeneration, these small, xylem-less strands did not regenerate the "intact" number of sieve tubes in an excised internode, even if it was treated apically with IAA. Attributing the results to the fact that the roots had been excised, the authors tested substances known to be produced by roots, concentrating on substituting cytokinins for the proximal roots while substituting IAA for the distal shoot. Zeatin or zeatin riboside increased sieve-tube regeneration to a level not significantly different from that in "intact" controls (Fig. 11 of HOUCK and LAMOTTE 1977). Kinetin or "2iP" (apparently N^6-isopentenyl-adenine) were not effective.

There are only a few other papers on hormonal effects on sieve-tube regeneration aside from those on cultured material. BENAYOUN et al. (1975) reported that IAA could replace distal leaves of *Cucumis* or *Coleus* in causing sieve-tube regeneration but provided no data on the point.

THOMPSON (1965) tested various chemicals, in addition to IAA, to determine the specificity of sieve-tube regeneration (his data have been graphed in Fig. 4–20 of JACOBS 1979). Auxins, including the herbicides 2,4-D and 2,4,5-T, increased sieve-tube regeneration. Other hormones tested were without effect at the single concentration used. GA_3 did not change to a statistically significant degree the number of sieve-tube strands regenerated, whether the GA_3 was added alone at several concentrations or with IAA. But because the mean decrease in strand number from adding GA_3 was 31%, there may well be a real effect of GA_3 that would show up if larger samples were used.

The smaller, xylem-less strands, whose sieve-tube regeneration was studied by HOUCK and LAMOTTE (1977), increased sieve-tube regeneration significantly when L-glutamic acid was added basally to excised internodes treated apically with IAA, but various other chemicals (L-proline, kinetin, GA_3, sucrose, and L-glutamine) were without effect.

4.2.1.3 Fibers

In his elegant study of hormonal control of the normal differentiation of phloem fibers (described below), ALONI (1976) pointed out that the fibers did not differentiate from the parenchymatous cells of the wound callus, even though sieve tubes and tracheary cells did. The differentiation of phloem fibers was always limited to the longitudinal vascular strands, never being found in the intervening parenchyma. In that sense, one could say that they do not regenerate; however, their regeneration has been seldom studied.

4.2.2 Differentiation During Normal Development in Shoots

4.2.2.1 Xylem

The role of hormones in normal tracheary differentiation has not been investigated as thoroughly as in regeneration. However, there are several quantitative studies confirming the hypothesis that auxin is the limiting factor for normal differentiation also. Using the same clone of *Coleus blumei* that had been used for regeneration studies, JACOBS and MORROW (1957) found a close parallel relation between the rates at which young leaves were producing diffusible auxin and tracheary cells (their Fig. 16). The "diffusible auxin", measured directly with the *Avena* curve bioassay, was later shown to be IAA, judging by location and color tests on paper chromatograms (SCOTT and JACOBS 1964). Comparison of the effectiveness of this endogenous IAA in causing procambial cells to differentiate into normal tracheary cells with the effectiveness of added IAA in causing tracheary regeneration revealed that 14 times as much IAA was required for regeneration as for normal differentiation. JACOBS and MORROW (1957) thought that this difference in effectiveness might explain why all the parenchyma cells of the stem did not routinely differentiate into xylem cells under the influence of the endogenous auxin. They surmised that it would be only when a vascular strand was cut that the level of IAA would rise in the adjoining parenchyma cells to the 14 × normal level required to touch off the transformation of parenchyma into tracheary cells.

WANGERMANN (1967) excised *Coleus* leaves and found that fewer tracheary cells differentiated in the trace to the excised leaf, thus confirming, but with convincing quantitative data, the much earlier report of JOST (1893) on *Phaseolus*. The entire effect of the leaf on normal tracheary differentiation she found was replaceable by 0.1 M "auxin" in lanolin. (The auxin's chemical nature was not specified.) Qualitative results along the same lines were provided for the fern *Osmunda* (STEEVES and BRIGGS 1960).

Autoradiographs prepared from thin sections of *Coleus* stems treated with tritiated IAA of high specific activity showed silver grains localized specifically over the secondary walls of the youngest differentiating tracheary cells (SABNIS et al. 1969). Although they presented no evidence as to what chemical the ^{3}H was attached to in the tracheary walls, the extensive prior evidence that IAA controlled tracheary regeneration and differentiation in that clone suggested that the ^{3}H was still with IAA. The fact that the specific localization appeared after only 3 h—the earliest time checked—also supported that view. Similar results were reported for *Coleus* by WANGERMANN (1970) and GEE (1972), who used ^{14}C-IAA.

4.2.2.2 Sieve-Tube Elements

Hormonal effects on the normal differentiation of primary sieve tubes in shoots are essentially uninvestigated.

4.2.2.3 Fibers

These cells are unusual in the amount of elongation they may undergo during differentiation [e.g., growing from 20 to 550,000 μm in *Boehmeria* (ALDABA 1927)]. The length and thickened walls of the fibers of several species are the basis of their economic value as in hemp, linen, and jute.

Adding GA_3 to intact plants increased the number of fibers in jute (SIRCAR and CHAKRAVERTY 1960) and gave longer and thicker-walled fibers in hemp (ATAL 1961) and other genera (STANT 1961, 1963).

SACHS (1972) began excision-substitution experiments on fiber differentiation, exploiting the occurrence in pea of strands of fibers (HAYWARD 1938) that ran in the cortex from the base of an internode up to and into the petiole attached at the node above. Excising a leaf before it was 25 μm long (along with the apical meristem) resulted in no differentiation of the fiber strand in the ensuing 2 weeks. The presumed stimulus from the young leaf (and/or apical meristem) could cross a graft union, judging by results of grafting a shoot-tip bearing two small leaf primordia into the side of a stem. IAA substituted for such young tissue was said to restore none of the fiber differentiation. From these results SACHS made the logical but too sweeping conclusion "auxin is not a component of the stimulus for fibre differentiation", rather than the more justified one that auxin was probably not the *sole* stimulus.

ALONI took up the problem of fiber differentiation and developed a physiological-anatomical approach, which, when combined with multiple-comparison statistics, resulted in two excellent papers (1976, 1979). Concentrating on the primary phloem fibers of *Coleus*, he used a technique of staining cleared material (ALONI and SACHS 1973) which allowed large samples to be processed efficiently.

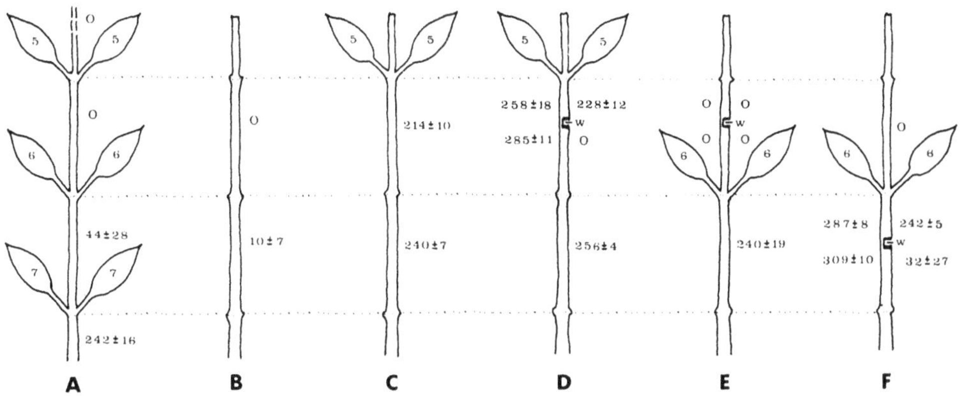

Fig. 4.10 A–F. Diagrams illustrating the effects of various treatments on the differentiation of phloem fibers in *Coleus* stems. *Numbers* beside the internodes show the average number of fibers (±standard error) in the adjacent half transverse section. The plants of diagram **A** were untreated. On the day the **A** plants were collected, the axillary shoots and indicated leaves of the other treatments were excised. Plants of **B–F** were harvested 2 weeks later. (ALONI 1976)

The cells that would differentiate much later into fibers were distinguishable up near the shoot apex as a group of fiber initial cells external to the primary sieve tubes. However, it was only many weeks later, when they were six or seven internodes below the apical bud (Fig. 4.10) and in internodes which had already developed cambial activity, that normal differentiation ensued, starting with the striking thickening of the wall that is characteristic of fibers. In these older internodes, approximately 500 differentiated phloem fibers were visible in a transverse section, with the cells being about 100 times as long as they were wide. (They averaged 2000 μm long in internode 7.)

Excision of the leaves reduced to zero the number of fibers differentiated. The presence of a single pair of mature leaves was sufficient to restore fiber differentiation to normal levels, but this inductive effect only operated on internodes below the leaf-pair (Fig. 4.10). Further evidence that the induction moved basipetally came from regeneration experiments in which a transverse slit was made two-thirds of the way through internode 5; fiber differentiation occurred everywhere except directly below the wound if the wound was kept open with parafilm (Fig. 4.10). If the wound was not kept open, fiber differentiation did occur directly below the wound, but the number of such cells was much smaller (ALONI 1976).

Substitution of IAA, GA_3, or both for all the leaves and for the stem above internode 5 revealed that IAA alone could restore about 13% of the normal number of phloem fibers in internode 5, GA_3 alone had no effect, but IAA and GA_3 added together at 0.05% in lanolin restored the full normal number of differentiated fibers (Fig. 4.11, ALONI 1979). Increasing the "concentration" of GA_3 in the IAA-GA lanolin produced longer fiber cells, as might be expected from the earlier literature reporting GA_3 effects on intact plants.

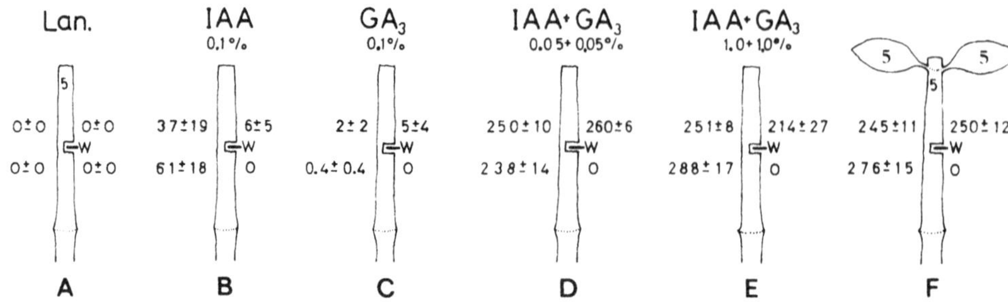

Fig. 4.11 A–F. The effect of IAA, GA_3, or both in replacing the effect of superjacent leaves in causing the differentiation of phloem fibers in *Coleus* stems. (ALONI 1979)

4.2.2.4 Cambium

The paucity of information about hormonal action on the regeneration of cambium is compensated for somewhat by the more extensive literature on the normal development of cambium. Since REINDERS-GOUWENTAK (1965) reviewed the pre-1960 literature, some important papers have been published. To the already known effects of auxin, they have added information on the action of later discovered hormones.

Before reviewing those papers I should point out several difficulties involved in studying cambium. "Cambial activity" in the strict sense refers to cell divisions. However, cambial activity is rarely measured by counting cell divisions in the cambial zone: doing so is too difficult. Instead, many investigators have resorted to counts of tracheary cells or even the width of total xylem as an indirect measure of cambial activity (e.g., AVERY et al. 1937). From the evidence (described above) that IAA controls the regeneration and differentiation of tracheary cells, it is obvious that such results, believed to reflect hormonal action on cambium, may instead be merely one more manifestation of hormone action on tracheary differentiation, with an increased number of total cells resulting secondarily from an effect of xylem differentiation on cambial activity. In cases of cambial reactivation in the spring, an increased number of xylem cells could result solely from the differentiation, without immediately prior division, of cells that have remained undifferentiated in the cambial zone over the winter. "Increased *width* of total xylem" or total phloem could, of course, be due to individual cells of wider diameter with no increase in cell number. These possible confounding factors are seldom sorted out in the literature.

The note by WAREING (1958), reporting no differentiated tracheary cells from added GA_3, some from IAA, but more when GA_3 was added with IAA, to excised sections of tree stems, was extended in later papers. For five genera, WAREING et al. (1964) found the maximal width of new secondary xylem and phloem tissue when IAA and GA were added together to excised pieces of stem (Fig. 4.12). The cells considered to be on the phloem side of the cambium did not differentiate into sieve tubes or companion cells, however, whether they resulted from GA alone or GA plus IAA. IAA added alone resulted in some tracheary cells but no phloem cells. Kinetin, whether added alone or with IAA, was said to be without effect or perhaps slightly inhibitory. The

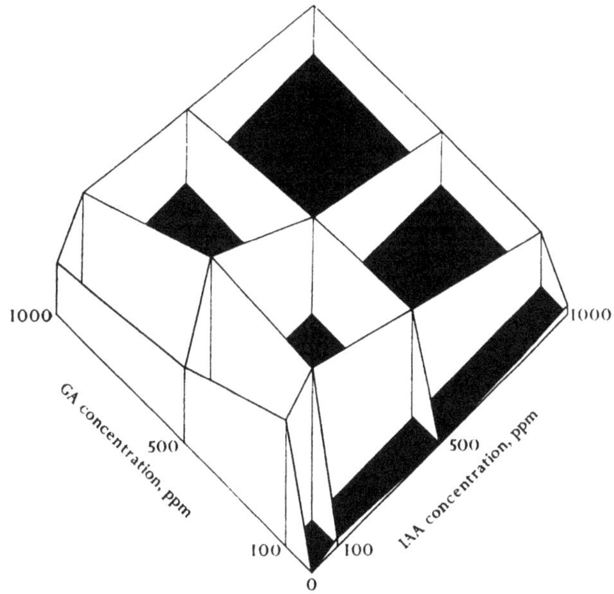

Fig. 4.12. The effect on the radial width of phloem produced of IAA, GA_3, or both, applied in lanolin to disbudded shoots of *Populus*. (DIGBY and WAREING 1966a)

data on IAA and GA were repeated in DIGBY and WAREING (1966a), with the added statement that in *Vitis* the newly produced phloem *did* contain differentiated sieve elements. (Whether there were more such differentiated cells from adding IAA or GA or both was not stated and presumably not known.)

Scrapings of the cambial area, earlier shown to contain auxin by direct bioassay (SÖDING 1940), were extracted with methanol and run on paper chromatograms. A zone with auxin activity and one with gibberellin activity were found in bioassays of the chromatograms (DIGBY and WAREING 1966b). The authors concluded that the relative amount of the two types of hormone in the cambial zone probably controls how many cells are produced to the phloem and xylem sides, with high GA and low IAA resulting in more cells to the secondary phloem.

Several other laboratories have reported that auxin can be obtained from the general cambial area, whether by extraction or diffusion (e.g., NITSCH and NITSCH 1965, SHEPHERD and ROWAN 1967, SHELDRAKE 1971, ZAJACZKOWSKI 1973, WODZICKI 1978, LITTLE et al. 1978). Authors of the first four papers, however, used the oat mesocotyl bioassay, which responds to all nine gibberellins tested as well as to auxins (NITSCH and NITSCH 1962). SHELDRAKE, in addition, did not cite the prior work of DIGBY and WAREING. Almost all the papers before the 1970's based their interpretations on growth responses in bioassays, often supplemented by information about location on chromatograms. LITTLE et al. (1978) provided more convincing identification of both IAA and abscisic acid (ABA) in the cambial region from combined gas chromatography-mass spectrometry.

Interest in other endogenous substances that might interact with IAA in stimulating cambial activity was natural in the light of evidence that adding a cambial extract along with auxin was more effective than auxin alone in stimulating cambial activity (SÖDING 1940, ZAJACZKOWSKI and WODZICKI 1975).

Cytokinins were found by bioassay of extracts of cambial scrapings (BOTTOMLEY et al. 1963) and of the "inner bark" of *Populus* trees (NITSCH and NITSCH 1965). The latter authors had guessed that *Populus* cambial zone would contain cytokinins because it was well known as a source of GAUTHERET'S (1959) cambial tissue cultures, which did not require added cytokinin for growth. Data of LETHAM (1967) support the view that carrot cambium contains more cytokinin than does phloem or xylem, as JACOBS (p 127, 1979) pointed out. ROGOZIŃSKA (1967), studying Scots pine, also reported cytokinin activity from extracts of "bark + cambium".

Several laboratories followed in the footsteps of DIGBY and WAREING (1966a) in adding various chemicals to de-budded stem sections, then checking the effect on cambial activity. Most measured or counted only secondary xylem as a measure of cambial activity, either ignoring the phloem entirely or finding it difficult to measure accurately (ROBARDS et al. 1969). DEMAGGIO (1966) reported that GA_3 increased the number of cells on the phloem side of pine cambium, but provided no convincing evidence (such as presence of sieve plates) that the cells were fully differentiated. For those who checked tracheary differentiation, it is perhaps not surprising—in view of all the evidence cited earlier that IAA normally limits tracheary regeneration—that auxins were the most generally effective class of chemical (ROBARDS et al. 1969, HEJNOWICZ and TOMASZEWSKI 1969, SHININGER 1971, HESS and SACHS 1972, ZAJACZKOWSKI 1973).

In some cases GA_3 or cytokinins increased the number of secondary xylem cells, particularly when combined with added auxin (HEJNOWICZ and TOMASZEWSKI 1969, HESS and SACHS 1972); in other cases they were without effect or even decreased the number (ROBARDS et al. 1969). Noting that earlier workers had shown that GA increased IAA transport (e.g., JACOBS and CASE 1965), as did cytokinin (e.g., BLACK and OSBORNE 1965), HEJNOWICZ and TOMASZEWSKI demonstrated that such increased transport of auxin was correlated with the increased formation of xylem from such hormonal combinations added to their woody stems. Sucrose also increased the number of xylem cells resulting from auxin (ROBARDS et al. 1969, ZAJACZKOWSKI 1973), an effect which similarly might be due to the increased IAA transport that is elicited by sucrose (SMITH and JACOBS 1968). The decrease in xylem area caused by the addition of abscisic acid (HESS and SACHS 1972) is likewise potentially explainable as due to abscisic acid's effect in decreasing IAA transport (CHANG and JACOBS 1973).

In short, the stimulating effects from adding GA or cytokinins along with auxin to de-budded stem segments are probably partly if not completely the result of increased transport of the auxin down the stem to the cambial region. The lack of such stimulation by added GA or cytokinin in other cases may then be explained on the basis of their being enough endogenous GA and cytokinin present so that neither is limiting auxin movement. Such a view raises the associated question: "Does GA or cytokinin have any direct effect—other than via auxin—on cambial activity (or on tracheary differentiation)?"

4.2.3 Control of Differentiation in Roots

By a comparison of the most distal level to which tracheary cells differentiated in cultured pea roots, TORREY (1953) concluded that IAA added to such cultures increased the rate at which tracheary cells differentiated, while drastically reducing the elongation of the roots. A similar stimulation of tracheary differentiation in bean roots on intact seedlings was obtained from GA_3 added to the nutrient solution (ODHNOFF 1963). For phloem differentiation in roots hormonal effects have been essentially uninvestigated. Cambial activity, however, has been shown to be stimulated in excised cultured roots if both an auxin and a cytokinin are added to the cut base (LOOMIS and TORREY 1964, TORREY and LOOMIS 1967, PETERSON 1973). Neither auxin nor cytokinin alone was sufficient. The requirement for both types of hormone for cambial activity in roots was considered tied to the transport requirement of the cytokinins: the added IAA caused much more movement of the cytokinin from radish root base toward the root tip and only when both hormones were on the *basal* end did cambial activity occur all along the root section (RADIN and LOOMIS 1974). A satisfying touch was added when RADIN and LOOMIS (1971) found three zones of endogenous cytokinin activity in extracts of radish roots from intact plants, two of the zones appearing at the time cambium was starting to form in the roots of intact seedlings. Eluates from each of these two zones stimulated cambial activity. The third zone of cytokinin activity did not appear until much later in radish development and did not stimulate cambium. Their interpretation was that in the intact plant the first two cytokinins moved down from the shoot and caused cambial activity to develop acropetally in the root, while the third cytokinin was apt to be formed in the root from which it would appear in the xylem exudate as reported by several earlier papers (see JACOBS 1979, for more details).

An early report of kinetin's action on stimulating cambial activity concerned the transition region between shoot and root on dark-grown pea (SOROKIN et al. 1962). Kinetin added alone or with auxins to the sucrose and penicillin solution bathing excised sections of the transition internode produced more normal-looking cambial activity and secondary xylem than did the auxins alone, judging by stained sections cut free-hand.

4.2.4 Evidence of Differentiation from Tissue Cultures

Many papers have been published on hormonal effects on tissue cultures (see reviews by TORREY et al. 1971, ROBERTS 1976, SHININGER 1979). The results have been diverse. One would expect some diversity, of course, from the reactions of different genera. One would also expect diversity from the various tissue origins of the cultures. In addition, in the stem of an intact plant, investigators might—with luck—discover that, of all the substances presumably required to differentiate a tracheary cell, only one was actually present in limiting amounts. This seems to be the case quite generally for auxin and tracheary cells. However, when one cuts out pieces for culturing that are smaller and smaller and less and less diverse histologically, it is only to be expected that one will often be left with a group of cells that now needs *many* substances added to permit cell differentiation. The substances that were available to the pith from the vascular tissue, for instance in a whole internode, will have to be supplied by the researcher culturing pith tissues per se. Such cell-division activity was demonstrated by JABLONSKI and SKOOG (1954) from vascular tissue and by BOTTOMLEY et al. (1963) more specifically from the internal phloem of tobacco.

Accordingly, unless results with tissue cultures are related in a meaningful way to the more intact organism they lose much of their potential value. An additional major weakness of most tissue-culture publications is that they omit completely information on the effects on sieve-tube differentiation of the hormonal treatments. Because sieve-tubes always form before tracheary cells in both normal development (ESAU 1965) and in most reported cases of regeneration (JACOBS 1970, ALONI and JACOBS 1977b), it is reasonable to expect them to form first in tissue cultures also. Among the few who reported on sieve-tube differentiation in cultures were WETMORE and RIER (1963) and JEFFS and NORTHCOTE (1966); both said that relatively high sucrose levels added with IAA increased the relative amount of phloem formed. ALONI (1980) reported that applying his clearing technique to callus cultures from five genera showed that sieve tubes did differentiate before tracheary cells, that low IAA levels resulted in sieve tubes but no tracheary cells, and that high sucrose levels increased callose on sieve plates but not the ratio of phloem to xylem.

Except for the fact that an auxin is typically required for tracheary differentiation in cultures, there is little consistency in the results of those concentrating on tissue cultures. It seems pointless, therefore, to list the varied results. References to the papers can be found in the reviews cited above.

4.3 General Discussion

A unified picture of vascular differentiation has arisen from the studies of regeneration and differentiation. After the discovery by LAMOTTE and JACOBS (1963), THOMPSON and JACOBS (1966), and THOMPSON (1967) that IAA controlled sieve-tube regeneration as well as tracheary regeneration, and that the sieve-tube development closely pre-figured—in both space and time—the development of the regenerating tracheary strands, it was natural to wonder if the *normal* development of these two vascular tissues was as different as classical anatomists believed. The classical view was that the first sieve tubes in vascular strand differentiated continuously and acropetally, whereas the first tracheary cells differentiated at an isolated locus near the leaf base (ESAU 1965). A critical appraisal of the classical literature made clear that not only were sample sizes painfully small by current physiological standards, but that collections were typically only made in the daytime. On the assumption that plants would not conveniently stop differentiating at 17.00 h, JACOBS and MORROW (1957, 1967) reinvestigated normal vascular differentiation, using larger samples collected throughout the 24 h. They found that sieve tubes differentiated first at an isolated locus near the leaf base and that later, and on the inner side of the trace, the first tracheary cell differentiated. (Most of the young leaves showing an isolated locus of sieve tube differentiation were indeed from the night-time collections.) In other words, the timing and the spatial pattern of differentiation of sieve tubes and tracheary cells were similar in both normal differentiation in the very young leaves and in regeneration in the older internode of the stem. For the latter, the earlier differentiation of sieve tubes was explicable by the observed fact that levels of IAA which were too low to cause tracheary regeneration were sufficient to cause the regeneration of sieve tubes (Fig. 4.9).

Many of the investigators who specialize in tissue culture have been interested in whether IAA is primarily limiting cell division, rather than tracheary differentiation (e.g., FOSKET 1970). In its general form this question is not very amenable to experimental treatment, other than to show that some treatments will increase tracheary differentiation more than they increase cell number (e.g., MINOCHA and HALPERIN 1974). It will become meaningful, however, when tissue culture specialists can take an isolated individual cell, already differentiated as a nontracheary cell, and induce it by hormonal treatment to change without cell division into a tracheary cell. Meanwhile, a more fruitful approach probably would be to determine in their cultures not merely tracheary cell[1] (or "wound vessel member") differentiation but also the differentiation of sieve tube members that both normal and regenerative development lead us to expect will precede and perhaps control the differentiation of tracheary cells.

The role of IAA in controlling the differentiation of tracheary cells (and to the much smaller extent that they have been studied, of sieve cells) seems well established. At least part of the role of other hormones (e.g., GA, ABA, cytokinins) in tissue differentiation is doubtless due to their demonstrated effects on the transport or available level of IAA. Whether they have more direct effects on cell differentiation is yet to be proven.

Finally, we will end with work of SACHS (1968a, b, 1969) which because of its originality and the fact that it applies to both roots and shoots does not fit neatly into any of our foregoing categories. He investigated the previously neglected question of what controls the path of the regenerating tracheary strands. In particular, he asked why the basipetally regenerating strands so often curve around the base of a wound and connect with the preexisting strand (as in Fig. 4.1). In a series of clever experiments with pea seedlings, he demonstrated that a newly regenerating tracheary strand tended to join a previously existing strand, particularly if the preexisting strand was well supplied with IAA. Auxin supplied to a pre-existing strand not only "attracted" the new strand, but changed the angle at which the new strand joined and decreased the number of plants that could be induced to form a new strand under the influence of exogenous IAA (SACHS 1968b, 1969).

[1] One of the biologists who devoted his life to tissue culture research pointed out that the "tracheary elements" of tissue culture do not correspond precisely with normal primary xylem and "only by tradition" is it justified to call their origin in tissue cultures "xylogenesis" (STREET 1978)

References

Aldaba VC (1927) The structure and development of the cell wall in plants. I. Bast fibers of *Boehmeria* and *Linum*. Am J Bot 14:16–24

Aloni R (1976) Polarity of induction and pattern of primary phloem fiber differentiation in *Coleus*. Am J Bot 63:877–889

Aloni R (1979) The role of auxin and gibberellin in differentiation of primary phloem fibers. Plant Physiol 63:609–614

Aloni R (1980) Role of auxin and sucrose in the differentiation of sieve and tracheary elements in plant tissue cultures. Planta 150:255–263

Aloni R, Jacobs WP (1977a) Polarity of tracheary regeneration in young internodes of *Coleus* (Labiatae). Am J Bot 64:395–403

Aloni R, Jacobs WP (1977b) The time course of sieve tube and vessel regeneration and their relation to phloem anastomoses in mature internodes of *Coleus*. Am J Bot 64:615–621

Aloni R, Sachs T (1973) The three-dimensional structure of primary phloem systems. Planta 113:345–353

Atal CK (1961) Effect of gibberellin on the fibers of hemp. Econ Bot 15:133–139

Avery GS, Burkholder PR, Creighton H (1937) Production and distribution of growth hormone in shoots of *Aesculus* and *Malus* and its probable role in stimulating cambial activity. Am J Bot 24:51–58

Benayoun J, Aloni R, Sachs T (1975) Regeneration around wounds and the control of vascular differentiation. Ann Bot 39:447–454

Beslow DT, Rier JP (1969) Sucrose concentration and xylem regeneration in *Coleus* internodes in vitro. Plant Cell Physiol 10:69–77

Black MK, Osborne DJ (1965) Polarity of transport of benzyladenine, adenine and indole-3-acetic acid in petiole segments of *Phaseolus vulgaris*. Plant Physiol 40:676–680

Bottomley W, Kefford NP, Zwar JA, Goldacre PL (1963) Kinin activity from plant extracts. I. Biological assay and sources of activity. Aust J Biol Sci 16:395–406

Chang YP, Jacobs WP (1973) The regulation of abscission and IAA by senescence factor and abscisic acid. Am J Bot 60:10–16

Delisle AL (1938) Morphogenetical studies in the development of successive leaves in *Aster*, with respect to relative growth, cellular differentiation, and auxin relationships. Am J Bot 25:420–430

DeMaggio AE (1966) Phloem differentiation: induced stimulation by gibberellic acid. Science 152:370–372

Digby J, Wareing PF (1966a) The effect of applied growth hormones on cambial division and the differentiation of the cambial derivatives. Ann Bot 30:539–548

Digby J, Wareing PF (1966b) The relationship between endogenous hormone levels in the plant and seasonal aspects of cambial activity. Ann Bot 30:607–622

Esau K (1965) Vascular differentiation in plants. Holt, Rinehart and Winston, New York

Fosket DE (1970) The time course of xylem differentiation and its relation to deoxyribonucleic acid synthesis in cultured *Coleus* stem segments. Plant Physiol 46:64–68

Gautheret RJ (1959) La culture des tissues végétaux. Masson et Cie, Paris

Gee H (1972) Localization and uptake of ^{14}C-IAA in relation to xylem regeneration in *Coleus* internodes. Planta 108:1–9

Goodwin RH (1937) The role of auxin in leaf development in *Solidago* species. Am J Bot 24:43–51

Hayward HE (1938) The structure of economic plants. MacMillan, New York

Hejnowicz A, Tomaszewski M (1969) Growth regulators and wood formation in *Pinus silvestris*. Physiol Plant 22:984–992

Hemberg T (1972) The effect of kinetin on the occurrence of acid auxin in *Coleus blumei*. Physiol Plant 26:98–103

Hess T, Sachs T (1972) The influence of a mature leaf on xylem differentiation. New Phytol 71:903–914

Houck DF, LaMotte CE (1977) Primary phloem regeneration without concomitant xylem regeneration: its hormone control in *Coleus*. Am J Bot 64:799–809

Jablonski JR, Skoog F (1954) Cell enlargement and cell division in excised tobacco pith tissue. Physiol Plant 7:16–24

Jacobs WP (1952) The role of auxin in differentiation of xylem around a wound. Am J Bot 39:301–309

Jacobs WP (1954) Acropetal auxin transport and xylem regeneration — a quantitative study. Am Nat 88:327–337

Jacobs WP (1956) Internal factors controlling cell differentiation in the flowering plants. Am Nat 90:163–169

Jacobs WP (1961) The polar movement of auxin in the shoots of higher plants: its occurrence and physiological significance. In: Klein RM (ed) Plant growth regulation. Iowa State Univ Press, Ames, pp 397–409

Jacobs WP (1970) Regeneration and differentiation of sieve-tube elements. Int Rev Cytol 28:239–273

Jacobs WP (1977) Polarity of indoleacetic acid in young *Coleus* stems. Plant Physiol 60:95–97

Jacobs WP (1979) Plant hormones and plant development. Cambridge Univ Press, New York

Jacobs WP, Case DB (1965) Auxin transport, gibberellin, and apical dominance. Science 148:1729–1731

Jacobs WP, McCready CC (1967) Polar transport of growth regulators in pith and vascular tissues of *Coleus* stems. Am J Bot 54:1035–1040

Jacobs WP, Morrow IB (1957) A quantitative study of xylem development in the vegetative shoot apex of *Coleus*. Am J Bot 44:823–842

Jacobs WP, Morrow IB (1967) A quantitative study of sieve tube differentiation in vegetative shoot apices of *Coleus*. Am J Bot 54:425–431

Jacobs WP, Danielson J, Hurst V, Adams P (1959) What substance normally controls a given biological process? II. The relation of auxin to apical dominance. Dev Biol 1:534–554

Jeffs RA, Northcote DH (1966) Experimental induction of vascular tissue in an undifferentiated plant callus. Biochem J 101:146–152

Jost L (1893) Über Beziehungen zwischen der Blattentwicklung und der Gefäßbildung in der Pflanze. Bot Z 51:89–138

LaMotte CE, Jacobs WP (1962) Quantitative estimation of phloem regeneration in *Coleus* internodes. Stain Technol 37:63–73

LaMotte CE, Jacobs WP (1963) A role of auxin in phloem regeneration in *Coleus* internodes. Dev Biol 8:80–98

Letham DS (1967) Regulators of cell division in plant tissues. V. A comparison of the activities of zeatin and other cytokinins in five bioassays. Planta 74:228–242

Little CHA, Heald JK, Browning G (1978) Identification and measurement of indoleacetic and abscisic acids in the cambial region by combined gas chromatography-mass spectrometry. Planta 139:133–138

Loomis RS, Torrey JG (1964) Chemical control of vascular cambium initiation in isolated radish roots. Proc Natl Acad Sci USA 52:3–11

McCready CC, Jacobs WP (1963) Movement of growth regulators in plants IV. Relationships between age, growth and polar transport in petioles of *Phaseolus vulgaris*. New Phytol 62:360–366

Minocha S, Halperin W (1974) Hormones and metabolites which control tracheid differentiation with or without concomitant effects on growth in cultured tuber tissue of *Helianthus tuberosus* L. Planta 116:319–331

Nitsch JP, Nitsch C (1962) Activités comparées de neuf gibbérellines sur trois tests biologiques. Ann Physiol Vég 4:85–97

Nitsch JP, Nitsch C (1965) Présence d'une phytokinine dans le cambium. Bull Soc Bot 112:1–10

Odhnoff C (1963) The effect of gibberellin and phenylboric acid on xylem differentiation and epidermal cell elongation in bean roots. Physiol Plant 16:474–483

Peterson RL (1973) Control of cambial activity in roots of turnip (*Brassica rapa*). Can J Bot 51:475–480

Radin JW, Loomis RS (1971) Changes in the cytokinins of radish roots during maturation. Physiol Plant 25:240–244

Radin JW, Loomis RS (1974) Polar transport of kinetin in tissues of radish. Plant Physiol 53:348–351

Reinders-Gouwentak CA (1965) Physiology of the cambium and other secondary meristems of the shoot. In: Ruhland W (ed) Handbuch der Pflanzenphysiologie XV/1. Springer, Berlin Göttingen Heidelberg, pp 1077–1105

Robards AW, Davidson E, Kidwai P (1969) Short-term effects of some chemicals on cambial activity. J Exp Bot 20:912–921

Roberts LW (1976) Cytodifferentiation in plants. Xylogenesis as a model system. Cambridge Univ Press, New York

Rogozińska JH (1967) The occurrence of cytokinins in Scots pine. Bull Acad Pol Sci 15:789–794

Sabnis DD, Hirshberg G, Jacobs WP (1969) Radioautographic analysis of the distribution of label from ^{3}H-indoleacetic acid supplied to isolated *Coleus* internodes. Plant Physiol 44:27–36

Sachs T (1968a) The role of the root in the induction of xylem differentiation in peas. Ann Bot 32:391–399

Sachs T (1968b) On the determination of the pattern of vascular tissue in peas. Ann Bot 32:781–790

Sachs T (1969) Polarity and the induction of organized vascular tissues. Ann Bot 33:263–275

Sachs T (1972) The induction of fibre differentiation in peas. Ann Bot 36:189–197

Scott TK, Jacobs WP (1963) Auxin in *Coleus* stems: limitation of transport at higher concentrations. Science 139:589–590

Scott TK, Jacobs WP (1964) Critical assessment of techniques for identifying the physiologically significant auxins in plants. In: Nitsch JP (ed) Régulateurs naturels de la croissance végétale. CNRS, Paris, pp 457–474

Sheldrake AR (1971) Auxin in the cambium and its differentiating derivatives. J Exp Bot 22:735–740

Shepherd KR, Rowan KS (1967) Indoleacetic acid in cambial tissue of radiata pine. Aust J Biol Sci 20:637–646

Shininger TL (1971) The regulation of cambial division and secondary xylem differentiation in *Xanthium* by auxins and gibberellin. Plant Physiol 47:417–422

Shininger TL (1979) The control of vascular development. Annu Rev Plant Physiol 30:313–337

Simon S (1908) Experimentelle Untersuchungen über die Differenzierungsvorgänge im Callusgewebe von Holzgewächsen. Jahrb Wiss Bot 45:351–478

Sinnott EW, Bloch R (1945) The cytoplasmic basis of intercellular patterns in vascular differentiation. Am J Bot 32:151–156

Sircar SM, Chakraverty R (1960) The effect of gibberellic acid on jute (*Corchorus capsularis*). Sci Cult 26:141–143

Smith CW, Jacobs WP (1968) The movement of IAA-^{14}C in the hypocotyl of *Phaseolus vulgaris*. In: Vardar Y (ed) The transport of plant hormones. Elsevier/North-Holland, Amsterdam, pp 48–59

Söding H (1940) Weitere Untersuchungen über die Wuchsstoffregulation der Kambiumtätigkeit. Z Bot 36:113–141

Sorokin HP, Mathur SN, Thimann KV (1962) The effects of auxins and kinetin on xylem differentiation in the pea epicotyl. Am J Bot 49:444–453

Stant MY (1961) The effect of gibberellic acid on fibre-cell length. Ann Bot 25:453–462

Stant MY (1963) The effect of gibberellic acid on cell width and the cell wall of some phloem fibers. Ann Bot 27:185–196

Steeves TA, Briggs WR (1960) Morphogenetic studies on *Osmunda cinnamomea* L. The auxin relationships of expanding fronds. J Exp Bot 11:45–67

Street HE (1978) Differentiation in cell and tissue cultures — regulation at the molecular level. In: Schütte HR, Gross D (eds) Regulation of developmental processes in plants. Fischer, Jena, pp 192–218

Thompson NP (1965) The influence of auxin on regeneration of xylem and sieve tubes around a stem wound. PhD Thesis, Princeton Univ. (No. 66–4017 from Univ Microfilms, Ann Arbor, Mich

Thompson NP (1967) The time course of sieve tube and xylem cell regeneration and their anatomical orientation in *Coleus* stems. Am J Bot 54:588–595

Thompson NP (1968) Polarity of IAA-^{14}C and 2,4-D-^{14}C transport and vascular regeneration in isolated internodes of peanut. In: Wightman F, Setterfield G (eds) Biochemistry and physiology of plant growth substances. Runge, Ottawa, pp 1205–1213

Thompson NP (1970) The transport of auxin and regeneration of xylem in okra and pea stems. Am J Bot 57:390–393

Thompson NP, Jacobs WP (1966) Polarity of IAA effect on sieve-tube and xylem regeneration in *Coleus* and tomato stems. Plant Physiol 41:673–682

Torrey JG (1953) The effect of certain metabolic inhibitors on vascular tissue differentiation in isolated pea roots. Am J Bot 40:525–533

Torrey JG, Loomis RS (1967) Auxin-cytokinin control of secondary vascular tissue formation in isolated roots of *Raphanus*. Am J Bot 54:1098–1106

Torrey JG, Fosket DE, Hepler PK (1971) Xylem formation: A paradigm of cyto-differentiation in higher plants. Am Sci 59:338–352

Wangermann E (1967) The effect of the leaf on differentiation of primary xylem in the internode of *Coleus blumei* Benth. New Phytol 66:747–754

Wangermann E (1970) Autoradiographic localization of soluble and insoluble ^{14}C from [^{14}C]indolylacetic acid supplied to isolated *Coleus* internodes. New Phytol 69:919–927

Wareing PF (1958) Interaction between indole-acetic acid and gibberellic acid in cambial activity. Nature 181:1744–1745

Wareing PF, Hanney CEA, Digby J (1964) The role of endogenous hormones in cambial activity and xylem differentiation. In: Zimmermann MH (ed) The formation of wood in forest trees. Academic Press, London New York, pp 323–344

Wetmore RH, Rier JP (1963) Experimental induction of vascular tissues in callus of angiosperms. Am J Bot 50:418–430

Wodzicki TJ (1978) Seasonal variation of auxin in stem cambial region of *Pinus silvestris* L. Acta Soc Bot Pol 47:225–231

Zajaczkowski S (1973) Auxin stimulation of cambial activity in *Pinus silvestris*. I. The differential cambial response. Physiol Plant 29:281–287

Zajaczkowski S, Wodzicki TJ (1975) Inhibition and requirement of natural stimulator for cambial xylem production in isolated stem segments of *Pinus silvestris*. Physiol Plant 33:71–74

5 Functions of Hormones at the Organ Level of Organization

A.W. NAYLOR

5.1 Introduction

Hormones, like other small molecules, have their effects first at the molecular level. The effects seen with the naked eye are far removed from the starting point. Many efforts have been made to account for the action of plant hormones. Massive doses of growth substances to the roots, stems, leaves, flowers, and fruits have elicited spectacular results, many of which have become of enormous commercial importance (NICKELL 1978, 1982, WITTWER 1978). But as far as basic physiology is concerned, one must be extremely cautious in equating reaction to large doses with reaction to physiologically normal amounts in the plant. Another common fallacy is to attribute growth behavior patterns to single hormones. While a single hormone may have overriding importance in a given instance, final expression is likely to be the result of the interaction of several hormones together with other internal and external factors. The analogy of an orchestra playing a composition is an apt one. Sometimes one instrument will have a solo role for a time with the other instruments playing supporting roles; later one or more of the supporting instruments assumes dominance. If the composition is to be played successfully, however, all instruments must be played in synchrony. The importance of consideration of growth from the standpoint of multiple factor interactions has also been stressed by SKOOG and MILLER (1957), HILL (1973), IVONIS (1981), RÜCKER (1982), LEOPOLD and NOODEN (Chap. 1, this Vol.), EVANS (Chap. 2, this Vol.), MATTHYSSE and SCOTT (Chap. 6, this Vol.).

In the final analysis, because of the enormous effects caused by little more than trace amounts of hormones, one must interpret hormone action in terms of enzyme regulators, genetic readout effectors or suppressors, membrane change effectors, nuclear division figure polarizers, regulators of microtubule formation and regulators of bioenergetic processes. The fact that amorphous masses of cells or groups of cells in tissue masses become polarized is fundamental to the establishment of the axis with a stem meristem at one end and a root meristem at the other.

If we are to understand phenomena such as tissue growth, cell differentiation, and organ development, we must first understand how spindle orientation or direction is regulated during nuclear division. Furthermore, we must understand how the cell cycle is started and stopped. The elucidation of these processes can probably be most readily achieved by using initially homogeneous populations of cells grown in axenic culture because such populations are those that can be most readily manipulated. Theoretically, if the details of the biochemical and bioelectric events relative to nuclear division, cytokinesis, and wall formation were known, it would be possible to narrow to a manageable degree the probable point or points of impact of hormones. For example, studies with cultured hamster cells have revealed a sequence of cell-cycle-dependent phosphorylations of histones that reflect specific changes in histone structure that are related to changes in function of chromatin. Among these has been the demonstration that a specific serine residue of the f1 histone is phosphorylated after cyclic AMP stimulation (HOHMANN et al. 1975). Thus, one of the effects of cyclic AMP on the regulation of cellular activity is achieved through phosphorylation of a histone which is accompanied by a change in conformation of DNA. Although the existence of cyclic-AMP in plant tissues has been a matter of debate (AMRHEIN 1977), unequivocal proof of its existence in *Zea mays* seedlings has been provided (JANISTYN 1981). To this was quickly added tobacco callus tissue, bean seedlings and immature corn kernels (JOHNSON et al. 1981). A need, however, still exists to show that c-AMP functions in plants. If it does, one may now envision a systematic progression of specific histone modifications necessarily occurring sequentially as chromatin functions in an orderly fashion during the cell cycle (GURLEY et al. 1977).

In actively dividing cells requiring cytokinin for growth, it is now clear that this hormone regulates processes associated with the G_2 phases or the transition from G_2 to mitosis (JOUANNEAU and TANDEAU DE MARSAC 1973). Auxin, on the other hand, stimulates incorporation of ^{3}H-thymidine into late G_1—early S-phase nuclei (SIMARD 1971, WARDELL 1975). Other discoveries of the roles of hormones in relation to the genetics of developmental regulation lie ahead. Many of the techniques required for making progress in this area of biochemical differentiation and development are becoming available.

Since the dawn of agriculture man has sought more and more control over the growth of his crop plants. As recently as 200 years ago 45% of the population was engaged in agriculture while today less than 5% are so employed, an eloquent tribute to man's increasing control of plant productivity. The race continues, however, between population growth and man's ability to increase the growth and reproduction of plants.

It is a basic operational tenet of the plant scientist that the more we know about the fundamental features of growth the better our chances of manipulating the rate of growth, the amount of growth, the time of reproductive activity and the amount of storage in storage organs. Two approaches are being used in our race with time: the empirical approach and the basic science approach. The basic discovery that plants produce hormones that play a regulatory role in growth, development, and reproduction has been seized upon by the agricultural chemical industry and exploited with spectacular results. Growth regulators can be considered the fastest growing component of the agricultural chemicals market. They already make up 5 to 10 percent of it.

Why the interest? Regulation of growth can be useful in many ways, as has been pointed out by NICKELL (1978). Some of these ways are:

- Promotion of rooting and the mass production of clones
- Simultaneous initiation or termination of dormancy of seeds, buds, and tubers

- Promotion or delay of flowering
- Chemical thinning of flowers
- Modification of sex expression
- Control of fruit set and fruit development
- Control of plant or organ size
- Regulation of the chemical composition of plants and fruit color
- Induction or prevention of fruit drop
- Prevention of post-harvest spoilage
- Increased plant resistance to pests
- Enhanced plant resistance to such environmental factors as temperature, water, and air pollution.

The impiricists work with both naturally produced and synthetic growth regulators, whereas the basic scientist is chiefly concerned with naturally produced growth regulators. By growth regulator is meant an organic compound other than nutrients that in low concentrations alters the physiological processes of plants. By this is meant that such compounds will alter the plants' life processes or structure in some beneficial way thereby enhancing yields, improving quality or facilitating harvesting. Some of these compounds are beneficial over only a narrow concentration range and may become lethal, as is found spectacularly for many of the herbicides.

In what follows we shall be concerned chiefly with naturally produced growth regulators—plant hormones. These are involved as correlating agents at all stages in the life cycle—from seed to seed. An integrative multifactor approach will be used because the single factor approach is neither adequate nor realistic.

5.2 Embryogeny— Establishment of Polarity[1]

There are numerous ways in which polarity can be established, even for a single zygote. Polarity becomes established whenever there is an unequal distribution of cellular constituents, ions, plastids, mitochondria, ribosomes, glyoxysomes, and Golgi bodies within a single cell. These differences lead to polarly oriented fluxes. Ion currents are involved in establishing developmental gradients (JAFFE 1981, SCHNEPF 1981). Unequal distribution can come about through unequal divisions of cells in which the organelles are uniformly distrubuted. This is accompanied by the setting up of metabolic gradients that are the essence of polarity. Light, temperature, pH, salts, metabolic fuels, hormones, and gravity can each have an overriding influence in the establishment of polarity. Often, the trajectory of the sperm to the egg nucleus serves this role. In seed plants pollen tubes grow along possible chemotropic factor gradients, not yet defined chemically, to the ovule where penetration and destruction of the synergid of the megagametophyte occurs. This is followed by rupture of the pollen tube and fertilization takes place (VAN DER PLUIJM 1964, CASS and JENSEN 1970).

[1] In the context of Chapters 5.2–5.4, Chapter 20 (STANGE, L.: Cellular Interactions During Early Differentiation) of Vol. 17 this Series, may also be of interest to the reader.

The mechanism whereby polarity is established is most easily investigated in single cell cultures. Much has been learned from studies with *Fucus* zygotes, *Acetabularia*, yeast, filaments of *Cladophora* and *Griffithsia* (reviewed by QUATRANO 1978; see also Chapters on polarity in Vol. 7 this Series, Chap. 5 and in Vol. 17 this series, Chap. 22). Work with *Fucus* zygotes has yielded the most information at both the whole cell and subcellular levels. Apolar zygotes of *Fucus* respond to a number of external gradients including light, gravity, pH, and physical pressure. For a number of hours after the imposition of one of these gradients the zygote's polar axis can be changed. This can be done with the same gradient from another direction or a different gradient. There is a time limit, however, and after a few hours the polar axis is irreversibly set. This is followed by the intracellular particles and external membrane "ion pumps" becoming asymmetrically distributed with certain factors accumulating at the point signaling the site of rhizoid formation (JAFFE 1981). Axis fixation appears to be associated with microfilament assembly or stability, in effect a cytoskeleton.

The microtubule band that presages the orientation of the spindle and cell plate begins development at the end of the S-phase or early in the G_2-phase (WICK et al. 1981). In actively dividing cells of several species requiring cytokinin the hormone apparently regulates processes in either the G_2-phase of the cell cycle or in the transition from G_2 to mitosis (reviewed by FOSKET 1977). However, a generalization cannot be made. In a careful analysis of results with a cytokinin-dependent suspension culture of soybean cells, cytokinin deprivation did not cause arrest at any particular point in the cell cycle (WANG et al. 1981). Cell arrest is extremely important. In its absence we have cancer. Trigonellin appears to be a naturally occurring cell-arrest factor. This substance is found to be abundant in cotyledons of dry seeds. It promotes preferential cell arrest in the G_2-phase during normal cell differentiation (TRAMONTANO et al. 1982).

External environmentally induced asymmetric localization of "ion pumps" appears to be achieved through the action of microfilaments. These "ion pumps", such as the Ca^+ pump, are involved in directing ions into the cytoplasm and, when in an asymmetric configuration, cause the generation of an electrical potential gradient across the cell. Localized influx of ions can serve to initiate enzyme activation, vesicle secretion associated with membrane and cell wall growth and spindle orientation, thereby amplifying intracellular polarity. The electrical gradient would inevitably result in redistribution of charged particles within the cell. Although asymmetric application of growth regulators is known to induce polarity, there are no known measurements showing the presence, let alone the cellular distribution of growth substances, in these cells.

In a very real sense, polarity is imposed on a zygote in the megagametophyte of seed plants. This is true because of the polarity of the megagametophyte within the ovule. Location of the egg at the micropyle and the antipodal cells in the chalazal end of the megasporangium provide a clearcut alignment. Shortly after fertilization the zygote shows a gradient in the density and distribution of organelles. A large vacuole is formed and the nucleus migrates to the chalazal end of the cell where it becomes surrounded by mitochondria and other cytoplasmic components. The first nuclear division is followed by formation of a transverse wall producing two distinctly different cells: a large highly vacuolated basal cell and a small cytoplasmatically dense terminal cell (SCHULZ and JENSEN 1968a, b). The basal cell divides immediately, giving rise through transverse divisions to the filamentous suspensor, while the apical cell, which will give rise to the embryo, is quiescent for a time. The first division of this cell is in a longitudinal plane. This is followed by a second division at right angles to the first, resulting in four cells. Each of these cells divides, producing an eight-celled embryo or octant. At this stage the potentiality for tissue and organ differentiation is established. This should be detectable through increasing differences in mRNA patterns. The upper tier of cells will give rise to the stem tip and cotyledons, while the lower tier will give rise to the hypocotyl and root. This becomes evident following the next essentially simultaneous periclinal division of the cells of the octant.

The cells of the 16-celled globular embryo or proembryo, as it is usually called, are coordinated in such a way that the outer cells form the protoderm and inner cells give rise to the procambium and ground meristem. This is the end of the globular stage although it continues to enlarge. Subsequent development involves the attainment of a new type of symmetry—bilateral in dicotyledous and bilateral in monocotyledons except for lack of one cotyledon. Differentiation becomes manifest in the "heart-shaped" and "torpedo-shaped" stages of the dicotyledonous embryos. One of the first signs of loss of radial symmetry occurs when the suspensor cells nearest the proembryo divide transversely, forming two cells. The daughter cell next to the proembryo is partitioned again, forming two cells. Derivatives of these two cells give rise to the root cortex, root cap, and root epidermis. Meanwhile, on the side of the globule the second sign of symmetry change is seen. The cells cease dividing opposite the pro-root meristem, while in dicotyledonous plants on either side of this quiescent area cells divide periclinally. But in monocots this division pattern is confined to one side of the quiescent area. For a time the most actively growing parts of the embryo are the cotyledon(s) and hypocotyl. Further divisions give rise to the torpedo stage of embryo development. Finally, in the depression at the base of the cotyledon(s), rapid cell divisions result in the development of an apical meristem that gives rise to leaf primordia. (see RAGHAVAN 1976 for more details.)

As the embryo matures, a procambial system differentiates throughout the hypocotyl and cotyledons. This system later gives rise to the components of the vascular system. This is foreshadowed by the development of differences in staining characteristics. Most commonly embryos have only differentiating xylem or phloem elements (BISALPUTRA and ESAU 1964).

In adventiously formed embryos, whether they be in the nucellus of an orange or grapefruit or free cells in nutrient culture, a suspensor is absent. However, in a group of cells destined to produce adventive embryos, some investigators think they can see a distinction between a dividing proembryonal cell and a nondividing suspensor-like cell. The primary distinction is the sharp increase in ribosomal content of the pro-embryonal cell, while there is no change in the suspensor-like cell (SUSSEX 1972). Recent findings in developmental physiology of the embryo have been reviewed by FELLENBERG (1982).

5.3 Hormonal Relations of the Developing Embryo

Commonly, the developing endosperm can export auxin and gibberellin to the developing embryo (MONULA 1978). Another potential source is the suspensor since this structure also possesses more auxin than the embryo (PRZYBYLLOK and NAGL 1977). Cytokinins and their derivatives also make their appearance during embryo development. When the embryo is at the heart-shaped stage, zeatin and 2-isopentenyladenine dominate in the suspensor. Zeatin-glucoside and zeatin-riboside reach their peaks during the cotyledonary stage. An important point is that cytokinin derivatives accumulate during the heart-shaped stage, while it is only later that nonstorage forms are found (LORENZI et al. 1978).

It has also been found that the suspensor of *Phaseolus coccineus* possesses high concentrations of gibberellins at the heart-shaped stage — 30 times as much as in the embryo. While some gibberellic acid might be imported into the suspensor, much of that present is synthesized in situ (CECCARELLI et al. 1979, 1981 a, b, c). As the embryo develops into the cotyledonary stage, gibberellin decreases in the suspensor while increasing in the embryo: the main gibberellin being GA_1 (ALPI et al. 1979).

Biochemical ontogeny is very striking in the developing embryo. Some essential substance(s) are produced in the suspensor that are exported to the young embryo. This can be easily demonstrated since removal of the suspensor, up to the heart-shaped stage results in cessation of embryo development. Sometime later the embryo becomes more self-sufficient (CIONINI et al. 1976, YEUNG and SUSSEX 1979). One of the deficiencies is in GA_3 production (YEUNG and SUSSEX 1979).

Following the heart-shaped stage, the amount of active cytokinins decreases rapidly while inactive forms increase (BURROWS and CARR 1970, NAGL 1974, LORENZI et al. 1978). This correlates well with the periods of cell division. Variation in amounts of cytokinins during embryonic development is important. In vitro experiments with young embryos show that cytokinin in too high concentration is associated with precocious cotyledonary differentiation (HACCIUS 1972, AMMIRATO 1977). Since such precocity can be avoided by changing the hormonal composition of the medium, it is evident that proportionality changes among the hormones are necessary for normal development.

It is assumed that IAA plays an important role in the very early stages of embryo development because in the complete absence of IAA there is abnormal embryo development (HACCIUS 1965). In the young seed the main source would be the developing endosperm and the suspensor. As the embryo matures auxin declines to very low levels (EEUWENS and SCHWABE 1975). However, the amount of bound auxin greatly increases (COHEN and BANDURSKI 1982) in the endosperm.

A multiple role is apparently played by ABA during growth of the seed. Throughout the growth period of the embryo the concentration of ABA is low (KING 1976, QUÉBÉDEAUX et al. 1976, HSU 1979, CHOINSKI et al. 1981), but as the embryo matures there is a rapid increase in amount. The increase is correlated with the synthesis and storage of food reserves (CHOINSKI et al.

1981). For example, vicilin production is stimulated in in vitro cultures of *Phaseolus* embryos in the presence of ABA (RUBENSTEIN et al. 1978). Furthermore, in some instances there is a correlation between the amount of ABA and the cessation of embryo growth in the maturing seed.

5.4 Vascular Development— Spatial Arrangement

One of the evolutionary innovations that have made it possible for plants and animals to attain large sizes, invade the land, and exist in the presence of considerable environmental stress was the development of vascular tissue. Much attention has been devoted to variations on the primary themes that have evolved through time, but our knowledge of the control of differentiation of these specialized cell types and tissues from ground parenchyma is extremely meager.

The problem is not easy to work on experimentally. Vascular development involves the differentiation of parenchyma cells in a developing organ into cell types very different from their neighbors. This occurs in highly symmetrical configurations within a matrix. Key questions for solution are what initiates, regulates, and maintains the spatial distribution of vascular cell formation within the stem, leaf, and root?

SHININGER (1979) sought to integrate the vast anatomical and the very limited physiological literature on vascularization, and showed a welcome trend toward experimentation. Differentiation is a continuous process in plants. As long as there is mitotic activity in the primary and secondary meristems differentiation takes place. New leaf, stem, and root tissues are formed. Some of the cells in a given cross section of the stem or root differentiate quickly, while others contain densely staining protoplasm with multiple tiny vacuoles that continue to divide. These can be described as meristemoids because they divide and produce specialized cell types and tissues. Meristemoids give rise to root hairs, guard cells of stomata and vascular strands. Attention was focused on such cells by BÜNNING in his 1952 and 1957 reviews. He pointed out that following unequal divisions of cells, the smaller of the two cells had the dynamic characteristics of a meristematic cell. Since such cells appeared in a highly patterned way, BÜNNING suggested that these dividing cells produced factors that prevented neighboring cells from also being meristematic. This led to his introducing the concept of mutual incompatibility. Such hypothetical factors presumably would be diffusible growth suppressants. This could account for the uniformity of placement of root hairs, stomates, etc. Experimental evidence in support of the concept that all epidermal cells of roots are potential producers of root hairs came from lateral incision experiments in which direct communication was interrupted between the stele and epidermis. Under these circumstances all cells of the epidermis developed root hair extensions (BEHRE 1929).

When one examines transverse sections from near the apex of a vascular plant's stem, one cannot help but be impressed by the beautiful symmetry of placement of the vascular bundles. In stems it appears that leaves play a key role in determining the location, kind, and amount of vascular bundle development. Therefore, phyllotaxies can be anticipated to be very important in determining stele types. The question of whether procambium differentiation in relation to leaf primordia is the initial stage of vascular tissue differentiation, or whether there is a stage in the process that precedes procambium has been carefully examined by STEEVES and SUSSEX (1972) and by GOODWIN (1978a). This is an important issue because it is crucial to interpretation of the relative influences of a terminal meristem and leaf primordia in regulating vascular differentiation. One could conclude that the terminal meristem plays a passive role in the process if the procambium initiated in the strands is associated with leaf primordia and the developing leaves determine the pattern of the vascular system. But if there is a stage preceding procambial initiation, this stage could be associated with the establishment of the basic pattern of the vascular system. If that is the case, the developing leaves have a subsequent modifying influence on the development of the vascular tissue. These questions have proven to be approachable through experimentation.

WARDLAW (1950, 1952) isolated the apical dome of shoots of five species of seed plants from their normal vasculature by incisions that left the apex connected to the older stem only by the pith. Micro-surgery was followed by basipetal differentiation of a cylinder of procambial-type tissue in the pith. As the apex produced leaf primordia, vascular strands developed basipetally in the procambium. Leaf primordia amputation experiments have also provided evidence supporting the central role of leaf primordia in determining both the location and amount of vascular strand development. For example, HELM (1931) found that when he removed very young leaf primordia of *Lysimachia punctata* and *Ricinus communis,* differentiation of vascular bundles failed to occur in the procambial tissue. NOZERAN et al. (1971) confirmed these findings with *Gleditsia triacanthos.* A further extension was made with isolated apices of *Geum chiloense.* When MCARTHUR and STEEVES (1972) removed leaf primordia as soon as they were recognizable, not only did vascular strands fail to produce a mature cylindrical stele but there was retention of an extended provascular cylinder. Replacement of the leaf primordia with IAA resulted in a more distinct provascular cylinder; and when both IAA and sucrose were applied together, a complete procambial cylinder developed below the provascular tissue. While no procambium could be detected in relation to the youngest leaf primordium, it was identifiable in all others. Another way of stating this is that in a sector of the apex occupied by a young leaf primordium, underlying provascular tissue is converted almost immediately to procambium. Thus, the barely delimited cylinder of meristematic tissue immediately below the meristem is considered the first stage of vascular differentiation and develops essentially as an activity of the terminal meristem. Subsequent differentiation to the procambial stage occurs only in relation to leaf primordia.

Hormone specialists generally agree that young leaves are a source of auxin and that the main plant hormone involved in vascular bundle induction and differentiation is auxin. In one of the few quantitative studies made on this point JACOBS and MORROW (1957) observed that the rate of production of xylem strands in *Coleus* petiole bases was directly correlated with the amount of auxin collectible from the bases.

A stimulation of cambial activity in trees by applied auxins was first reported by SÖDING (1936, 1938). Because of incomplete replacement of the stem tip by auxin it was concluded that other factors must also be involved. At least one of these is gibberellin. DIGBY and WAREING (1966) demonstrated interaction of auxin and gibberellic acid in three genera *Acer, Robinia,* and *Populus*. In *Populus* both IAA and GA applied alone to disbudded shoots induced some cambial activity, but a combination of the two showed a tremendous synergism, the sum of the resulting growth being many times greater than the sum of the individual responses to the single hormones. The fates of the products of cambial activity are clearly under hormone control. With IAA alone, fully differentiated xylem is produced. With GA alone, cambial products on the xylem side do not differentiate into mature xylem. Depending upon the concentrations of the combinations of GA and IAA, variable degrees of differentiation are obtained. Maximum production of fully differentiated xylem occurred with a GA/IAA ratio of 100 ppm/500 ppm. In contrast, no phloem differentiation takes place in the presence of IAA alone, but GA alone does considerably promote it. Indole acetic acid in low concentrations increases the action of GA in promoting phloem development but at higher concentrations becomes antagonistic; the most favorable GA/IAA ratio for phloem development being 500 ppm/100 ppm. As expected, the two hormones have their main controlling action in separate phases of the cell's development and for as yet unexplained reasons the patterns of action and interaction are different in the incipient xylem and phloem cells.

5.5 Regeneration

The ability of plant tissue to regenerate whole plants from smaller and smaller bits of tissue has been a fascinating subject for experimentation for a very long time. Support for the hypothesis that even extremes in differentiation of cells do not necessarily lead to loss in capacity to develop into a whole new plant grew out of tissue slice experiments. Totipotency, as a concept, rests on the premise that differentiation is not a result of irreversible changes in the genome, as the cell undergoes changes in form and function. There could be no proof of the totipotency concept, however, until tissue culture techniques had reached an advanced state. Fundamental to proving totipotency is to achieve regeneration of whole plants from single cells or protoplasts of numerous origins. In order to do this, single living cells from each of the plant's organs must be obtainable in culture, preferably in axenic suspension-culture or through cell wall digestion procedures leading to the production of protoplasts (COCKING 1972).

For definitive testing of the totipotency concept, single cells or protoplasts must be picked axenically from the culture medium and placed on a suitable growing medium; then through manipulation of the mineral, vitamin, amino acid, and hormonal components of the medium cause development of a tissue (usually callus) that gives rise to embryoids that in turn develop roots, stems, and leaves. The first investigators to successfully meet the demands of these rigorous criteria strongly supporting the concept of totipotency were BRAUN (1959) and VASIL and HILDEBRANDT (1965), who regenerated flowering tobacco plants from single cells. They were followed by TAKEBE et al. (1971), who regenerated normal whole *Nicotiana tabacum* plants from protoplasts. Regeneration of whole plants from cells of monocotyledonous plants proved more difficult to achieve. Eventually, however, this was done and the list continues to grow, though not as rapidly as for dicotyledonous plants (VASIL 1982).

Plant tissues that have served as a source of cells for callus production include virtually all of those containing living cells. Even guard cells of the epidermis, meosphyll cells, microsporocytes, and pollen have been induced to form callus followed by differentiation of embryoids. The most striking variable among cells from these many sources is the lag period before active growth is induced. (See REINERT and BAJAJ 1977 for procedures.) In each instance, however, large populations of cells are employed and only a few proceed through all the stages to produce normal plants. In other words, the processes of selection are at work and it cannot be said that all cells are totipotent.

Development of a completely defined culture medium for the successful establishment of organ and tissue culture has been one of the major achievements of plant physiologists in this century. However, culture conditions favoring isolated root or callus growth have rarely proven suitable for organogenesis. Seemingly, each tissue type requires a different formulation, depending upon whether the investigator's objective is undifferentiated parenchyma production or organogenesis. A standard basal medium consists of macro- and micronutrients, vitamins, a carbon source, organic growth factors including amino acids, and plant hormones. Ten of the most successful basal nutrient media developed during the mid-portion of the twentieth century were tabulated by NARAYANASWAMY (1977).

Differentiation of organs from callus tissue has been brought about for a large number of species by manipulation of the nutrient and hormonal components of the culture media. Almost invariably regeneration occurs more readily on a medium simpler than is required for callus or tissue culture. Each tissue seems to have its own requirements and sometimes as yet undefined adjuvants play a critical role in organogenesis. Furthermore, the practitioner of the art of organogenesis must take into account such factors as the origin of the callus, its genotype and age, endogenous hormone levels, and several physical factors when designing his experiments (REINERT et al. 1977).

The transmission of "organizing factors" from already organized tissue was demonstrated by WETMORE and SOROKIN (1955) when xylogenesis occurred in callus tissue following grafting of a bud into it. While complete mimicry of the bud has not been achieved, xylogenetic effects can be induced with a combination of a disaccharide containing a glucosyl radical at the nonreducing end, IAA (ROBERTS 1969) and benzylaminopurine (MINOCHA and HALPERIN 1974). In contrast, GA_3 and ABA inhibit tracheid differentiation (MINOCHA and HALPERIN 1974).

Subsequent to the successful production of plantlets in cultures of tobacco callus by SKOOG and MILLER (1957) through the manipulation of auxin and kinetin concentrations, an even better nutrient medium was introduced by MURASHIGE and SKOOG (1962). Improvement was achieved by increasing the levels of the major elements as well as B, Mn, and Zn among the minor elements, raising the sucrose level to 3% and adding myo-inositol. Marked success has been achieved with asexual test tube propagation of plants of both economic and noneconomic importance including ornamentals, food crops, cereals, bulb plants, and tree species (KARNOSKY 1981). Often the procedure is: first, obtain callus formation: second, transfer to a medium formulated to induce bud formation; third, make another transfer to a root inducing medium; fourth, transfer plantlets to a medium favorable for seedling growth; and fifth, transfer the young plants to soil. Tables including plants so generated through 1977 have been compiled by NARAYANSWAMY (1977). These lists are being added to monthly. However, there are still a great many plants for which regeneration is as yet only partial, that is, only roots or shoots have been generated, not whole plantlets.

Results from the in vitro experiments are steadily revealing the number of factors that are involved in the regeneration process. Sometimes the deficiency is in the amount of available auxin or kinetin-like substances and then again the deficiency may be in a particular form of nitrogenous product or metabolic intermediate that changes with time. This point was effectively made by BUTENKO et al. (1967), who noted that mineral salts should be present in high concentration for embryo initiation. Addition of nitrogen, either inorganic (KNO_3, NH_4NO_3) or organic (amino acids and amides) facilitates embryogenesis (TAZAWA and REINERT 1969) but the molarity requirements are not a variable. Glutamine (5 mM) added to White's basal medium which contain 3.2 mM inorganic nitrogen as well as amino acids induced embryogenesis, while higher concentrations of KNO_3 (40 mM) were necessary to achieve the same results. Embryogenesis, however, could not be achieved with nitrate as a sole source of nitrogen at any concentration; yet nitrate was necessary for development at advanced stages. The reason for this lies in the ontogenetic development of enzyme spectra in the growing embryo – nitrate reductase being produced relatively late.

WHITE (1939) was the first to note that the degree of submergence of a callus in liquid culture could be correlated with the appearance of buds. He guessed that an oxygen gradient was involved. This view has been supported experimentally by KESSEL and CARR (1972), who found with carrot tissue that a reduction in the available oxygen favored shoot formation while an increased oxygen gradient promoted rooting. It would probably prove fruitful to determine to what extent hormonal balances are affected by oxygen gradients.

Temperature, photoperiod, light intensity, pH, sugar concentration and quality of the substrate are among the factors that have been shown to play determining roles in organogenesis and embryogenesis. How these environmental factors mediate their effects remains to be determined. Their impact on the metabolic and hormonal balances initiating organogenesis and the differentiation process is a fruitful field for investigation.

Especially annoying impediments in the way of exploiting tissue culture as a tool for morphogeneticists, practical plant propagationists and molecular biologists is the apparent loss of capacity for organogenesis presently attributed to age of tissue, senescence, nutrition, and ploidy changes. Many tissues display a high potential for organogenesis or embryogenesis when first put in culture but with subculture become more and more difficult to stimulate to produce buds or roots (TORREY 1967, MURASHIGE and NAKANO 1967). The loss of morphogenetic potential in a tissue may be due to genetic or physiological changes as subculturing occurs. Systematic adjustments in oxygen gradients, inorganic and organic components of the culture medium, pH, hormonal factors, temperature, and light will probably lead to success in obtaining regeneration if the problem is one of physiological change. While it is generally best to avoid using tissue that has been in culture for a long time if one is attempting regeneration, the outlook is not completely bleak. Calli that had been continuously subcultured through 30 passages (PADMANABHAN et al. 1974) showed that tissue cultures from leaves of tomato still had organogenic capacity on a massive scale.

When the change in cultured material is genetic, especially when euploidy, or aneuploidy, or chromosome aberrations have occurred, the chances of restoration to morphogenetic capacity are low but if the change with time is only metabolic, then the chances for restoration are good. Addition or substitution of phytohormones, adjustments in nutrient constituents and other environmental factors may very well lead to organogenesis. In his review on the cytogenetics of differentiation in tissue and cell culture D'AMATO (1977) concluded that the only cell line in the axis of higher plants that is endowed with genetic stability is the meristematic cell line and this is restricted to the apical dome including a few leaf primordia. Employment of apical meristem cultures, apparently, is the only procedure that ensures genetic uniformity for in vitro propagation.

As a generalization, it appears true that as cells proliferate in callus or suspension culture, there is a high probability of chromosome variation between cells and of the emergence of cell lines with chromosome complements quite different from the parent plant tissue (BAYLIS 1980). However, since cultures from some species (BAYLIS 1980) seem to retain chromosome complements of the parent plant through many transfers, there appears to be a factor or factors that can ensure uniformity in the chemistry and physics of the cell cycle. It is highly desirable that we learn what these factors are. Knowledge of how to ensure stability of the chromosomal system in culture is urgent because of the desirability of somatic hybridization procedures and genetic manipulations employed in bioengineering.

The importance of the interplay of hormonal factors in the determination of the character of stems, whether they be leafy and photosynthetic, rhizomatomacious and horizontal growing, or leaflets and bulbous storage reservoirs has been well illustrated with *Solanum tuberosum*. BOOTH (1959, 1963) and KUMAR and WAREING (1972) showed that removal of the apical bud was followed, as expected, by the upper lateral buds growing into leafy shoots. Applications of indoleacetic acid to the cut-surface of the stump partially reimposes inhibition of lateral bud growth. Application of gibberellic acid alone promotes extension of the axillary shoots. Simultaneous application of auxin and gibberellin, how-

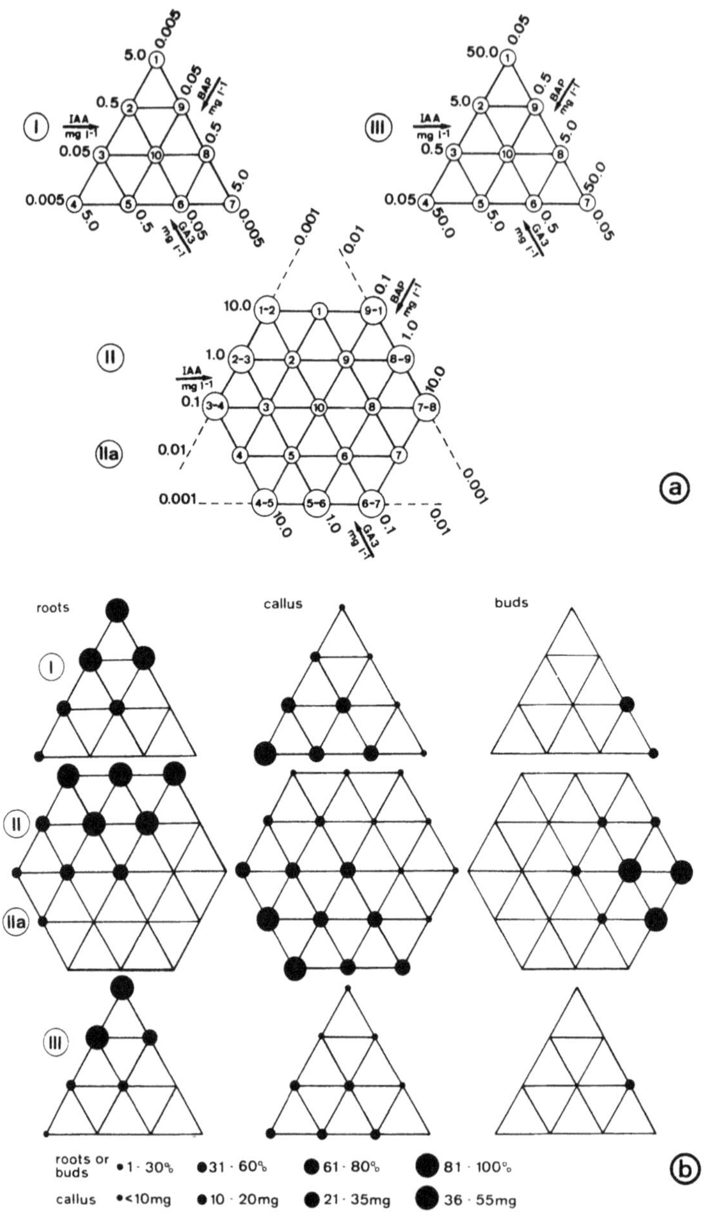

Fig. 5.1 a, b. Organ and callus differentiation on *Digitalis purpurea* leaf explants cultured under the simultaneous influence of IAA, GA_3 and BAP. **a** Graphs of the combinations of growth regulators put together according to coordinate triangles. *Roman numerals* indicate the various concentration ranges (series): 0.005 to 5.0 mg l^{-1} (*I*); 0.001 to 10.0 mg l^{-1} (*II* and *IIa*); 0.05 to 50.0 mg l^{-1} (*III*). *Arabic numerals* indicate the quantitative proportions of the growth substances (groups *1* to *10* and *1–2* to *9–1*). In series *I, II, III* four different concentrations are distributed in regular intervals. In series *IIa* the concentration of one growth substance is very low, whereas the concentrations of the other two partners vary in their active range in opposite direction. **b** Induction of roots and buds and growth of callus tissue. (After RÜCKER 1982, p 144)

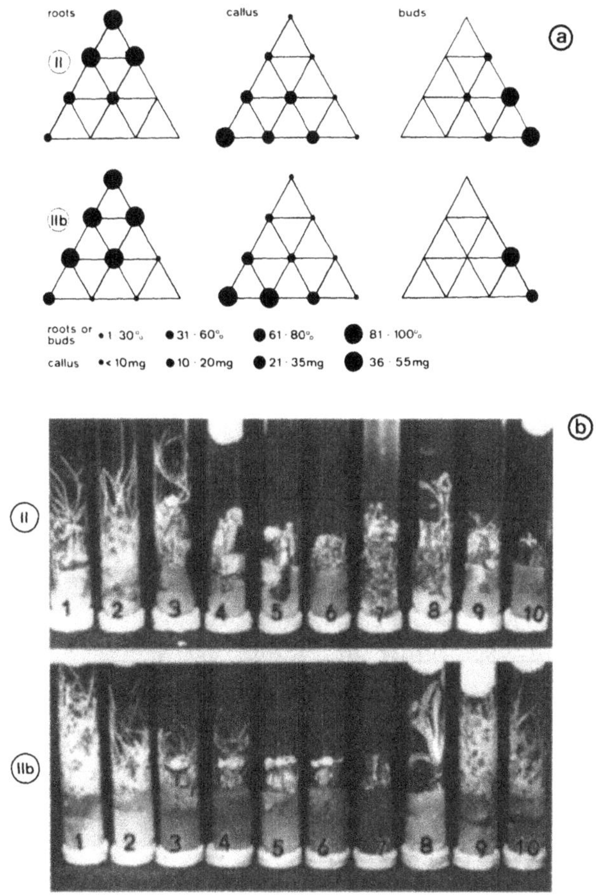

Fig. 5.2 a, b. Root, bud, and callus formation of leaf explants of *Digitalis purpurea* under the influence of high (*II*) and low (*IIb*) N-concentrations comparatively represented. In both series *II* and *IIb* the concentrations of the growth regulators are identical to groups marked with the same *arabic* numeral (for concentrations see Fig. 5.1 a). **a** Induction of roots and buds and callus growth under the influence of high and low N-content. **b** Representative examples for the formation of roots, buds and callus in the leaf explants. (After RÜCKER 1982, p 147)

ever, is followed by the axillary buds producing scale-like leaves on thin, elongated, horizontally growing stems (stolons) typical of stems bearing tubers belowground, but these were well above the surface. Thus IAA and GA individually induce only small quantitative changes in the otherwise normal shoots, but in combination divert the bud into a vastly different pathway of development including ability to respond to gravity. If the induced rhizome growing tips are treated with kinetin, a dramatic change in pattern of growth occurs. The joint control by IAA and GA coming from the tip of the stump is neutralized and the growing tip of the rhizome quickly assumes an upright, leafy growth form.

The action of serial combinations of IAA, GA_3, and another cytokinin BAP (benzylaminopurine) on leaf explants of *Digitalis purpurea* provides more precise information about the vectorial nature of tissue response to hormones (RÜCKER 1982). Execution of the three factor experiments was by means of the triangle method employed earlier in the development of "ideal" nutrient solutions (HAMNER et al. 1942). The results when depicted on triangular graphs (see Fig. 5.1) permit quick determination of the concentrations of the growth substances that induce differentiation of roots, buds or callus by simultaneous action of the several hormones (see Fig. 5.2). Root formation responded very sensitively to stimulating amounts of IAA and to the inhibitory influence of BAP. High GA_3 concentrations stimulated or inhibited root formation in accordance with the concentration of IAA. Bud production was dependent upon the presence of both BAP and IAA. Although GA_3 was not necessary for the initiation of buds, it proved to have an enhancing effect over a narrow low concentration range and a strong depressing effect at only slightly higher concentrations. High IAA concentrations inhibited callus formation. The nitrogen content of the medium was of definite importance. Low nitrogen tended to promote root development but decreased bud formation; callus production was, however, relatively insensitive to the N-concentration. It is probable that RÜCKER'S (1982) model for triple combinations of growth substances with different nitrogen sources will be used on a wide scale in the future and will lead to success in regenerating plantlets from tissues that have never before yielded to experimentation.

5.6 Flower Initiation

One of the most dramatic morphogenetic phenomena known in plants occurs when the vegetative meristem turns from the production of leaves to the generation of floral primordia. Cells of the apical meristem-producing leaf primordia have a highly ordered sequence of divisions that produce the characteristic phyllotaxy. Suddenly, signals causing the production of flower primordia result in new orientations of spindles and division figures that produce sepal, petal, stamen, and pistil initials. The meristem is totally transformed. Rarely is there a gradual transition from the vegetative to the reproductive phase. Accompanying the sharp change in behavior of the apical meristem one should expect changes in physiology—changes specifically associated with flower primordia induction. Indeed, one can reasonably refer separately to the physiology of flower induction, flower development, fertilization (fruit set), fruit development, and seed development. Each of these phases has distinct internal and often external environmental requirements. The importance of considering reproduction in plants from a multiphasic viewpoint was first clearly enunciated by KLEBS (1918). Flowering, alone, can be separated into (1) initiation of changes in the apex leading from production of leaf primordia to floral primordia; (2) differentiation of floral parts; (3) growth and final maturation of all parts;

and (4) anthesis. Though these stages grade into one another, recognition that they exist can markedly improve the precision of interpretation of experiments on reproduction.

Early investigators concerned with reproduction among plants, whether they be fungi, algae, ferns, or seed plants, were directed toward an understanding of the effects of various environmental factors upon reproduction as a whole. Distinctions were not made between flower initiation and flower development. Water, light, temperature, and nutritional factors received the greatest emphasis during the first part of the twentieth century. This was followed by the direction of attention by numerous investigators to biochemical balances correlated with the development of fertile flowers into fruits. This line of investigation was enormously stimulated by the findings of KRAUS and KRAYBILL (1918). Discovery of the phenomenon of photoperiodism (GARNER and ALLARD 1920, 1923) made it possible to study with precision chemical changes associated with floral induction. With the increasing acceptance of the concept of phytohormones as correlating agents that followed upon WENT's (1928) development of a biological method for measuring plant hormones, research became focused on growth accelerators and inhibitors potentially involved in reproduction in plants.

JULIUS SACHS (1888) was among the first to conduct experiments on floral development. The results of his investigations led him to formulate the hypothesis of a "Blütenbildender Stoff" (flower-forming substance) generated in the leaves and transported to buds in leaf axils and there induce flower formation. He came to this conclusion following his discovery, with three species, that flowers were differentiated on portions of stems kept in total darkness when other parts of the plant, with leaves, were exposed to the light. No clear-cut distinction could be drawn, however, between the possibilities the flower promoting factor was photosynthate or a special hormone.

Early in the twentieth century G. KLEBS carried out numerous studies in an attempt to discover the combination of environmental factors that promoted fruitfulness vs. vegetativeness. By 1918, KLEBS had formulated a set of generalizations that were of practical use in controlling flowering. These factors were light duration and intensity, water supply, temperature, and nutrients. As a result of his careful observations on plants grown from seed to maturity, KLEBS (1913) introduced the concept of "ripeness to flower". This phrase describes the physiological and biochemical balance when the plant will flower upon exposure to an appropriate environmental stimulus. This may or may not coincide with the end of the juvenile phase. Plants requiring low temperature vernalization illustrate this point. Such plants may require some weeks of growth before the juvenile phase is passed and the plants become sensitive to low temperatures. Following the vernalization period no flower initials are formed, but the plant is now ripe to flower and will do so when subjected to appropriate temperatures and/or photoperiods.

Meanwhile, in another German laboratory, H. FISCHER (1905, 1916) was studying the effects of available mineral nutrients, particularly nitrogen, light intensity, and CO_2 concentration on flowering. In 1905 he suggested that the crucial combination of factors was a "balance of carbohydrate over nitrogen". His general observations led him to conclude that when photosynthesis was vigorous in proportion to available nitrogen, flowering occurred; and when nitrogen was high in relation to carbohydrate, flowers would not appear. Whether flower primordia were initiated and did not develop was not determined. By 1916 FISCHER had convinced himself that flowering is based on a definite "ratio" between the carbohydrate and nitrogen present in the plant. This was an unfortunate expression because it implied a mathematical precision that does not exist. Nevertheless, this criticism does not mean there is no important relationship between nitrogen and carbohydrate levels and flower development following floral initiation. There is.

Neither KLEBS nor FISCHER had chemical evidence to use in support of their conclusions. Their reasoning, however, inspired KRAUS and KRAYBILL (1918) to make the necessary chemical determinations and correlate their findings with anatomical and morphological responses. For the first time, chemical data supported the concept of a relationship between the composition of a plant (tomato) and its degree of fruitfulness. KRAUS and KRAYBILL's experiments were concerned almost entirely with development of flowers into fruits. Differentiation of flower primordia was not dealt with at all. The authors were aware that flower primordia differentiation, flower development, and fruit set are separate phenomena. In the summary of their paper KRAUS and KRAYBILL (1918) indicated a fear there would be misinterpretation of their findings when they said "... the study was undertaken in connection with the fruit setting problem ..." and "The conditions for the initiation of floral primordia, and even blooming, are probably different from fruit setting". Despite the clarity of their claim that they were dealing only with the physiology of fruit setting, their results have been widely misquoted and misused in the literature on floral initiation.

Floral initiation is best studied at the microscopic level. The fact is, floral initiation often occurs and is unrecorded because abortion takes place or there is failure of the flower initials to develop to the macroscopic level where detection is easy with the unaided eye. There are many pieces of evidence that this is true. One was provided by NAYLOR (1941) with the photoperiodically sensitive cocklebur plant (*Xanthium pennsylvanicum*). He found that even when the plants were severely deficient in nitrogen or phosphorus or potassium they differentiated floral primordia after the leaves received an appropriate photoperiodic treatment. However, floral development did not occur. If dissection of stem apices had not been carried out, floral initiation would have gone undetected and the plants would have been recorded as vegetative.

Flower initiation can be viewed in at least two ways. One can make the assumption there is a basic tendency to produce flowers following a certain juvenile period and that failure to do so is the result of a neutralizing inhibitor(s) issuing from the leaves. It then follows that flowering occurs once the inhibitor or inhibitors are removed or neutralized. An alternative assumption is that plants remain vegetative until flower-promoting factors are produced and exported from the leaves to the growing apex following synchronization of environmental components. These hypotheses have been debated by GREGORY (1936), VON DENFFER (1950a, b), EVANS (1975) and LANG et al. (1977) and remain an unsettled but thought-provoking issue.

Regardless of how one views the phenomenon of floral initiation, it is clear that when a plant changes from the vegetative to the reproductive phase many previously quiescent genes must be activated in a large number of cells in localized areas of the shoot apex. On the morning following a single long photoinductive cycle, specific patches of cells subjacent to the apical cells of the shoot apex of *Lolium temulentum* show that they have received a specific signal. This is indicated by an upsurge in the production of RNA (CHERRY and VON HAYSTEE 1965, KNOX and EVANS 1968). These are the cells destined to give rise to the spikelets. Analogous findings have been reported for the long-day plant *Sinapis alba* (BERNIER 1971). Soon after the activated cells begin to synthesize protein they go into mitosis. The exact nature of the signal(s) and what determines which cells will be target cells is yet to be determined. (For a review, see VINCE-PRUE 1975, pp 231–244.) Experimental data thus far accumulated can be interpreted to mean that a floral stimulus acts at least partially at the transcriptional level but action at the translation level has not been excluded. Once started, activation of other genes follows in an orderly sequence almost as if a single regulatory gene controls the whole gene sequence involved in flower production.

Since growth regulators have the potential of altering the expression of the genome, investigators interested in the reproductive physiology of plants have quickly tested new introductions to see if they will affect the time or number of flowers initiated. As a result of testing of natural and synthetic growth regulators, a number have been shown to modify flowering behavior. Among these compounds, the gibberellins are particularly important from both the theoretical and practical standpoints.

At present, the gibberellins are the only group of chemicals known to evoke flowering in a wide range of plant species. However, the effect is almost entirely restricted to long-day and biennial plants that grow as rosettes in short days and flowering is signalled by elongation of the stem (see VINCE-PRUE 1975 for extensive review.) Gibberellins have also figured in some spectacular examples of shortening of the juvenile phase in trees—particularly in species of Cupressaceae and Taxodiaceae (see PHARIS and KUO 1977 for review).

Promotion of early flowering in gymnosperms by the application of gibberellins (GA_1, GA_2, GA_3, GA_4, GA_5, GA_7, GA_9, GA_{14}, and GA_{24}) had been almost (HASHIZUME 1967) confined to members of the Cupressaceae and Taxodiaceae families until the late 1970's, when mixtures of gibberellins less polar than GA_3 began to be employed. A combination of GA_4/GA_7 proved to be effective in overcoming juvenility and irregularity in cone production in a number of members of the Pinaceae including some commercially important forest tree species: Douglas fir (ROSS 1976, ROSS and PHARIS 1976, PURITCH et al. 1979), Sitka spruce (TOMPSETT 1977, 1978, TOMPSETT and FLETCHER 1979, TOMPSETT et al. 1980, PHILIPSON 1983), and *Pinus taeda* (ROSS and GREENWOOD 1979). How one should relate these topical application results to untreated plants is not clear, because bioassays indicate that although growth-regulator levels in young shoots change greatly with stage of development, the levels are similar in good and poor cone-producing clones (MCMULLAN 1980). Clearly, more precise measurements are needed of the spectrum of natural gibberellins in association with initiation of reproduction in gymnosperms.

Similar experiments to those done with gymnosperms involving applications of gibberellins to leaves and buds of woody angiosperms such as apple, pear, black currant, grape, *Syringa*, and *Fuchsia* have been negative (WAREING and PHILLIPS 1981). However, growth retardants such as chlorocholine chloride (CCC) and 2,4-dichlorobenzyltributyl phosphonium chloride (Phosphon-D), which apparently block the gibberellin biosynthesis pathway, promote flowering in apples, pears, and azaleas.

The amount of GA found to cause flowering ranges from as little as 3 μg per plant in *Lolium temulentum* (EVANS 1964) to 100 μg or more in other species. Furthermore, the application may have to be repeated at frequent intervals. Only about 50% of *Sinapis alba* plants, for example, initiated flowers after 200 days of repeated applications—the total amount of GA per plant reaching 180 μg (EL TANNIR-LOMBA et al. 1962). Although the number of long-day species yielding to GA treatment has been steadily extended, there are still a number that have not been induced to flower. Part of the problem is to supply the correct gibberellin or combination of gibberellins. More than 50 gibberellins have been isolated, but very few are available in quantity for experimentation. When MICHNIEWICZ and LANG (1962) tested nine gibberellins on the long-day plant *Silene armeria*, flowering was obtained only with GA_7. At ten times the dose, GA_3 did not induce flowering. Furthermore, they found that both GA_4 and GA_7 were considerably more active than GA_3 in inducing *Crepis parviflora* to flower.

Although it has been relatively easy to induce bolting and flowering in rosette-forming long-day plants with gibberellins, caulescent long-day plants have proven to be singularly refractory. Evidence is accumulating which indicates that in at least some species two gibberellins are involved in the flowering of rosette forms, one being responsible for stem elongation, while the second influences flower initiation (STODDART 1962).

After more than 25 years of experimentation with the gibberellins, it is now clear that they play only a part in the control of flowering. They are not synonymous with florigen. Normally, the gibberellins do not induce flowering in absolute short-day plants, but there are a few quantitative short-day plants that have responded by flowering (WITTWER and BUKOVAC 1958, HESLOP-HARRISON 1969, NANDA et al. 1969). Possibly the effect of the growth regulator is to reduce the length of the juvenile phase of development in this class of photoperiodically sensitive plants.

It is now clear that flower initiation requires the presence of several factors. Gibberellins can only promote or induce flower initials when this hormone is the one limiting factor for flower initiation. Since gibberellic acid is known to influence several cellular processes, it is possible that in floral initiation it fulfills different functions in different photoperiodic response groups. Since floral induction does not appear to be necessarily correlated with levels of endogenous GA's, the effect of exogenously supplied GA may be artifactual and not a physiological equivalent. The conclusion that must be reached at this time is that GA causes flower initiation indirectly through the production of other flower-promoting factors.

The search for a suitable system for the study of the biochemistry of flower initiation led NITSCH and NITSCH (1967a, b) to employ tissue culture techniques for that purpose. Progress in the study of flower initiation at the biochemical level had been hampered by lack of a satisfactory means for assaying individual substances or mixtures or plant extracts with potential for chemical induction of flowering. Another important feature of tissue culture is that it provides a means for determining the contributions of different organs to the initiation and development of flowers. The NITSCHES and their colleagues were successful in inducing flowering in vitro with photoperiodic species by keeping the stock plants vegetative under noninductive environments, excising pieces of tissue from them, planting these in culture tubes, and finally, subjecting the cultured explants to inductive treatments. This technique has been employed successfully to induce flower formation on root tissues of a long-day species *Cichorium intybus*, leaf tissue of the short-day plant *Streptocarpus nobilis* and internode sections of *Plumbago indica* L., clone "Angker" another short-day plant (reviewed by C. NITSCH 1968). The action of auxins, gibberellins, cytokinins, ABA, amino acids, and ethylene on induction and development of flowers in vitro have been most intensively studied with 7-mm stem segments from vegetative vigorously growing *Plumbago indica* plants. NITSCH and NITSCH's (1967a) K nutrient solution solidified with agarose, served as the basal medium. Test compounds were added to this medium singly and in combination.

Vegetative *Plumbago indica* stem segments were responsive to photoperiod in vitro after 4 weeks in culture (NITSCH and NITSCH 1967b). By that time vegetative buds with one or two small leaves had been generated. Flower initiation took place on short photoperiods; but on long photoperiods only vegetative buds were formed. A minimum level of 8%–10% sucrose was necessary for flower initiation. Sucrose could be replaced by maltose and to some extent by cellobiose but not by mannitol or lactose. Auxins and gibberellins inhibited formation of flower buds. Thymine or its precursor, orotic acid, enhanced production of flower buds when adenine and kinetin were also in the medium. Following embryonic bud initiation several amino acids, as well as glutamine and asparagine, tended to reduce flower bud development at 3×10^{-4} M or above. In contrast urea, at an equivalent concentration, clearly promoted flower bud development. Both cis and trans-isomers of ABA stimulated flower bud production when the cultures were exposed to short photoperiods but were without visible effects on long photoperiods. Perhaps most significant of all, C. NITSCH (1968) found that by combining zeatin (10^{-6} M), adenine (10^{-4} M) and thymine (10^{-4} M), she could obtain flower initiation and development in 20% of her cultures even when she maintained her explants on nonphotoinductive cycles. Another significant finding was that ethylene and its metabolic precursor, methionine, stimulated flowering in *Plumbago indica* stem segments under noninductive photoperiods (C. NITSCH 1968, NITSCH and NITSCH 1969). Previously, pineapple (*Ananas sativus* L.) had been considered the outstanding exception to the generalization that short-day plants do not respond to chemical applications to flower.

COUSSON and TRAN THAN VAN (1981) made the remarkable discovery that by using precisely defined physical and chemical conditions it is possible to cause 1 × 10 mm pieces of explant tissue from the basal part of the inflorescence of *Nicotiana tabacum* cv. Samsun—a short-day plant—to proceed directly to the production of flower buds. From 5 to 25 flower initials were obtained on 95% to 100% of their explants cultured on 4.0 mm "pseudo-calic" glass beads. This high degree of success was achieved by culturing the tissue at 24 °C under 50 Wm^2 of continuous light and using a MURASHIGE and SKOOG (1962) macro- and micro-nitrient medium containing myoinositol (100 mg l^{-1}), thiamin-HCl (0.4 mg l^{-1}) and glucose (30 g l^{-1}) and equimolar concentrations of IBA and kinetin (2×10^{-7} or 3×10^{-7} or 5×10^{-7} M) and precisely adjusted pH values of 5.6 or 6.0 or 7.5 respectively. Neither IBA nor kinetin provided a sufficient condition to trigger flower initiation, but pH was of crucial importance. COUSSON and TRAN THAN VAN (1981) concluded that since light, temperature, and pH strongly affect de novo flower initiation from a thin layer of cells, determination of the electrical profile is called for in elucidating the physicochemical processes triggering flower initiation. Furthermore, it will be important to investigate the presumed pH-mediated uptake of IBA. Although these investigators were successful in obtaining flower initiation in the absence of leaves, it must be borne in mind that the tissue they used as starting material had already been preconditioned to flower. Nevertheless, theirs is a major achievement.

5.7 Flower Development

Once sufficient signals for floral initiation have been received by the receptive cells of the stem meristems, the genetic apparatus is reprogrammed to produce floral structures. There is a teeter-point, however, between vegetative and reproductive development; and it is at this juncture that a good deal of experimentation has been directed in an effort to determine if minerals, photosynthate, temperature or known hormones or their presumed antagonists have a role in floral initiation and development. If insufficient signals for floral initiation are received in the buds, a mixture of leaves and floral parts will be generated. In photoperiodically sensitive plants requiring more than one photoinductive cycle for floral phyllody initiation or incomplete induction can be avoided by supplying the requisite number of photoperiods. A considerable body of literature now exists concerning the control of flower morphogenesis and in particular of sex expression by environmental, chemical, and hormonal influences. An extensive review of this field has been given by NAYLOR (1952) and HESLOP-HARRISON (1972). Environmental and nutritional aspects have been summarized by NIGHTINGALE (1948) and SCHWABE (1963) and hormonal aspects have been considered in depth by ZEEVART (1978).

Flower development is nearly always arrested or the primordia are aborted when there is an acute carbohydrate shortage. Although the most obvious correlation is with lack of basic substrate required in synthesizing new protoplasm and cell walls, other less obvious factors may also be involved. In the temperate zones plants are exposed to daily alternations of relatively high and low levels of carbohydrate, and inevitably this rhythm is reflected in the morphogenetic centers. Ordinarily, a high proportion of the daily assimilate goes to the young growing parts of the plant. It may be speculated that plant hormones accompany the assimilates. At least there is a high correlation when auxins are applied topically. Except when there is a surplus of carbohydrate, there inevitably is intraplant competition for carbohydrate between centers of growth (WILLIAMS 1960), regions of storage and use in respiration. Developing flowers are centers of growth intensively involved in the use of carbohydrate and nitrogen in the production of new cells. If a large number of floral initials have been generated, competition for the limited substrates is bound to occur among the developing flowers. This competition will be conditioned by the nearness of the several sinks to the source or sources of substrate, efficiency of the transport systems, area of leaf surface irradiance levels, and possibly the presence of the suspected assimilate mobilizing hormone, brassinolide (KRIZEK and MANDAVA 1983).

While low irradiance levels often limit the number of flowers developed, duration of the light period perhaps more frequently dominates in the floral development phase. In any event, an unsuitable carbohydrate nitrogen balance under either long or short photoperiods can limit the number of flowers that reach the pollination stage of development.

EL HINNAWY (1956) has reviewed the relation of macronutrient nutrition to flower initiation. He concluded that variations in macronutrient levels may greatly influence the time of appearance of floral buds, the rate of subsequent floral development, the number of flowers produced and the stamen—pistil ratios; but there was no alteration in initiation of floral primordia in photoperiodically sensitive plants.

Although extreme nitrogen deficiency does not prevent the short-day plant *Xanthium strumarium* from initiating flower primordia (NAYLOR 1941) nitrogen nutrition does have a marked effect on floral development. NEIDLE (1939) found that fully photoperiodically induced *Xanthium* plants receiving an abundant supply of nitrogen developed flowers more rapidly than the plants that were becoming increasingly nitrogen-deficient. Subsequent to photoinduction, staminate head production was favored by high nitrogen supply and long days. Abortion of developing pollen was greatest on the plants on short photoperiods. Regardless of day length during floral development, the plants receiving an abundance of nitrogen had many more fruits than those given a low supply. Nevertheless, seed production was greatest in the burrs produced by the plants receiving both high nitrogen and long photoperiods. This was true because of the high frequency of parthenocarpy under short photoperiods. High nitrogen levels during short-day photoperiodic induction also accelerates flowering in *Perilla* (CHAILAKHYAN 1945) and *Kalanchoe blossfeldiana* (RÜNGER 1961).

There is now strong evidence that iron is very important in floral development in *Xanthium* (SMITH et al. 1957). When Fe was supplied iron deficient plants after six photoinductive cycles, normal staminate inflorescence development occurred but the pistillate inflorescences were abnormal and reduced in number. If Fe was not added to the −Fe plants following photoinduction, many of them died. It is not possible at this time to say if Fe is directly or indirectly involved. Most probably, however, the effects are indirect and reflective of the importance of Fe in cytochromes, ferredoxin, nitrate reductase, and sulfite reductase as well as several other enzymes that are essential in cellular function.

In general, a high level of available nitrogen promotes female relative to male flower part development, while low available nitrogen is associated with the reverse. This generalization has been strikingly demonstrated with monoecious and andromonoecious species—TIEDJENS (1928) with *Cucumis sativus* and BRANTLEY and WARREN (1960) with *Cucumis melo*. Furthermore, there are numerous reports of corresponding effects of nitrogen nutrition on sexual expression in dioecious and hermaphrodite species. HOWLETT'S (1936, 1939) work with *Lycopersicon esculentum* provides evidence that strongly supports this view. In the perfect flowers of tomato, maximum length of the pistil in relation to the stamens was obtained with the highest level of nitrogen. In addition, in short photoperiods high nitrogen was associated with stamen abortion and pollen sterility.

The pistil is the last of the floral structures to be differentiated during flower formation. In some dioecious species, such as *Lychnis dioica*, pistil initiation is linked to the presence of a sex chromosome (WESTERGAARD 1958). In many species, however, sexual expression is more subject to modification by environmental factors acting on metabolic processes. This aspect has been studied frequently in monoecious members of the Cucurbitaceae. In these species, as in other monoecious types, light and temperature are particularly potent factors influencing the production of stamens and/or pistils.

Unusually long photoperiods and high temperatures repress, whereas short photoperiods and low temperatures promote pistillate flower development in species such as *Cucurbita pepo* and *Cucumis sativus* (NITSCH et al. 1952). These observations were confirmed and extended considerably by MATSUO et al. (1968). Comparable observations on other monoecious plants were compiled earlier by LOEHWING (1938). Irradiance level together with photoperiod can be of crucial importance. In *Zea mays* staminate flowers failed to appear in winter in the Columbus, Ohio area but pistillate flowers continued to be produced (SCHAFFNER 1930). It may be significant that endogenous changes in gibberellin-like substances accompany sex reversal of the apical inflorescence of corn (ROOD et al. 1980). *Ambrosia trifida* responds to progressively shortened photoperiod by producing increasing numbers of pistillate flowers (MANN 1942). Temperature also affects sexuality in *Ambrosia*. Lowering the temperature to 15–20 °C leads to increasing numbers of pistillate flowers, while increasing the temperature to 35 °C is correlated with proportionally more staminate flowers (JONES 1947).

How do photoperiod and temperature bring about their effects on sex expression? A definite answer is not known. Since it is known, however, that ethylene treatment stimulates the development of pistillate flowers in cucumbers (MININA and TYLKINA 1947, SIMS and GLEDHILL 1969, RUDICH et al. 1969), it is hypothesized that under short photoperiod and chilling temperature conditions metabolism of methionine is tipped toward ethylene generation. It is now feasible to test this hypothesis and it should be tested.

Very probably in those instances where application of auxins has been correlated with increased pistillate flower production the effect is indirect. BURG and BURG (1966), for example, report that auxins stimulate ethylene production. Perhaps once again this is the factor most directly involved in morphogenesis following chemical treatment. Staminate flower initiation, however, appears to be a different matter. In certain flowering plants staminate flower production is now associated with elevated concentrations of gibberellins (GALUN 1959). This view is strongly supported by the finding that gibberellin can induce anther and pollen development in a stamen-less tomato mutant (PHATAK et al. 1966). Morphactin (ethyl ester of 2-chloro-fluorenol-9-carboxylic acid) also greatly stimulates the production of staminate flowers (BOSE and NITSCH 1970). Studies are overdue on the interaction of gibberellins, morphactin, and cytokinins in stamen initiation and development.

Cytokinins probably play a role in sex expression. It may be that in certain monecious species cytokinins are below a critical level for differentiation of perfect flowers. Topical applications of cytokinin to a known staminate flowered grapevine 3 weeks before anthesis induced the plant to produce perfect flowers (NEGI and OLMO 1966). An additional example is to be found in the production of male flowers by *Mercurialis annua* following treatment with cytokinins (DURAND 1969). Furthermore, soaking of *Luffa acutangula* seed in a N-benzyladenine solution promoted early pistillate flower production (BOSE and NITSCH 1970). In this connection it may be important to note that cytokinins have been reported to induce ethylene formation in plant tissues (FUCHS and LIEBERMAN 1968).

While little more than inferences can be made at this time, it appears that in the development of floral organs auxins, gibberellins, cytokinins, morphactins and ethylene play significant but, as yet, undefined roles preceding the development of tissues giving rise to fruits.

5.8 Fruit Setting and Development

In nature, the fruit-setting period is accompanied by a high degree of uncertainty. Climatic factors such as late frosts or a very wet or a hot, dry period during a brief pollination period may practically eliminate fruit and seed production for a crop year. Because of the importance of the pollination–fertilization–fruit development phase in the life cycle of angiosperms, a great deal of attention has been devoted to it. Many internal and external factors impinge upon it.

Nitrogen, among all the elements, is most frequently limiting to fruitfulness. Water deficiency may also block fruit setting and development. Once fruit development is underway, competition for water may be partially responsible for the abscission of immature fruits. Some malformations of strawberries and raspberries have been correlated with a deficiency of water during the pollination–fertilization period. Much emphasis has been placed on hormones and growth regulators in fruit production. The hormone work has been reviewed since 1970 by NITSCH (1971), WELLENSIEK (1973), MOORE and ECKLUND (1975), COOMBE (1976) and GOODWIN (1978b).

Ordinarily fruit growth is relatively rapid. Mobilization of food in the growing fruit is often faster than it can be supplied by photosynthesis. Thus physiological balances in other parts of the plant may be drastically altered. Although

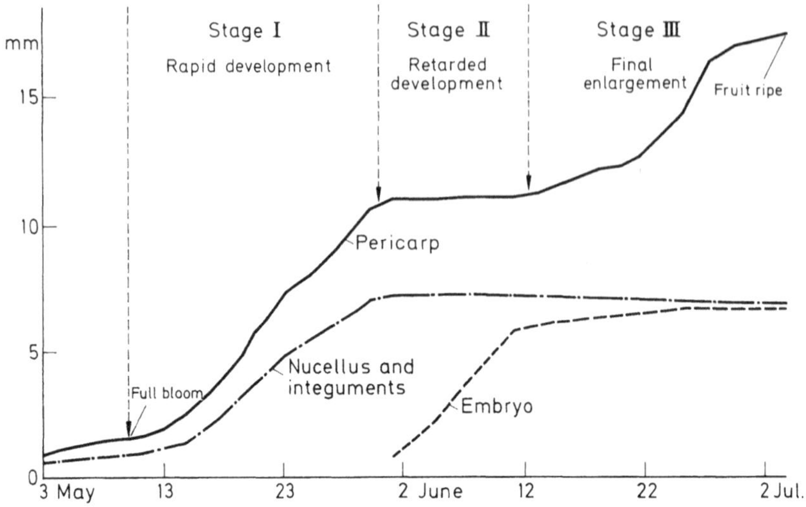

Fig. 5.3. Growth stages in development of fruit and seed components of sour cherry from pre-bloom to fruit ripening. (After TUKEY and YOUNG 1939)

fruits may differ greatly with respect to their morphological origins (see COOMBE 1976), from the standpoint of physiology there is merit in defining fruits, as does NITSCH (1952), as structures that surround ovules, development of these structures being closely linked with development of the ovules. This definition applies to such dissimilar structures as the floral axis of pineapple, receptacle of strawberry and apple, and the syconium of fig. It even applies to seedless fruits, since ovules are initially present but fail to develop.

The coordination of growth of the ovary and seed is so programmed that the pericarp begins growth a little before or at the same time as the nucellus and integuments and often many days before the zygote begins to grow into the young embryo. The growth curves for fruits are commonly of two types: sigmoid curves, as for bean, pea, and tomato, or double sigmoid curves, as for peach, plum, and cherry (LEOPOLD and KRIEDEMANN 1975). For our purposes, the sour cherry will serve as a model. In a detailed analysis of the growth of several parts of the fruit and seed of *Prunus acida* TUKEY and YOUNG (1939) recorded growth rates of the ovary and seed parts (Fig. 5.3). Their study also related cell division and enlargement to overall development of the ovary wall (Fig. 5.4). It is notable that cell division in the fleshy exocarp ceased shortly after the first period of rapid development, whereas cell enlargement was biphasic and continued to the time of ripening. Rapid enlargement of the cells of the pericarp for about 3 weeks following full bloom was succeeded by essentially no increase for 2 weeks and then there was a second sigmoid like growth period lasting 3 weeks. During this latter phase of growth, the size of the individual cells increased remarkably—up to 25 times in diameter from the size at full bloom. Rapid growth of the embryo was accompanied by little or no growth of the pericarp. But once full growth of the embryo was achieved, enlargement

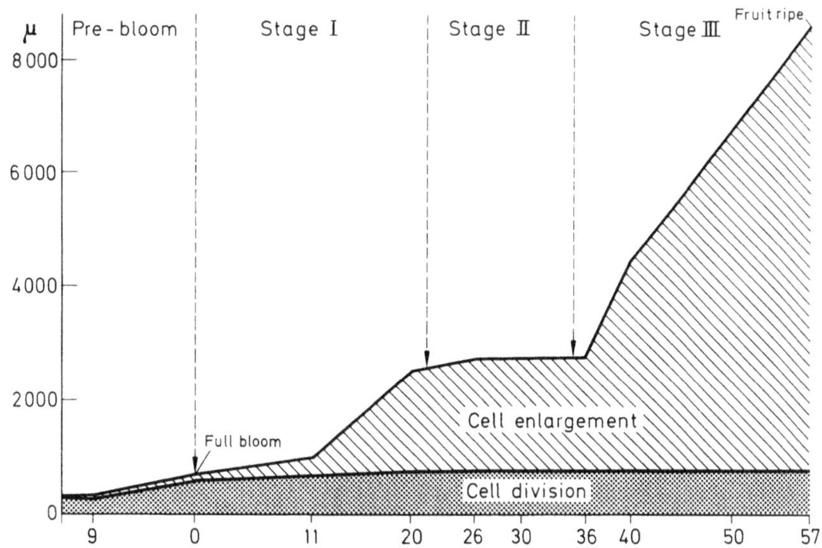

Fig. 5.4. Increase in thickness of ovary wall of sour cherry, from pre-bloom to fruit ripening, showing amount due to increase in number of cells and amount due to enlargement of cells already present. *Slanting lines* size changes due to cell enlargement. *Stippling* size changes due to cell division. (After TUKEY and YOUNG 1939)

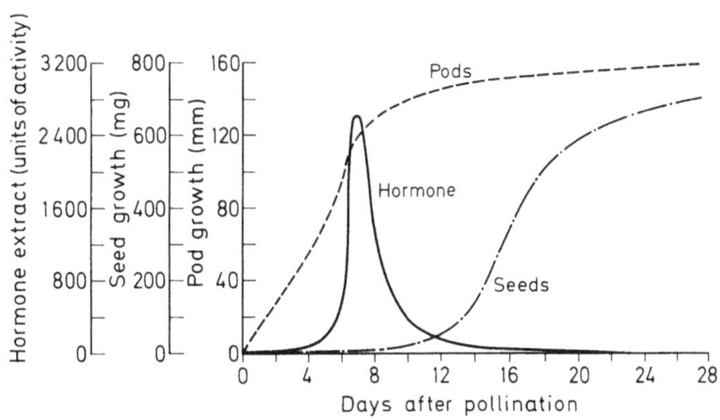

Fig. 5.5. Growth of bean pods and seeds in terms of length and fresh weight, respectively, compared with amount of extractable cell-elongating hormones obtained from the seeds at intervals following pollination. (After MITCHELL et al. 1951)

of the cells of the pericarp was resumed. What is the relationship of these well-delineated growth phases to endogenous growth factors?

Important observations on the timing of production of a gibberellin-like substance were made by MITCHELL et al. (1951) on Black Valentine bean during fruit and seed development (Fig. 5.5). While pod development began shortly after pollination, the gibberellin-like growth factor was barely detectable until 4 days after pollination. The growth factor reached its peak for activity approximately 4 days later, shortly before the end of the "grand-period-of-growth" of the pods. Interestingly, it was when the growth factor was at the peak of its activity that growth in size of the seed began.

The physiological significance of auxin in pollen has been questioned by BARENDSE et al. (1970) because of the difficulty in demonstrating its presence in the pollen of numerous species. BARENDSE et al. (1970) were able to show in *Petunia hybrida* the presence of auxins in the pollen and style and gibberellins in both of these tissues plus the ovary. Both hormones were present in all three tissue types of the three lily species investigated. Nevertheless, only one significant correlation could be drawn from their data. Gibberellins seemed to be of importance for the initial stages of pollen germination. BARENDSE et al. (1970) further concluded that since different hormones are continually interacting with one another, it is premature to propose a specific role for any one of them in the fertilization process.

Unless pollination occurs, or some other stimulus is provided, growth of the ovary usually stops; but if pollination takes place, ovary growth normally continues. The exact causes are not known. It has been speculated that the cessation of growth is due to the combination of an undersupply of growth-promoting hormones and an oversupply of growth inhibitors. Clear-cut data, however, are lacking. Fortunately, a pertinent time period for detailed investigation of this problem has been established for tomato and gherkin (NITSCH 1951). Crucial events taking place shortly after pollination seem to involve more than the known hormones. A short time interval is spotlighted as critical for continuation of growth because it is possible to grow the ovaries of these plants to maturity on a simple mineral/sugar medium if taken 2 days after pollination, whereas ovaries transferred to the same medium at the time of pollination stay alive but will not grow. Some doubt has been thrown on the possibility that the known hormones are directly involved. BAJAJ and COLLINS (1968) tested the effects of gibberellic acid, indolebutyric acid, naphthalene acetic acid, and kinetin, as well as the presence of a calyx on the formation of strawberry fruits under aseptic, test-tube conditions. The size, shape, and general growth of the fruits was found to be dependent on the number of fertilized ovules. Furthermore, the fruits tended to be undersized when the calyx was removed. However, when high concentrations ranging from 500 to 6000 ppm of indolebutyric acid, 2-naphthoxyacetic acid and 1-naphthaleneacetic acid prepared in a free-flowing lanolin emulsion were used on just-opening flowers of strawberries grown in the greenhouse, receptacles grew very well (THOMPSON 1964). The nucellus of each unfertilized ovule also grew. THOMPSON (1964) thought the growth substance treatment provided only an initial stimulus which perhaps was expressed as promotion of the development of the nucellus, and the subsequent growth of the strawberry depended upon the fruit tissues producing their own growth substances or attracting them from the parent plant.

Chloroform extracts of pollen can replace many of the initial effects of pollen on ovary growth. Whatever the factor(s) may be, they are nonspecific, for they will cause "fruit set" in unrelated genera and even unrelated families (GUSTAFSON 1937). Once beyond the critical period following pollination and fertilization, many fruits seem to be self-sufficient for plant hormones.

Soon after the introduction of synthetic auxins it was discovered that they would induce parthenocarpic development of fruits in many species (Table 5.1).

Table 5.1. Parthenocarpic development of fruit induced by growth substances

Plant	Auxins	Gibberellins	Kinins	Acid	Combination of IAA, GA, K	References
Pepper	×					Gustafson (1937)
Cucurbits	×					Gustafson (1937)
Solanaceae	×					Gustafson (1937)
Begonia	×					Gustafson (1942)
Blackberries	×					Gustafson (1942)
Cymbidium	×					Gustafson (1942)
Figs	×					Gustafson (1942)
Gladiolus	×					Gustafson (1942)
Grape	×					Gustafson (1942)
Holly	×					Gustafson (1942)
Primula	×					Gustafson (1942)
Pear	×					Gustafson (1942)
Snapdragon	×					Gustafson (1942)
Avocado	×					Stewart and Hield (1951)
Muskemelon	×					Elassar et al. (1974)
Oil palm	×					Thomas et al. (1973)
Rosa rugosa	×					Jackson (1968)
Strawberry	×					Thompson (1964)
Apple		×				Davison (1960)
Almond		×				Crane et al. (1960)
Apricot		×				Crane et al. (1960)
Peach		×				Crane et al. (1960)
Blueberry		×				Barker and Collins (1965)
Fig		×				Crane (1965)
Delaware grape		×				Clore (1965)
Grapefruit		×				Randhawa et al. (1964)
Mandarin		×				Randhawa et al. (1964)
Guava		×				Teaotia et al. (1961)
Pear		×				Gil et al. (1972)
Plum		×				Jackson (1968) Marlangeon (1973)
Pereskia aculeata		×				Sachar (1962)
Rosa sherardii		×				Jackson and Blundell (1966)
Tomato		×				Wittwer et al. (1957)
Fig			×			Crane (1965)
Grape			×			Weaver et al. (1966)
Muskmelon			×			Jones (1965)
Rosa sherardii				×		Jackson (1968c)
Blueberry					×	Mainland and Eck (1968)
Cherry					×	Rebeiz and Crane (1961) Nyeki (1973)
Apple					×	Kotob and Schwabe (1971)
Mango					×	Chacko and Singh (1969)

This was particularly true for those possessing a large number of ovules per ovary. Only about 20% of horticultural crops respond to auxins, indicating that factors other than auxins must also affect fruit set. Later, when gibberellins, particularly GA_4 and GA_7, were tested, it was found that they were effective not only on species responsive to auxin but also on flowers where the ovary contained only one or a few ovules. From Table 5.1, one may draw the generalization that auxins are most effective in inducing parthenocarpy in multiseeded fruits, while gibberellins seem to be most effective when there are relatively few ovules. Cytokinins are active on fig and grape and even abscisic acid has been effective on *Rosa sherardii* (G.A.D. JACKSON 1966, 1968). Morphactin applications have also been found to increase both the total number and weight of tomato fruits (LUKASIK 1975). In a number of species—notably blueberry, cherry, apple, and mango—a combination of auxins, gibberellins, and cytokinins is necessary to achieve full fruit development.

Since the pattern of hormone activity in the developing fruits of seeded and naturally parthenocarpic fruit is similar and only one phytohormone treatment is necessary in the chemically induced parthenocarpic fruits, it seems most probable some phytohormones act as inducing agents starting the filling of "metabolic sinks", while others like abscisic and jasmonic acids stop protoplasmic growth (DATHE et al. 1981). Thus the role of plant hormones in fruit development remains obscure.

In the commercial production of fruit, the ability to control the number of fruit per tree is of great importance. Sometimes it is desirable to increase fruit set while at others it is imperative that fruit thinning be practiced. Chemical treatments have been devised to satisfy both needs. It is now common practice to use growth regulators to control fruit set as well as to influence the size, shape, and rate of maturation.

Shape is often used as a special identifying characteristic. An excellent example of this is found in the Red Delicious variety of apple. This apple is valued for its flavor, aroma, and keeping qualities, and is readily recognized by its characteristic elongated appearance and pronounced calyx lobes. Under some environmental conditions, however, the tendency is for the Red Delicious fruit to be nearly spherical. This can be avoided by spraying the trees between the time of full bloom and petal fall with a combination of gibberellin A_4 and A_7 and 6-benzyladenin. The resulting fruits are increased in size and have the typical elongate appearance of the Red Delicious apple, and are therefore more readily marketable (GREENHALGH and GODLEY 1976). These and many other aspects of the present state of the art of growth regulation with agricultural chemicals have been ably treated by WEAVER (1972) and NICKELL (1982).

5.9 Seed Development

While studies on the development of embryos from undifferentiated cells in tissue culture have been very useful in revealing some of the information needed to understand the impact of localized environmental factors required to probe some of the complexities of seed development, there is good reason to conclude we are only in a position to ask good questions—questions that can be answered through experimentation. Seed development involves both growth and differentiation. Typically, the parts of the ovule that show the greatest amount of DNA replication, RNA activity and accompanying metabolic activity are the embryo and endosperm followed by the integuments. The endosperm, of course, provides the embryo with its immediate environment.

The special role of endosperm in supporting embryo growth was delineated by VAN OVERBEEK et al. (1941), who demonstrated that the liquid endosperm of coconuts would support the growth and development of *Datura* embryos up to the point where young embryos produce enough of the requisite growth factors, including amino acids, vitamins and hormones that are essential until the embryos become dependent only upon a supply of sugar. Following this work, liquid endosperm became a regular constituent of embryo culture media and remained so until the major growth factors it supplied were identified. In this regard it should be emphasized that the rates of biochemical pathway differentiation vary greatly from species to species. This accounts for differences in requirements for amino acids, vitamins, and hormones in fully defined culture media by different species.

In an attempt to learn what liquid endosperm provided that was essential for embryo development, it was soon learned that up to 85% of the dry weight of coconut milk is carbohydrate with sorbitol making up the bulk of that. Next in importance are two other sugar alcohols—myo-inositol and scyllo-inositol. Glucose is the major sugar. Inositol is an intermediate in the formation of galacturonic acid—a cell wall component—and may also function like a vitamin. During seed maturation inositol is transformed into phytin, a mixed potassium, magnesium, calcium, and phosphorous salt.

Liquid endosperm contains a number of nitrogenous compounds including proteins, amides, free amino acids, and nucleic acids. In some instances these might be used directly and in others they are used only after further degradation. While it has been found that very young embryos must be nurtured in a liquid medium, older embryos develop most quickly on a semi-solid substrate. Initiation of root primorida seems to be facilitated by a liquid environment, while shoot development has a lesser liquid requirement. The normal progression in the seed, of course, is from liquid to cellular endosperm. Embryos that are most easily cultured can use glucose or sucrose as the sole carbon source and NO_3 as the nitrogen source. Apparently biochemical pathway differentiation occurs very early in these species (REINERT et al. 1977).

The hormonally active fraction of liquid endosperms contains several kinds of growth substances: indoleacetic acid, gibberellins, cytokinins, and several as yet uncharacterized substances. The probable role of these undefined correlating agents must be investigated. Relatively few studies have been made by single investigators or research teams in which the several known growth regulators have been assayed for from single samples taken during the full period of ovary wall growth and seed development. Interactions must be occurring since auxins, gibberellins, cytokinins, and abscisic acid are all simultaneously present in the seed (EEUWENS and SCHWABE 1975, RADLEY 1976). Almost certainly they are

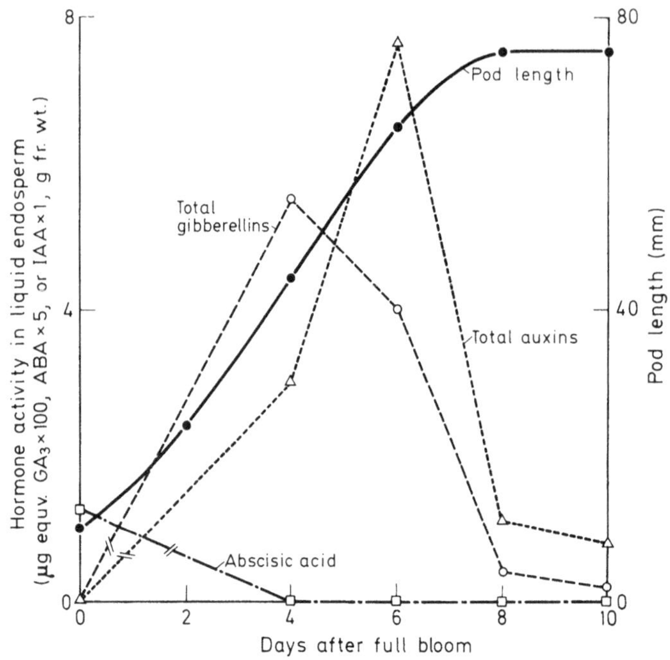

Fig. 5.6. Elongation of the pod wall of pea in relation to hormone activity in the liquid endosperm of the seed. (EEUWENS and SCHWABE 1975)

acting in concert. In *Pisum sativum* (EEUWENS and SCHWABE 1975) and wheat (RADLEY 1976) and doubtless, other plants as well, the amounts of the individual substances differ markedly during the course of fruit and seed development (Fig. 5.6 and 5.7).

It can be predicted from tissue culture studies that auxins must be increasing in amount as the embryo initiates the root primordium and we might expect a surge of cytokinins as the stem primordium is initiated (SKOOG and MILLER 1957). However, clean-cut correlations have not been made in naturally developing seeds.

General conclusions that can be drawn at this time are that the gibberellins synthesized in developing seeds are, indeed, involved in both seed and fruit growth; and as maturation occurs, there is a buildup of conjugated gibberellins and gibberellin derivatives—"bound forms". These persist and are potentially available at the time of germination. It is tempting to think that the hexapyranosyl and glycosyl esters of gibberellin are made available during germination. FRYDMAN and MACMILLAN (1973) and YAMANE et al. (1975) have shown with radioactive gibberellin that during germination of *Phaseolus vulgaris* there was little liberation of gibberellin but, on the contrary, a further accumulation of label occurs in glucosides. Gibberellins synthesized in developing seeds are highly correlated with both seed and fruit growth. This aspect of gibberellic acid metabolism has been reviewed by MOORE (1979).

Cytokinin levels also change during seed development. In *Zea mays* (MILLER 1967) and *Triticum aestivum* (WHEELER 1972), for example, it was found that

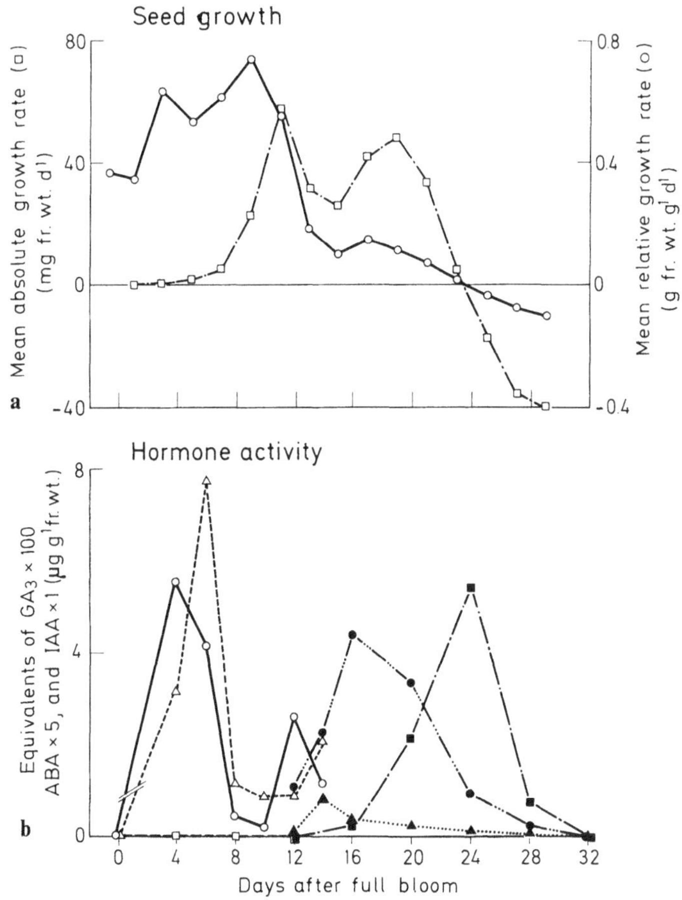

Fig. 5.7 a, b. Changes in **a** absolute and relative growth rates of the developing pea seed in relation to **b** extractable hormone activity in the liquid endosperm and embryo. Symbols in **b**: total gibberellins, ○——○ liquid endosperm, ●——● embryo; total auxins; △——△ liquid endosperm, ▲——▲ embryo; total abscisic acid; □——□ liquid endosperm, ■——■ embryo. (EEUWENS and SCHWABE 1975)

cytokinin levels reached a peak 11 days after pollination and declined almost back to base level during the next 10 days. An interesting correlation has been found in *Pisum arvense* (field pea) during seed development (BURROWS and CARR 1970). One major and two minor cytokinin peaks coincide with the maximum volume of the endosperm and with the two periods of rapid growth of the whole seed and embryo. The possibility is certainly strong that cytokinins are involved in growth and development of embryos, fruits, and seeds. How they are triggered and at what cytochemical level they act is yet to be determined.

Seeds are a rich source not only of gibberellins and cytokinins but also of auxins. Relatively few generalizations can be made about auxin content, stage of endosperm development, and growth rate of embryos. During growth of apple seeds one finds two peaks of auxin accumulation (LUCKWILL et al.

1969). The first of these coincides with the change from the free nuclear to cellular stage of endosperm development. As the embryo grows at the expense of the cellular endosperm there is an increase in auxin and a second peak is achieved. In wheat, auxin production begins about 3 weeks after anthesis and reaches a peak about 1 week before the embryo reaches maximum fresh weight. As the seed matures, the auxin content declines to nearly zero (WHEELER 1972). In contrast to these two examples, the auxin level in pea (*Pisum sativum*) endosperm reaches its maximum level within a week after pollination. During the following week there is marked decline followed by a short increase (EEUWENS and SCHWABE 1975). The one generalization about auxins in seeds that seems to hold is that free auxin declines to a very low level as seeds mature (BEWLEY and BLACK 1978). Quite possibly, a major portion of the free auxin is converted to bound forms and ester conjugates, S-acyl anhydrides, acyl anhydrides, and amide conjugates. *Zea mays* and *Avena sativa* can serve as models for such conversions. BANDURSKI and his associates are adding rapidly to our knowledge of the bound auxins: where they are distributed in the seed and patterns of turnover and usage during germination (see review COHEN and BANDURSKI 1982). Thus far, however, comparable information is not available for any other plant hormone.

It is commonly thought that arrest of embryo growth in the seed and prevention of precocious germination is under the control of abscisic acid. Through surgical removal of parts of the ovary wall, seed coats, and endosperm it has been found that growth inhibitors are present in different tissues in accordance with species. Several inhibitors have been characterized, but the one most intensively studied is abscisic acid. Once again there is no consistent pattern followed during seed development. In wheat, abscisic acid reaches a maximum level 3 weeks after anthesis and then proceeds to decline to its lowest level at seed maturity (MCWHA 1975, RADLEY 1976). At that point, very little is present. The story is rather different in peas. Abscisic acid is virtually undetectable until the beginning of the third week following pollination, thereafter there is a sharp rise in production until the 24th day. A sharp decline follows to near absence at maturity (EEUWENS and SCHWABE 1975). These seeds—wheat and pea—do not have a dormant period. Unfortunately, very few analogous measurements have been made during development of seeds that do undergo a period of dormancy. For the sake of future experimentation it is useful to know that seeds of many species do possess appreciable quantities of abscisic acid at maturity. This fact has fueled the current dogma that ABA is the controlling factor in limiting embryo growth and prevention of vivipary. The issue should not be considered closed, however, until more measurements are available.

5.10 Seed Dormancy — Germination Inhibitors

Most seeds are dormant at the time of fruit maturity. In those rare instances where this is not the case, germination may occur in the fruit—a condition known as vivapary. When that occurs, it is usually the result of failure of the embryo to go into a quiescent state. Arrested growth of the embryo can result from the accumulation of one or more growth inhibitors, reduction in water supply, reduced food supply, or reduced oxygen availability.

Dormancy may persist from a few days to scores of years. In crop plants, it is highly desirable that endogenous causes of seed dormancy do not persist much beyond the time of harvest. Once the water content of the seed has dropped below 15%, as is usually the case in storage, germination will not occur. In the instance of crop seeds, it is highly desirable that they germinate uniformly when put under favorable water, temperature, oxygen, and light conditions. This is true because the time of harvest is closely linked to the time of germination. Consequently, plant breeders select for uniformity in germination.

It is a very different matter when one is considering natural vegetation. In nature, nonuniformity in germination can be highly advantageous to persistence in a niche, because extreme weather conditions can wipe out an entire population of uniformly flowering plants—as in annuals. Staggering of germination definitely increases the chances of survival in a given area.

Among the several causes of dormancy are presence of growth inhibitors, lack of growth hormones, impermeability of seeds to water, mechanically resistant seed coats, impermeability to oxygen, immature embryos, and dormant embryos. Emphasis in this account will be placed on the nature of embryo dormancy.

Embryos in some seeds, even though fully mature, will not grow when placed under environmental conditions favorable for germination. Sometimes this is due to the presence of inhibitors in one or more of the tissues surrounding the embryo or in the embryo proper (EVENARI 1949). In other instances, there is subthreshold production or absence of growth-promoting substances.

A very large number of germination inhibitors have been listed (WAREING 1965). These substances have been found in extracts from every plant organ including roots, stems, leaves, fruits and seeds. Many of these compounds prevent germination without affecting the seeds irreversibly and when dissipated through leaching or oxidative processes permit germination. In watery fruits such as melons and tomatoes, just as for seeds from saline environments, osmotic inhibition is presumed. Osmotic inhibition is inferred because when the seeds are removed from such environments and placed in water they germinate. With tomatoes, however, fractionation of the juice has revealed the presence of at least three inhibitors (STEWART and NAYLOR 1953). High osmotic pressure may be due to sugars, sodium chloride, or other substances. This aspect of germination inhibition, as well as others, has been thoroughly reviewed by MAYER and POLJAKOFF-MAYBER (1982).

Naturally occurring metabolic inhibitors may accumulate during embryo growth. In many such instances germination occurs only after considerable leaching with water. Since intensive metabolic activity accompanies germination, metabolic poisons are particularly effective germination inhibitors. Excellent examples are found among the respiratory inhibitors cyanide, dinitrophenol, azide, fluoride, and hydroxylamine. Each of these inhibits germination at concentrations inhibitory to respiration. But, as with many other poisons, at certain concentrations cyanide and dinitrophenol cause some embryos to break dormancy. Furthermore, it is interesting to note that cyanide at low concentrations is actively metabolized by seeds (TAYLORSON and HENDRICKS 1973).

Naturally occurring phenolic compounds such as catechol, resorcinol, pyrogallol, salicylic, gallic, ferulic, caffeic, and coumaric acids all serve as germination inhibitors. Indeed, some of the selective herbicides such as the substituted phenols and cresols also serve as germination inhibitors. Because of the widespread occurrence of coumarin in plants and its demonstrated inhibitory effects at levels found in fruits and some seeds, it is considered to be a natural germination inhibitor (LERNER et al. 1959). In time, of course, coumarin is metabolized in resting seeds to a level below inhibition.

Without doubt, the best-documented naturally occurring germination inhibitor is ABA. Presence of ABA has been reported in an impressive number of fruits and seeds (SONDHEIMER et al. 1968, ADDICOTT and LYONS 1969, MILBORROW 1974). The concentration of ABA required to bring about its inhibitory effects varies, depending upon the species, between 5–100 ppm. Abscisic acid appears to interact with gibberellic acid and kinetin in controlling germination; but it must also be remembered that ABA is often accompanied in the seed by other growth inhibitors. For example, studies by WEBB and WAREING (1972) showed by means of chromatography that one or more inhibitors other that ABA must be involved in dormancy in *Acer pseudoplatanus*.

Significantly, ABA additions have been shown to increase osmotic inhibition (DÖRFFLING 1970). But the crucial measurements of natural ABA increase with osmotic pressure increase have not been made. A comprehensive review of the role of phytohormone in seed dormancy has been published by KHAN and TAO (1978).

The importance of cytokinins in the germination process has been clarified to a marked degree. KHAN (1975) emphasized the importance of the cytokinins in overcoming dormancy and dovetailing with the action of gibberellins in the early stages of germination. Gibberellins still appear to be the most potent of the hormones in substituting for the embryo in starting the sequence of metabolic processes associable with germination (JONES 1973, VARNER 1974). In grains, the aleurone cells, in response to gibberellins, both synthesize and secrete hydrolases such as alpha-amylase, protease, and ribonuclease (BENNETT and CHRISPEELS 1972), while secretion but not synthesis of B-gibberellin is associated with the presence of gibberellin (TAIZ and JONES 1970). Since synthesis of new proteins is initiated, one must infer that specific m-RNA molecules are being either generated or activated. Inasmuch as these proteins do not all appear simultaneously, it is not clear if gibberellin is targeting something that starts an entrained series of events or if it is serving as a multi-targeting hormone.

TREWAVAS (1976) provided an excellent figure summarizing what was known of the time sequence of biochemical events accompanying germination as reflected in barley aleurone cells subjected to gibberellic acid treatment and simultaneously supplied with $^{32}PO_4$ (Fig. 5.8). The electron microscope has also been useful in revealing that within minutes following the start of seed hydration there is a remarkable growth in endoplasmic reticulum in the cells. This, of course, is indicative of aroused protein and lipid synthesis because these are

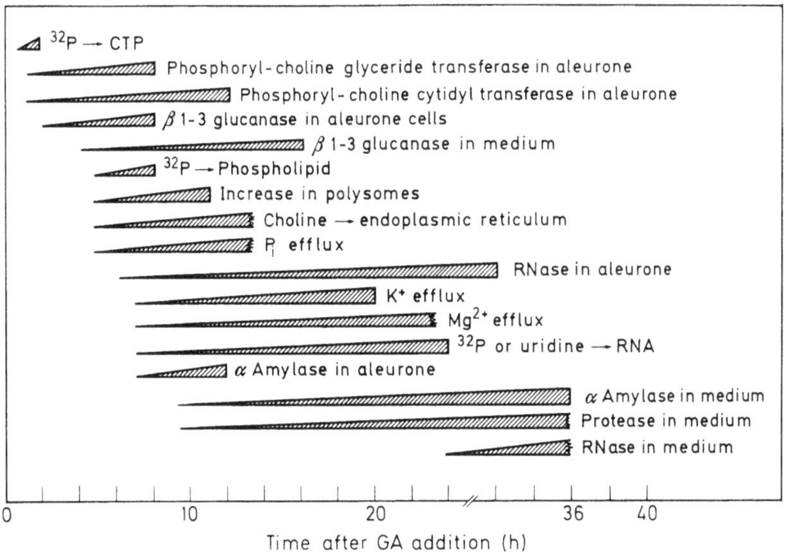

Fig. 5.8. Metabolic changes which occur in barley aleurone cells after gibberellic acid treatment. *Left point of triangle* time at which the metabolic change is initiated; *apex* time at which the metabolic change is at its maximum. *Triangles with serrated edges* time of maximal metabolic change was not determined. (After TREWAVES 1976)

the primary components of the membranes. Messenger RNA reading for the enzymes involved in the synthesis of these membrane components must already be present in the cytoplasm or the requisite m-RNA's are synthesized de novo.

While a prominent role for ABA in dormancy seems clear, the search for other inhibitors must continue. A decline in ABA does not necessarily guarantee release from dormancy. This has proven to be true for lettuce and *Fraxinus americana* (SONDHEIMER et al. 1974, BRAUN and KHAN 1975) and for *Pyrus communis* and *Prunus persica* (DENNIS et al. 1978, BONAMY and DENNIS 1977). Furthermore, it has been found that *Acer saccharinum* seeds possess high levels of ABA yet show no dormancy (RUDNICKI and SUZKA 1969).

Future work on dormancy will profit from simultaneous studies on decrease in levels of inhibitors and inhibitor-related processes.

5.11 Conclusions

The relative importance of the several known plant hormones in cell division and as correlating agents serving in the differentiation of the zygote into an embryo, seedling and adult plant is becoming reasonably well established at the qualitative level. There is abundant evidence the quantitative stage has arrived. Analytical procedures are now sufficiently sensitive for detection of physiologically significant amounts of known plant hormones in both their active and "storage" forms. The new technology must be employed to reinvestigate much of what is now taken as dogma. When this is done, the field of plant hormones will have reached the primary biochemical level.

Since there is increasing recognition that hormonal effects are essentially of a vectorial rather than linear nature, it is essential that cells and tissues be analyzed for more than one hormone at a time. This will call for an increasing use of new experimental designs. Hormonal action takes place in a matrix and this needs to be recognized when measurements are being made and reported.

It is to be expected that lesser known naturally produced growth accelerators and retardants will be experimented with increasingly. Fortunately, the day of fixation on "the" plant growth hormone(s) is over and acceptance of the philosophy of many hormones acting in concert is becoming increasingly accepted. Very probably, there are some important plant hormones yet to be described.

The tools of molecular biology are now sufficiently sharp to begin to determine the mechanism of action of plant hormones. Physiologists have done fairly well in providing guide lines for fruitful biochemical investigations to be made. In this regard many models are already available from animal hormone investigations. For example, the concept of the second messenger in hormonal action in plants must be taken into account and exploited. There are hints throughout this chapter that there are significant blank spots in our information about signals that serve as hormone evocators. It is important that answers to these questions be supplied soon.

Students of the problem of differentiation will be well served when it is determined what the controlling factors are in spindle orientation preceding nuclear and cell division. Special signals must be involved causing an apical meristem to produce leaves in a pattern we call phyllotaxy. Furthermore, this same terminal meristem can react to another set of signals causing it to become a multimeristem that is used up in the production of numerous flowers as in the sunflower. Learning the role(s) of known and perhaps yet to be discovered hormones in this messenger–evocator scenario will be an exciting phase in plant physiology.

References

Addicott F, Lyon JL (1969) Physiology of abscisic acid and related substances. Annu Rev Plant Physiol 20:139–164

Alpi A, Lorenzi R, Cionini P, Bennici A, D'Amato F (1979) Identification of gibberellin A_1 in the embryo suspensor of *Phaseolus coccineus*. Planta (Berl) 147:225–228

Ammirato P (1977) Hormonal control of somatic embryo development from cultured cells of caraway. Interactions of abscisic acid, zeatin, and gibberellic acid. Plant Physiol (Bethesda) 59:579–586

Amrhein N (1977) The current status of cyclic AMP in higher plants. Annu Rev Plant Physiol 28:123–132

Bajaj YPS, Collins WB (1968) Some factors affecting the in vitro development of strawberry fruits. Proc Am Soc Hortic Sci 93:326–333

Barendse GWM, Rodrigues Pereira AS, Berkers PA, Driessen FM, van Eyden-Emons A, Linskens HF (1970) Growth hormones in pollen, styles and ovaries of *Petunia hybrida* and of *Lilium* species. Acta Bot Neerl 19:175–186

Barker WG, Collins WB (1965) Parthenocarpic fruit set in the lowbush blueberry. Proc Am Soc Hortic Sci 87:229–233

Bayliss MW (1980) Chromosomal variations in plant tissues in culture. In: Vasil IK (ed) Perspectives in plant cell and tissue culture. Supplement 11 A, International review of cytology. Academic New York, pp 113–144

Behre K (1929) Physiologische und zytologische Untersuchungen über Drosera. Planta (Berl) 7:208–306

Bennett PA, Chrispeels MJ (1972) *De novo* synthesis of ribonuclease and β-1,3-glucanase by aleurone cells of barley. Plant Physiol (Bethesda) 49:445–447

Bernier G (1971) Structural and metabolic changes in the shoot apex in transition to flowering. Can J Bot 49:803–819

Bewley JD, Black M (1978) Physiology and biochemistry of seeds. Development, germination and growth, vol I. Springer, Berlin Heidelberg New York, pp 88–101

Bisalputra T, Esau K (1964) Polarized light study of phloem differentiation in embryo of *Chenopodium album*. Bot Gaz 125:1–7

Bonamy PA, Dennis FG, Jr (1977) Abscisic acid levels in seeds of peach. I. Changes during maturation and storage. J Am Soc Hortic Sci 102:23–26

Booth A (1959) Some factors concerned in the growth of stolons in potato. J Linn Soc Lond Bot 56:166–169

Booth A (1963) The role of growth substances in the development of stolons. In: Ivins JD, Milthorpe IL (eds) The growth of the potato. Butterworth, London, pp 99–113

Bose TK, Nitsch JP (1970) Chemical alteration of sex expression in *Luffa acutangula*. Physiol Plant 23:1206–1211

Brantley BB, Warren GF (1960) Sex expression and growth in muskmellon. Plant Physiol (Bethesda) 35:741–745

Braun AC (1959) A demonstration of the recovery of the crown gall tumor cell with the use of complex tumors of single-cell origin. Proc Natl Acad Sci USA 45:932–938

Braun J, Khan AA (1975) Endogenous abscisic acid levels in germinating and nongerminating lettuce seed. Plant Physiol (Bethesda) 56:731–733

Bünning E (1952) Morphogenesis in plants. Surv Biol Prog 2:105–140

Bünning E (1957) Polarität und inaquale Teilung des pflanzlichen Protoplasten. Protoplasmatologia 8:1–86

Burg SP, Burg EA (1966) The interaction between auxins and ethylene and its role in plant growth. Proc Natl Acad Sci USA 55:262–269

Burrows WJ, Carr DJ (1970) Cytokinin content of pea seeds during their growth and development. Physiol Plant 23:1064–1070

Butenko RG, Strogonov BP, Babaeva JA (1967 Somatic embryogenesis in carrot tissue cultures under conditions of high salt concentration in the medium. Dokl Akad Nauk SSSR 175:1179–1181

Cass DD, Jensen WA (1970) Fertilization in barley. Am J Bot 57:62–70

Ceccarelli N, Lorenzi R, Alpi A (1979) Kaurene and kaurenol biosynthesis in cell-free system of *Phaseolus coccineus* suspensor. Phytochemistry (Oxf) 18:1657–1658

Ceccarelli N, Lorenzi R, Alpi A (1981a) Kaurene metabolism in cell-free extracts of *Phaseolus coccineus* suspensors. Plant Sci Lett 21:325–332

Ceccarelli N, Lorenzi R, Alpi A (1981b) Gibberellin biosynthesis in *Phaseolus coccineus* suspensor. Z Pflanzenphysiol 102:37–44

Ceccarelli N, Lorenzi R, Alpi A (1981c) Kaurene biosynthesis in intact *Phaseolus coccineus* suspensors. Experientia (Basel) 37:478

Chacko EK, Singh RN (1969) Induction of parthenocarpy in mango (*Mangifera indica* L.) using plant growth regulators. Hort Science 4:121–123

Chailakhyan MKh (1945) Flowering in different plant species as a response to nitrogenous food. Dokl Acad Sci USSR 47:146–149

Cherry JH, van Haystee R (1965) Comparison of messenger RNA in photoperiodically induced and non-induced *Xanthium* buds. Science (Wash DC) 150:1450–1453

Choinski JS Jr, Trelease RN, Doman DC (1981) Control of enzyme activities in cotton cotyledons during maturation and germination. III. In-vitro embryo development in the presence of abscisic acid. Planta (Berl) 152:428–435

Cionini P, Bennici A, Alpi A, D'Amato F (1976) Suspensor, gibberellin and in vitro development of *Phaseolus coccineus* embryos. Planta (Berl) 131:115–117

Clore WJ (1965) Responses of Delaware grapes to gibberellin. Proc Am Soc Hortic Sci 87:259–263

Cocking EC (1972) Plant cell protoplasts – isolation and development. Annu Rev Plant Physiol 23:29–50

Cohen JD, Bandurski RS (1982) Chemistry and physiology of the bound auxins. Annu Rev Plant Physiol 33:403–430

Coombe BG (1976) The development of fleshy fruits. Annu Rev Plant Physiol 27:207–227

Cousson A, Tran Thanh Van K (1981) In vitro control of de novo flower differentiation from tobacco thin cell layers cultured on a liquid medium. Physiol Plant 51:77–84

Crane JC (1965) The chemical induction of parthenocarpy in the Calimyrna fig and its physiological significance. Plant Physiol (Bethesda) 40:606–610

Crane JC, Primer PE, Campbell RC (1960) Gibberellin-induced parthenocarpy in *Prunus*. Proc Am Soc Hortic Sci 75:129–137

D'Amato A (1977) Cytogenetics of differentiation in tissue. In: Reinert J, Bajaj YPS (eds) Plant cell, tissue, and organ culture— applied and fundamental aspects. Springer, Berlin Heidelberg New York, pp 343–357; 442–464

Dathe W, Rönsch H, Preiss A, Schade W, Sembdner G, Schreiber K (1981) Endogenous plant hormones of the broad bean, *Vicia faba* L. (—)-jasmonic acid, a plant growth inhibitor in pericarp. Planta (Berl) 153:530–535

Davison RM (1960) Fruit-setting in apple using gibberellic acid. Nature (Lond) 188:681–682

Denffer D von (1950a) Blühhormon oder Blühhemmung? Neue Gesichtspunkte zur Physiologie der Blütenbildung. Naturwissenschaften 37:296–301

Denffer D von (1950b) Blühhormon oder Blühhemmung? Neue Gesichtspunkte zur Physiologie der Blütenbildung. Naturwissenschaften 37:317–321

Dennis FG Jr, Martin GC, Gaskin P, MacMillan J (1978) Hormones in pear seeds. II. Levels of abscisic acid, dihydrophaseic acid, and their metabolites in relation to seed dormancy in several *Pyrus* species. J Am Soc Hortic Sci 103(3):314–317

Digby J, Wareing PF (1966) The effect of applied growth hormones on cambial division and the differentiation of cambial derivatives. Ann Bot 30:539–548

Dörffling K (1970) Abscisinsäure und Keimungshemmung in der Tomatenfrucht. Planta (Berl) 93:243–256

Durand B (1969) Sélection de génotypes mâles de *Mercurialis annua* L. (2n = 16) en fonction de leurs sensibilités aux cytokinines. CR Acad Sc, Paris 268, série D, 2049–2051

Eeuwens CJ, Schwabe WW (1975) Seed and pod wall development in *Pisum sativum* L. in relation to extracted and applied hormones. J Exp Bot 26:1–14

Elassar G, Rudich J, Palevitch D, Kedar N (1974) Induction of parthenocarpic fruit development in cucumber by growth regulators. Hort. Science 9:238–239

El Hinnawy EL (1956) Some aspects of mineral nutrition and flowering. Meded. Landbouwhogeschool, Wageningen, Nederland 56:1–51

El Tannir Lomba J, Parmentier A, Bouillienne R (1962) Action comparative de la gibbérelline et de l'A.I.A. sur les plantes de jours longs et de jours courts cultivées en diverses photoperiodes et en conditions constantes dans le phytotron de Liège. Proc 16th Int Hortic Congr 5:641–648

Evans LT (1964) Inflorescence initiation in *Lolium temulentum* L. V. The role of auxins and gibberellins. Aust J Biol Sci 17:10–23

Evans LT (1975) Daylength and the flowering of plants. Benjamin, Menlo Park, California, p 122

Evenari M (1949) Germination inhibitors. Bot Rev 15:153–194

Fellenberg G (1982) Developmental physiology. Fortschr Bot 44:205–221

Fischer H (1905) Über die Blütenbildung in ihrer Abhängigkeit vom Licht und über die blütenbildenden Substanzen. Flora 94:478–490

Fischer H (1916) Zur Frage der Kohlensäureernährung der Pflanze. Gartenflora 65:232–237

Fosket DE (1977) The regulation of the plant cell cycle by cytokinin. In: Rost TL, Gifford EM Jr (eds) Mechanism and control of cell division. Dowden, Hutchinson and Ross, Stroudburg, PA, pp 62–91

Frydman VM, MacMillan J (1973) Identification of gibberellins A_{20} and A_{29} in seed of *Pisum sativum* cv Progress No 9 by combined gas chromatography-mass spectrometry. Planta (Berl) 115:11–15

Fuchs Y, Lieberman M (1968) Effects of kinetin, IAA, and gibberellin on ethylene production and their interactions in growth of seedlings. Plant Physiol (Bethesda) 43:2029–2036

Galun E (1959) Effects of gibberellic acid and naphthaleneacetic acid on sex expression and some morphological characters in the cucumber plant. Phyton (Horn) 13:1–8

Garner WW, Allard HA (1920) Effect of the relative length of day and night and other factors of the environment on growth and reproduction in plants. J Agric Res 18:553–606

Garner WW, Allard HA (1923) Further studies in photoperiodism the response of the plant to relative length of day and night. J Agric Res 23:871–920

Gil GF, Martin GC, Griggs WH (1972) Fruit set and development in the pear: Extractable endogenous hormones in parthenocarpic and seeded fruits. J Am Soc Hortic Sci 97:731–735

Goodwin PB (1978a) Phytohormones and growth and development of organs of the vegetative plant. In: Letham DS, Goodwin PB, Higgins TJV (eds) Phytohormones and related compounds: A comprehensive treatise, Vol II. Elsevier/North Holland, pp 31–174

Goodwin PB (1978b) Phytohormones and fruit growth. In: Letham DS, Goodwin RB, Higgins TJV (eds) Phytohormones and related compounds: A comprehensive treatise, vol 2. Elsevier/North Holland, Amsterdam, pp 175–214

Greenhalgh WJ, Godley GL (1976) Studies on fruit shape in apples: a survey of factors influencing the development of "typiness" of the cultivar Red Delicious. Aust J Exp Agric Anim Hus 16:592–595

Gregory FG (1936) The effect of length of day on flowering of plants. Sci Hortic (Wye, Kent) 4:143–154

Gurley LR, Walters RA, Hildebrand CE, Ratliff RL, Hohmann PG, Tobey RA (1977) Sequential biochemical events related to cell proliferation. In: Rost TL, Gifford EM (eds) Mechanisms and control of cell division. Dowden, Hutchinson and Ross, Stroudsburg, PA, pp 3–43

Gustafson FG (1937) Parthenocarpy induced by pollen extracts. Am J Bot 24:102–107

Gustafson FG (1942) Parthenocarpy: Natural and artificial. Bot Rev 8:599–654

Haccius BZ (1965) Haben "Gewebekulturen-Embryonen" einen Suspensor? Ber Dtsch Bot Ges 78(11):11–21

Haccius B (1972) Primärblatt-ähnliche Fiederkotyledonen nach Cytokinin-Behandlung junger Embryonen von *Eranthis hiemalis*. Beitr Biol Pflanz 48:301–311

Hamner KC, Lyon CB, Hamner CL (1942) Effect of mineral nutrition on the ascorbic acid content of the tomato. Bot Gaz 103:586–616

Hashizume H (1967) Experimental induction of female flowers in young Japanese larch (*Larix leptolepis* Gordon). J Jpn For Soc 49:405–408

Helm J (1931) Untersuchungen über die Differenzierung der Sproßscheitelmeristeme von Dikotylen unter besonderer Berücksichtigung des Procambiums. Planta (Berl) 15:105–191

Heslop-Harrison J (1969) *Cannabis sativa* L. In: Evans LT (ed) The induction of flowering. Cornell Univ. Press, Ithaca. NY, pp 205–226

Heslop-Harrison J (1972) Sexuality in angiosperms. In: Steward FC (ed) Plant physiology: A treatise, Vol VIC. Academic, New York, pp 133–289

Hill TA (1973) Endogenous plant growth substances. Edward Arnold, London

Hohmann P, Robey RA, Gurley LR (1975) Cell-cycle-dependent phosphorylation of serine and threonine in Chinese hamster cell F1 histones. Biochem Biophys Res Commun 63:126–133

Howlett FS (1936) The effect of carbohydrate and of nitrogen deficiency upon microsporogenesis and the development of the male gametophyte in the tomato, *Lycopersicum esculentum* Mill. Ann Bot (Lond) 50:767–803

Howlett FS (1939) The modification of flower structure by environment in varieties of *Lycopersicum esculentum*. J Agric Res 58:79–117

Hsu F (1979) Abscisic acid accumulation in developing seeds of *Phaseolus vulgaris* L. Plant Physiol (Bethesda) 63:552–556

Ivonis IY (1981) Natural growth regulators in opening buds of weeping birch (*Betula pendula* Roth). Sov Plant Physiol 28(32):416–422

Jackson DI (1968) Gibberellin and growth in stone fruits. Induction of parthenocarpy in plum. Aust J Biol Sci 21:1103–1106

Jackson GAD (1968) Hormonal control of fruit development, seed dormancy and germination with particular reference to *Rosa*. Soc Chem Ind (Lond) Monogr 30:127–156

Jackson GAD, Blundell JB (1966) Effect of dormin on fruit-set in *Rosa*. Nature (Lond) 212:1470–1471

Jacobs WP, Morrow IB (1957) A quantitative study of xylem development in the vegetative shoot apex of *Coleus*. Am J Bot 44:823–842

Jaffe LF (1981) The role of ion currents in establishing developmental gradients. In: Schweiger HG (ed) International cell biology 1980–1981. Second International Congress on Cell Biology, Berlin (West), Aug. 31–Sept. 5, 1980. Springer, Berlin Heidelberg New York, pp 507–511

Janistyn B (1981) Gas chromatographic, mass- and infrared-spectrometric identification of cyclic adenosine-3′; 5-monophosphate (c-AMP) in maize seedlings (*Zea mays*). Z Naturforsch 36C:193–196

Johnson LP, McLeod JK, Parker CW, Letham DS, Hunt NH (1981) Identification and quantitation of adenosine-3,5′-cyclic monophosphate in plants using gas chromatography-mass spectrometry and high-performance liquid chromatography. Planta (Berl) 152:195–201

Jones CM (1965) Effects of benzyladenine on fruit set in muskmellon. Proc Am Soc Hortic Sci 87:335–340

Jones KL (1947) Studies on *Ambrosia* IV. Effects of short photoperiod and temperature on sex expression. Am J Bot 34:371–377

Jones RL (1973) Gibberellins: Their physiological role. Annu Rev Plant Physiol 24:571–598

Jouanneau JP, Tandeau de Marsac N (1973) Stepwise effects of cytokinin activity and DNA synthesis upon mitotic cycle events in partially synchronized tobacco cells. Exp Cell Res 77:167–174

Karnosky DP (1981) Potential for forest tree improvement via tissue culture. Bioscience 31:114–120

Kessell RHJ, Carr AH (1972) The effect of dissolved oxygen concentration on growth and differentiation of carrot tissue. J Exp Bot 23:996–1007

Khan AA (1975) Primary, preventive and permissive roles of hormones in plant systems. Bot Rev 41:391–420

Khan AA, Tao K-L (1978) Phytohormones, seed dormancy and germination. In: Letham DS, Goodwin PB, Higgins TJV (eds) Phytohormones and related compounds: A comprehensive treatise, vol 2. Elsevier/North Holland, Amsterdam, pp 371–422

King RW (1976) Abscisic acid in developing wheat grains and its relation to grain growth and maturation. Planta (Berl) 132:43–51

Klebs G (1913) Über das Verhältnis der Aussenwelt zur Entwicklung der Pflanze. Sber Akad Wiss Heidelberg. Ser B No 5

Klebs G (1918) Über die Blütenbildung von *Sempervivum*. Flora (Jena) 111/112:128–151, Festschrift Stahl Jena

Knox RB, Evans LT (1968) Inflorescence initiation in *Lolium temulentum* L. XII. An autoradiographic study of evocation in the shoot apex. Aust J Biol Sci 21:1083–1094

Kotob MA, Schwabe WW (1971) Induction of parthenocarpic fruit in Cox's orange pippin apples. J Hortic Sci 46:89–93

Kraus EJ, Kraybill HR (1918) Vegetation and reproduction with special reference to the tomato. Oregon Agric Exp St Bull 149

Krizek DT, Mandava NB (1983) Influence of spectral quality on the growth response of intact bean plants to brassinosteroid, a growth-promoting steroidal lactone. II. Chlorophyll content and partitioning of assimilate. Physiol Plant 57:324–329

Kumar D, Wareing PF (1972) Factors controlling stolon development in the potato plant. New Phytol 71:639–648

Lang A, Chailakhyan MK, Frolova IA (1977) Promotion and inhibition of flower formation in a day neutral plant in grafts with a short-day plant and a long-day plant. Proc Natl Acad Sci USA 74:2412–2416

Leopold AC, Kriedemann PE (1975) Plant growth and development, 2nd edn. McGraw-Hill, New York, pp 318–319

Lerner HR, Mayer AM, Evenari M (1959) The nature of the germination inhibitors present in dispersal units of *Zygophyllum dumosum* and *Trigonella arabica*. Physiol Plant 12:245–250

Loehwing WF (1938) Physiological aspects of sex in angiosperms. Bot Rev 4:581–625

Lorenzi R, Bennici A, Cionini P, Alpi A, D'Amato F (1978) Embryo-suspensor relations in *Phaseolus coccineus:* Cytokinins during seed development. Planta (Berl) 143:59–62

Luckwill LC, Weaver P, MacMillan J (1969) Gibberellins and other growth hormones in apple seeds. J Hortic Sci 44:413–424

Lukasik S (1975) Some observations upon the influence of morphactin IT3456 applied into the soil on the yield of tomatoes. Acta Agrobot 28:43–55

McArthur ICS, Steeves TA (1972) An experimental study of vascular differentiation in *Geum chiloense* Balbis. Bot Gaz 133:276–287

McMullan EE (1980) Effect of applied growth regulators on cone production in Douglas-fir and relation of endogenous growth regulators to cone production capacity. Can J For Res 10:405–414

McWha J (1975) Changes in abscisic acid levels in developing grains of wheat (*Triticum aestivum* L.). J Exp Bot 26:823–827

Mainland CM, Eck P (1968) Induced parthenocarpic fruit development in highbush blueberry. Proc Am Soc Hort Sci 92:284–289

Mann LK (1942) Effects of photoperiod on sex expression in *Ambrosia trifida*. Bot Gaz 103:780–787

Marlangeon RC (1973) Tres dosis de acido giberelico a 75 ppm inducen partenocarpia en ciruelo e incrementan el cuaje en flores libremente polinizadas. Phyton (Horn) 31:95–99

Matsuo E, Minohara Y, Takanashi S (1968) Studies on the growth behaviour of cucumber in controlled environments. III. Effect of different photoperiodic combinations of various kinds of light on the growth and sex differentiation of cucumber. J Jpn Soc Hortic Sci 37:328–332

Mayer AM, Poljakoff-Mayber A (1982) The germination of seeds, 3rd edn. Pergamon, Oxford

Michniewicz M, Lang A (1962) Effect of nine different gibberellins on stem elongation and flower formation in cold requiring and photoperiodic plants grown under non-inductive conditions. Planta (Berl) 58:549–563

Milborrow BV (1974) The chemistry and physiology of abscisic acid. Annu Rev Plant Physiol 25:259–307

Miller CO (1967) Cytokinins in *Zea mays*. Ann NY Acad Sci 144:251–257

Minina EG, Tylkina LG (1947) Physiological study of the effect of gases upon sex differentiation in plants. Compt Rend Acad URSS II, 55:165–168

Minocha SC, Halperin W (1974) Hormones and metabolites which control tracheid differentiation with or without concomitant effects on growth, in cultured tuber tissue of *Helianthus tuberosus* L. Planta (Berl) 116:319–331

Mitchell JW, Skaggs DP, Anderson WP (1951) Plant growth-stimulating hormones in immature bean seeds. Science (Wash DC) 114:159–161

Monula MAK (1978) Gibberellin-like substances in parts of the developing barley grain. Physiol Plant 44:268–272

Moore TC (1979) Biochemistry and physiology of plant hormones. Springer, Berlin Heidelberg New York, pp 99–101; 114–116; 120–127

Moore TC, Ecklund PR (1975) Role of gibberellins in the development of fruits and seeds. In: Krishnamorthy HN (ed) Gibberellins and plant growth. Wiley Eastern, New Delhi, pp 145–182

Murashige T, Nakano R (1967) Chromosome complement as a determinant of the morphogenetic potential of tobacco cells. Am J Bot 54:963–970

Murashige T, Skoog F (1962) A revised medium for rapid growth and bioassays with tobacco tissue cultures. Physiol Plant 15:473–497

Nagl W (1974) The *Phaseolus* suspensor and its polytene chromosomes. Z Pflanzenphysiol 73:1–44

Nanda KK, Krishnamorthy HN, Toky KL, Lata K (1969) Effect of gibberellin A_3, A_{4+7} and A_{13} and of (−)-kaurine on flowering and extension growth of *Impatiens balsamina* under different photoperiods. Planta (Berl) 86:134–141

Narayanaswamy S (1977) Regeneration of plants from tissue cultures: Applied and fundamental aspects. In: Reinert J, Bajaj YPS (eds) Plant cell, tissue, and organ culture. Springer, Berlin Heidelberg New York, pp 179–248

Naylor AW (1941) Effects of nutrition and age upon rate of development of terminal staminate inflorescences of *Xanthium pennsylvanicum*. Bot Gaz 103:342–353

Naylor AW (1952) Physiology of reproduction in plants. Surv of Biol Prog 2:259–300

Negi SS, Olmo HP (1966) Sex conversion of a male *Vitis vinifera* L. by a kinin. Science (Wash DC), 152:1624–1625

Neidle EK (1939) Nitrogen nutrition in relation to photoperiodism in *Xanthium pennsylvanicum*. Bot Gaz 100:607–618

Nickell LG (1978) Plant growth regulators. Chem Eng News 56(41):18–34

Nickell LG (1982) Plant growth regulators: Agricultural uses. Springer, Berlin Heidelberg New York

Nightingale GT (1948) The nitrogen nutrition of green plants. II. Bot Rev 14:185–221

Nitsch C (1968) Effects of growth substances on the induction of flowering of a short-day plant in vitro. In: Wightman F, Setterfield G (eds) Biochemistry and physiology of plant growth substances. Runge, Ottawa, Canada, pp 1385–1398 (Proc 6th Int Conf Plant Growth Substances)

Nitsch C, Nitsch JP (1967a) The induction of flowering in vitro in stem segments of *Plumbago indica* L. I. The production of vegetative buds. Planta (Berl) 72:355–370

Nitsch C, Nitsch JP (1967b) The induction of flowering in vitro in stem segments of *Plumbago indica* L. II. The production of reproductive buds. Planta (Berl) 72:371–384

Nitsch C, Nitsch JP (1969) Floral induction in a short-day plant, *Plumbago indica* L. by 2-chloroethane phosphonic acid. Plant Physiol (Bethesda) 44:1747–1748

Nitsch JP (1951) Growth and development in vitro of excised ovaries. Am J Bot 38:566–577

Nitsch JP (1952) Plant hormones in the development of fruits. Q Rev Biol 27:33–57

Nitsch JP (1971) Perennation through seeds and other structures: fruit development. In: Steward FC (ed) Plant physiology: A treatise. Academic, New York, pp 413–501

Nitsch JP, Kurtz E Jr, Liverman J, Went FW (1952) The development of sex expression in cucurbit flowers. Am J Bot 39:32–43

Nozeran R, Bancilhon L, Neville P (1971) Intervention of internal correlations in the morphogenesis of higher plants. Adv Morphog 9:2–66

Nyeki J (1973) Fruit set promoted by chemical induction in Pandy sour cherry. Acta Agron Acad Sci Hung 22:207–209

Padmanabhan V, Paddock EF, Sharp WR (1974) Plantlet formation from *Lycopersicon esculentum* leaf callus. Can J Bot 52:1429–1432

Pharis RP, Kuo CG (1977) Physiology of gibberellins in conifers. Can J For Res 7:299–325

Phatak SC, Wittwer SH, Honma S, Bukovac MJ (1966) Gibberellin induced anther and pollen development in a stamen-less tomato mutant. Nature (Lond) 209:635–636

Philipson JJ (1983) The role of gibberellin $A_{4/7}$, heat and drought in the induction of flowering in Sitka spruce. J Exp Bot 34:291–302

Przybyllok T, Nagl W (1977) Auxin concentration in the embryo and suspensors of *Tropaeolum majus*, as determined by mass fragmentation (single ion detection). Z Pflanzenphysiol 84:463–465

Puritch GS, McMullan EE, Meagher MD, Simmons CS (1979) Hormonal enhancement in Douglas-fir grafts and seedlings. Can J For Res 9:193–200

Quatrano RS (1978) Development of cell polarity. Annu Rev Plant Physiol 29:487–510

Québédeaux B, Sweetser P, Rowell J (1976) Abscisic acid levels in soybean reproductive structures during development. Plant Physiol (Bethesda) 58:363–366

Radley M (1976) The development of wheat grain in relation to endogeneous growth substances. J Exp Bot 27:1009–1021

Raghavan V (1976) Experimental embryogenesis in vascular plants. Academic, London

Randhawa GS, Dhuria HS, Nair PKR (1964) A note on gibberellin-induced parthenocarpy in citrus. Indian J Hortic 21:171–172

Rebeiz CA, Crane JC (1961) Growth regulator-induced parthenocarpy in the Bing cherry. Proc Am Soc Hortic Sci 78:69–75

Reinert J, Bajaj YPS (eds) (1977) Applied and fundamental aspects of plant cell, tissue and organ culture. Springer, Berlin Heidelberg New York, p 803

Reinert J, Bajaj YPS, Zbell B (1977) Aspects of organization: organogenesis, embryogenesis, cytodifferentiation. In: Street HE (ed) Plant tissue and cell culture, 2nd edn. University California Press, Berkeley and Los Angeles, pp 389–427

Roberts LW (1969) The initiation of xylem differentiation. Bot Rev 35:201–250

Rood S, Pharis RP, Major DJ (1980) Changes in endogenous gibberellin-like substances with sex reversal of the apical inflorescence of corn. Plant Physiol (Bethesda) 66:793–796

Ross SD (1976) Differential flowering responses by four year old Douglas fir seedlings and equi-sized two-year old grafts to gibberellins and auxin. Acta Hortic 56:163–167

Ross SD, Greenwood MS (1979) Promotion of flowering in the Pinaceae by gibberellins. II. Grafts of mature and immature *Pinus taeda*. Physiol Plant 45:207–210

Ross SD, Pharis RP (1976) Promotion of flowering in the Pinaceae by gibberellins. 1. Sexually mature, nonflowering grafts of Douglas fir. Physiol Plant 36:182–186

Rubenstein J, Phillips R, Green C, Gengenbach B (1978) The plant seed. Academic Press, New York

Rücker W (1982) Kombinierter Einfluss von Indolessigsäure, Gibberellin und Benzylaminopurin auf Kallus- und Organdifferenzierung an Blattexplantaten von *Digitalis purpurea*. Z Pflanzenphysiol 107:141–151

Rudich J, Halevy AH, Kedar N (1969) Increase in femaleness of three cucurbits by treatment with ethrel, an ethylene-releasing compound. Planta (Berl) 86:69–76

Rudnicki R, Suszka B (1969) Abscisic acid in non-dormant seeds of silver maple (*Acer saccharinum*). Bull Acad Pol Sci Ser V, Sci Biol 17:325–331

Rünger W (1961) Über den Einfluss der Stickstoffernährung und der Temperatur während der Langtag- und Kurztagperioden auf die Blütenbildung von *Kalanchoe blossfeldiana*. Planta (Berl) 56:517–529

Sachar RC (1962) Comparative effects of gibberellin and indole compounds on the induction of parthenocarpy in sexually incompatible *Pereskia aculeata*. Am J Bot 49:913–917

Sachs J (1888) Über die Wirkung der ultravioletten Strahlen auf die Blüthenbildung. Arb Bot Instit Würzburg 3:372:388

Schaffner JH (1930) Sex reversal and the experimental production of neutral tassels in *Zea mays*. Bot Gaz 90:279–298

Schnepf E (1981) Polarity and gradients in tip growing plant cells. In: Schweiger HG (ed) International cell biology 1980–1981. 2nd International Congress on Cell Biology, Berlin (West), Aug. 31–Sept. 5, 1980. Springer, Berlin Heidelberg New York, pp 485–488

Schulz SR, Jensen WA (1968a) *Capsella* embryogenesis: the early embryo. J Ultrastruct Res 22:376–392

Schulz SR, Jensen WA (1968b) *Capsella* embryogenesis: the egg, zygote and young embryo. Am J Bot 55:807–819

Schwabe WW (1963) Morphogenetic responses to climate. In: Evans LT (ed) Environmental control of plant growth. Academic Press, New York, pp 311–336

Shininger TL (1979) The control of vascular development. Annu Rev Plant Physiol 30:313–337

Simard A (1971) Initiation of DNA synthesis by kinetin and experimental factors in tobacco pith in vitro. Can J Bot 49:1541–1549

Sims WL, Gledhill BL (1969) Ethrel effects on sex expression and growth development in pickling cucumbers. Calif Agric 23:4–6

Skoog F, Miller CO (1957) Chemical regulation of growth and organ formation in plant tissue cultured in vitro. Symp Soc Exp Biol 11:118–131

Smith HJ, McIllrath WJ, Bogorad L (1957) Some effects of iron deficiency on flowering in *Xanthium*. Bot Gaz 118:174–179

Söding H (1936) Über den Einfluss von Wuchstoff auf das Dickenwachstum der Bäume. Ber Dtsch Bot Ges 54:291–304

Söding H (1937) Wuchsstoff und Kambiumtätigkeit der Bäume. Jahrb Wiss Bot 84:639–670

Sondheimer E, Tzou DS, Galson EC (1968) Abscisic acid levels and seed dormancy. Plant Physiol (Bethesda) 43:1443–1447

Sondheimer E, Galson EC, Tinelli E, Walton DC (1974) The metabolism of hormones during seed germination and dormancy. IV. The metabolism of (S)-2-^{14}C-abscisic acid in ash seed. Plant Physiol (Bethesda) 54:803–808

Steeves TA, Sussex IM (1972) Patterns in plant development. Prentice Hall, Englewood Cliffs

Stewart JW, Naylor AW (1953) Isolation of a seed germination inhibitor from tomato juice. Proceedings of the American Society of Plant Physiologists. University of Florida, Gainesville, FL., September 1953, p 23

Stewart WS, Hield HZ (1951) Effects of growth-regulating sprays on fruit set in avacado. Proc Am Soc Hortic Sci 58:53–56

Stoddart JL (1962) Effect of gibberellin on a nonflowering genotype of red clover. Nature (Lond) 194:1063–1064

Sussex IM (1972) Somatic embryos in long-term carrot tissue cultures: Histology, cytology, and development. Phytomorphology 22:50–59

Taiz L, Jones RL (1970) Gibberellic acid, β-1,3-glucanase and the cell walls of barley aleurone layers. Planta (Berl) 92:73–84

Takebe I, Lahib G, Melchers G (1971) Regeneration of whole plants from isolated mesophyll protoplasts of tobacco. Naturwissenschaften 58:318–320

Taylorson RB, Hendricks SB (1973) Promotion of seed germination by cyanide. Plant Physiol (Bethesda) 52:23–27

Tazawa M, Reinert J (1969) Extracellular and intracellular chemical environments in relation to embryogenesis in vitro. Protoplasma 68:157–173

Teaotia SS, Pandey IC, Mathur RS (1961) Gibberellin induced parthenocarpy in guava (*Psidium guajava* L.). Curr Sci (Bangalore) 30:312

Thomas RL, Seth AK, Chau KW, Ooi SC (1973) Induced parthenocarpy in the oil-palm. Ann Bot (Lond) 37:447–452

Thompson PA (1964) The effect of applied growth substances on development of the strawberry fruit. I. Induction of parthenocarpy. J Exp Bot 15:347–358

Tiedjens VA (1928) Sex ratios in cucumber flowers affected by different conditions of soil and light. J Agric Res 36:721–746

Tompsett PB (1977) Studies of growth and flowering in *Picea sitchensis* (Bong.) Carr. 1. Effects of growth regulator applications to mature scions on seedling rootstocks. Ann Bot 41:1171–1178

Tompsett PB (1978) Studies of growth and flowering in *Picea sitchensis* (Bong.) Carr. 2. Initiation and development of male, female and vegetative buds. Ann Bot 42:889–900

Tompsett PB, Fletcher AM (1979) Promotion of flowering on mature *Picea sitchensis* by gibberellin and environmental treatments. The influence of timing and hormonal concentration. Physiol Plant 45:112–116

Tompsett PB, Fletcher AM, Arnold GM (1980) Promotion of cone and seed production in Sitka spruce by gibberellin application. Ann Appl Biol 94:421–429

Torrey JG (1967) Morphogenesis in relation to chromosomal constitution in long-term plant tissue cultures. Physiol Plant 20:265–275

Tramontano WA, Hartnett CM, Lynn DG, Evans LS (1982) Relationship between trigonelline concentration and promotion of cell arrest in G2 in cultured roots of *Pisum sativum*. Phytochemistry (Oxf) 21:1201–1206

Trewavas AJ (1976) Plant growth substances. In: Bryant JA (ed) Molecular aspects of gene expression in plants, chap. 7. Academic Press, New York

Tukey HB, Young JO (1939) Histological study of the developing fruit of sour cherry. Bot Gaz 100:723–749

van der Pluijm JE (1964) An electron microscopic investigation of the filiform apparatus in the embryo sac of *Torenia fournieri*. In: Linskens HF (ed) Physiology and fertilization. North-Holland, Amsterdam, pp 8–16

van Overbeek J, Conklin ME, Blakeslee AF (1941) Factors in coconut milk essential for growth and development of very young *Datura* embryos. Science (Wash DC) 94:350–351

Varner JE (1974) Gibberellin control of a secretory tissue. In: Runeckles VC, Sondheimer E, Walton DC (eds) The chemistry and biochemistry of plant hormones: Recent advances in phytochemistry. Academic Press, New York, pp 123–130

Vasil IK, Hildebrandt AC (1965) Differentiation of tobacco plants from single, isolated cells in microculture. Science (Wash DC) 150:889–892

Vasil IK (1982) Plant cell culture and somatic cell genetics of cereals and grasses. In: Vasil IK, Scowcroft WR, Frey KJ (eds) Plant improvement and somatic cell genetics. Academic, New York, pp 179–203

Vince-Prue D (1975) Photoperiodism in plants. McGraw Hill, London, p 444

Wang TL, Everett NP, Gould AR, Street HE (1981) Studies on the control of the cell cycle in cultured plant cells. III. The effects of cytokinin. Protoplasma 106:23–35

Wardell WL (1975) Rapid initiation of thymidine incorporation into deoxyribonucleic acid in vegetative tobacco stem segments treated with indole-3-acetic acid. Plant Physiol (Bethesda) 56:171–176

Wardlaw CW (1950) The comparative investigation of apices of vascular plants by experimental methods. R Soc London Phil Trans B 234:583–602

Wardlaw CW (1952) The effect of isolating the apical meristem in *Echinopsis, Nuphar, Gunnera,* and *Phaseolus*. Phytomorphology 2:240–242

Wareing PF (1965) Endogenous inhibitors in seed germination and dormancy. In: von Ruhland W (ed) Handbuch der Pflanzenphysiologie, vol 1512. Springer, Berlin Heidelberg New York, pp 909–924

Wareing PF, Phillips IDJ (1981) Growth and differentiation in plants. 3rd edn. Pergamon, Oxford (see p 256)

Weaver RJ (1972) Plant growth substances in agriculture. Freeman, San Francisco

Weaver RJ, Van Overbeek J, Pool RM (1966) Effect of kinins on fruit set and development in *Vitis vinifera*. Hilgardia 37:181–201

Webb DP, Wareing PF (1972) Seed dormancy in *Acer*: Endogenous germination inhibitors and dormancy in *Acer pseudoplatanus* L. Planta (Berl) 104:115–125

Wellensiek SJ (ed) (1973) Symposium on growth regulators in fruit production. International Society for Horticultural Science. Acta Hortic 34:1–507

Went FW (1928) Wuchsstoff und Wachstum. Recl Trav Bot Neerl 25:1–116

Westergaard M (1958) The mechanism of sex determination in dioecious flowering plants. Adv Genet 9:217–281

Wetmore RH, Sorokin S (1955) On the differentiation of xylem. J Arnold Arbor Harv Univ, 36:305–317

Wheeler AW (1972) Changes in growth-substance contents during growth of wheat grains. Ann Apl Biol 72:327–334

White PR (1939) The controlled differentiation in a plant tissue culture. Bull Torrey Bot Club 66:507–513

Wick SM, Seagull RW, Osborn M, Weber K, Gunning BES (1981) Immunofluorescence microscopy of organized micro arrays in structurally stabilized meristematic plant cells. J Cell Biol 89:685–690

Williams RF (1960) The physiology of growth in the wheat plant. 1. Seedling growth and the pattern of growth at the shoot apex. Aust J Biol Sci 13:401–428

Wittwer SH (1978) Phytohormones and chemical regulators. In: Letham DS, Goodwin PB, Higgins TJV (eds) Phytohormones and related compounds: A comprehensive treatise, vol II, chap 13. Elsevier/North Holland, Amsterdam
Wittwer SH, Bukovac MJ (1958) The effects of gibberellin on economic crops. Econ Bot 12:213–255
Wittwer SH, Bukovac MJ, Sell HM, Weller LE (1957) Some effects of gibberellin on flowering and fruit setting. Plant Physiol (Bethesda) 32:39–41
Yamane H, Murofushi N, Takahashi N (1975) Metabolism of gibberellins in maturing and germinating bean seeds. Phytochemistry (Oxf) 14:1195–1200
Yeung EC, Sussex IM (1979) Embryogeny of *Phaseolus coccineus:* The suspensor and the growth of the embryo proper in vitro. Z Pflanzenphysiol 91:423–433
Zeevart JAD (1978) Hormones and flower formatin. In: Letham DS, Goodwin PB, Higgins TJV (eds) Phytohormones and related compounds: A comprehensive treatise, vol II. Elsevier/North Holland Biomedical, Amsterdam, pp 291–327

6 Functions of Hormones at the Whole Plant Level of Organization

A.G. MATTHYSSE and T.K. SCOTT

6.1 Introduction

This chapter is intended to provide an overview of the functions of the five major plant growth hormones in the intact plant: auxin (IAA), gibberellins (GA), cytokinins (K), ethylene, and abscisic acid (ABA). The effects of these hormones have been described in detail on numerous occasions over the last 10 years (see general references at the end of this chapter). This review utilizes these previous reviews as well as including a survey of the literature published between January 1977 and January 1980. The most recent and surely the most detailed treatment of the subject is to be found in *Phytohormones and Related Compounds: A Comprehensive Treatise.* D.S. LETHAM et al., Volumes 1 and 2, 1978. The intent of the present review is to provide a more general overview and to consider overall patterns of regulatory controls by hormones in the whole plant.

The preceding chapters in this Encyclopedia volume deal more directly with analyses of hormonal effects at levels ranging from the cell to the organ. The subject matter of this chapter— hormones and the whole plant— necessarily requires a description of phenomena rather than an analysis of effects.

The present review is intended to facilitate the rapid acquisition of the major aspects of our current knowledge of plant hormones: their sites of synthesis, their routes of transport, their functions in the whole plants, and how these relate to environmental stimuli. To condense this information into a readily utilizable form we have chosen to construct diagrams showing the role of hormones in the "typical" whole plant (Figs. 6.1–6.8, p 236–243). We have not included specialized or unique features of specific groups of plants, nor have we included information about areas which are only partly understood and which may well be clarified in the next few years. Where controversy exists over the role of plant hormones we have presented that information which appears to apply to the majority of species studied. Further, in some cases, responses are quite different for different species; these are indicated on the diagrams by "some plants". This species specificity in itself may be significant and is discussed further below.

Sources for the diagrams are listed at the end of this chapter. General reviews are listed first followed by a listing of specific sources for each diagram. Text references are listed last. The 8 diagrams (figures) are grouped together at the end of this chapter.

6.2 Some Perspectives on the Nature of Hormones

The traditional definition of hormone—a chemical regulator which acts as a messenger, is synthesized in one location, and is transported to another location at which it has a specific effect at low concentrations—best summarizes the existing experimental data and the conceptual viewpoint of both plant and animal endocrinologists. (For a critique of this definition see Chap. 1, this Vol.). In plants, in contrast to animals, the lack of endocrine glands or of identifiable specialized endocrine cells which synthesize a hormone(s) has led to difficulties in determining exact sites of hormone synthesis. At best, general locations of synthesis have been determined for IAA and GA in the apical meristem and young leaves, cytokinins and GA in the root, ABA in the root and root cap, and ABA in wilting leaves. Hormones are also synthesized by flowers, fruits, and germinating seeds but the exact sites of synthesis are, in general, less clear in these cases. IAA, about which there is the most information, is synthesized in anthers, growing pollen tubes, ovaries, and embryos (Fig. 6.1).

The existence of a closed circulatory system in higher animals provides the organism with an easy and efficient route for the transport of hormones from the site of synthesis to the target tissues. In plants some hormones appear to be transported directly in the vascular tissue: for example, cytokinin, GA, and ABA move from the root to the shoot in the xylem; GA moves out of young leaves in the phloem; and ABA is transported out of wilting leaves in the phloem (Fig. 6.1). However, auxin is not transported directly in the vascular tissue, but instead appears to be transported in cells associated with the phloem (Fig. 6.1). Ethylene poses a special problem in that it is a diffusable gas. However, its precursor, 1-aminocyclopropane-1-carboxylic acid (ACC), is transported from the root to the shoot in the xylem. Therefore, using the traditional concept of a hormone as a translocated chemical messenger, ACC may be more aptly considered to be a hormone than ethylene.

In plants it is difficult to define precise target organs, tissues, or cells for the various hormones. This is particularly apparent if one looks at the effects of each individual hormone as shown in Figs. 6.2 through 6.6. Similar difficulties are encountered with some animal hormones, for example adrenalin. In fact there is no theoretical reason nor does it follow from the definition of a hormone why each hormone must have only one target tissue or cell. This point is described in detail in Chapter 1, this Volume.

A conspicuous feature of hormone action in higher animals is the use of a second messenger in the target tissue's response to a hormone. In higher plants this possibility has received less attention. An example of a second messenger in higher plants may well be the transport of ACC from waterlogged roots to the leaves where it is converted into ethylene which in turn produces the physiological response (BRADFORD and YANG 1980). In this case ACC appears to be acting as the hormone and ethylene as the second messenger which is produced locally and has local effects. Other cases in which second messengers may be involved include the stimulation of local ethylene production by IAA (OSBORNE 1978), the local mobilization of Ca^{+2} by IAA (PENNY and PENNY

1978), and the regulation of local flowering stimulators or inhibitors by GA (ZEEVAART 1978). Because of the inherent difficulties in the transport of a diffusable gas, ethylene may generally act as a second messenger which is produced locally in response to a hormonal signal from elsewhere in the plant. As would be expected for a second messenger, ethylene effects are often restricted to the local region of its production (LIEBERMAN and KUNISHI 1972).

Very few processes in the higher plant or animal appear to be regulated by a single hormone. Instead the synergistic and antagonistic actions of two or more hormones are generally used to provid multiple controls and balanced responses to a variety of internal and external signals (see Chap. 1 and 2, this Vol.). Hormones regulate the organism at a variety of levels ranging from responses to the external environment, growth, maturation, and physiological homeostasis to cellular differentiation and biochemical modulations. In this chapter only the first three of these levels are considered. Hormonal effects on cellular differentiation are considered in Chapters 2, 4, and 5, this Volume. Biochemical aspects of hormone action are considered in Volume 9, this Series.

6.3 A General View of the Role of Hormones in the Whole Plant

If one had to make a broad generalization regarding the roles of the five major plant hormones, it would be that auxin and cytokinin are the shoot and root hormones respectively and that gibberellin, ethylene, and abscisic acid are involved in both long-term seasonal and environmental responses of the plant. Short-term responses may have more diverse control mechanisms. Short-term tropistic responses in shoots with few exceptions are regulated principally by auxin, while gravitropic responses in the root are regulated principally by abscisic acid and other inhibitors. The short-term control of water balance is also regulated by abscisic acid and ethylene.

Auxin is produced in shoot meristems and expanding leaves. Cytokinins are synthesized in root tips (Fig. 6.1). They are thus transported in opposite directions in the main body of the plant. Their mechanisms of transport are quite different (Chap. 3, this Vol.). They each participate in the imposition of apical dominance; auxin in the shoot and cytokinin in the root. They also each promote branching in the opposite portion of the plant. In this connection, the regulation of tuber formation may be regarded as a special case of stem lateral branch formation since auxin inhibits and cytokinin stimulates the production of tubers. Auxin and cytokinin each participate, in the absence of the other, in stimulating the formation of missing portions of the plant, i.e., auxin-induced adventitious rooting of stems and cytokinin-induced formation of shoots in tissue culture. In addition auxin and cytokinin each stimulate cambial activity in the portion of the plant in which they are synthesized (Figs. 6.2 and 6.3).

Long-term seasonal responses in which gibberellin is involved seem to be those in which growth and synthetic processes are promoted and the surrounding environment is generally favorable. For example, gibberellin stimulates seed germination, cell division and elongation, bolting, juvenile characteristics of leaves and stems, outgrowth of lateral buds, flowering in a number of species, and finally both fruit growth and fruit set (Fig. 6.4). Further, gibberellin tends to inhibit processes which if set into motion would result in the formation of storage organs, dormancy, and senescence.

In contrast, long-term seasonal responses in which abscisic acid is involved seem to be those in which growth and synthetic processes are inhibited and the surrounding environment is generally unfavorable. For example, abscisic acid stimulates bud and tuber dormancy, leaf abscission and senescence, and fruit abscission and senescence and ripening. In general, abscisic acid is inhibitory to growth and to seed germination (Fig. 6.5).

Ethylene often has inhibitory effects but, like abscisic acid, is involved in long-term responses which are principally concerned with changes in the developmental stages of the plant. The processes effected include the whole life cycle from seed germination, vegetative growth, flowering, fruiting, and ultimately to senescence. Ethylene is not solely an inhibitory hormone; its effects on processes such as seed germination and adventitious root formation are usually stimulatory. The formation and action of ethylene in some, but not all, cases is linked to the continuous presence of indole acetic acid. The classic example is the inhibition of elongation growth by high concentrations of auxin which result in the production of ethylene (Fig. 6.6).

Short-term responses to environmental stimuli are mediated principally by auxin and abscisic acid. Other, less well-characterized inhibitors may play a significant role as well. Gravitropic (Fig. 6.7) and phototropic (Fig. 6.8) responses in shoots often result from lateral auxin transport and the resulting stimulation of growth. Gravitropic responses in roots result principally from lateral redistribution of abscisic acid (and/or other inhibitors) and the resulting inhibition of growth. These tropic responses have generally been measured using seedling plants in which one would expect relatively rapid responses to changing environmental stimuli. Changes in water balance also require rapid adjustments for plants of any age to survive, and adaptation to water stress generally involves increases in abscisic acid especially in leaves, which in turn causes the closure of the stomata. Abscisic acid synthesis in roots is increased by either water stress or flooding (Fig. 6.7). Flooding also results in increased transport of 1-aminocyclopropane-1-carboxylic acid (ACC), an ethylene precursor, in the xylem to the leaves where its conversion to ethylene results in wilting.

Many of the regulatory phenomena described above involve the use of hormones to communicate between plant parts and to integrate the responses of one part of the plant with another, often over long distances. An example of a rapid integrated response in which transfer of information over some distance is involved is the response of leaves to changes in water balance (stress or flooding) at the root. The most-studied of these long-distance phenomena is the much slower interaction and integration of the growth and differentiation of the plant as a whole. It is a matter of necessity that the root meristem and the shoot meristem communicate with one another and that the apical meristems communicate their status to presumptive lateral meristems. Auxin plays a major role in carrying information from the shoot apex to regulate the growth of shoot lateral branches and lateral roots; cytokinin serves as a carrier from the root apex as it regulates the growth of lateral roots and lateral branches of the shoot. In addition to these well-documented interactions, there exists the possibility that communicative and integrative processes may involve, in some cases, the regulation of the synthesis and transport of one hormone by another. Such a possibility is raised by the apparent increase in auxin synthesis in response to low levels of abscisic acid (TAL et al. 1979).

As yet the effects of hormones on the amount and transport of other hormones in the intact plant are not sufficiently understood to allow generalizations to be made.

Integration and communication between the root and the shoot, as well as the response to changes in the availability of water, are common problems that all land angiosperms must solve. Thus it is not surprising that these phenomena present the clearest picture of hormonal interactions since most species appear to have adopted a common solution. At the other extreme, problems of the regulation of flowering and fruiting cannot be solved similarly by different species of plants since it is these processes, among others, that are important in the separation of species. Even plants which respond to the same environmental stimuli may have evolved from ancestors who responded to different stimuli, and thus even these plants may not show the same hormonal controls. Therefore, the extreme variation found in the hormonal regulation of flowering responses should not be surprising. This variation may include both those hormones which promote flowering and the interaction between these hormones and others which may inhibit flowering.

In *Oenothera lamarkiana* gibberellin can replace the cold requirement for flowering; in *Oenothera biennis* gibberellin cannot replace cold treatment. In *Lactuca dentata* and *sativa* gibberellin promotes short-day flowering of these long day plants; in *Lactuca scariola,* also a long-day plant, gibberellin promotes elongation but has no effect on flowering under short days (ZEEVAART 1978). There are many cases of species of plants belonging to the same genus in which one species flowers under short days and another species flowers under long days, e.g., *Chrysanthemum morifolium* (SDP), *C. maximum* (LDP); *Lemna paucicostata* (SDP), *L. perpusilla* (SDP), *L. gibba* (LDP); *Nicotiana tabacum* var. Maryland mammouth (SDP), var. Havana A (LDP), *N. sylvestris* (LDP) (VINCE-PRUE 1975). Thus the flowering stimulus may be at least as variable as ecological niches. Somewhere in the middle between species uniformity and species diversity are those regulatory processes which are broadly similar in all species but which exhibit species-specific variations. For example, during the stratification of seeds of *Acer saccharum,* cytokinin and gibberellin increase followed by a decrease in cytokinin, then a decrease in gibberellin, and then germination (WEBB et al. 1973a). In seeds of *Acer pseudoplatanus* there is a gradual decrease in cytokinin level and no change in gibberellin level during stratification (WEBB et al. 1973b). In addition to stratification, other processes which would be expected to show species-specific variation include germination, which must occur when the environment becomes favorable; dormancy, which must occur when the environment becomes unfavorable; water balance, which must depend on the niche the plant occupies; and the degree of apical dominance, which determines the plant habit. In general these processes are regulated in the same direction by the same hormones in all plants, but the exact degree and timing of such responses depends on the species.

6.4 Some Considerations Regarding Experimental Design

Although it is virtually impossible to do ideal experiments with intact plants, certain considerations should be kept in mind in all experiments. Manipulations of whole plants are at best disruptive. Specifically, cutting a plant or plant part will inevitably produce a wound response and in addition it will interrupt hormonal signals which would have been received in the intact plant. Replacing plant parts with variously applied hormones may lead to the presence of the hormones in unnatural places, and to unusual metabolism of these hormones both at the cut surface and within the plant. In addition exogeneously applied hormones may increase an endogenous pool to the point of producing hormone imbalance and metabolic disturbance.

An increasingly important consideration is the unequivocal identification and quantification of hormones extracted from the plant. Clearly, to determine when a hormone is performing *as* a hormone requires that distinctions between active, accessible, and inactive, less accessible, or stored forms of the hormones should be made where possible. Long-term experiments require careful analysis of the processes and kinetics involved, and the careful consideration of the possibility of any intermediate steps. In addition, the presence of microorganisms, as contaminants, is likely to affect both long-term and short-term processes, and, therefore, should be avoided, if possible (LIBBERT and RISCH 1969).

6.5 Some Considerations Regarding Research Strategies

The general view concerning hormones in the whole plant presented above suggests that there is a base of information adequate to serve as a foundation for further research in this field. However, because of the low level of funding and relatively small number of researchers in this field, research strategies become more important than in many other areas. A survey of the literature suggests that one source of considerable confusion is the large number of species studied (for example, the species index of the most current comprehensive review of the field cites work done with more than 500 species, LETHAM et al. 1978). Since each species occupies its own ecological niche, one would expect each species to have some unique responses to the environment and these responses are likely to be hormonally mediated. Indeed it would appear that much of the literature supports this view. It is our opinion that it would be advisable to concentrate research efforts on a relatively small number of species to further our basic understanding of hormonal physiology of plants. The advantages of selecting a few model systems are abundant and this strategy has been successful in other areas of biology. A minimum number of four species comes to mind: a monocot, a dicot without symbiotic nitrogen fixation, a dicot with symbiotic nitrogen fixation, and a perennial fruit-producing species. Each of these species should be agriculturally significant.

It seems that with some effort, major laboratories could come to an agreement on the species which would be identified as principal objects of research efforts. Obviously there are the special cases in which a particular species of organism is ideally suited for a specific study. However, in general, it would seem that only when we have a comprehensive understanding of the interaction of hormonal processes in a few plants will it be practicable to expand and extend our knowledge to a wider diversity of species about which we have only fragmentary knowledge at present.

A second strategy, the study of diseases of plants, has been especially helpful in the past and can be expected to continue to be useful (Chap. 5, Vol. 4, this Series 1976). The original discovery of the gibberellins is the most striking example of the usefulness of studying the diseased plant as an aid to our understanding of the healthy plant (KUROSAWA 1926). Other examples include the study of infections of leaves by the fungus *Omphalia flavida* in elucidating the role of auxin in the prevention of abscission (SEQUEIRA and STEEVES 1954) and the study of infections of stems by *Corynebacterium fasciens* in elucidating the role of cytokinins in the promotion of shoot branching (THIMANN and SACHS 1966). These examples provide evidence that some diseases do involve relatively simple hormonal interactions between the pathogen and the host; such diseases provide a useful way to analyze the hormonal physiology of the host. Although not all diseases can be expected to be of this type, the study of carefully chosen diseases will continue to aid our understanding of the regulatory roles of the plant hormones.

Another strategy which, although seldom exploited, has yielded considerable information regarding hormonal control mechanisms with the expenditure of relatively little effort is the use of mutants. Two prominent examples of this are the use of dwarf mutants in studies of the role of gibberellins (BRIAN and HEMMING 1955) and the use of the flacca mutant of tomato in studies of the role of abscisic acid (TAL et al. 1979). There is now available to the hormone physiologist a vast array of mutants which are known to have or can be suspected of having regulatory defects. In addition mutants can be induced in higher plants by chemical treatments or by irradiation. Such mutants could be screened (or even in some cases, hopefully, selected) for properties of interest to the hormone physiologist. Some mutations, such as the lack of cytokinin or auxin synthesis, would probably be lethal, but it may be possible to obtain temperature-sensitive mutants of these types. Other types, such as those which are insensitive to excessive amounts of hormones, may be selected directly. With such mutants in hand, the physiologists should more easily identify the sites of action, the roles, and the interactions of the five plant hormones.

6.6 Conclusion

A review of research on the hormonal physiology of the whole plant reveals that there exists a body of knowledge sufficiently compelling to warrant a general conceptual model of hormonal regulation. For example, we now know with some certainty the sites of synthesis for four of the five major plant hormones, and the broad outlines of how they are distributed in the whole plant. The major roles of auxin, cytokinin, and gibberellic acid in the vegetative plant seem well established. The roles of ethylene in the regulation of fruit ripening and expansion versus elongation growth are also well established. However, the role(s) of abscisic acid, with the exception of the response to water stress, seems less clear. There also exist areas in which considerable effort has been expended, but about which no general conclusions can yet be made. For example, the role of hormones in the flowering and in the germinating seed continue to be obscure. Although some hormonal interactions are relatively well understood, the details of most hormonal interactions remain unclear. The understanding of such interactions is undoubtedly crucial to completing our picture of the role of hormones in the whole plant.

There is reason for optimism regarding further progress in this field. There exist old techniques which may be refined and new techniques which have yet to be fully exploited, whose application should result in additional knowledge. One may perceive that progress to date has come slowly and in stages. However, this need not continue to be the case. In light of the current world food crisis, workers should be encouraged to move forward boldly building on current knowledge and recently acquired data to advance the field as rapidly as possible.

References

General Review

Audus LJ (1972) Plant growth substances, 3rd edn. Interscience, New York London

Carr DJ (ed) (1972) Plant growth substances 1970. Springer, Berlin Heidelberg New York

Fredrick JF (ed) (1967) Plant growth regulators. Ann NY Acad Sci Vol 144, Art 1

Hillman JR (ed) (1978) Isolation of plant growth substances. Cambridge Univ Press, Cambridge London New York Melbourne

Jacobs WP (1979) Plant hormones and plant development. Cambridge Univ Press, Cambridge London New York Melbourne

Kaldewey H, Vardar Y (eds) (1972) Hormonal regulation in plant growth and development. Verlag Chemie, Weinheim

Kudrov T, Ivanova I, Karanov Ed (eds) (1977) Plant growth regulators. Bulgarian Acad Sci, Sofia

Leopold AC, Kriedemann PE (1975) Plant growth and development, 2nd edn. McGraw-Hill, New York

Letham DS, Goodwin PB, Higgins TJV (eds) (1978) Phytohormones and related compounds— a comprehensive treatise Vols 1 and 2. Elsevier/North-Holland, Amsterdam

Moore TC (1979) Biochemistry and physiology of plant hormones. Springer, New York Heidelberg Berlin

Pilet PE (ed) (1977) Plant growth regulation. Springer, Berlin Heidelberg New York
Plant Growth Substances 1973 (1974) Hirokawa Publ, Tokyo
Scott TK (ed) (1979) Plant regulation and world agriculture. Plenum, New York
Skoog F (ed) (1980) Plant growth substances 1979. Springer, Berlin Heidelberg New York
Symposia of Society for Experimental Biology XXXI (1977) Integration of activity in the higher plant. Cambridge Univ, Cambridge
Thimann KV (1972) The natural plant hormones. In: Steward FC (ed) Plant physiology vol VI B. Academic Press, London New York, pp 3–332
Thimann KV (1977) Hormone action in the whole life of plants. Univ Massachusetts Press, Amherst
Torrey JG, Clarkson DT (eds) (1975) The development and function of roots. Academic Press, London New York
Vardar Y (ed) (1968) The transport of plant hormones. Elsevier/North-Holland, Amsterdam
Wareing PF, Phillips IDJ (1978) The control of growth and differentiation in plants, 2nd edn. Pergamon, Oxford
Weaver RJ (1972) Plant growth substances in agriculture. Freeman, San Francisco
Wightman F, Setterfield G (1968) Biochemistry and physiology of plant growth substances. Runge, Ottawa
Wilkins MB (ed) (1969) The physiology of plant growth and development. McGraw-Hill, New York

Sites of Hormone Synthesis and Transport (Fig. 6.1)

Coombe BG (1976) The development of fleshy fruits. Annu Rev Plant Physiol 27:207–228
Eeuwens CJ, Schwabe WW (1975) Seed and pod wall development in *Pisum sativum* L. in relation to extracted and applied hormones. J Exp Bot 26:1–14
Elisezer J, Morris DA (1979) Effects of temperature and sink activity on the transport of ^{14}C-labelled indol-3yl-acetic acid in the intact pea plant (*Pisum sativum* L.). Planta 147:216–224
Feldman L (1979) Cytokinin biosynthesis in roots of corn. Planta 145:315–321
Goldsmith MHM (1977) The polar transport of auxin. Annu Rev Plant Physiol 28:439–478
Goodwin PB (1978) Phytohormones and growth and development of organs of the vegetative plant. In: Letham DS, Goodwin PB, Higgins TJV (eds) Phytohormones and related compounds— A comprehensive treatise, Vol II. Elsevier/North Holland, Amsterdam, pp 31–174
Goodwin PB (1978) Phytohormones and fruit growth. In: Letham DS, Goodwin, PB, Higgins TJV (eds) Phytohormones and related compounds— A comprehensive treatise, Vol II. Elsevier/North Holland, Amsterdam, pp 175–214
Goodwin PB, Gollnow BI, Letham DS (1978) Phytohormones and growth correlations. In: Letham DS, Goodwin PB, Higgins TJV (eds) Phytohormones and related compounds— A comprehensive treatise, Vol II. Elsevier/North Holland, pp 215–250
Graebe JE, Ropers HJ (1978) Gibberellins. In: Letham DS, Goodwin PB, Higgins TJV (eds) Phytohormones and related compounds— A comprehensive treatise, Vol I. Elsevier/North Holland, Amsterdam, pp 107–204
Jones RL (1973) Gibberellins: Their physiological role. Annu Rev Plant Physiol 24:571–598
Letham DS (1978) Cytokinins. In: Letham DS, Goodwin PB, Higgins TJV (eds) Phytohormones and related compounds— A comprehensive treatise, Vol I. Elsevier/North Holland, Amsterdam, pp 205–264
Milborrow BV (1978) Abscisic acid. In: Letham DS, Goodwin PB, Higgins TJV (eds) Phytohormones and related compounds— A comprehensive treatise, Vol I. Elsevier/North Holland, Amsterdam, pp 295–348

Radley M (1976) The development of wheat grain in relation to endogenous growth substances. J Exp Bot 27:1009–1021

Rowntree RA, Morris DA (1979) Accumulation of ^{14}C from exogenous labelled auxin in lateral root primordia of intact pea seedlings (*Pisum sativum* L.). Planta (Berl) 144:463–466

Schneider EA, Wightman F (1978) Auxins. In: Letham DS, Goodwin PB, Higgins TJV (eds) Phytohormones and related compounds— A comprehensive treatise, Vol I. Elsevier/North Holland, Amsterdam, pp 29–106

Schneider EA, Wightman F (1974) Metabolism of auxin in higher plants. Annu Rev Plant Physiol 25:487–513

Scott TK (1972) Auxins and roots. Annu Rev Plant Physiol 23:235–258

Sheldrake AR (1973) The production of hormones in higher plants. Biol Rev 48:509–559

Torrey JG (1976) Root hormones and plant growth. Annu Rev Plant Physiol 27:435–459

Van Staden J, Smith AR (1978) The synthesis of cytokinins in excised roots of maize and tomato under aseptic conditions. Ann Bot 42:751–753

Indole Acetic Acid (Fig. 6.2)

Aloni R (1979) Role of auxin and gibberellin in differentiation of primary phloem fibers. Plant Physiol (Bethesda) 63:609–614

Eeuwens CJ, Schwabe WW (1975) Seed and pod wall development in *Pisum sativum* L. in relation to extracted and applied hormones. J Exp Bot 26:1–14

Goodwin PB (1978) Phytohormones and growth and development of organs of the vegetative plant. In: Letham DS, Goodwin PB, Higgins TJV (eds) Phytohormones and related compounds— A comprehensive treatise, Vol II. Elsevier/North Holland, Amsterdam, pp 31–174

Goodwin PB (1978) Phytohormones and fruit growth. In: Letham DS, Goodwin PB, Higgins TJV (eds) Phytohormones and related compounds— A comprehensive treatise, Vol II. Elsevier/North Holland, Amsterdam, pp 175–214

Goodwin PB, Gollnow BI, Letham DS (1978) Phytohormones and growth correlations. In: Letham DS, Goodwin PB, Higgins TJV (eds) Phytohormones and related compounds— A comprehensive treatise, Vol II. Elsevier/North Holland, Amsterdam, pp 215–250

Khan AA, Tao KL (1978) Phytohormones, seed dormancy and germination. In: Letham DS, Goodwin PB, Higgins TJV (eds) Phytohormones and related compounds— A comprehensive treatise, Vol II. Elsevier/North Holland, Amsterdam, pp 371–422

McGlasson WB, Wade DL, Adato I (1978) Phytohormones and fruit ripening. In: Letham DS, Goodwin PB, Higgins TJV (eds) Phytohormones and related compounds— A comprehensive treatise, Vol II. Elsevier/North Holland, Amsterdam, pp 447–494

Noodén LD, Leopold AC (1978) Phytohormones and the endogenous regulation of senescence and abscission. In: Letham DS, Goodwin PB, Higgins TJV (eds) Phytohormones and related compounds— A comprehensive treatise, Vol II. Elsevier/North Holland, Amsterdam, pp 329–370

Osborne DS (1978) Ethylene. In: Letham DS, Goodwin PB, Higgins TJV (eds) Phytohormones and related compounds— A comprehensive treatise, Vol I. Elsevier/North Holland, Amsterdam, pp 265–294

Patrick JW (1979) Auxin-promoted transport of metabolites in stems of *Phaseolus vulgaris* L. J Exp Bot 114:1–13

Radley M (1976) The development of wheat grain in relation to endogenous growth substances. J Exp Bot 27:1009–1021

Reinhold L (1978) Phytohormones and the orientation of growth. In: Letham DS, Goodwin PB, Higgins TJV (eds) Phytohormones and related compounds— A comprehensive treatise, Vol II. Elsevier/North Holland, Amsterdam, pp 251–290

Rubinstein B, Nagao MA (1978) Lateral bud outgrowth and its control by the apex. Bot Rev 42:83–113

Sachs TA (1975) The control of differentiation of vascular networks. Ann Bot 39:197–204

Saunders P (1978) Phytohormones and bud dormancy. In: Letham DS, Goodwin PB, Higgins TJV (eds) Phytohormones and related compounds— A comprehensive treatise, Vol II. Elsevier/North Holland, Amsterdam, pp 423–446

Schneider EA, Wightman F (1978) Auxins. In: Letham DS, Goodwin PB, Higgins TJV (eds) Phytohormones and related compounds— A comprehensive treatise, Vol I. Elsevier/North Holland, Amsterdam, pp 29–106

Scott TK (1972) Auxins and roots. Annu Rev Plant Physiol 23:235–258

Torrey JG (1976) Root hormones and plant growth. Annu Rev Plant Physiol 27:435–459

Zeevaart JAD (1978) Phytohormones and flower formation. In: Letham DS, Goodwin PB, Higgins TJV (eds) Phytohormones and related compounds— A comprehensive treatise, Vol II. Elsevier/North Holland, Amsterdam, pp 291–328

Cytokinin (Fig. 6.3)

Eeuwens CJ, Schwabe WW (1975) Seed and pod wall development in *Pisum sativum* L. in Relation to extracted and applied hormones. J Exp Bot 26:1–14

Goodwin PB (1978) Phytohormones and growth and development of organs of the vegetative plant. In: Letham DS, Goodwin PB, Higgins TJV (eds) Phytohormones and related compounds— A comprehensive treatise, Vol II. Elsevier/North Holland, Amsterdam, pp 31–174

Goodwin PB (1978) Phytohormones and fruit growth. In: Letham DS, Goodwin PB, Higgins TJV (eds) Phytohormones and related compounds— A comprehensive treatise, Vol II. Elsevier/North Holland, Amsterdam, pp 175–214

Goodwin PB, Gollnow BI, Letham DS (1978) Phytohormones and growth correlations. In: Letham DS, Goodwin PB, Higgins TJV (eds) Phytohormones and related compounds— A comprehensive treatise, Vol II. Elsevier/North Holland, Amsterdam, pp 215–250

Hall RH (1973) Cytokinins as a probe of developmental processes. Annu Rev Plant Physiol 24:415–444

Khan AA, Tao KL (1978) Phytohormones, seed dormancy and germination. In: Letham DS, Goodwin PB, Higgins TJV (eds) Phytohormones and related compounds— A comprehensive treatise, Vol II. Elsevier/North Holland, Amsterdam, pp 371–422

Letham DS, Higgins TJV, Goodwin PB, Jacobsen JV (1978) Phytohormones in retrospect. In: Letham DS, Goodwin PB, Higgins TJV (eds) Phytohormones and related compounds— A comprehensive treatise, Vol I. Elsevier/North Holland, Amsterdam, pp 1–28

Noodén LD, Leopold AC (1978) Phytohormones and the endogenous regulation of senescence and abscission. In: Letham DS, Goodwin PB, Higgins TJV (eds) Phytohormones and related compounds— A comprehensive treatise, Vol II. Elsevier/North Holland, Amsterdam, pp 329–370

Phillips IDJ (1975) Apical dominance. Annu Rev Plant Physiol 26:341–367

Saunders P (1978) Phytohormones and bud dormancy. In: Letham DS, Goodwin PB, Higgins TJV (eds) Phytohormones and related compounds— A comprehensive treatise, Vol II. Elsevier/North Holland, pp 423–446

Webb DP, van Staden J, Wareing PF (1973) Seed dormancy in *Acer*. Changes in endogenous cytokinins, gibberellins and germinations inhibitors during the breaking of dormancy in *Acer saccharum*. Marsh J Exp Bot 24:105–116

Webb DP, van Staden J, Wareing PF (1973) Seed dormancy in *Acer*. Changes in endogenous germination inhibitors, cytokinins, and gibberellins during the breaking of dormancy in *Acer pseudoplatanus* L. J Exp Bot 24:741–750

Zeevaart JAD (1978) Phytohormones and flower formation. In: Letham DS, Goodwin PB, Higgins TJV (eds) Phytohormones and related compounds— A comprehensive treatise, Vol II. Elsevier/North Holland, Amsterdam, pp 291–328

Gibberellic Acid (Fig. 6.4)

Eeuwens CJ, Schwabe WW (1975) Seed and pod wall development in *Pisum sativum* L. in relation to extracted and applied hormones. J Exp Bot 26:1–14

Goodwin PB (1978) Phytohormones and growth and development of organs of the vegetative plant. In: Letham DS, Goodwin PB, Higgins TJV (eds) Phytohormones and related compounds— A comprehensive treatise, Vol II. Elsevier/North Holland, Amsterdam, pp 31–174

Goodwin PB (1978) Phytohormones and fruit growth. In: Letham DS, Goodwin PB, Higgins TJV (eds) Phytohormones and related compounds— A comprehensive treatise, Vol II. Elsevier/North Holland, Amsterdam, pp 175–214

Goodwin PB, Gollnow BI, Letham DS (1978) Phytohormones and growth correlations. In: Letham DS, Goodwin PB, Higgins TJV (eds) Phytohormones and related compounds— A comprehensive treatise, Vol II. Elsevier/North Holland, Amsterdam, pp 215–250

Graebe JE, Ropers HJ (1978) Gibberellins. In: Letham DS, Goodwin PB, Higgins TJV (eds) Phytohormones and related compounds— A comprehensive treatise, Vol I. Elsevier/North Holland, Amsterdam, pp 107–204

Jones RL (1973) Gibberellins: Their physiological role. Annu Rev Plant Physiol 24:571–598

Khan AA, Tao KL (1978) Phytohormones, seed dormancy and germination. In: Letham DS, Goodwin PB, Higgins TJV (eds) Phytohormones and related compounds— A comprehensive treatise, Vol II. Elsevier/North Holland, Amsterdam, pp 371–422

Kumor S, Datta KS, Nauda KK (1977) Gibberellic acid causes flowering in the short-day plants *Panicum miliaceum* L., *P. miliare* Lamk., and *Setaria italica* (L.), P. Beauv. Planta (Berl) 134:95–96

McGlasson WB, Wade DL, Adato I (1978) Phytohormones and fruit ripening. In: Letham DS, Goodwin PB, Higgins TJV (eds) Phytohormones and related compounds— A comprehensive treatise, Vol II. Elsevier/North Holland, Amsterdam, pp 447–494

Mulligan DR, Patrick JW (1979) Gibberellic acid-promoted transport of assimilated in stems of *Phaseolus vulgaris* L. Planta (Berl) 145:233–238

Noodén LD, Leopold AC (1978) Phytohormones and the endogenous regulation of senescence and abscission. In: Letham DS, Goodwin PB, Higgins TJV (eds) Phytohormones and related compounds— A compehensive treatise, Vol II. Elsevier/North Holland, Amsterdam, pp 329–370

Radley M (1976) The development of wheat grain in relation to endogenous growth substances. J Exp Bot 27:1009–1021

Saunders P (1978) Phytohormones and bud dormancy. In: Letham DS, Goodwin PB, Higgins TJV (eds) Phytohormones and related compounds— A comprehensive treatise, Vol II. Elsevier/North Holland, Amsterdam, pp 423–446

Taylorson RB, Hendricks SB (1977) Dormancy in seeds. Annu Rev Plant Physiol 28:331–354

Webb DP, Van Staaden J, Wareing PF (1973) Seed dormancy in *Acer*. Changes in endogenous cytokinins, gibberellins, and germination inhibitors during the breaking of dormancy in *Acer saccharum* Marsh. J Exp Bot 24:105–116

Webb DP, Van Staaden J, Wareing PF (1973) Seed dormancy in *Acer*. Changes in endogenous germination inhibitors, cytokinins, and gibberellins during the breaking of dormancy in *Acer pseudoplatanus* L. J Exp Bot 24:741–750

Zeevaart JAD (1978) Phytohormones and flower formation. In: Letham DS, Goodwin PB, Higgins TJV (eds) Phytohormones and related compounds— A comprehensive treatise. Vol II. Elsevier/North Holland, Amsterdam, pp 291–328

Abscisic Acid (Fig. 6.5)

Eeuwens CJ, Schwabe WW (1975) Seed and pod wall development in *Pisum sativum* L. in relation to extracted and applied hormones. J Exp Bot 26:1–14

Glinka Z (1977) Effects of abscisic acid and of hydrostatic pressure gradient on water movement through excised sunflower roots. Plant Physiol (Bethesda) 59:933–935

Goodwin PB (1978) Phytohormones and growth and development of organs of the vegetative plant. In: Letham DS, Goodwin PB, Higgins TJV (eds) Phytohormones and related compounds— Comprehensive treatise, Vol II. Elsevier/North Holland, Amsterdam, pp 31–174

Goodwin PB (1978) Phytohormones and fruit growth. In: Letham DS, Goodwin PB, Higgins TJV (eds) Phytohormones and related compounds— A comprehensive treatise, Vol II. Elsevier/North Holland, Amsterdam, pp 175–214

Karmoker JL, Van Steninck RFM (1978) Stimulation of volume flow and ion flux by abscisic acid in the excised root system of *Phaseolus vulgaris* L. cv. Redland Pioneer. Planta (Berl) 141:37–43

Kefeli VI, Kadyrov ChSh (1971) Natural growth inhibitors, their chemical and physiological properties. Annu Rev Plant Physiol 22:185–196

McGlasson WB, Wade DL, Adato I (1978) Phytohormones and fruit ripening. In: Letham DS, Goodwin PB, Higgins TJV (eds) Phytohormones and related compounds— A comprehensive treatise, Vol II. Elsevier/North Holland, Amsterdam, pp 447–494

Milborrow BV (1974) The chemistry and physiology of abscisic acid. Annu Rev Plant Physiol 25:259–307

Milborrow BV (1978) Abscisic acid. In: Letham DS, Goodwin PB, Higgins TJV (eds) Phytohormones and related compounds— A comprehensive treatise, Vol I. Elsevier/ North Holland, Amsterdam, pp 295–348

Noodén LD, Leopold AC (1978) Phytohormones and the endogenous regulation of senescence and abscission. In: Letham DS, Goodwin PB, Higgins TJV (eds) Phytohormones and related compounds— A comprehensive treatise, Vol II. Elsevier/North Holland, Amsterdam, pp 329–370

Osborne DS (1978) Ethylene. In: Letham DS, Goodwin PB, Higgins TJV (eds) Phytohormones and related compounds— A comprehensive treatise, Vol I. Elsevier/North Holland, Amsterdam, pp 265–294

Radley M (1976) The development of wheat grain in relation to endogenous growth substances. J Exp Bot 27:1009–1021

Tal M, Imber D, Erez A, Epstein E (1979) Abnormal stomatal behavior and hormonal balance in flacca, a wilty mutant of tomato. V. Effect of abscisic acid on indoleacetic acid metabolism and ethylene evolution. Plant Physiol (Bethesda) 63:1044–1048

Taylorson RB, Hendricks SB (1977) Dormancy in seeds. Annu Rev Plant Physiol 28:331–354

Torrey JG (1976) Root hormones and plant growth. Annu Rev Plant Physiol 27:435–459

Wareing PF, Saunders PF (1971) Hormones and dormancy. Ann Rev 22:261–288

Webb DP, Van Staaden J, Wareing PF (1973) Seed dormancy in *Acer*. Changes in endogenous cytokinins, gibberellins and germination inhibitors during the breaking of dormancy in *Acer saccharum* Marsh. J Exp Bot 24:105–116

Webb DP, Van Staaden J, Wareing PF (1973) Seed germination in *Acer*. Changes in endogenous germination inhibitors, cytokinins, and gibberellins during the breaking of dormancy in *Acer pseudoplatanus* L. J Exp Bot 24:741–750

Weyers JDB, Hillman JR (1979) Sensitivity of *Commelina* stomata to abscisic acid. Planta (Berl) 146:623–628

Wright STC (1978) Phytohormones and stress phenomena. In: Letham DS, Goodwin PB, Higgins TJV (eds) Phytohormones and related compounds— A comprehensive treatise, Vol II. Elsevier/North Holland, Amsterdam, pp 495–536

Ethylene (Fig. 6.6)

Abeles FB (1972) Biosynthesis and mechanism of action of ethylene. Annu Rev Plant Physiol 23:259–292

Abeles FB (1972) Ethylene in plant biology. Academic Press, London New York

Burg SP (1962) The physiology of ethylene formation. Annu Rev Plant Physiol 13:265–302

Goodwin PB (1978) Phytohormones and growth and development of organs of the vegetative plant. In: Letham DS, Goodwin PB, Higgins TJV (eds) Phytohormones and related compounds— A comprehensive treatise. Vol II. Elsevier/North Holland, Amsterdam, pp 31–174

Goodwin PB (1978) Phytohormones and fruit growth. In: Letham DS, Goodwin PB, Higgins TJV (eds) Phytohormones and related compounds— A comprehensive treatise. Vol II. Elsevier/North Holland, Amsterdam, pp 175–214

Kahn AA, Tao KL (1978) Phytohormones, seed dormancy and germination. In: Letham DS, Goodwin PB, Higgins TJV (eds) Phytohormones and related compounds— A comprehensive treatise. Vol II. Elsevier/North Holland, Amsterdam, pp 371–422

Lieberman M (1975) Biosynthesis and regulatory control of ethylene in fruit ripening. A review. Physiol Veg 13:489–499

Lieberman M (1979) Biosynthesis of action of ethylene. Annu Rev Plant Physiol 30:533–591

McGlasson WB, Wade DL, Odato I (1978) Phytohormones and fruit ripening. In: Letham DS, Goodwin PB, Higgins TJV (eds) Phytohormones and related compounds— A comprehensive treatise. Vol II. Elsevier/North Holland, Amsterdam, pp 447–493

Morgan PW (1972) Effect on ethylene physiology. In: Audus LJ (ed) Herbicides: Physiology, biochemistry, and ecology. Academic Press, London New York, pp 255–280

Noodén LD, Leopold AC (1978) Phytohormones and the endogenous regulation of senescence and abscission. In: Letham DS, Goodwin PB, Higgins TJV (eds) Phytohormones and related compounds— A comprehensive treatise. Vol II. Elsevier/North Holland, Amsterdam, pp 329–369

Osborne DJ (1978) Ethylene. In: Letham DS, Goodwin PB, Higgins TJV (eds) Phytohormones and the related compounds— A comprehensive treatise. Vol I. Elsevier/North Holland, Amsterdam, pp 265–294

Phan CT (1971) L'ethylène métabolisme et activité métabolique. Masson et Cie, Paris

Pratt HK, Goeschl JD (1969) Physiological roles of ethylene in plants. Annu Rev Plant Physiol 20:541–584

Reinholt L (1978) Phytohormones and the orientation of growth. In: Letham DS, Goodwin PB, Higgins TJV (eds) Phytohormones and the related compounds— A comprehensive treatise. Vol II. Elsevier/North Holland, Amsterdam, pp 251–289

Spencer MS (1969) Ethylene in nature. Fortschr Chem Org Naturst 27:31–80

Wright STC (1978) Phytohormones and stress phenomena. In: Letham DS, Goodwin PB, Higgins TJV (eds) Phytohormones and the related compounds— A comprehensive treatise. Vol II. Elsevier/North Holland, Amsterdam, pp 495–536

Yang SF (1974) The biochemistry of ethylene: biogenesis and metabolism. Recent Adv Phytochem 7:131–164

Yang SF, Adams DO (1980) The biosynthesis of ethylene. In: Stumpf PK, Conn EE (eds) The biochemistry of plants— A comprehensive treatise. Vol 4, Lipids: Structure and function. Academic Press, New York, pp 163–175

Zeevaart JAD (1978) Phytohormones and flower formation. In: Letham DS, Goodwin PB, Higgins TJV (eds) Phytohormones and the related compounds— A comprehensive treatise. Vol II. Elsevier/North Holland, Amsterdam, pp 291–327

Environment Responses (Gravity, Temperature, Water), (Fig. 6.7)

Barendse GWM, Gilissen HAM (1977) The diffusion of gibberellins into agar and water during early germination of *Pharbitis nil* Choisy. Planta (Berl) 137:169–175

Goodwin PB, Gollnow BI, Letham DS (1978) Phytohormones and growth correlations. In: Letham DS, Goodwin PB, Higgins TJV (eds) Phytohormones and related compounds— A comprehensive treatise. Vol II. Elsevier/North Holland, Amsterdam, pp 215–250

Graebe JE, Ropers HJ (1978) Gibberellins. In: Letham DS, Goodwin PB, Higgins TJV (eds) Phytohormones and related compounds— A comprehensive treatise. Vol I. Elsevier/North Holland, Amsterdam, pp 107–204

Hiller LK, Kelly WC, Powell LE (1979) Temperature interactions with growth regulators and endogenous gibberellin-like activity during seed stalk elongation in carrots. Plant Physiol (Bethesda) 63:1055–1061

Hoad GV (1978) Effect of water stress on abscisic acid levels in white lupin (*Lupinus alba* L.) fruit, leaves, and phloem exudate. Planta (Berl) 142:387–390

Hsiao TC (1973) Plant responses to water stress. Annu Rev Plant Physiol 24:519–570

Jones RL (1973) Gibberellins: Their physiological role. Annu Rev Plant Physiol 24:571–598

Juniper BE (1976) Geotropism. Annu Rev Plant Physiol 27:385–406

Karmoker JL, Van Steveninck RFM (1978) Stimulation of volume flow and ion flux by abscisic acid in the excised root systems of *Phaseolus vulgaris* L. cv. Redland Pioneer. Planta (Berl) 141:37–43

Khan AA, Tao KL (1978) Phytohormones, seed dormancy and germination. In: Letham DS, Goodwin PB, Higgins TJV (eds) Phytohormones and related compounds— A comprehensive treatise, Vol II. Elsevier/North Holland, Amsterdam, pp 371–422

Letham DS (1978) Cytokinins. In: Letham DS, Goodwin PB, Higgins TJV (eds) Phytohormones and related compounds— A comprehensive treatise, Vol I. Elsevier/North Holland, Amsterdam, pp 205–264

Milborrow BV (1978) Abscisic acid. In: Letham DS, Goodwin PB, Higgins TJV (eds) Phytohormones and related compounds— A comprehensive treatise, Vol I. Elsevier/North Holland, Amsterdam, pp 295–348

Osborne DS (1978) Ethylene. In: Letham DS, Goodwin PB, Higgins TJV (eds) Phytohormones and related compounds— A comprehensive treatise, Vol I. Elsevier/North Holland, Amsterdam, pp 265–294

Raschke K(1975) Stomatal action. Annu Rev Plant Physiol 26:309–340

Reinhold L (1978) Phytohormones and the orientation of growth. In: Letham DS, Goodwin PB, Higgins TJV (eds) Phytohormones and related compounds— A comprehensive treatise, Vol II. Elsevier/North Holland, Amsterdam, pp 251–290

Schneider EA, Wightman F (1978) Auxins. In: Letham DS, Goodwin PB, Higgins TJV (eds) Phytohormones and related compounds— A comprehensive treatise, Vol I. Elsevier/North Holland, Amsterdam, pp 29–106

Scott TK (1972) Auxins and roots. Annu Rev Plant Physiol 23:235–258

Taylorson RB, Hendricks SB (1977) Dormancy in seeds. Annu Rev Plant Physiol 28:331–354

Torrey JG (1976) Root hormones and plant growth. Annu Rev Plant Physiol 27:435–459

Wareing PF, Saunders PF (1971) Hormones and dormancy. Annu Rev Plant Physiol 22:261–288

Wilkins MB (1977) Gravity and light-sensing guidance systems in primary roots and shoots. In: Symposia of society for experimental biology XXXI. Integration of activity in the higher plant. Cambridge University Press, Cambridge, pp 275–336

Wilkins MB (1979) Growth control mechanisms in gravitropism. In: Haupt W, Feinleib ME (eds) Encyclopedia of Plant Physiology, New Series, Vol 7. Springer, Berlin Heidelberg New York, pp 601–626

Wright STC (1978) Phytohormones and stress phenomena. In: Letham DS, Goodwin PB, Higgins TJV (eds) Phytohormones and related compounds— A comprehensive treatise, Vol II. Elsevier/North Holland, Amsterdam, pp 495–536

Environmental Response (Light), (Fig. 6.8)

Dennison DS (1979) Phototropism. In: Haupt W, Feinleib ME (eds) Encyclopedia of Plant Physiology, New Series, Vol 7. Springer, Berlin Heidelberg New York, pp 506–556

Graebe JE, Ropers HJ (1978) Gibberellins. In: Letham DS, Goodwin PB, Higgins TJV (eds) Phytohormones and related compounds— A comprehensive treatise, Vol I. Elsevier/North Holland, Amsterdam, pp 29–106

Henson IE, Wareing PF (1977) Cytokinins in *Xanthium strumarium* L.: Some aspects of the photoperiodic control of endogenous levels. New Phytol 78:35–45

Jones RL (1973) Gibberellins: Their physiological role. Annu Rev Plant Physiol 24:571–598

Khan AA, Tao KL (1978) Phytohormones, seed dormancy and germination. In: Letham DS, Goodwin PB, Higgins TJV (eds) Phytohormones and related compounds— A comprehensive treatise, Vol II. Elsevier/North Holland, Amsterdam, pp 371–422

Letham DS (1978) Cytokinins. In: Letham DS, Goodwin PB, Higgins TJV (eds) Phytohormones and related compounds— A comprehensive treatise, Vol I. Elsevier/North Holland, Amsterdam, pp 205–264

Milborrow BV (1978) Abscisic acid. In: Letham DS, Goodwin PB, Higgins TJV (eds) Phytohormones and related compounds— A comprehensive treatise, Vol I. Elsevier/North Holland, Amsterdam, pp 295–348

Osborne DS (1978) Ethylene. In: Letham DS, Goodwin PB, Higgins TJV (eds) Phytohormones and related compounds— A comprehensive treatise, Vol I. Elsevier/North Holland, Amsterdam, pp 265–294

Reinhold L (1978) Phytohormones and the orientation of growth. In: Letham DS, Goodwin PB, Higgins TJV (eds) Phytohormones and related compounds— A comprehensive treatise, Vol II. Elsevier/North Holland, Amsterdam, pp 251–290

Schneider EA, Wightman F (1978) Auxins. In: Letham DS, Goodwin PB, Higgins TJV (eds) Phytohormones and related compounds— A comprehensive treatise, Vol I. Elsevier/North Holland, Amsterdam, pp 29–106

Suttle JC, Zeevaart JAD (1979) Stem growth, flower formation, and endogenous gibberellins in a normal and a dwarf strain of *Silene armeria*. Planta (Berl) 145:175–180

Torrey JG (1976) Root hormones and plant growth. Annu Rev Plant Physiol 27:435–459

Vince-Prue D (1975) Photoperiodism in plants. McGraw-Hill, London

Wareing PF, Saunders PF (1971) Hormones and dormancy. Annu Rev Plant Physiol 22:261–288

Wilkins MB (1977) Gravity and light-sensing guidance systems in primary roots and shoots. In: Symposia of society for experimental biology XXXI. Integration of activity in the higher plant. Cambridge University Press, Cambridge, pp 275–336

Zeevaart JAD (1976) Physiology of flower formation. Annu Rev Plant Physiol 27:321–348

Zeevaart JAD (1978) Phytohormones and flower formation. In: Letham DS, Goodwin PB, Higgins TJV (eds) Phytohormones and related compounds— A comprehensive treatise, Vol II. Elsevier/North Holland, Amsterdam, pp 291–328

Text References

Bradford KJ, Yang SF (1980) Xylem transport of 1-aminocyclopropane-1-carboxylic acid, an ethylene precursor, in waterlogged tomato plants. Plant Physiol (Bethesda) 65:322–326

Brian PW, Hemming HG (1955) The effect of gibberellic acid on shoot growth of pea seedlings. Physiol Plant 8:669–681

Kurosawa E (1926) Experimental studies on the secretion of *Fusarium heterosporum* on rice plants. Trans Nat Hist Soc Formosa 16:213–227

Libbert E, Risch H (1969) Interactions between plants and epiphytic bacteria regarding their auxin metabolism. V. Isolation and identification of the IAA-producing and -destroying bacteria from pea plants. Physiol Plant 22:51–58

Lieberman M, Kunishi AT (1972) Thoughts on the role of ethylene in plant growth and development. In: Carr DJ (ed) Plant growth substances 1970. Springer, Berlin Heidelberg New York, pp 549–560

Osborne DS (1978) Ethylene. In: Letham DS, Goodwin PB, Higgins TJV (eds) Phytohormones and related compounds— a comprehensive treatise Vol I. Elsevier/North-Holland, pp 265–294

Penny P, Penny D (1978) Rapid responses to phytohormones. In: Letham DS, Goodwin PB, Higgins TJV (eds) Phytohormones and related compounds— a comprehensive treatise Vol II. Elsevier/North-Holland, Amsterdam, pp 537–597

Sequeira L, Steeves TA (1954) Auxin inactivation and its relation to leaf drop caused by the fungus *Omphalis flavida*. Plant Physiol (Bethesda) 29:11–16

Tal M, Imber D, Erez A, Epstein E (1979) Abnormal stomatal behavior and hormonal balance in flacca, a wilty mutant of tomato. V. Effect of abscisic acid on indoleacetic acid metabolism and ethylene evolution. Plant Physiol (Bethesda) 63:1044–1048

Thimann KV, Sachs T (1966) The role of cytokinins in the "fasciation" disease caused by *Corynebacterium fasciens*. Am J Bot 53:731–739

Vince-Prue D (1975) Photoperiodism in plants. McGraw-Hill, London

Webb DP, Van Staaden J, Wareing PF (1973a) Seed dormancy in *Acer*. Changes in endogenous cytokinins, gibberellins and germination inhibitors during the breaking of dormancy in *Acer saccharum* Marsh. J Exp Bot 24:105–116

Webb DP, Van Staaden J, Wareing PF (1973b) Seed dormancy in *Acer*. Changes in endogenous germination inhibitors, cytokinins and gibberellins during the breaking of dormancy in *Acer pseudoplatanus* L. J Exp Bot 24:741–750

Zeevaart JA (1978) Phytohormones and flower formation. In: Letham DS, Goodwin PB, Higgins TJV (eds) Phytohormones and related compounds—a comprehensive treatise Vol II. Elsevier/North-Holland, Amsterdam, pp 291–328

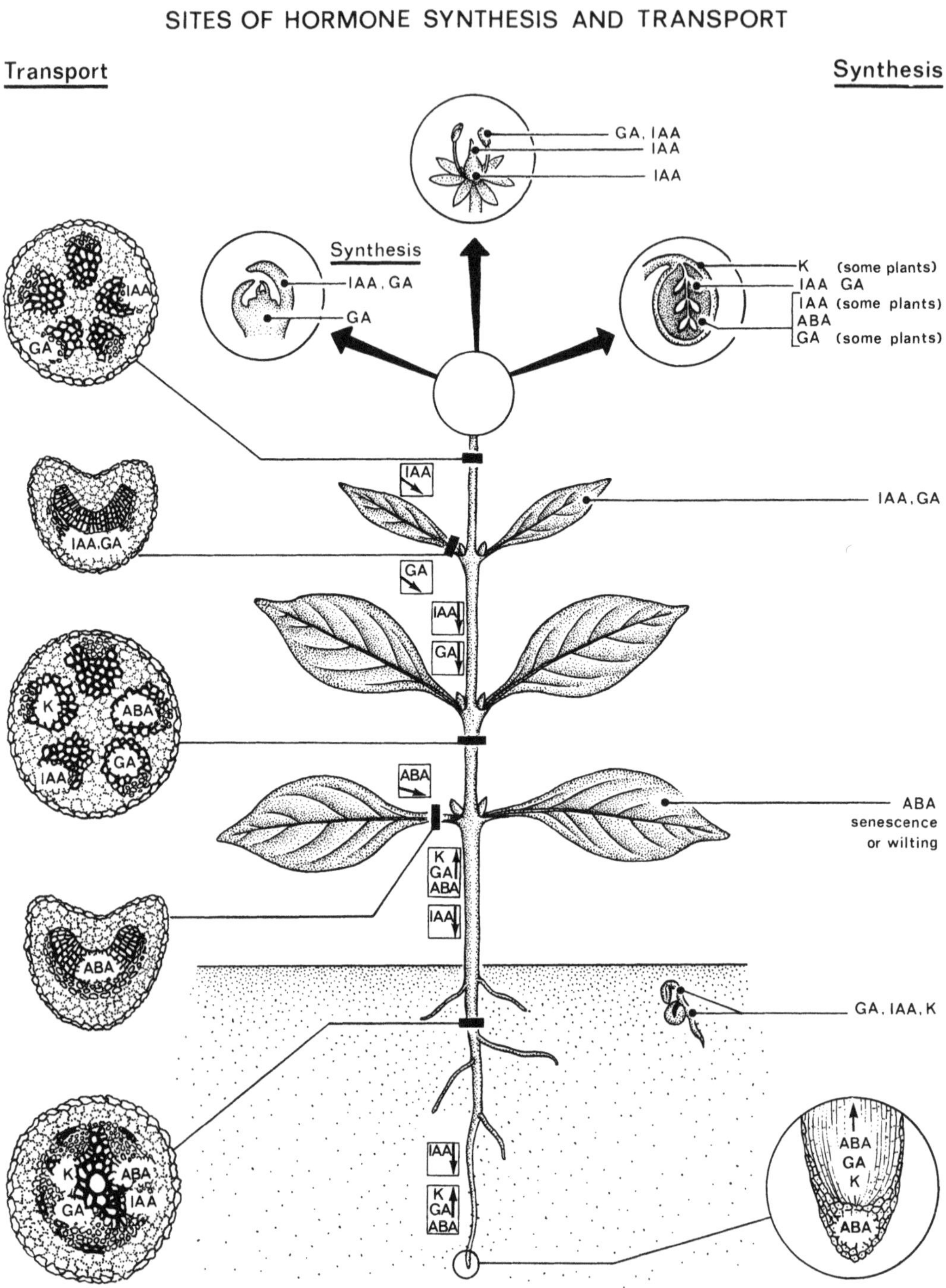
SITES OF HORMONE SYNTHESIS AND TRANSPORT
Transport
Synthesis
GA, IAA
IAA
IAA
Synthesis
IAA, GA
GA
K (some plants)
IAA GA
IAA (some plants)
ABA
GA (some plants)
IAA
GA
IAA
IAA, GA
IAA, GA
GA
IAA
GA
K
ABA
GA
IAA
ABA
ABA
senescence
or wilting
K
GA
ABA
IAA
ABA
GA, IAA, K
IAA
K
GA
ABA
K
ABA
IAA
GA
ABA
GA
K
ABA

Fig. 6.1

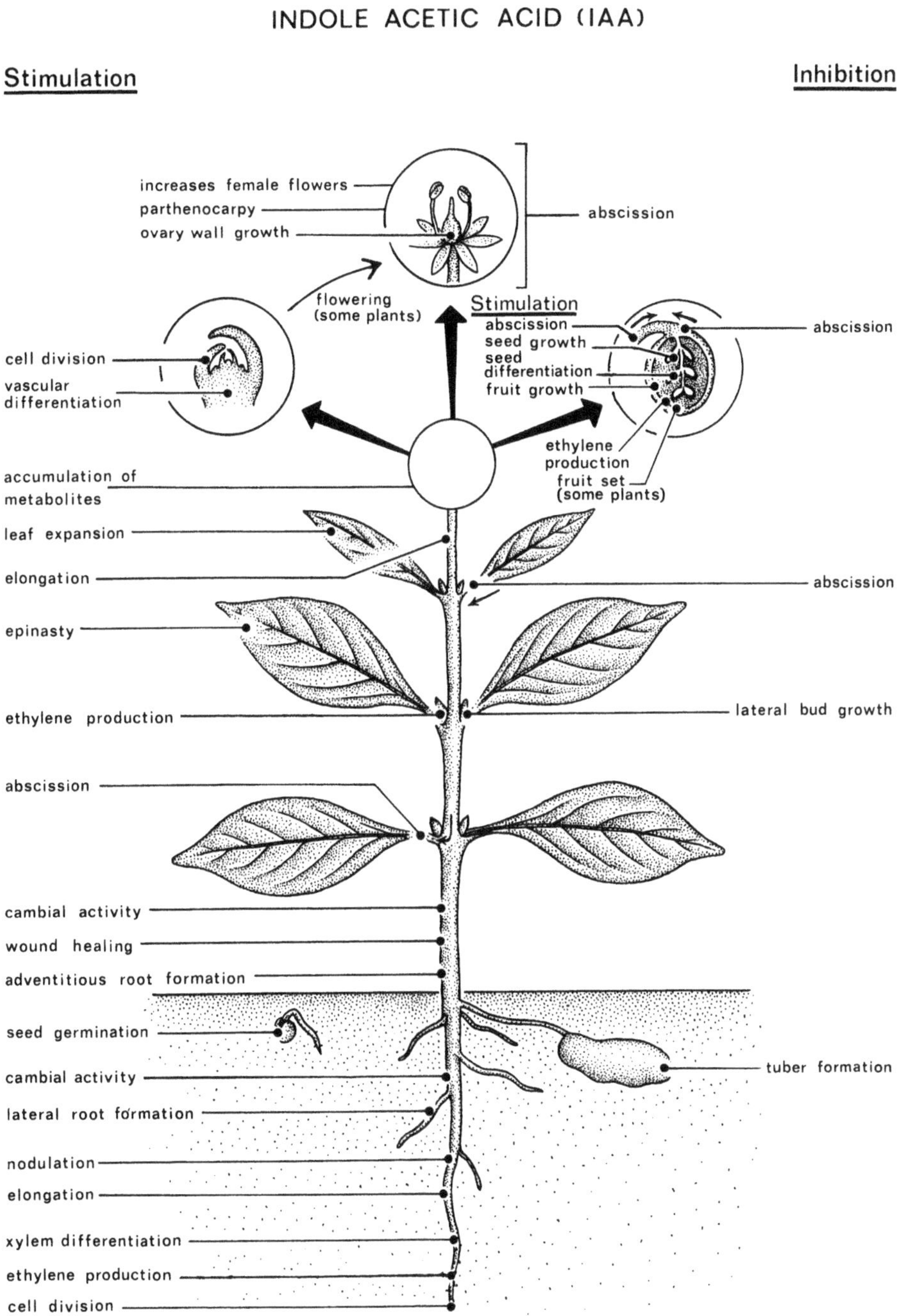
INDOLE ACETIC ACID (IAA)
Stimulation
Inhibition
increases female flowers
parthenocarpy
ovary wall growth
abscission
flowering (some plants)
cell division
vascular differentiation
Stimulation
abscission
seed growth
seed differentiation
fruit growth
abscission
ethylene production
fruit set (some plants)
accumulation of metabolites
leaf expansion
elongation
abscission
epinasty
ethylene production
lateral bud growth
abscission
cambial activity
wound healing
adventitious root formation
seed germination
tuber formation
cambial activity
lateral root formation
nodulation
elongation
xylem differentiation
ethylene production
cell division

Fig. 6.2

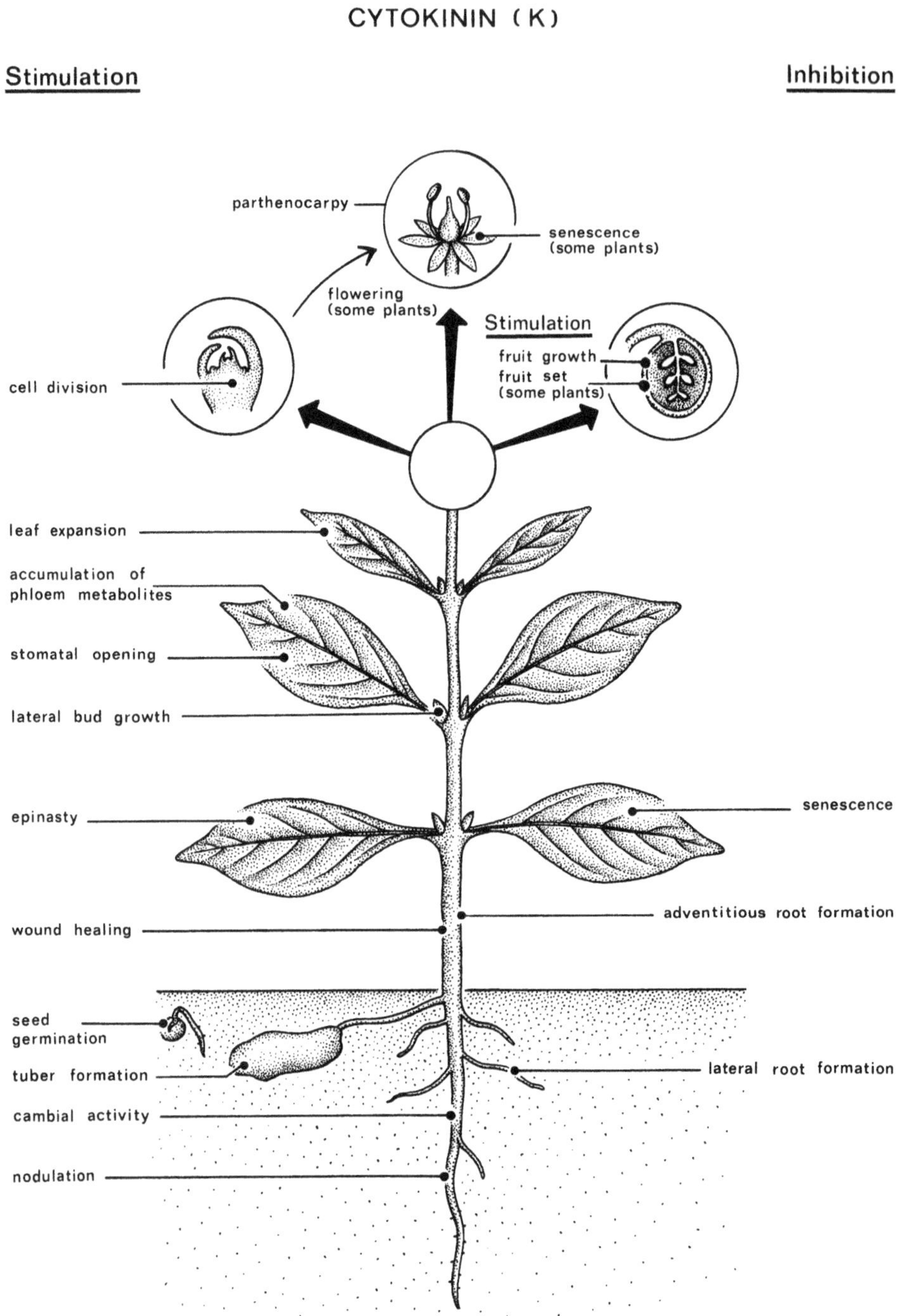
CYTOKININ (K)
Stimulation
Inhibition
parthenocarpy
senescence (some plants)
flowering (some plants)
Stimulation
fruit growth
fruit set (some plants)
cell division
leaf expansion
accumulation of phloem metabolites
stomatal opening
lateral bud growth
epinasty
senescence
adventitious root formation
wound healing
seed germination
tuber formation
cambial activity
nodulation
lateral root formation

Fig. 6.3

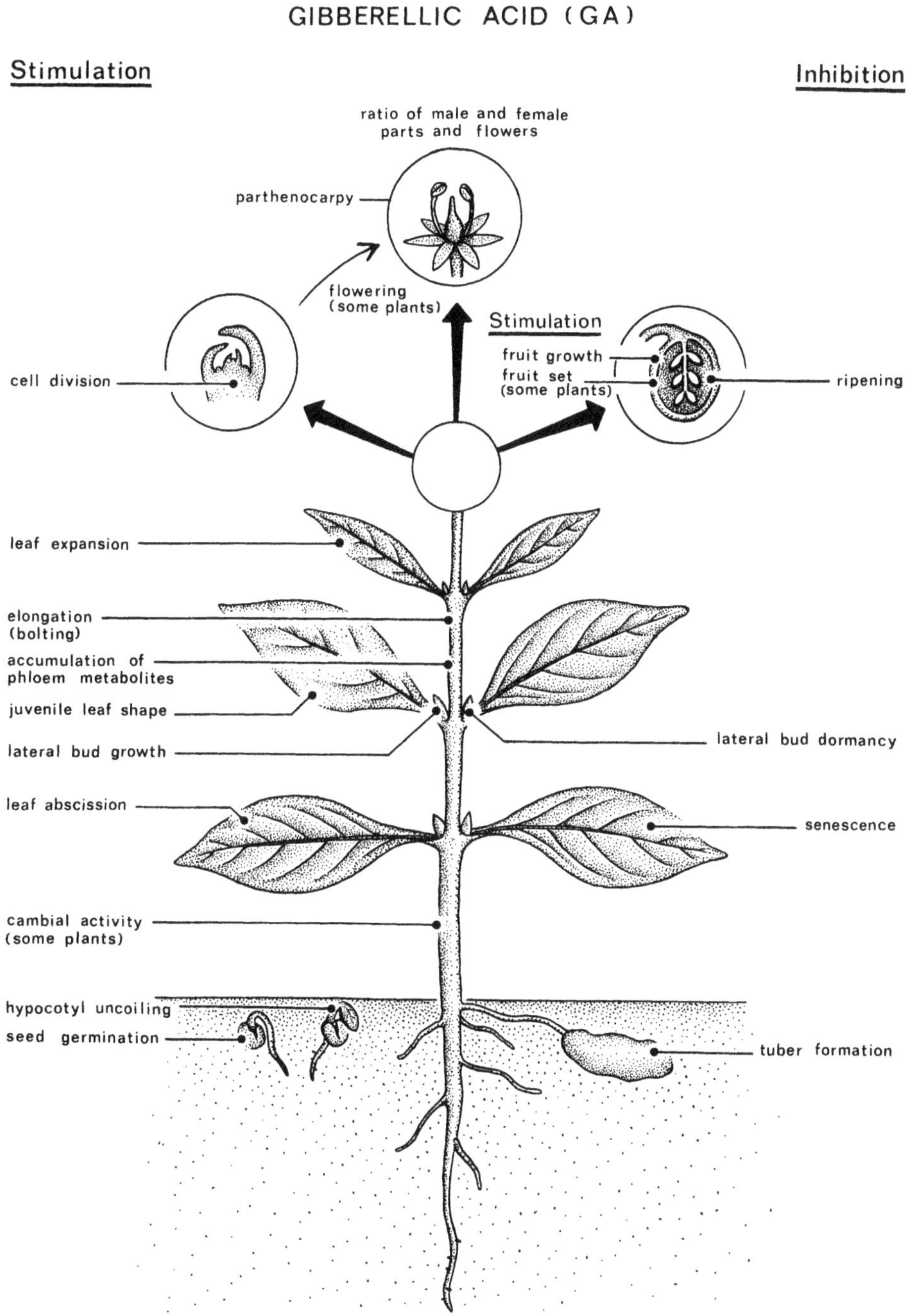
GIBBERELLIC ACID (GA)
Stimulation
Inhibition
ratio of male and female parts and flowers
parthenocarpy
flowering (some plants)
cell division
Stimulation
fruit growth
fruit set (some plants)
ripening
leaf expansion
elongation (bolting)
accumulation of phloem metabolites
juvenile leaf shape
lateral bud growth
lateral bud dormancy
leaf abscission
senescence
cambial activity (some plants)
hypocotyl uncoiling
seed germination
tuber formation

Fig. 6.4

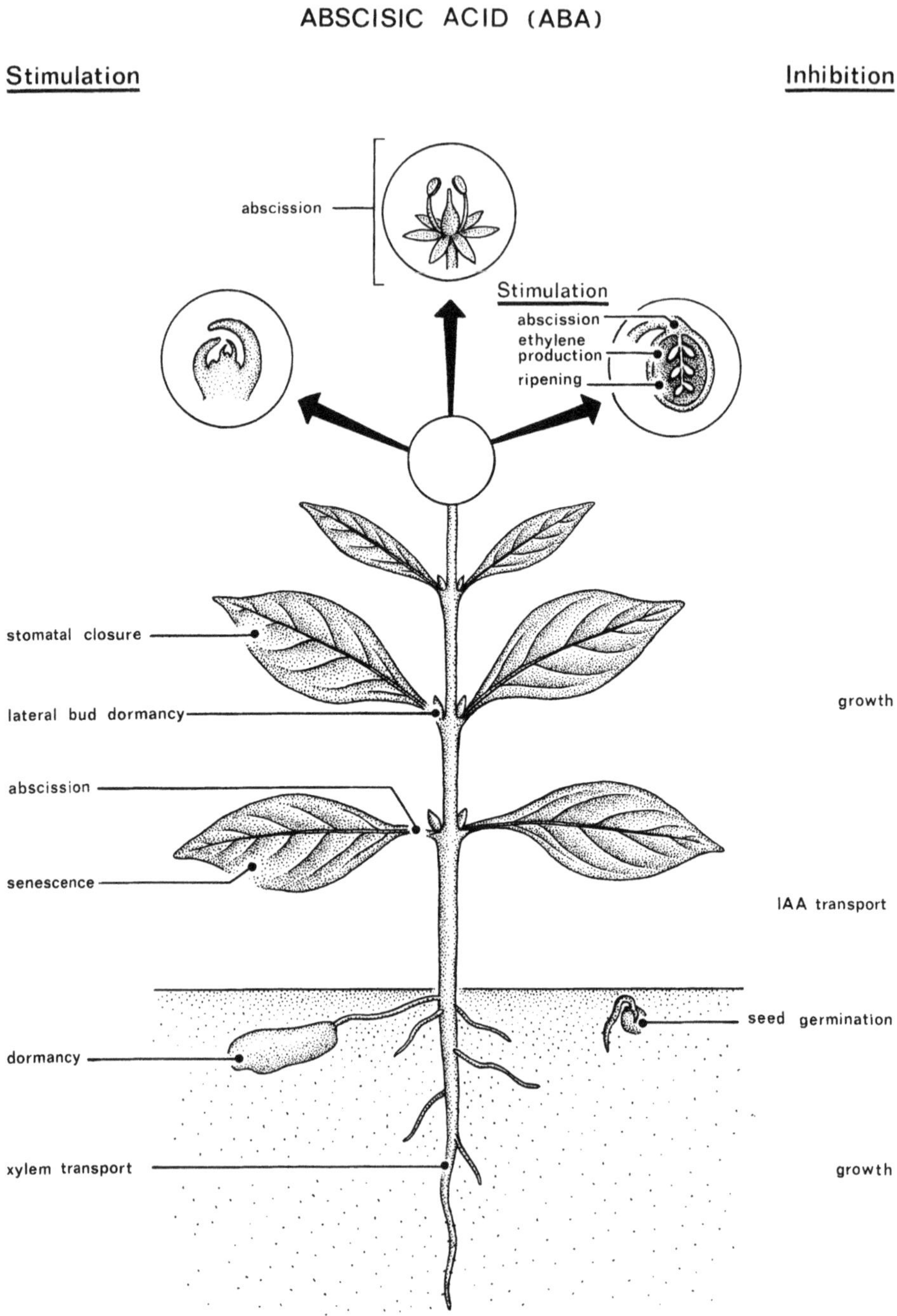
ABSCISIC ACID (ABA)
Stimulation
Inhibition
abscission
Stimulation
abscission
ethylene production
ripening
stomatal closure
lateral bud dormancy
abscission
senescence
dormancy
xylem transport
growth
IAA transport
seed germination
growth

Fig. 6.5

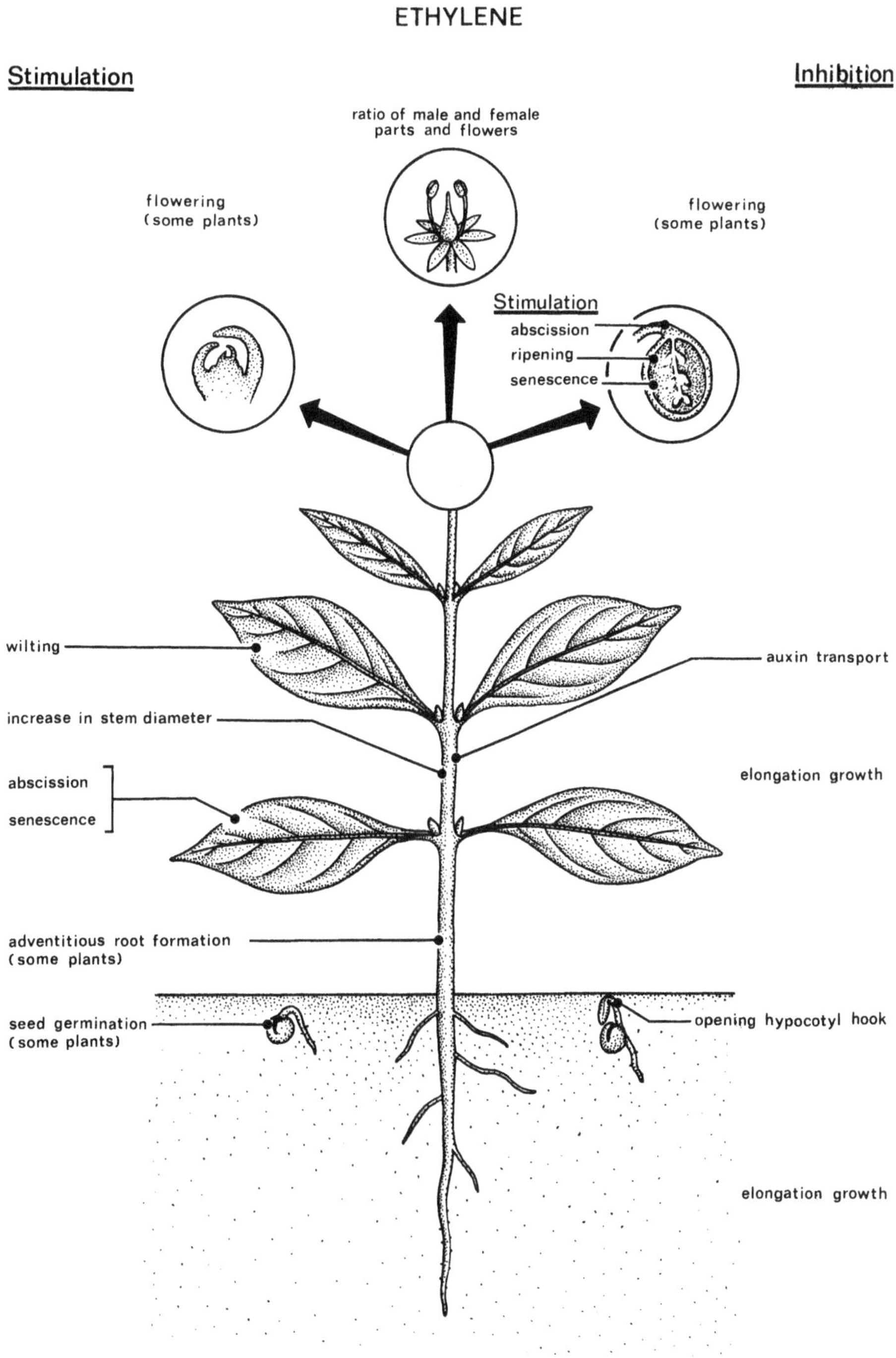
ETHYLENE
Stimulation
Inhibition
ratio of male and female parts and flowers
flowering (some plants)
flowering (some plants)
Stimulation
abscission
ripening
senescence
wilting
increase in stem diameter
abscission
senescence
adventitious root formation (some plants)
seed germination (some plants)
auxin transport
elongation growth
opening hypocotyl hook
elongation growth

Fig. 6.6

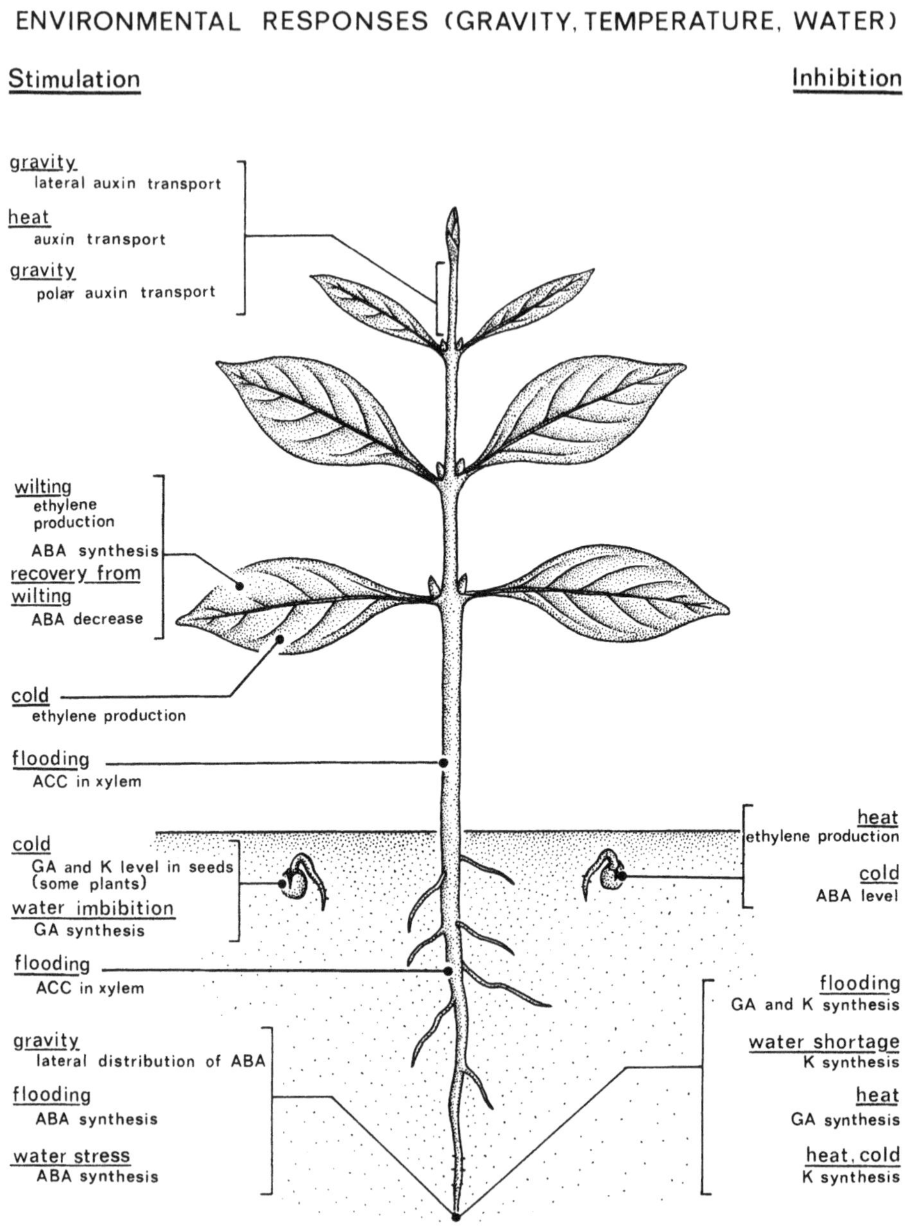
ENVIRONMENTAL RESPONSES (GRAVITY, TEMPERATURE, WATER)
Stimulation
Inhibition
gravity
lateral auxin transport
heat
auxin transport
gravity
polar auxin transport
wilting
ethylene production
ABA synthesis
recovery from wilting
ABA decrease
cold
ethylene production
flooding
ACC in xylem
cold
GA and K level in seeds (some plants)
water imbibition
GA synthesis
flooding
ACC in xylem
gravity
lateral distribution of ABA
flooding
ABA synthesis
water stress
ABA synthesis
heat
ethylene production
cold
ABA level
flooding
GA and K synthesis
water shortage
K synthesis
heat
GA synthesis
heat, cold
K synthesis

Fig. 6.7

ENVIRONMENTAL RESPONSES (LIGHT)

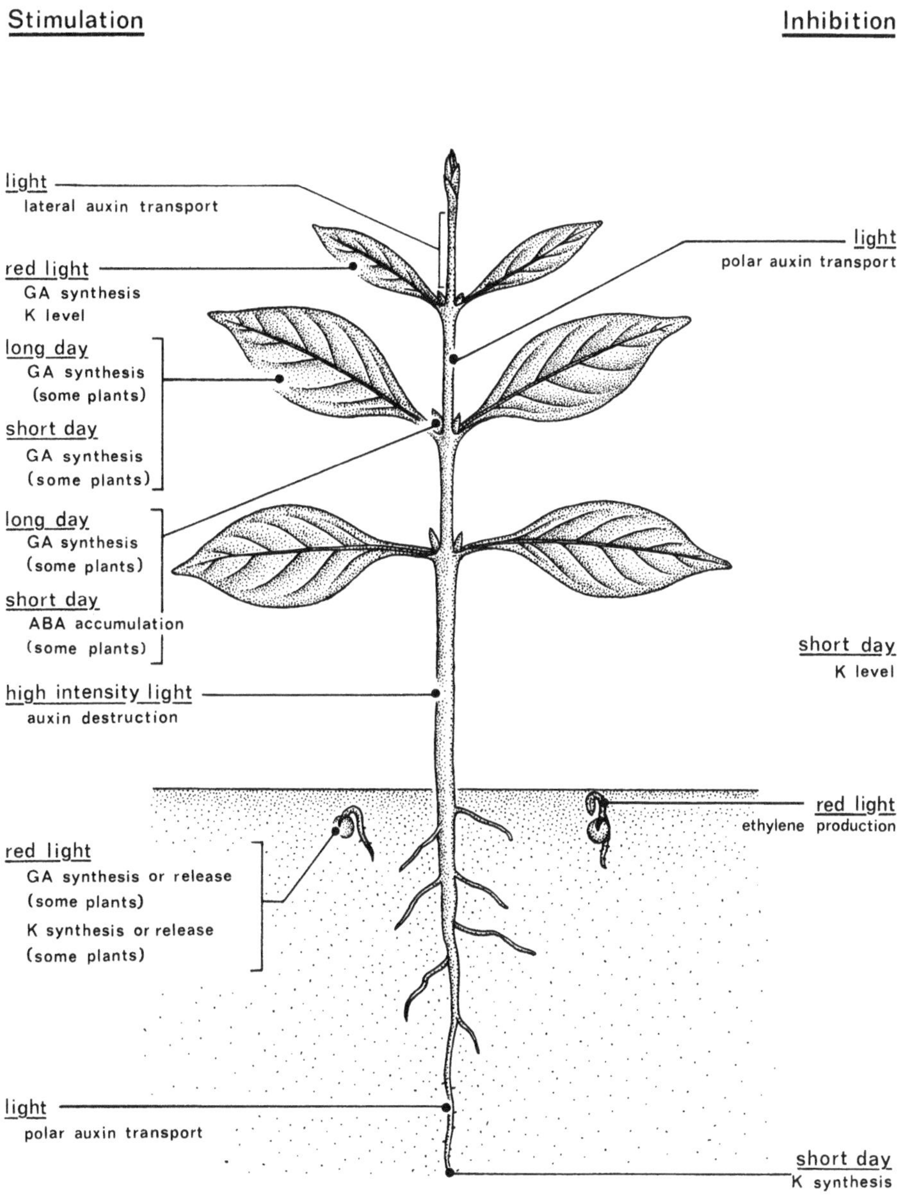

Fig. 6.8

7 Auxin Waves and Plant Morphogenesis

S. ZAJĄCZKOWSKI, T.J. WODZICKI, and J.A. ROMBERGER

7.1 Introduction

Realization of any developmental pattern requires spatial and temporal ordering of morphogenic events into a hierarchic system of processes occurring at all levels of organization of the plant body. Developmental ordering results from operation of an analogous hierarchic system of information transfer, of which the morphogenic action of plant hormones is generally agreed to be an important part. This distinctive aspect of the physiological role of phytohormones still remains, however, among the least understood and most intriguing problems of plant biology.

Most contemporary theories and hypotheses that purport to explain the role of phytohormones in plant morphogenesis relate their mode of action to spatial or temporal changes in molecular concentration. These explanations generally conform with classical gradient-field theories (CHILD 1941) which postulate that the differentiation of a cell is a consequence of its position within fields of concentration gradients of morphogenically active substances that are somehow maintained in the bodies of multicellular organisms. Critical studies by WOLPERT (1969) allowed distinction between systems specifying cell position within the developing organ and the processes of interpretation (or realization) of the positional information at a subcellular level. Within limitations determined by the cell genome, availability of metabolites, and developmental history, positional information determines the nature of molecular differentiation that the cell will undergo. In this concept, the specification of positional information precedes and is independent of its interpretation. Positional information, as discussed by WOLPERT (1969), may be specified by concentration gradients of morphogens or by other physical gradients generated by the rhythmic processes inherent in living cells (GOODWIN and COHEN 1969, HEJNOWICZ 1975). The idea of positional information implies the presence of a system of virtual coordinates with respect to which the cell can sense its position within the organ. Thus, an important condition for specification of cell position is establishment of a polarity in the multicellular system. The polarity in this case means a direction with respect to which the cell position is determined.

7.2 Polar Transport of Auxin and Positional Information

Processes of specification and realization of the morphological pattern occur in an order imposed by what may be called a hierarchic system of regulation.

Some of the general attributes of such systems operating in higher plants can be deduced theoretically on the basis of experimental data available in the literature. These attributes are summarized below:

1. A distinctive feature of higher plants is axial polarity of the whole body. This implies that the control systems also have an analogous polarity. Such systems can be described mathematically by application of the principles of vector and tensor calculus.
2. Among known plant hormones, only auxin exhibits basipetally polar transport. Auxin transport has a striking polarity which mirrors the structural and functional axial polarity of the plant body. Thus, even though other phytohormones are known to cause morphogenic effects, a prominent (and probably dominant) role in the spatial and temporal control of plant morphogenesis must be ascribed to auxin [1], and especially to the still obscure processes comprising the mechanisms of its basipetal transport.
3. The actual control system allows propagation of morphogenic information over long distances with seemingly equal precision in all directions. Consequently, in spite of apparent coupling with the transport of auxin, the control system cannot depend directly upon the messages carried basipetally by the auxin molecules themselves.
4. The actual system must be capable of translating various signals accompanying diverse environmental and physiological situations into a universal (or at least organismal) language specifying the positional information that can be perceived and interpreted by plant cells throughout their life span.

7.3 The Origin of a New Concept

A new concept of a morphogenic control system accommodating some of the principles of WOLPERT's theory of positional information is presented in the following pages. This concept has emerged from the results of a long series of experiments on the role of phytohormones, especially that of auxin, in xylogenesis. Our interest for many years has been focused on the function of plant hormones in regulation of seasonal cambial activity and the differentiation of secondary xylem in forest trees. Experimental results were obtained by extraction, purification, and bioassay by standard techniques (WODZICKI 1964, 1965, 1968, 1971, WODZICKI and WODZICKI 1973, 1980), application of synthetic hormones to trees growing under natural conditions (WODZICKI and ZAJĄCZKOWSKI 1974, WODZICKI et al. 1982, PORANDOWSKI et al. 1982), and by variations of sterile culture techniques in isolated stem segments. The latter allowed fully controlled application of synthetic and natural substances (BROWN and WODZICKI 1969, ZAJĄCZKOWSKI 1969, 1973, 1978, ZAJĄCZKOWSKI and WODZICKI 1975, ZAJĄCZKOWSKI and ROMBERGER 1978). The results clearly indicated a dominant role of auxin in control of these processes. Even so, no consistent correlation could be found between the concentration of auxin in the cambial region and the morphogenic phenomena that were studied. A seasonal pattern,

[1] Cf. also: L. STANGE, Cellular actions during early differentiation (Chapter 20, Vol. 17 of this Encyclopedia).

which to some extent reflected changes in xylem formation, was found only when basipetal efflux of natural auxin from the cambial region of pine stems was investigated.

In this chapter we first discuss the experimental evidence that led us to evolve new concepts of the role of auxin in plant morphogenesis. Some of the evidence has already been published in a series of technical papers (ZAJĄCZKOWSKI and WODZICKI 1978a, WODZICKI et al. 1979, WODZICKI and WODZICKI 1981, ZAKRZEWSKI 1983). An outline of the proposed model of regulation of morphogenesis by auxin has also been published previously (ZAJĄCZKOWSKI and WODZICKI 1978b, ZAJĄCZKOWSKI et al. 1983). Some of the theoretical and experimental aspects of the model have been presented and discussed at several meetings and conferences (ZAJĄCZKOWSKI and WODZICKI 1978c, 1980, WODZICKI 1980a, b, ZAJĄCZKOWSKI 1980). However, some previously unpublished material is included here.

7.4 Some Experimental Evidence

Large pieces of stem collected from mature forest trees (*Pinus silvestris* L., *Picea abies* (L.) Karst., *Larix decidua* Mill., *Acer pseudoplatanus* L., *Quercus robur* L.) were brought into the laboratory and resectioned into tissue blocks comprising phloem, the cambial region, and the most recent secondary xylem. After the dead bark was removed and the mature xylem had been trimmed to leave only the 1 to 2 mm closest to the cambial zone, the final dimensions of the tissue blocks were: 20 to 30 mm in width, 3 to 5 mm in thickness, and from 5 to 160 mm in length, as appropriate to the specific experiment. Strips of agar gel, plain or supplemented with synthetic growth regulators, were applied to the apical and basal ends of the stem tissue blocks for various times from 10 to 100 min. The strips were then bioassayed by a slightly modified Went's *Avena* (cv. Seger I) coleoptile curvature test (FUNKE 1939) either directly or after the tissue blocks had been subdivided into shorter (typically 6 mm long) sections. In the latter case, the amount of stimulation of curvature of *Avena* coleoptiles ascribable to basipetal efflux of auxin was tested after 6 to 10 min. This stimulation, presumably elicited by natural auxin collected in the agar receiver strips, was expressed in terms of equivalent amounts of authentic IAA, for which calibration curves were prepared.

Preliminary experiments with pine tissue blocks of a range of lengths (5, 10, 20, 40, 80, and 160 mm), to which strips of plain (unsupplemented) agar gels were applied for 60 min, revealed an initially unexpected pattern of stimulation of curvature that was evoked by the auxin collected from the cambial region. The stimulation evoked by the auxin collected in agar gel strips at the basal ends of the tissue blocks declined rapidly with increasing length of the sections from 5 to about 20 mm (Fig. 7.1 A). Further increase in length of the sections beyond 20 mm had little effect on basipetal translocation of auxin. The yield of auxin from the basal ends was always greater than from the apical ends, which confirmed the polarity of the auxin efflux. A decline with increasing section length in the amount of auxin translocated in the acropetal direction to agar receiver blocks was also observed. The amount of auxin collected decreased gradually as the length of the sections increased from 5 to 160 mm.

The decreasing diffusion of auxin to agar from increasingly longer sections of tissue was already observed in the experiments of De la Fuente and Leopold (1966). However, they interpreted the decline as being due to the binding of some of the auxin by the tissues themselves. In this view, the longer the route of transport through the tissue, the greater the amount of auxin fixed. Such an explanation, however, was inadequate for experiments concerning transport of endogenous auxin in the stem cambial region, which auxin presumably was initially uniformly distributed along the length of the tissue blocks (Wodzicki 1978). Length of cambial fusiform cells, in which most of the auxin seemingly is transported (Nix and Wodzicki 1974), averaged 3.3 mm in the pine tissue blocks that were studied. As restriction of basipetal efflux of auxin occurred only if the length of the tissue segment was reduced to some minimum (about 20 mm), a new working hypothesis became attractive: In the shorter stem blocks the degree of cell integration into a supracellular (organ level) system controlling polar transport of auxin is so incomplete that there is no effective control of the efflux of auxin to agar. This working hypothesis was then tested by experiments, some of which are summarized below.

Both ends of pine stem tissue blocks (72 to 80 mm long) were first placed in contact with agar gels for 10 min, then the tissue blocks were cut into sections 6 mm long, and fresh agar strips were immediately applied to both ends for 6 min, after which the auxin that diffused into the agar from the basal ends was bioassayed.

It appeared that the stimulation of coleoptile curvature ascribed to the auxin collected in agar from a series of successive short segments that were initially adjacent varied in a way such that an oscillating pattern was revealed (Fig. 7.1 B). If such oscillation in space is interpreted as a wave-like change moving along the stem, then the wavelengths observed were about 20 mm. The "amplitudes" of variations in efflux were equivalent to as much as 80 $\mu g\, l^{-1}$ of synthetic IAA. Thus, restriction of the auxin efflux observed in the above-mentioned preliminary experiments occurred only when the length of tissue blocks increased to equal the demonstrated wavelength. This supports the hypothesis that integration of cells into a supracellular system plays a significant role in a mechanism controlling polar transport of endogenous auxin in the cambial region of the pine stem. Similar experiments were performed with more than 50 pine stem tissue blocks collected from different trees during various times of year. Oscillatory patterns were revealed in more than 95% of the experiments. The average wavelength of the oscillations was 20 mm. The range was 15 to 40 mm. The shortest wavelengths and the highest amplitudes of efflux occurred at the time of spring initiation of cambial activity. Similar oscillatory patterns of auxin basipetal efflux were found in experiments with tissue blocks collected from other tree species of four genera: *Larix, Picea, Acer* (Wodzicki and Wodzicki 1981), and *Quercus* (Zakrzewski 1983). The wavelengths and amplitudes also were similar to those found in pine.

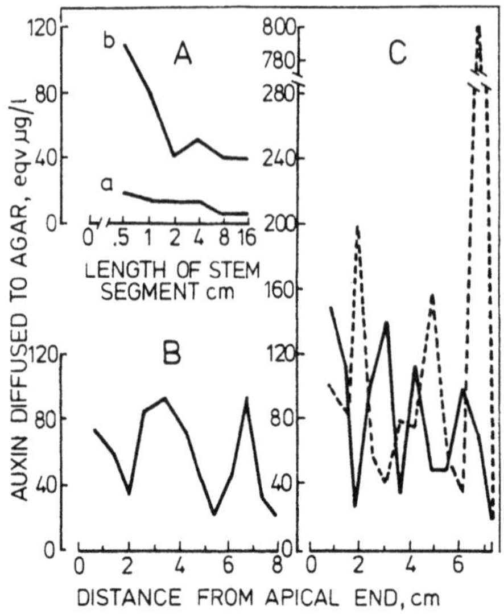

Fig. 7.1 A–C. Amount of endogenous auxin diffused to agar gels from the cambial regions of pine stem sections during 6 min. **A** Apical (*a*) or basal (*b*) efflux of auxin from sections of different lengths. **B** and **C** Basal efflux from successive sections 5 or 6 mm long. *Solid lines* no pretreatment; *broken lines* pretreatment by apical application of exogenous IAA for 100 min prior to sectioning and positioning for 6-min basal efflux test. (**A** and **B** adapted from ZAJĄCZKOWSKI and WODZICKI 1978a; **C** adapted from WODZICKI et al. 1979)

Oscillatory phenomena associated with polar transport of synthetic auxin have also been reported in *Avena* and *Zea* coleoptiles (NEWMAN 1963, HERTEL and FLORY 1968, SHEN-MILLER 1973a, b), thus the oscillations of basipetal efflux of endogenous auxin found in our experiments may reflect a more general property of the mechanism involved in polar transport of auxin.

Oscillatory phenomena could be basic to the functioning of efficient systems of spatial and temporal control of morphogenesis. Such systems would inherently have a high information-transmitting capacity. These points have already been discussed by WINFREE (1967), GOODWIN and COHEN (1969), PAVLIDIS (1971), and HEJNOWICZ (1975). Involvement of oscillatory phenomena in the control of morphogenesis seems especially likely in larger organisms— those having dimensions in centimeters (CRICK 1971)—because the time required for establishment of concentration gradients by diffusion alone over long distances becomes extremely long (CRICK 1970, MUNRO and CRICK 1971).

The stem tissues of large trees can serve as excellent systems in which to study mechanisms of long-distance morphogenic control. Tissues of the cambial regions of tree stems have a high degree of structural and physiological uniformity and thus also are excellent models in which to study polar transport of auxin and its possible involvement in the transmission of morphogenic information. In theory, the functioning of a morphogenic field depends upon its competence to transmit relevant information within its domain and to modulate the parameters of such transmission in response to factors that affect growth and differentiation.

The validity of this thinking and its applicability to model systems comprising tissue blocks from tree stems was tested in another series of experiments. The immediate objective was to determine the effects of exogenously added indole-3-acetic acid (IAA), and other substances known to affect morphogenic processes in plants, upon the oscillating systems previously demonstrated to be present in the cambial regions of such tissue blocks. Strips of agar gels containing synthetic growth regulators were applied to the apical (and sometimes the basal) ends of experimental stem tissue blocks prepared as described previously. Concentrations of the applied substances (separately or combined) were as follows: indole-3-acetic acid (IAA), 0.1 mg l^{-1}; gibberellic acid (GA), 1.0 mg l^{-1}; zeatin (ZEA), 0.1 mg l^{-1}; abscisic acid (ABA), 0.1, 1.0, or 10 mg l^{-1}; triiodobenzoic acid (TIBA), 10.0 mg l^{-1}. After they had served as hormone sources for 20 to 100 min at 20° C, the agar gels were discarded and the tissue blocks were cut into 6-mm segments. As in the previous experiments, strips of unsupplemented agar gels were immediately applied to both ends of the short segments. After 10 min, the amounts of auxin that had diffused from the basal ends of the tissue blocks were bioassayed. It appeared that apical supply of synthetic IAA increased the amplitude of oscillation of auxin efflux along the entire 72 mm length of the tissue blocks. This was particularly well manifested at the time of spring initiation of cambial activity, when the increase over the control value of auxin efflux in several instances exceeded 100% of the synthetic auxin that was supplied (Fig. 7.1 C).

The reported velocity of the polar transport of auxin in a variety of plant tissues has usually ranged between 10 and 15 mm h^{-1} (GOLDSMITH 1969, 1977, THIMANN 1972). Thus, the increase of amplitude after 100 minutes of IAA treatment, which we observed as far as 70 mm from the place of application, was probably not due to translocation of the applied synthetic auxin. This supposition was confirmed experimentally. In these tests ^{14}C-IAA was applied to apical ends of longer pine tissue blocks. Bioassays for hormone activity were then combined with determinations of radioactivity in receiver strips of agar gel in contact with the basal ends of short serial segments cut from the longer blocks (WODZICKI and BROWN unpublished experiments).

When TIBA, an inhibitor of polar transport of auxin, was added, the amplitude of auxin efflux was greatly reduced. In fact, in some of the stem blocks TIBA application resulted in the complete disappearance of the wave-like pattern (Fig. 7.2 A). Supplementing the IAA in agar gels with ABA produced effects similar to those obtained with IAA + TIBA. The effects of both inhibitors were visible along the whole 66 mm length of the stem tissue blocks even when the application period was shortened to 20 min. If we consider that the velocity of basipetal translocation of ABA is estimated to be 20 to 35 mm h^{-1} (MILBORROW 1978), and that of TIBA to be 2.5 mm h^{-1} (THOMSON et al. 1973), it seems improbable that the observed modulations could be due to a direct action of molecules of the applied substances. Our observations that the modulating effects of IAA, ABA, and TIBA are apparently translocated faster than their molecules are consistent with the hypothesis that the oscillatory system involved in polar transport of auxin may function in control of plant morphogenesis analogously to the signalling mechanisms discussed by CRICK (1971). This means that the population of cells participating in the polar transport of auxin functions as an integrated supracellular system of coupled oscillators, and that the role of apical meristematic centers can be compared to that of the pacemaker regions described by GOODWIN and COHEN (1969).

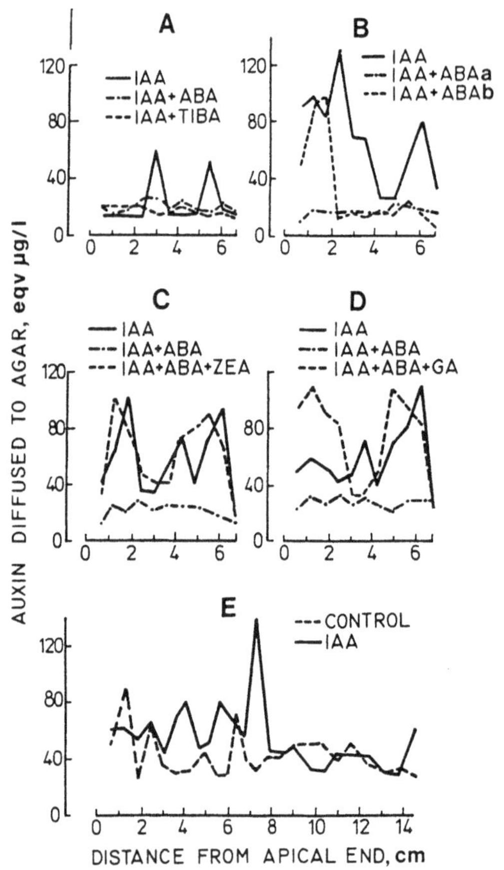

Fig. 7.2 A–E. Interactions of IAA with ABA, GA, TIBA, and zeatin (ZEA), alone or in combination, as evidenced by effects upon the wave-like pattern of auxin efflux from the basal ends of successive 6-mm-long pine stem sections. Protocol as in Fig. 7.1. Test substances were applied for 100 min prior to sectioning and positioning for 10-min auxin basal efflux period. Test substances were applied apically only in **A**, **C**, **D**, and **E**, but both apically (*a*) and basally (*b*) in **B**. Controls had no exposure to exogenous substances. (**A**, **B**, **C**, and **D** adapted from Wodzicki and Wodzicki 1981; **E** from Wodzicki et al. 1979)

The effects of ABA upon the wave-like pattern of the basipetal efflux of auxin were also observed when the inhibitor was applied for 100 min to the basal ends of tissue blocks (66 mm long). In these tests, however, suppression of the wave amplitudes was observed only at distances up to about 40 mm from the basal end to which ABA was applied (Fig. 7.2 B). This implies that the supracellular oscillatory system, in spite of its polarity, is capable of transmitting signals in both the basipetal and the acropetal direction. The velocity of signal transmission in the acropetal direction, however, appears to be slower.

Application of zeatin in agar gels to the apical ends of stem segments before sectioning them produced some increase of the auxin efflux. The most interesting action of both cytokinin and gibberellic acid, however, was prevention of the depressive effect of abscisic acid upon the wave-like pattern of the basipetal efflux of auxin (Fig. 7.2 C, D).

The results of experimental application of synthetic substances to stem tissue blocks showed that the oscillatory system involved in polar transport of auxin is sensitive both to growth stimulators and inhibitors, and that it reacts to them by modulation of at least the amplitude of the oscillations. Since the observed oscillations are translocated in space, the oscillatory system involved in the polar transport of auxin can be analyzed as a moving wave. As the basipetal transport of auxin in plant tissues is a function of the activities of individual polar cells (LEOPOLD 1961, GOLDSMITH 1969), the wave-like pattern found in these experiments arises from oscillations within these cells and the wave propagation must be basipetal. Because the oscillators in an individual cell are coupled with those in contiguous cells, any modulation of wave parameters can also be transmitted acropetally. Because the wavelength is several times longer than the length of cambial fusiform initial cells, there must be a phase difference between oscillations in longitudinally contiguous cells. If the wave is assumed to be sinusoidal then the relations between the wavelength (λ), length of cambial fusiform cells (L), period of oscillation within cells (T), and period of delay between phases of oscillation in longitudinally contiguous cells (Δt), can be expressed by the formula (ZAJĄCZKOWSKI and KASPRZAK in preparation):

$$\frac{L}{\lambda}=\frac{\Delta t}{T}.$$

The velocity of the basipetal signal (wave) propagation (V_s) is described by:

$$V_s=\frac{\lambda}{T} \quad \text{or} \quad V_s=\frac{L}{\Delta t}.$$

The relation between the velocity of signal propagation and the velocity of polar transport of auxin molecules (V_m), then, is as follows:

$$V_s=V_m\left(\frac{T}{\Delta t}+1\right) \quad \text{or} \quad V_s=V_m\left(\frac{\lambda}{L}+1\right).$$

For the functioning of the supracellular oscillatory system described at least two interacting subsystems in each cell are required. The auxin molecules must be under control of one of the subsystems for a period of time equalling Δt and under control of the other subsystem for a period T. When the experimental data were used to calculate velocity of signal propagation in cambium, it was found to be about 70 mm h^{-1}; the period of cell oscillations was about 20 min; and the phase shift (or lag) between contiguous cells was about 3 min. The calculated velocity of signal propagation appeared to be similar to that found in experiments with IAA applied to the apical ends of tissue blocks 144 mm long (Fig. 7.2 E). The calculated period of oscillation is compatible with the reported oscillation of electric potential differences which develop after application of synthetic auxin to other plant tissues (NEWMAN 1963, HERTEL and FLORY 1968, POLEVOY 1972). This compatibility may indicate an involvement of the system responsible for control of H^+ secretion across the plasmalemma or transport of other ions. Thus, the measured auxin wave may form in conjunction with oscillations in ionic transport at the cellular level. A model of hormonal control of plant morphogenesis, using the principles of the above-described oscillatory system, is discussed in the next section.

7.5 The Proposed Model

Let us assume the theoretically most simple case: that polar transport of auxin occurs primarily within a single layer of cells in the cambial zone, the envelope of cambial initials. The observed wave-like pattern can then be interpreted as a moving wave in two-dimensional space (Fig. 7.3 A). If the direction of propagation is parallel to the stem axis, the wave fronts are perpendicular to the stem axis (Fig. 7.3 B a). The vectors of wave propagation, which are orthogonal to the wave fronts, form a vector field parallel to the stem axis. In nature a variety of agents may modify the orientation of the vectors of wave propagation. Agents such as auxin sources, auxin acceptors, processes and factors affecting auxin transport (including other growth regulators), branching of the stem axis, structural differences between fusiform and ray initials in the cambium, etc., may bend the wave fronts locally and cause the vectors of wave propagation to deviate from the axial direction. The vector field can thus acquire a complex pattern. The areas of left and right inclinations of vectors (Fig. 7.3 Bb) form a domain pattern. This invisible domain pattern in the vector field probably has some similarity to the sometimes visible orientational domains of the cambial fusiform initials (Hejnowicz 1967, 1968, Hejnowicz and Romberger 1973, 1979).

Close similarity between behavior of the postulated auxin waves and the mapped patterns of cell orientation raises the question of whether the vectors of auxin wave propagation could constitute a morphogenic "map" carrying positional information to cambial fusiform cells. An apparent difficulty is that we have not yet confronted the truly vast difference between migration rates of orientational domains (Hejnowicz and Romberger 1973, 1979) and those of the auxin wave system as discussed here. Whatever controls the movement of orientational domains is probably several hierarchial levels removed from any positional signal system emanating from the shoot apices — or may depend upon a separate, but operationally similar, system. This lack of understanding is, however, not an argument against the possibility that auxin wave vector fields could constitute a dynamic positional information map on an organismal scale. Realization of the information encoded in the auxin vectors could be accomplished, for example, by asymmetric anticlinal divisions, locally variable intrusive growth, and modification of the competitive status of variously oriented initial cells. The intensity of these processes, at least to some extent, might also be encoded in other characteristics of the auxin wave (such as amplitude and frequency) the variation of which need not necessarily affect the pattern of the vector field.

Plant physiologists generally agree that polar movement of auxin in plant organs depends upon the inherent polarity of the plant body (Leopold 1961, Thimann 1972). Reorientation processes (primarily non-randomly oriented divisions and unequal intrusive growth) in the spiral or zig-zag bridges of tissues in ring-barked sections of stem, such as described by Brown and Cormack (1937), Sachs (1969), Kirschner et al. (1971), and Thair and Steeves (1976), indicate that the cells of the cambial regions recognize the direction of auxin

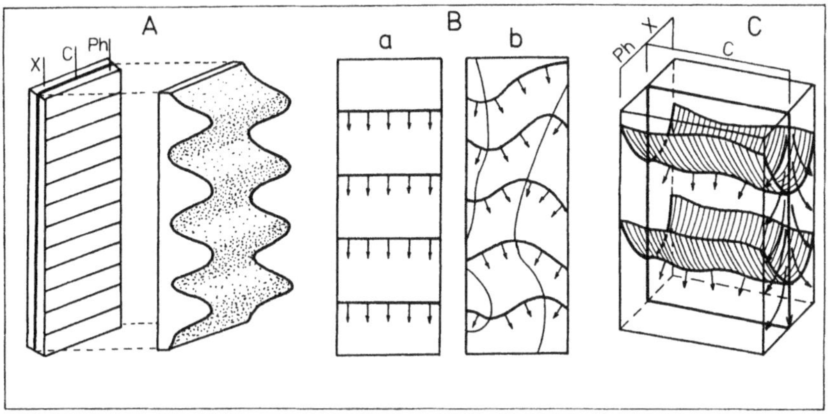

Fig. 7.3 A–C. Representations of auxin waves. **A** Two-dimensional wave of variation in amount of auxin efflux from a serial set of short stem segments simulating experiments referred to in Figs. 7.1 and 7.2. *C* cambium; *Ph* phloem; *X* xylem. **B** Morphogenic maps depicting (in **a**) regular isophasic fronts of auxin waves (*heavy lines*) and (in **b**) domains of right and left inclination of vectors of wave propagation. Domain boundaries (*lighter lines*) may be quite irregular. The vectors (*arrows*) of wave propagation are orthogonal to the wave fronts. The drawings represent tangential views of the cambium from outside the circumference of a stem segment. **C** Three-dimensional diagram of auxin wave fronts (*shaded, curved surfaces*) and vectors of wave propagation (*arrows*) in the secondary tissues of the stem. *Short arrows* represent vectors within the cambial envelope. *Trajectories of the longer arrows* drawn in the right-hand (radial) face of the block represent the angles between the vectors of auxin wave propagation and the axial direction at various distances from the cambium. *C* cambium; *Ph* phloem; *X* xylem. (**A** and **B** adapted from ZAJĄCZKOWSKI and WODZICKI 1978a; **C** from ZAJĄCZKOWSKI and WODZICKI 1978b)

transport. The oscillatory phenomena involved in the polar transport of auxin may constitute a specific mechanism, by means of which the positional information controlling these morphogenic cellular processes is specified.

So far, this discussion has been confined to the behavior of auxin waves in two-dimensional space. In real stems, however, polar transport of auxin is not restricted to the cambial envelope. It also occurs in the zone of differentiating xylem and phloem (NIX and WODZICKI 1974). Consequently three-dimensional aspects of the auxin waves must be considered. As reviewed by GOLDSMITH (1969), polarity of auxin transport in the meristematic tissues of the stem gradually decreases as the derivative cells progress through their differentiation. Thus, bending of auxin waves can be expected near the xylem and phloem boundaries of the cambial region. Such bending would result in curvature of the auxin wave fronts (isophasic surfaces in the three-dimensional aspect) as indicated in Fig. 7.3C. As the vectors of wave propagation are orthogonal to the wave fronts, a three-dimensional vector field is formed in which the theoretical trajectories of the vectors form straight lines in the cambium (as indicated in the schematic radial view of the tissue in Fig. 7.4 Aa). Everywhere else the trajectories are curved and deviate increasingly from the axial direction as the distance from the cambium increases.

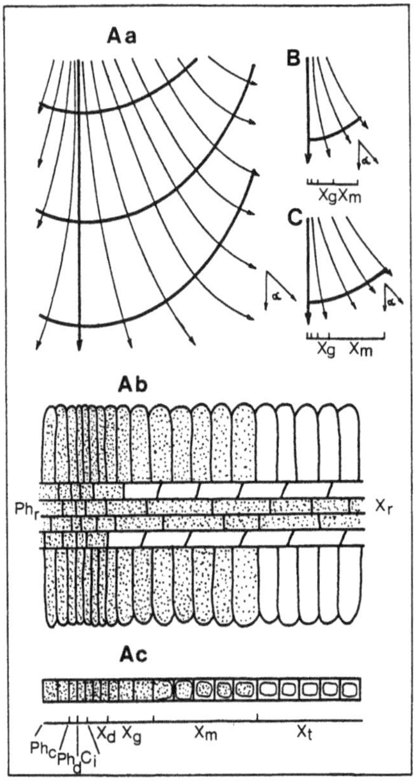

Fig. 7.4 A–C. A model of control of morphogenesis by vectors of auxin wave propagation in the final phase (autolysis) of axial tracheid maturation in secondary xylem of the stem. **A** Trajectories of vectors of auxin wave propagation (**a**) in the cambial region and in differentiating secondary tissues as seen in radial (**b**) and transverse (**c**) sections. The breakdown of the cytoplasm is initiated when some critical angle (α) between the vector of auxin wave propagation and the cell axis is attained. **B** and **C** Vector trajectories associated with a model of regulation of earlywood and latewood differentiation. Ph_c conducting phloem; Ph_d dividing phloem; C_i cambial initial; X_d dividing xylem; X_g enlarging xylem; X_m maturing xylem; X_t mature xylem (nomenclature according to WILSON et al. 1966), Ph_r phloem ray; X_r xylem ray. (Adapted from ZAJĄCZKOWSKI and WODZICKI 1978b)

We now return to the hypothesis that auxin waves can specify morphogenic positional information. As an example, we will use tissue of the stem cambial region of a coniferous tree species (Fig. 7.4). The meristematic cells of the cambial zone include not only the cambial initials, but also those phloem and xylem cells still capable of division (WILSON et al. 1966). For simplicity we shall here ignore the ray initials and their derivatives. The axial xylem derivatives, after undergoing their last divisions in the cambial zone, enter the stage of enlargement growth. The next stage is maturation, during which the secondary cell wall is deposited. This stage ends with the autolysis of the protoplast (WODZICKI 1971). As new xylem cells are produced, the position of a given differentiating cell becomes increasingly more distant from the cambium.

As seen in a schematic radial view (Fig. 7.4Aa, b, c), the continuous straight line of vector trajectories passes through one layer of cells in the cambial zone. This continuous noncurving vector defines the region in which there is no radial deviation of the vectors of auxin wave propagation from the cell axes. That distinctive region of the auxin wave vector field may be recognized as the envelope of cambial initials because after each periclinal division of an initial the axis of only one of the newly formed cells can really be coincident with the nondeviating vector direction. The direction and magnitude of the small deviation of the vector from the cell axis, such as occurs in at least half of the daughter cells, may determine the developmental fate of the cell so that it becomes either a xylem or a phloem derivative. During the course of axial element differentiation the radial deviation between the direction of the vector of wave propagation and the cell axis gradually increases. It is smallest for the cells constituting the cambial zone. Theoretically it is nil in the initials themselves. As the deviation increases, the cells lose their ability to divide. They then begin rapid radial growth. Further increase of the vector deviation from the cell axial direction is accompanied by the process of secondary wall deposition. The greatest deviation is reached at the boundary between maturing and mature xylem where autolysis of the protoplast terminates the process of tracheid differentiation (Fig. 7.4 B, C).

The coincidence between increasing deviation of vectors of the auxin wave from a cell's axis and the sequence of morphogenic events during axial tracheid differentiation suggests that the three-dimensional vector field, which exists when auxin waves are propagated through stems, can specify positional information to differentiating cambial derivatives. Realization of this information may depend upon expression of specific segments of the cell genome and also may depend upon availability of metabolites. It should be noted that in the postulated system the same angle of vector deviation from the cell axis (in the radial plane) may occur at various distances from the cambial initials depending upon the degree of curvature of the auxin wave fronts. Theoretically, if the velocity of the wave increases, a greater curvature of the wave fronts due to boundary effects or inhomogeneities of the medium through which the wave is propagated may be expected. If that were the case, any critical vector angle coinciding with breakdown of vacuolar membranes and initiation of autolysis of protoplasts in differentiating tracheids (Wodzicki and Brown 1973) should occur closer to the cambium. Thus, factors increasing the velocity of the auxin wave should result in a narrower zone of differentiating tracheids (as is seen in spring and early summer), whereas inhibitors should allow development of the wider zone characteristic of latewood formation. This offers at least a partial explanation of the seasonal variation of the width of the zone of maturing xylem (Wodzicki 1971, Wodzicki and Wodzicki 1980).

Positional information specified by configuration of the vector fields of the auxin waves in the zone of differentiating tracheids may also be a factor in controlling orientation of microfibrils in the successive cell wall layers, S_1, S_2, and S_3 (Frey-Wyssling 1959, Wardrop 1964a). Changes may result from increasing structural complexity of the cellular medium in which the wave is propagated. Circular deposition of fibrils around the primary pitfields permeated by plasmodesmata (Liese 1965, Schmid 1965) may be a manifestation of vector rotation around an obstacle within a vector field. Such behavior is common for various vector fields irrespective of their physical nature. A possible example of an analogous rotation at the supracellular level is the orientation of cambial fusiform cells producing ellipsoidal patterns of secondary tissues in the regions of the main stem around lateral branches. The discussion concerning the cambial regions of coniferous trees can be extended with only slight modifications to xylogenesis in angiosperms, including formation of vessels.

The polarity of a plant organ is a resultant of the polarities of its individual cells. During organogenesis specific regions of the plant body become specialized (differentiated) in the direction of various limited functions. This differentiation is accompanied by a gradual diminution of the polarity of the individual cells. Thus, the polarity of cells in the remaining meristematic regions eventually becomes dominant. These regions also become the main routes of polar transport of auxin and are the only regions in which the auxin wave vectors are parallel with the axes of the cells. This is especially true of the cambial initial cells.

Organ polarity and the organ axis tend to parallel each other, probably because maintenance of the polarity then requires the least energy. The same thermodynamic principle should apply to relations between the polarities of individual cells and the vectors of auxin wave propagation. There are only very limited regions of the plant body in which the condition of parallelism can be fulfilled and maintained. These regions are the meristems. The implication is that cells outside the meristematic regions can be maintained in the undifferentiated (structurally and functionally unspecialized, but highly polarized) condition only at high energy cost. Outside the meristem, processes that reduce maintenance energy expense would thus be favored. The processes of cell differentiation and the gradual loss of polarity apparently contribute to the reduction of energy requirements. The cells that spend little or no energy on compensating for the vector deviation are those that remain undifferentiated and meristematically active. Thus differences in energetic balance of the cells resulting from various amounts of energy being spent for compensation of the deviation of the vector of auxin waves from the direction of the main polarity axis might provide positional information with a potential for controlling developmental processes.

The postulated ability of cells to recognize the direction of vectors of auxin wave propagation, and adjustment of cell axes in meristems to those vectors, also provides a basis for a hypothetical mechanism that could control tropisms and coordinated growth of the whole plant. Let us assume that an organ is exposed to a gradient of an external factor affecting oscillations in the individual cells of an organ in which the polar transport of auxin is active. If the direction of the gradient does not coincide with the organ axis, there will be differences between oscillations on opposite sides of the organ and, due to cell coupling, a lateral phase shift can be expected. This could cause radial deviation of the vectors of the auxin wave from the polarity axis of the cells in those organ sectors mediating the main polar transport of auxin (Fig. 7.5A).

If the above were the case, no continuous trajectory of the vectors of auxin wave propagation could be maintained in the zone of major polar transport of auxin unless additional energy were spent on counteracting the deviational effects of the external factor. Differential growth of cells, and the consequent asymmetric growth of the organ, restores the latter to a position such that the vectors of wave propagation in the zone of major polar transport of auxin again coincide with the organ polarity. Thus a state of minimum energy requirement is restored. Figure 7.5 A, B illustrates this type of regulation in the case of phototropic and geotropic responses. The proposed mechanisms of tropistic

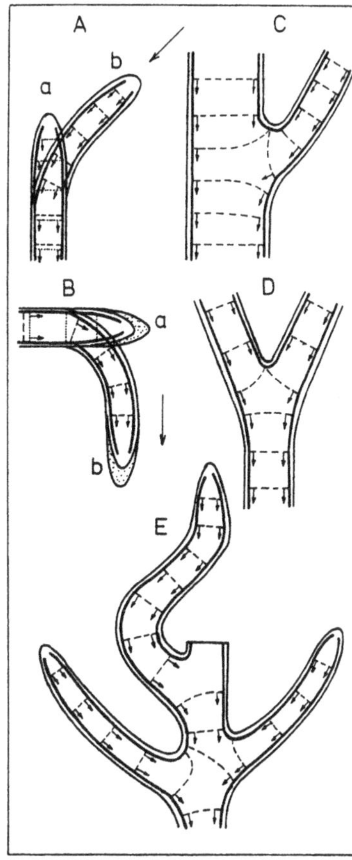

Fig. 7.5 A–E. Models of regulation of shoot and root morphogenesis by vectors of auxin wave propagation (*small arrows*). *Dotted lines* are theoretical tangents to isophasic surfaces of auxin waves in the zone of major polar transport of auxin in the plant organ (*heavy lines*) prior to any adjustment due to the effects of external stimuli. *Broken lines* are similar tangents adjusted for the effects of external stimuli and/or the interaction of different auxin waves in the branching region. **A** and **B** tropic responses in the apical meristem regions. **A** Phototropism in the shoot. **B** Geotropism in the root. Organs are represented before (**a**) and after (**b**) morphogenic response. *Long arrows* represent vectors of the stimuli: light or gravity. **C** and **D** effects of interactions of auxin waves in intersecting axes. The main stem dominates in **C**. In **D** there is co-domination. **E** theoretical fields of the vectors of auxin wave propagation in a regenerating shoot after decapitation. (Adapted from ZAJĄCZKOWSKI and WODZICKI 1978b)

responses advanced by various authors may be considered as also deriving whatever validity they have from the principle of deviation of the vectors of auxin wave propagation.

Attainment and maintenance of the least energy state of the system requires that cell and organ polarities be adjusted to the pervasive environmental factor, which is gravity. If we consider only a single higher plant stem axis in which an auxin wave is propagated, the organ polarity commonly coincides with the vector of gravity. The auxin waves propagated in the various shoot segments of a branched plant body, however, interact at the nodes of branch insertion (Fig. 7.5 C). Such interaction essentially means a change of the phase difference of the laterally adjacent cellular oscillators in that region and the imposed tendency of deviation of the auxin wave vectors from the direction of gravity. As mentioned earlier, modulations of auxin waves can be transmitted acropetally and can serve as a major source of information for realization of morphogenic processes of the interacting shoot in a coordinated growth. The deviation of the vectors caused by wave interaction decreases with increasing distance from the branching region due to damping effects and because the effects of gravity and moments of force of the branch weight become increasingly dominant.

If there were no interaction between the auxin waves in branches and in the main stem, the branches would grow generally parallel to that stem. We propose that adjustment of the organ axes to the configuration of the auxin wave vectors in the zone of major polar transport of auxin results in deviation of the branch axes from the direction of gravity.

The principles of the mechanism outlined above can be applied to explain various types of branching that may derive from different quantitative relations between the parameters of the auxin waves moving along the organ axes. For example, equal deviations of vectors in two interacting shoots may produce dichotomy (Fig. 7.5 D). In extreme cases domination by a main axis may completely inhibit propagation of auxin waves in lateral axes and result in suppression of lateral buds. Experimentally imposed horizontal positioning of the main (dominant) stem would be expected to produce radial deviation of the vectors toward the direction of gravity. Consequently lateral axes inserted along the lower part of the main stem should be under greater suppression than they would have been if the axis had remained in the upright position. Conversely, partial release from suppression would be expected in the lateral axes inserted along the upper side of the inclined stem. This would explain differential growth of lateral axes (buds) on the opposite sides of the stem such as is known to occur in various cases of gravimorphism.

As mentioned previously, the meristematically active shoot apex, which is producing auxin, most probably is the organizing center of the auxin wave. If one branch of a shoot is decapitated, the auxin wave moving down from its apex disappears and its interaction with the waves of other shoots fades away. Under these circumstances the auxin wave in the intact shoot closest to the point of decapitation is affected most strongly by gravity, which means that the vectors of wave propagation in the zones of major polar transport of auxin tend to parallel the direction of gravity. Thus the processes of shoot reorientation are initiated (Fig. 7.5 E). This phenomenon in the case of interaction between a decapitated main stem and its lateral branches is commonly known as release from apical dominance. Young plant organs can adjust themselves to the vector of auxin wave propagation by unequal rates of cell division and elongation on opposite sides of the organ. Parts of organs with secondary structure usually adjust organ position by localized formation of reaction tissues, which commonly are modified xylem (Wardrop 1964b, 1965). The mechanism of formation of the latter in relation to control by auxin waves has already been discussed in some detail (Zajączkowski and Wodzicki 1978b).

7.6 Further Implications

As an extension of the preceding discussion about the probable consequences of interactions between auxin waves in branched shoot systems, a control mechanism capable of coordinated growth of all plant parts (axes) can be proposed. Vector fields in the zones of major polar transport of auxin in axes of various orders can be envisioned as morphogenic fields specifying positional information

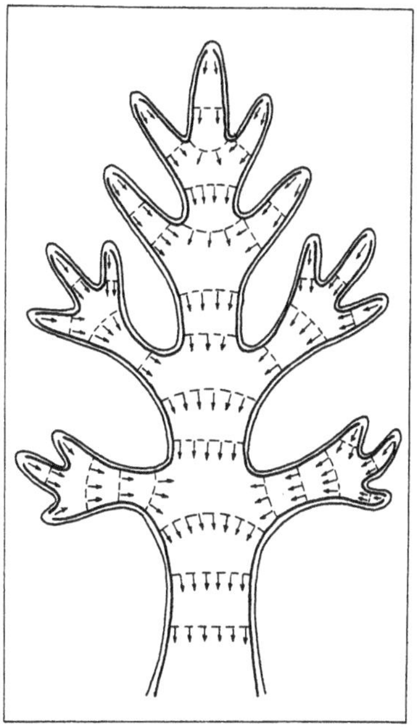

Fig. 7.6. Two-dimensional simplification of the three-dimensional field of the vectors of auxin wave propagation (*arrows*) in the aerial part of a plant with a continuous cambial envelope (*heavy black line*). *Broken lines* represent the theoretical tangents to isophasic surfaces adjusted for the effects of external stimuli and interactions of colliding auxin waves in regions of branching. (Adapted from ZAJĄCZKOWSKI and WODZICKI 1978b)

within developing plant organs (Fig. 7.6). Adjustment of the morphogenesis of plant organs in response to configuration of such vector fields again represents a tendency to default to the state that requires the least energy to maintain. This concept offers a universal mechanism of hierarchical control of plant morphogenesis by which the responses to practically all morphogenically active agents described in the literature could be mediated.

In our opinion, the concepts advanced herein, though they need further elaboration and refinement, have the potential to accommodate much of the accumulated information about the varied regulatory effects on plant morphogenesis produced by the several phytohormones and by some other morphogenically active substances.

Acknowledgments. The authors express their gratitude for facilities and support provided by the Warsaw Agricultural University (Poland), the School of Forest Resources, University of Georgia (U.S.A.), and Harvard Forest, Harvard University (U.S.A.), where the several parts of the experimental and theoretical research discussed herein were accomplished. The generous support of the U.S. Department of Agriculture, Forest Service, through the PL-480 Program is especially acknowledged. The junior author also is grateful for support from the InterAcademy Exchange Program between the U.S.A. and Poland, which enabled him to work in Poland in 1981.

References

Brown AB, Cormack RGH (1937) Stimulation of cambial activity, locally in the region of application and at distances in relation to a wound, by means of heteroauxin. Can J Res Sec C 15:433–441

Brown CL, Wodzicki TJ (1969) A simple technique for investigating cambial activity and differentiation of cambial derivatives. For Sci 15:26–29

Child CM (1941) Patterns and problems of development. Univ Chicago Press, Chicago

Crick FHC (1970) Diffusion in embryogenesis. Nature 225:420–422

Crick FHC (1971) The scale of pattern formation. In: Davies DD, Balls M (eds) Control mechanisms of growth and differentiation. Symp Soc Exp Biol 25:429–438

de la Fuente RK, Leopold AC (1966) Kinetics of polar auxin transport. Plant Physiol 41:1481–1484

Frey-Wyssling A (1959) Die pflanzliche Zellwand. Springer, Berlin Göttingen Heidelberg

Funke H (1939) Über den Nachweis kleiner Wuchsstoffmengen. Jahrb Wiss Bot 88:373–388

Goldsmith MHM (1969) Transport of plant growth regulators. In: Wilkins MB (ed) The physiology of plant growth and development. McGraw-Hill, New York London, pp 125–162

Goldsmith MHM (1977) The polar transport of auxin. Annu Rev Plant Physiol 28:439–478

Goodwin BC, Cohen MH (1969) A phase-shift model for the spatial and temporal organization of developing systems. J Theor Biol 25:49–107

Hejnowicz Z (1967) Some observations on the mechanism of orientation movement of woody stems. Am J Bot 54:684–689

Hejnowicz Z (1968) The structural mechanism involved in the changes of grain in timber. Acta Soc Bot Pol 37:347–365

Hejnowicz Z (1975) A model for a morphogenetic map and clock. J Theor Biol 54:345–362

Hejnowicz Z, Romberger JA (1973) Migrating cambial domains and the origin of wavy grain in xylem of broadleaved trees. Am J Bot 60:209–222

Hejnowicz Z, Romberger JA (1979) The common basis of wood grain figures is the systematically changing orientation of cambial fusiform cells. Wood Sci Tech 13:89–96

Hertel R, Flory R (1968) Auxin movement in corn coleoptiles. Planta 82:123–144

Kirschner H, Sachs T, Fahn A (1971) Secondary xylem reorientation as a special case of vascular tissue differentiation. Isr J Bot 20:184–198

Leopold AC (1961) The transport of auxin. In: Ruhland W (ed) Encyclopedia of plant physiology Vol 14. Springer, Berlin Göttingen Heidelberg, pp 671–682

Liese W (1965) The fine structure of bordered pits in softwoods. In: Côté WA Jr (ed) Cellular ultrastructure of woody plants. Syracuse Univ Press, Syracuse, pp 271–290

Milborrow BV (1978) Abscisic acid. In: Letham DS, Goodwin PB, Higgins TJV (eds) Phytohormones and related compounds: A comprehensive treatise Vol 1. Elsevier/North-Holland, Amsterdam Oxford New York, pp 295–347

Munro M, Crick FHC (1971) The time needed to set up a gradient: detailed calculations. In: Davies DD, Balls M (eds) Control mechanisms of growth and differentiation. Symp Soc Exp Biol 25:439–453

Newman IA (1963) Electric potentials and auxin translocation in *Avena*. Austr J Biol Sci 16:629–646

Nix LE, Wodzicki TJ (1974) The radial distribution and metabolism of IAA-^{14}C in *Pinus echinata* stems in relation to wood formation. Can J Bot 52:1349–1355

Pavlidis T (1971) Populations of biochemical oscillators as circadian clocks. J Theor Biol 33:319–338

Polevoy VV (1972) Physiology and biochemistry of auxin effects on the growth of plant cells (in Russian). In: Mekhanyzmy reghulyatornykh processov, Trudy Peterghofskogho Byologhicheskogo Instituta, 21:191–207 (Izdatyelstvo Leningradskogho Universiteta, Leningrad)

Porandowski J, Rakowski K, Wodzicki TJ (1982) Apical control of xylem formation in pine stems. II. Responses of differentiating tracheids. Acta Soc Bot Pol 51:187–201

Sachs T (1969) Polarity and induction of organized vascular tissues. Ann Bot 33:263–275

Schmid R (1965) The fine structure of pits in hardwoods. In: Côté Jr (ed) Cellular ultrastructure of woody plants. Syracuse Univ Press, Syracuse, pp 291–304

Shen-Miller J (1973a) Rhythmicity in the basipetal transport of indolacetic acid through coleoptiles. Plant Physiol 51:615–619

Shen-Miller J (1973b) Rhythmic differences in the basipetal movement of indoleacetic acid between separated upper and lower halves of geotropically stimulated corn coleoptiles. Plant Physiol 52:166–170

Thair BW, Steeves TA (1976) Response of the vascular cambium to reorientation in patch grafts. Can J Bot 54:361–373

Thimann KV (1972) The natural plant hormones. In: Steward FC (ed) Plant physiology: A treatise Vol VI B. Physiology of growth and development. Academic Press, London New York, pp 3–332

Thomson K-S, Hertel R, Müller S (1973) 1-N-naphthylphthalamic acid and 2,3,5-triiodobenzoic acid: In-vitro binding to particulate cell fractions and action on auxin transport in corn coleoptiles. Planta 109:337–352

Wardrop AB (1964a) The structure and formation of the cell wall in xylem. In: Zimmermann MH (ed) The formation of wood in forest trees. Academic Press, London New York, pp 87–134

Wardrop AB (1964b) The reaction anatomy of arborescent angiosperms. In: Zimmermann MH (ed) The formation of wood in forest trees. Academic Press, London New York, pp 405–456

Wardrop AB (1965) The formation and function of reaction wood. In: Côté WA Jr (ed) Cellular ultrastructure of woody plants. Syracuse Univ Press, Syracuse, pp 371–390

Wilson BF, Wodzicki TJ, Zahner R (1966) Differentiation of cambial derivatives: Proposed terminology. For Sci 12:438–440

Winfree AT (1967) Biological rhythms and the behavior of populations of coupled oscillators. J Theor Biol 16:15–42

Wodzicki T (1964) Photoperiodic control of natural growth substances and wood formation in larch (*Larix decidua* D.C.). J Exp Bot 15:584–599

Wodzicki TJ (1965) Annual ring of wood formation and seasonal changes of natural growth-inhibitors in larch. Acta Soc Bot Pol 34:117–151

Wodzicki TJ (1968) On the question of occurrence of indole-3-acetic acid in *Pinus silvestris* L. Am J Bot 55:564–571

Wodzicki TJ (1971) Mechanism of xylem differentiation in *Pinus silvestris* L. J Exp Bot 22:670–687

Wodzicki TJ (1978) Seasonal variation of auxin in stem cambial region of *Pinus silvestris* L. Acta Soc Bot Pol 47:225–231

Wodzicki TJ (1980a) Oscillatory-wave mechanism of polar transport of auxin and plant morphogenesis (in Polish). In: II Ogólnopolska Konferencja: Mechanizmy Regulacji Morfogenezy Roślin. Warsaw Agric Univ, Rogów, pp 67–74

Wodzicki TJ (1980b) Control of cambial activity. In: Little CHA (ed) Control of shoot growth in trees. Proc Workshop IUFRO Working Parties on Xylem and Shoot Growth Physiol (Fredericton, NB, Canada, July 1980), pp 173–183

Wodzicki TJ, Brown CL (1973) Organization and breakdown of the protoplast during maturation of pine tracheids. Am J Bot 60:631–640

Wodzicki TJ, Wodzicki AB (1973) Auxin stimulation of cambial activity in *Pinus silvestris*. II. Dependence upon basipetal transport. Physiol Plant 29:288–292

Wodzicki TJ, Wodzicki AB (1980) Seasonal abscisic acid accumulation in stem cambial region of *Pinus silvestris* and its contribution to the hypothesis of a late-wood control system in conifers. Physiol Plant 48:443–447

Wodzicki TJ, Wodzicki AB (1981) Modulation of the oscillatory system involved in polar transport of auxin by other phytohormones. Physiol Plant 53:176–180

Wodzicki TJ, Zajączkowski S (1974) Effect of auxin on xylem tracheid differentiation in decapitated stems of *Pinus silvestris* L. and its interaction with some vitamins and growth regulators. Acta Soc Bot Pol 43:129–148

Wodzicki TJ, Wodzicki AB, Zajączkowski S (1979) Hormonal modulation of the oscillatory system involved in polar transport of auxin. Physiol Plant 46:97–100

Wodzicki TJ, Rakowski K, Starck Z, Porandowski J, Zajączkowski S (1982) Apical control of xylem formation in pine stems. I. Auxin effects and availability of assimilates. Acta Soc Bot Pol 51:187–201

Wolpert L (1969) Positional information and the spatial pattern of cellular differentiation. J Theor Biol 25:1–47

Zajączkowski S (1969) Xylem formation in isolated stem segments of *Pinus silvestris* L. Acta Soc Bot Pol 38:671–675

Zajączkowski S (1973) Auxin stimulation of cambial activity in *Pinus silvestris* I. The differential cambial response. Physiol Plant 29:281–287

Zajączkowski S (1978) Cambial activity and the complex of natural growth regulators in *Pinus silvestris* L. (in Polish with English summary) Zesz Nauk Szk Gł Gospod Wiejsk A R 107, Warsaw

Zajączkowski S (1980) Functioning of the morphogenetic field associated with the polar transport of auxin (in Polish). Proc: II Ogólnopolska konferencja— Mechanizmy regulacji morfogenezy roślin. Warsaw Agric Univ, Rogów, pp 75–78

Zajączkowski S, Romberger JA (1978) Polarity of xylem formation in isolated stem segments of *Pinus silvestris*. Physiol Plant 44:175–180

Zajączkowski S, Wodzicki TJ (1975) Inhibition and requirement of natural stimulator for cambial xylem production in isolated stem segments of *Pinus silvestris*. Physiol Plant 33:71–74

Zajączkowski S, Wodzicki TJ (1978a) On the question of stem polarity with respect to auxin transport. Physiol Plant 44:122–126

Zajączkowski S, Wodzicki TJ (1978b) Auxin and plant morphogenesis— A model of regulation. Acta Soc Bot Pol 47:233–243

Zajączkowski S, Wodzicki TJ (1978c) Auxin control of plant morphogenesis— A model of regulation. Abstracts, Inaugural Meet Fed Eur Soc Plant Physiol, Edinburgh, p 398

Zajączkowski S, Wodzicki TJ (1980) Plant morphogenetic field as an oscillatory system coupled to the phenomena of polarity and auxin stimulation. Abstracts, Fed Eur Soc Plant Physiol 2nd Congr, Santiago de Compostela, Spain, pp 729–730

Zajączkowski S, Wodzicki TJ, Bruinsma J (1983) A possible mechanism for whole-plant morphogenesis. Physiol Plant 57:306–310

Zakrzewski J (1983) Hormonal control of cambial activity and vessel differentiation in *Quercus robur*. Physiol Plant 57:537–542

Author Index

Page numbers in *italics* refer to the references

Aasheim T, Iversen T-H *126*
Abe H, Uchiyama RM, Sato R, Muto S 58, *62*
Abeles FB 13, *17*, 29, 60, *62*, *231*
Abeles FB, Lonski J 32, *62*
Abeles FB, Rubinstein B 29, *62*
Abeles FB, see Craker LE 6, *18*, 32, *65*
Abrol BK, Audus LJ 122, *126*
Acaster MA, see Kende H 61, *70*
Acton GJ, Murray PB 31, 32, *62*
Adams D, see Ohlrogge JB 11, *21*
Adams DO, Yang SF 29, *62*
Adams DO, see Yang SF *232*
Adams DO, see Yu Y-B 29, *79*
Adams P, see Jacobs WP 7, *19*, 152, *169*
Adams PA, Montague MH, Tepfer M, Rayle DL, Ikuma H, Kaufman PB 49, *62*
Adams PA, see Kaufman PB 27, 40, 45, 48, 49, *70*
Adato I, see McGlasson WB *228*, *230*, *231*
Addicott FT 101, *126*
Addicott FT, Lynch RS 16, *17*
Addicott FT, Lyon JL 23, *62*, 206, *208*
Addicott FT, see Robinson BJ 117, *143*
Aharoni N 33, *62*
Aharoni Y, see Goldschmidt EE 26, *67*
Aitchison PA, Yeoman MM 37, *62*
Albaum HG, Kaiser S, Nestler HA 115, *126*
Aldaba VC 160, *168*
Allard HA, see Garner WW 187, *211*
Aloni R 159, 160, 161, 162, 166, *168*, *228*
Aloni R, Jacobs WP 152, 155, 166, *168*
Aloni R, Sachs T 160, *168*
Aloni R, see Benayoun J 153, 158, *168*
Alpi A, Lorenzi R, Cionini P, Bennici A, D'Amato F 177, *208*
Alpi A, see Ceccarelli N 177, *209*
Alpi A, see Cionini P 177, *210*
Alpi A, see Lorenzi R 177, *213*
Altman A, Goren R 42, *62*
Altman A, Mozes R 119 *126*
Altman A, see Mozes R 119, *140*
Alvim R, Hewett EW, Saunders PF 120, 121, 122, *126*
Ammirato P 177, *208*
Amrhein N 10, *17*, 173, *208*
Anderson WP, see Mitchell JW 197, *213*
Andreae WA, Good NE 83, *126*
Andreae WA, Venis MA, Jursic F, Dumas T 24, 29, 53, *62*
Anker L 27, 28, 48, 50, *62*
Anonymous 42, *62*
Antcliff AJ, see Skene KGM 121, *144*
Antoszewski R, Lis E 100, *126*
Apelbaum A, Burg SP 41, *62*
Archer SA 53, 60, *62*
Arias I, Williams PM, Bradbeer JW 34, *62*
Arisz WH 114, *126*
Arnold GM, see Tompsett PB 189, *216*
Asakawa Y, Tamari K, Inoue K, Kaji J 119, *127*
Asmundson CM, see Shindy WW 119, *144*
Atack AV, see Sargent JA 24, *75*
Atal CK 160, *168*
Atkin RK, Barton GE, Robinson DK 34, *62*, 120, *127*
Atmar VT, Throneberry GO, Kuehn GD 61, *62*
Atsmon D, Lang A, Light EN 120, *127*
Audus LJ 4, 5, 9, 12, 13, *17*, 48, *62*, *226*
Audus LJ, see Abrol BK 122, *126*
Audus LJ, see Da Cruz GS 97, 100, 104, 116, 118, *130*
Audus LJ, see Kundu KK 54, *70*
Audus LJ, see Yeomans LM 82, 91, 125, *148*
Augier H 58, 59, *62*, *63*
Aung LH, see Wright RD 56, *79*
Aurich D, see Sembdner G 119, 120, *144*
Avanzi MG 43, *63*
Avery GS, Burkholder PR, Creighton H 162, *168*

Babaeva JA, see Butenko RG 182, *209*
Back A, Richmond AE 26, *63*
Back A, see Guttman R 36, *68*
Backus GE, Schrank AR 101, *127*
Bahre R 57, *63*
Bajaj YPS, Collins WB 198, *208*
Bajaj YPS, see Reinert J 181, 200, *215*
Baker DA, see Hall SM 121, *133*

Baker LR, see Rudich J 124, *143*
Baker RA, see Keitt GW 112, *137*
Balasimha D, Ram G, Tewari MN 24, *63*
Ball NG, Dyke IJ 55, *63*
Ballarin Denti A, see Marré E 102, *139*
Banas A, see Kentzer T 59, *70*
Bancilhon L, see Nozeran R 179, *214*
Bandurski RS 101, *127*
Bandurski RS, Schulze A 31, *63*
Bandurski RS, see Cohen JD 177, 204, *210*
Bandurski RS, see Epstein E 101, 122, *131*
Bandurski RS, see Hall PL 101, 122, *133*
Bandurski RS, see Nowacki J 8, *20*, 101, 122, *141*
Bandurski RS, see Ueda M 101, 122, *145*
Banerjee AB, see Guha AK 60, *68*
Banerji D, Laloraya MM 56, *63*
Banko TJ, Boe AA 120, *127*
Barea JM, Navarro E, Palomares A, Montoya E 36, 59, *63*
Barendse GWM 119, *127*
Barendse GWM, Gilissen HAM *232*
Barendse GWM, Rodrigues Pereira AS, Berkers PA, Driessen FM, Eyden-Emons A van, Linskens HF 198, *208*
Barker WG, Collins WB 199, *209*
Barkley GM, Evans ML 48, *63*
Barlow HWB, see Fulford RM 100, 101, *132*
Barlow PW 38, 39, 41, 43, 47, *63*
Barlow PW, Macdonald PDM 38, *63*
Barry AJ 101, *127*
Barry AJ, see Morris DA 100, 101, 113, *140*
Barton GE, see Atkin RK 34, *62*, 120, *127*
Baruah P, see Sharma PK 39, *75*
Basler E 100, *127*
Basler E, McBride R 100, *127*
Basrur PK, see Basrur SV 39, *63*
Basrur SV, Fletcher RA, Basrur PK 39, *63*
Batra MW, Edwards KL, Scott TK 51, *63*, 86, 100, 101, 115, 118, *127*
Batt S, Venis MA 116, *127*
Batt S, Wilkins MB, Venis MA 116, 117, *127*
Bayer MH 122, *127*
Bayliss MW 183, *209*
Bayliss MW, see Gould AR 35, *68*
Beasley CA, Ting IP 57, *63*
Beasley CA, see Dhindsa RS 57, *66*
Beever JE, Woolhouse HW 120, *127*
Beevers L, see Chin T-Y 28, 30, 42, *65*
Behl R, see Hartung W 92, 118, 126, *133, 134*
Behre K 178, *209*
Bellandi DM, Dörffling K 119, 125, *127*
Bellandi DM, see Dörffling K 91, 100, 119, 125, *131*
Benayoun J, Aloni R, Sachs T 153, 158, *168*
Benes J, see Chvojka L 119, *129*
Bennett JH, see White KL 25, *78*
Bennett PA, Chrispeels MJ 206, *209*
Bennici A, see Alpi A 177, *208*
Bennici A, see Cionini P 177, *210*
Bennici A, see Lorenzi R 177, *213*
Bentley-Mowat JA, Reid SM 58, 59, *63*
Ben-Zioni A, see Itai C 120, 121, *135*
Berg AR, see Gagianas AA 112, 113, *132*
Berg AR, see Salminen SO 25, *75*
Bergmann L 38, *63*
Berkers PA, see Barendse GWM 198, *208*
Berliner MD, see Wood NL 58, *78*
Bernhard SA, see Rosenberry TL 25, *75*
Bernier G 189, *209*
Berrie AMM, Robertson J 34, *64*
Beslow DT, Rier JP 153, *168*
Beth K, see Thimann KV 58, *77*
Bewley JD, Black M 204 *209*
Bewley JD, see Durley RC 23, *66*
Beyer A 81, *127*
Beyer E, see Morgan PW 28, *73*
Beyer EM 112, 113, *127*
Beyer EM, Johnson AL, Sweetser PB 112, *127*
Beyer EM Jr 14, *17*
Beyer EM Jr, Morgan PW 97, 107, *127*
Beyer EM, Morgan PW 11, *17*, 28, *64*
Biale J 6, *17*
Biro RL, Hunt ER Jr, Erner Y, Jaffe MJ 34, *64*
Bisalputra T, Esau K 176, *209*
Bittner S, see Goldschmidt EE 30, *67*
Bjerknes CA, see Van't Hof J 37, *78*
Black M, Vlitos AJ 31, *64*
Black M, see Bewley JD 204, *209*
Black MK, Osborne DJ 122, 124, *127*, 164, *168*
Black MK, see Osborne DJ 92, 101, 122, 123, 124, *141*
Blackman GE, see Little ECS 101, *139*
Blakely LM, Rodaway SJ, Hollen LB, Croker SG 52, *64*
Blakeslee AF, see Overbeek J van 200, *217*
Bloch R, see Sinnott EW 149, *170*
Blumenfeld A, Gazit S 42, *64*
Blundell JB, see Jackson GAD 199, *212*
Boe AA, see Banko TJ 120, *127*
Boeken G, Oostveldt P van 40, 47, *64*

Bogorad L, see Smith HJ 193, *216*

Bollag JM, Galun E 101, *127*

Bonamy PA, Dennis FG, Jr 207, *209*

Bonnemain JL 84, 100, 101, 117, 118, *127, 128*

Bonnemain JL, Bourbouloux A 98, 100, *128*

Bonnemain JL, see Bourbouloux A 84, 98, 100, 104, 117, 118, *128*

Bonnemain J-L, see Couillerot J-P 120, *130*

Bonnemain J-L, see Everat-Bourbouloux A 98, 100, *131*

Bonner H, see Bonner J 5, *17*

Bonner J, Bonner H 5, *17*

Bonner J, English J Jr 5, *17*

Bonnett HT Jr, Torrey JG 82, 92, 125, *128*

Booth A 183, *209*

Bopp M 57, *64*, 113, *128*

Bornman CH, see Staden J van 23, 56, *77*

Bose JC 101, *128*

Bose TK, Nitsch JP 195, *209*

Böttger M 101, *128*

Böttger M, see Dörffling K 91, 100, 119, 124, 125, *131*

Bottomley W, Kefford NP, Zwar JA, Goldacre PL 164, 166, *168*

Bouchet M, see Gaspar Th 52, *67*

Boudet AM, see Marigo G 112, *139*

Bouiellenne-Walrand M, see Gaspar Th 52, 53, *67*

Bouillienne R, see El Tannir Lomba J 190, *210*

Bourbouloux A, Bonnemain JL 84, 98, 100, 104, 117, 118, *128*

Bourbouloux A, Bonnemain JL, Vasseur J 84, 98, 100, 117, 118, *128*

Bourbouloux A, see Bonnemain JL 98, 100, *128*

Bowen MR, Hoad GV 120, *128*

Bowen MR, Wareing PF 100, 119, *128*

Bowen MR, Wilkins MB, Cane AR, McCorquodale I 104, 117, 118, 125, 126, *128*

Bowen MR, see Hoad GV 13, *19*, 121, *134*

Bowen MR, see Lenton JR 120, *138*

Boyer JS, see Meyer RF 47, *72*

Boyle WS, Evans JO 44, *64*

Boysen Jensen P 81, *128*

Bradbeer JW, see Arias I 34, *62*

Bradbeer JW, see Ross JD 34, *75*

Bradford KJ, Dilley DR 33, *64*

Bradford KJ, Yang SF 8, *17*, 220, *234*

Bralczyk J, Wielgat B, Wasilewska-Dabrowska LD, Kleczdowski K 59, *64*

Brantley BB, Warren GF 194, *209*

Braun AC 9, 17, *18*, 181, *209*

Braun J, Khan AA 207, *209*

Brauner L 81, 101, *128*

Breen CM, see Staden J van 58, *77*

Breen JJ, see Dedolph RR 103, *130*

Brenneman T, see Goldsmith MHM 100, 101, 117, *132*

Brian PW 57, 59, 61, *64*

Brian PW, Hemming HG 48, 56, *64*, 225, *234*

Briant RE, see Morris DA 51, *73*, 100, 117, *140*

Briarty LG, see Gunning BES 14, *19*

Bridges IG, Hillman JR, Wilkins MB 118, *128*

Briggs WR 31, *64*

Briggs WR, Steeves TA, Sussex IM, Wetmore RH 91, *128*

Briggs WR, see Scott TK 7, *21*, 31, *75*, 83, 91, 98, 99, 109, 122, *143*

Briggs WR, see Steeves TA 91, *145*, 160, *170*

Briggs WR, see Vanderhoef LN 9, *22*, 31, *78*

Brossard D, Tepper H 100, *128*

Brossard D, see Tepper HB 118, *145*

Broughton WJ 47, *64*

Brown AB, Cormack RGH 252, *259*

Brown BT, Johansen O, Katekar GF, Sasse WH 112, *128*

Brown BT, Johansen O, Sasse WHF 112, *128*

Brown CL, McAlpine RG, Kormanik PP 15, *18*

Brown CL, Wetmore RH 99, 113, 114, *129*

Brown CL, Wodzicki TJ 245, *260*

Brown CL, see Wodzicki TJ 255, *261*

Brown HT, Morris GH 4, *18*

Brown R, see Yeoman MM 46, *79*

Browning G 120, 121, *129*

Browning G, see Little CHA 163, *169*

Bruinsma J, see Zajaczkowski S 246, *262*

Bryant JA 35, *64*

Buggeln RG 58, *64*

Buggeln RG, Craig JS 58 *64*

Bukovac MJ, see Goren R 28, *68*

Bukovac MJ, see Hayashi F 13, *19*

Bukovac MJ, see Phatak SC 195, *214*

Bukovac MJ, see Rasmussen HP 101, *142*

Bukovac MJ, see Wittwer SH 190, 199, *218*

Bünning E 178, *209*

Bünning E, Reisener HJ, Weygand F, Simon H, Klebe JF 84, *129*

Burg EA, see Burg SP 10, 13, *18*, 28, 29, *64*, 194, *209*

Burg SP 24, 41, *64*, *232*

Burg SP, Burg EA 10, 13, *18*, 28, 29, *64*, 194, *209*

Burg SP, Dijkman MJ 29, *64*

Burg SP, see Apelbaum A 41, *62*

Burg SP, see Chadwick AV 24, 29, 53, *65*

Burg SP, see Kang BG 29, 41, *69*

Burkholder PR, see Avery GS 162, *168*

Burkiewicz K, see Kentzer T 59, *70*

Burris RH, see Dravnieks DE 27, *66*

Burrows WJ, Carr DJ 121, *129*, 177, 203, *209*
Burrows WJ, see Carr DJ 121, *129*
Burrows WJ, see Reid DM 120, *143*
Burström H 47, 51, 52, 53, *64*
Butcher DN, Street HE 52, *64*
Butenko RG, Strogonov BP, Babaeva JA 182, *209*
Butts JS, see Fang SC 101, *131*

Campbell RC, see Crane JC 199, *210*
Cande WZ, Goldsmith MHM, Ray PM 114, *129*
Cande WZ, Ray PM 113, 114, *129*
Cane AR, see Bowen MR 104, 117, 118, 125, 126, *128*
Cane AR, see Wilkins MB 91, 102, 103, *147*
Canny MJ 87, 88, *129*
Carlucci AF, see Provasoli L 58, 59, *74*
Carr AH, see Kessell RHJ 182, *212*
Carr DJ *226*
Carr DJ, Burrows WJ 121, *129*
Carr DJ, Reid DM 13, *18*, 120, 121, *129*
Carr DJ, Reid DM, Skene KGM 120, *129*
Carr DJ, see Burrows WJ 121, *129*, 177, 203, *209*
Carr DJ, see Drake G 91, 114, *131*
Carr DJ, see Reid DM 120, *143*
Carraway KL, see Skoog F 38, *76*
Case DB, see Jacobs WP 11, *19*, 164, *169*
Cass DD, Jensen WA 174, *209*
Cataldo DA, see Goldsmith MHM 100, 101, 117, *132*
Ceccarelli N, Lorenzi R, Alpi A 177, *209*
Chacko EK, Singh RN 199, *209*
Chadwick AV, Burg SP 24, 29, 53, *65*
Chailakhyan MK, see Lang A 188, *213*
Chailakhyan MKh 193, *209*
Chailakhyan MKh, Khlopenkova LP, Khazhakyan KhK 119, *129*
Chailakhyan MKh, see Lang A 5, *19*
Chakraverty R, see Sircar SM 160, *170*
Chalutz E, Lieberman M 60, *65*
Chang KD, see Muir RM 31, *73*
Chang YP, Jacobs WP 164, *168*
Chapman RW, Vance BD 58, *65*
Chapman RW, see Morris RO 120, *140*
Chatson KB, see Constabel F 41, *65*
Chau KW, see Thomas RL 199, *216*
Chen J, Meeuse BJD 5, *18*
Chenou E, Sossountzov L, Lefebvre M-F 100, 117, *129*
Cherry JH, Haystee R van 189, *209*
Child CM 244, *260*
Chin T-Y, Beevers L 28, 30, *65*
Chin T-Y, Lockhart JA 119, *129*
Chin T-Y, Meyer MM Jr, Beevers L 42, *65*
Ching TM, Fang SC 84 *129*
Choinski JS Jr, Trelease RN, Doman DC 177, *209*
Cholodny N 81, 125, *129*
Chrispeels MJ, see Bennett PA 206, *209*
Christie AE, Leopold AC 82, 85, 112, 113, 114, 115, *129*
Chvojka L, Stohr J, Hejmova L, Benes J 119, *129*
Cionini P, Bennici A, Alpi A, D'Amato F 177, *210*
Cionini P, see Alpi A 177, *208*
Cionini P, see Lorenzi R 177, *213*
Clark WG 83, 112, *129*
Clarkson DT, see Torrey JG *227*
Cleland CF 5, *18*
Cleland CF, see Phillips DA 121, 122, *142*
Cleland R 6, 16, *18*
Cleland R, Thompson M, Rayle DL, Purves WK 27, 28, 48, 49, *65*
Cleland RE 9, 15, *18*, 102, *130*
Cleland RE, Prins H, Harper R, Higinbotham N 102, *130*
Cline MG, see Rehm MM 23, 48, *75*
Clor MA 124, *130*
Clore WJ 199, *210*
Clowes FAL 38, 39, *65*
Cocking EC 180, *210*
Cohen D, Robinson JB, Paleg LG 124, *130*
Cohen ID, see Epstein E 101, 122, *131*
Cohen JD, Bandurski RS 177, 204, *210*
Cohen MH, see Goodwin BC 244, 248, 249, *260*
Colby SR 25, *65*
Colclasure GC, see Krul WR 91, *138*
Collins WB, see Bajaj YPS 198, *208*
Collins WB, see Barker WG 199, *209*
Conklin ME, see Overbeek J van 200, *217*
Conrad H, Saltman P, Eppley R 59, *65*
Considine PJ, Patching JW 60, *65*
Constabel F, Kurz WGW, Chatson KB, Kirkpatrick JW 41, *65*
Cook JC Jr, see Grove MD 5, *18*
Cooke RJ, Kendrick RE 31, *65*
Cooke RJ, Saunders PF 31, *65*
Cooke RJ, Saunders PF, Kendrick RE 31, *65*
Coombe BG 16, *18*, 195, 196, *210*, *227*
Corcoran MR, see Jacobson A 26, *69*
Cordes W 101, *130*
Corgan JN, Peyton C 32, *65*
Cormack RGH, see Brown AB 252, *259*

Cote P, see Walton DC 33, *78*
Coudron CA, see Zimmerman DC 5, *22*
Couillerot J-P, Bonnemain J-L 120, *130*
Cousson A, Tran Thanh Van K 191, 192, *210*
Craig JS, see Buggeln RG 58, *64*
Craker LE, Abeles FB 6, *18*
Craker LE, Abeles FB, Shropshire W 32, *65*
Crane JC 199, *210*
Crane JC, Primer PE, Campbell RC 199, *210*
Crane JC, see Rebeiz CA 199, *215*
Creanor J, see Mitchison JM 36, *73*
Creighton H, see Avery GS 162, *168*
Crick FHC 248, 249, *260*
Crick FHC, see Munro M 248, *260*
Croes AF, Dodemont HJ, Stumm C 43, *65*
Croker SG, see Blakely LM 52, *64*
Crozier A, Reid DM 119, *130*
Crozier A, see Reid DM 34, *75*, 120, *143*
Cummins WR, Kende H, Raschke K 119, *130*
Curtis RW 60, *65*
Curtis RW, see Ilag L 53, 60, *69*
Cutler H, see Phillips IDJ 5, *21*

Da Cruz GS, Audus LJ 97, 100, 104, 114, 116, 118, *130*
Dale JE, see Felippe GM 56, *67*
D'Amato A 183, *210*
D'Amato F, see Alpi A 177, *208*
D'Amato F, see Cionini P 177, *210*
D'Amato F, see Lorenzi R 177, *213*
Danielson J, see Jacobs WP 7, *19*, 152, *169*
Darimont E, Gaspar Th, Hofinger M 54, 55, *65*
Darwin C 81, *130*
Darwin F, see Darwind C 4, *18*
Darwind C, Darwin F 4, *18*
Dathe W, Rönsch H, Preiss A, Schade W, Sembdner G, Schreiber K 200, *210*
Datta KS, see Kumor S *230*
Davenport TL, Jordan WR, Morgan PW 97, *130*
Davenport TL, Morgan PW, Jordan WR 97, 101, *130*
Davey JE, Staden J van 7, *18*, 121, 122, *130*
Davey JE, see Staden J van 120, 121, *146*
Davidson D, MacLeod RD 40, *65*
Davidson D, see Friedberg SH 37, *67*
Davidson E, see Robards AW 164, *170*
Davidson O, see MacLeod RD 40, *72*
Davies CR, see Seth AK 11, *21*, 30, *75*, 124, *144*
Davies FS, Mau SC, Nooden LD 13, *18*
Davies LJ, Rappaport L 119, *130*
Davies PJ 93, 103, 113, 117, *130*
Davies PJ, Doro JA, Tarbox AW 51, *65*, 100, 118, *130*
Davies PJ, Mitchell EK 118, *130*
Davies PJ, Rubery PH 115, *130*
Davies PJ, see Mitchell EK 104, 118, *140*
Davis W, see Wareing PF 33, 54, *78*
Davison RM 120, *130*, 199, *210*
Davison RM, Young H 13, *18*, 120, *130*
Day BE 101, *130*
Dayanadan P, Hebard FV, Kaufman PB 48, 55, *65*
De Muth PJ, see Jacobs WP 126, *136*
Dedolph RR, Breen JJ, Gordon SA 103, *130*
Dedolph RR, see Naqvi SM 103, 114, *141*
DeGroote D, Muir RM 55, *66*
DeLa Torre C, see Gonzalez-Fernandez A 39, *67*
Delisle AL 150, *168*
DeMaggio AE 164, *168*
Denffer D von 188, *210*
Dennis FG Jr, Martin GC, Gaskin P, MacMillan J 207, *210*
Dennis FG, Jr, see Bonamy PA 207, *209*
Dennison DS *233*
Dhaliwal AS, Dhaliwal GK, Moloney RJ 58, *66*
Dhaliwal GK, see Dhaliwal AS 58, *66*
Dhindsa RS, Beasley CA, Ting IP 57, *66*
Dhuria HS, see Randhawa GS 199, *215*
Dietrich J, see Walker GWR 43, *78*
Diez JL, see Gonzalez-Fernandez A 39, *67*
Digby J, Wareing PF 46, *66*, 163, 164, *168*, 180, *210*
Digby J, see Wareing PF 162, *171*
Dijkman MJ, see Burg SP 29, *64*
Dilley DR, see Bradford KJ 33, *64*
Dilley DR, see Mayak S 29, *72*
Dimalla GG, Staden J van 28, *66*
Dimalla GGA, see Staden J van 121, *146*
Dodemont HJ, see Croes AF 43, *65*
Dohrmann U, Hertel R, Kowalik H 116, *131*
Dohrmann U, see Ray PM 117, *142*
Dolk HE 81, 88, *131*
Doman DC, see Choinski JS Jr 177, *209*
Dorée M, see Guern J 119, *133*
Dörffling K 206, *210*
Dörffling K, Bellandi DM, Böttger M, Lückel H, Menzer U 91, 100, 119, 125, *131*
Dörffling K, Böttger M 124, *131*
Dörffling K, see Bellandi DM 119, 125, *127*

Dörfler M, see Köhler K-H 33, *70*
Doro JA, see Davies PJ 51, *65*, 100, 118, *130*
Dostal R 4, *18*
Downing RD, see Khan AA 26, *70*
Drake G, Carr DJ 91, 114, *131*
Dravnieks DE, Skoog F, Burris RH 27, *66*
Driessen FM, see Barendse GWM 198, *208*
Du Buy HG, Olson RA 114, *131*
Dubucq M 55, *66*
Dugue N, see Miginiac E 98, 100, 117, 118, *140*
Dumas T, see Andreae WA 24, 29, 53, *62*
Dumbroff EB, see Enu-Kwesi L 34, *66*
Durand B 195, *210*
Durley RC, Bewley JD, Railton ID, Pharis RP 23, *66*
Durnam DJ, Stuart DA, Jones RL 47, 49, *66*
Dyer AF 45, *66*
Dyke IJ, see Ball NG 55, *63*

Eck P, see Mainland CM 199, *213*
Ecklund PR, see Moore TC 195, *214*
Edelmann J, see Schildknecht H 5, *21*
Edwards KL, Goldsmith MHM 115, *131*
Edwards KL, Scott TK 51, *66*
Edwards KL, see Batra MW 51, *63*, 86, 100, 101, 115, 118, *127*
Edwards ME, Miller JH 41, *66*
Eeuwens CJ, Schwabe WW 177, 200, 202, 203, 204, *210*, *227*, *228*, *229*, *230*
Eib M, see Gordon SA 84, *133*
Eilati SK, see Goldschmidt EE 26, *67*
El-Antably HMM, Larsen P 54, *66*
Elassar G, Rudich J, Palevitch D, Kedar N 199, *210*
Elema J, see Kögl F 27, *70*
El Hinnawy EL 193, *210*
Eliasson L 29, *66*, 100, *131*
Eliezer J, Morris DA 84, 100, 101, 114, 115, 116, *131*
Elisezer J, Morris DA *227*
Elliott MC 51, *66*
Elliott MC, Hall JF, Weston GD 41, *66*
Elliott MC, see Martin HV 100, *139*
El-Saidi MT 124, 126, *131*
El Tannir Lomba J, Parmentier A, Bouillienne R 190, *210*
Engelbrecht L, see Mothes K 124, *140*
English J Jr, see Bonner J 5, *17*
Enu-Kwesi L, Dumbroff EB 34, *66*
Eppley R, see Conrad H 59, *65*
Epstein E, Cohen ID, Bandurski RS 101, 122, *131*
Epstein E, see Tal M 223, 225, *231*, *235*
Erez A, see Tal M 223, 225, *231*, *235*
Erner Y, see Biro RL 34 *64*
Ernest LC, Valdovinos JG 28, *66*
Esashi Y, see Goto N 11, *18*, 24, 31, 50, *68*
Esau K 16, *18*, 166, *168*
Esau K, see Bisalputra T 176, *209*
Eschrich W 100, 101, 117, *131*
Etherton B 102, *131*
Evans JO, see Boyle WS 44, *64*
Evans LS, Tramontano WA 5, *18*
Evans LS, Van't Hof J 42, 43, *66*
Evans LS, see Tramontano WA 175, *217*
Evans LT 188, 190, *210*
Evans LT, see Knox RB 189, *212*
Evans ML 24, 48, 51, 52, 54, *66*
Evans ML, Schmitt MR 28, *67*
Evans ML, Simon M, Vesper MJ 28, *67*
Evans ML, see Barkley GM 48, *63*
Evans ML, see Gougler J 52, 54, *68*
Evans ML, see Hertel R 84, 123, 124, *134*
Evans ML, see McBride R 29, 52, 53, *72*
Evans ML, see Mulkey TJ 51, 53, *73*
Evenari M 205, *210*
Evenari M, see Lerner HR 206, *213*
Even-Chen Z, see Goldschmidt EE 30, *67*
Everat-Bourbouloux A, Bonnemain J-L 98, 100, *131*
Everett NP, see Wang TL 175, *217*
Eyden-Emons A van, see Barendse GWM 198, *208*

Faber E-R 82, 125, *131*
Fahn A, see Kirschner H 252, *260*
Fan V, see Filner B 102, *131*
Fang SC, Butts JS 101, *131*
Fang SC, see Ching TM 84, *129*
Feldman L *227*
Feldman LJ 38, 51, 54, *67*
Felippe GM, Dale JE 56, *67*
Fellenberg G 176, *210*
Ferroni A, see Marré E 102, *139*
Field RJ, Peel AJ 100, 119, *131*
Filner B, Hertel R, Steele Ch, Fan V 102, *131*
Fischer H 187, 188, *210*
Fitting H 81, *131*
Fletcher AM, see Tompsett PB 189, *216*
Fletcher RA, Zalik S 31, *67*, 101, *131*
Fletcher RA, see Basrur SV 39, *63*
Flippen-Anderson JL, see Grove MD 5, *18*
Flory R, see Hertel R 90, 92, 95, 96, 103, 107, 111, 112, 113, 122, *134*, 248, 251, *260*
Fonnesbech M 52, 56, *67*
Forman M, see Robinson BJ 117, *143*
Forrence LE, see Gaither DH 23, 54, *67*

Fosket DE 167, *168*, 175, *211*
Fosket DE, Short KC 37, 43, *67*
Fosket DE, see Short KC 36, *76*
Fosket DE, see Torrey JG 165, *171*
Fox JE, Weis JS 124, *131*
Frankland B, Wareing PF 7, *18*, 28, *67*
Fredrick JF *226*
Frenkel C 6, *18*
Frenkel C, see Janes HW 32, *69*
Frey-Wyssling A 255, *260*
Friedberg SH, Davidson D 37, *67*
Fries D, see Gaspar Th 52, *67*
Frissel MJ, see Veen H 116, 125, *146*
Frolova IA, see Lang A 5, *19*, 188, *213*
Fry SC, Wangermann E 111, 122, *131*
Frydman VM, MacMillan J 202, *211*
Fuchs Y, Lieberman M 10, *18*, 29, *67*, 195, *211*
Fuchs Y, see Gertman E 29, *67*
Fuente RK de la, Leopold AC 11, *18*, 90, 91, 92, 93, 94, 96, 103, 105, 107, 114, 115, 123, *130, 131*, 247, *260*
Fuente RK de la, see Leopold AC 114, 122, 123, *138*
Fujii T, see Kondo N 31, *70*
Fujii T, see Skoog F 38, *76*
Fujii T, see Suzuki T 54, *76*
Fujii T, see Yamaki T 100, *148*
Fulford RM, Quinlan JD, Lacey HJ, Barlow HWB 100, 101, *132*
Fuller GL, see Kriedemann PE 33, *70*
Funke H 246, *260*
Furuya M, see Imaseki H 32, *69*
Furuya M, see Wada M 46, *78*

Gagianas AA, Berg AR 112, 113, *132*
Gagianas AA, see Salminen SO 25, *75*
Gaither DH, Lutz DH, Forrence LE 23, 54, *67*
Gallori E, see Russo VEA 60, *75*
Galson EC, see Sondheimer E 206, 207, *216*
Galston AW, Hillman WS 83, *132*
Galston AW, see Okerse R 25, *74*
Galun E 195, *211*
Galun E, see Bollag JM 101, *127*
Gamburg KZ, Osharova LM 39, *67*
Garcia-Martinez JL, see Ohlrogge JB 11, *21*
Gardener G, see Shaw S 98, *144*
Gardner GM, see Kende H 117, *137*
Garner WW, Allard HA 187, *211*
Gaskin P, see Dennis FG Jr 207, *210*
Gaspar Th, Bouchet M, Fries D 52, *67*
Gaspar Th, Xhaufflaire A 24, 27, 54, *67*
Gaspar Th, Xhaufflaire A, Bouiellenne-Walrand M 52, 53, *67*
Gaspar Th, see Darimont E 54, 55, *65*
Gausman HW, see Morgan PW 11, *20*, 28, *73*
Gauthert RJ 164, *168*
Gavrilova NS, see Svarinskaya RA 40, *76*
Gayadin AP, see Konings H 118, *138*
Gazit S, see Blumenfeld A 42, *64*
Gee H 117, *132*, 160, *168*
Gee H, Greyson RI 100, *132*
Geissler AE, see Katekar GF 112, *137*
Gengenbach B, see Rubenstein J 178, *215*
Gentile AC, Klein RM 60, *67*
Gepstein S, Thimann KV 32, *67*
Gerdes I, see Libbert E 124, *139*
Gertman E, Fuchs Y 29, *67*
Gibbons GSB, Wilkins MB 54, *67*
Gil GF, Martin GC, Griggs WH 199, *211*
Gilbert SF, see Jacobs M 85, 115, *135*
Gilissen HAM, see Barendse GWM *232*
Gillespie B, Thimann KV 84, *132*
Gimenez-Martin G, see Gonzalez-Fernandez A 39, *67*
Ginkel U, see Kaldewey H 91, 92, 93, 94, 103, 113, *136*, *137*
Gledhill BL, see Sims WL 194, *216*
Glinka Z *231*
Godlewski M 40, *67*
Godley GL, see Greenhalgh WJ 200, *211*
Goebel K 4, *18*
Goeschl JD, Pratt HK 32 *67*
Goeschl JD, see Pratt HK 41, *74*, *232*
Goethem G van, see Oostveldt P van 47, *77*
Goldacre PL, see Bottomley W 164, 166, *168*
Goldschmidt EE, Aharoni Y, Eilati SK, Riov JW, Monselise SP 26, *67*
Goldschmidt EE, Goren R, Even-Chen Z, Bittner S 30, *67*
Goldsmith MHM 9, *18*, 84, 85, 86, 87, 94, 95, 96, 97, 102, 103, 107, 109, 111, 114, 115, 116, 122, 125, *132*, *227*, 249, 251, 253, *260*
Goldsmith MHM, Cataldo DA, Karn J, Brenneman T, Trip P 100, 101, 117, *132*
Goldsmith MHM, Goldsmith TH 116, *132*
Goldsmith MHM, Goldsmith TH, Martin MH 96, 107, 116, *132*
Goldsmith MHM, Ray PM 96, 114, 115, *132*
Goldsmith MHM, Thimann KV 84, 91, 94, *132*
Goldsmith MHM, Wilkins MB 84, *132*
Goldsmith MHM, see Cande WZ 114, *129*
Goldsmith MHM, see Edwards KL 115, *131*

Goldsmith MHM, see Greenwood MS 91, 103, 111, 122, *133*
Goldsmith MHM, see Little CHA 114, *139*
Goldsmith MHM, see Sussman MR 117, *145*
Goldsmith TH, see Goldsmith MHM 96, 107, 116, *132*
Goldthwaite J, see Manos PJ 50, *72*
Gollnow BI, see Goodwin PB *227*, *228*, *229*, *230*, *232*
Gonzalez-Fernandez A, Diez JL, Gimenez-Martin G, DeLa Torre C 39, *67*
Good JEG, see Wareing PF 28, 34, *78*
Good NE, see Andreae WA 83, *126*
Goodwin BC, Cohen MH 244, 248, 249, *260*
Goodwin PB 178, 195, *211*, *227*, *228*, *229*, *230*, *231*, *232*
Goodwin PB, Gollnow BI, Letham DS *227*, *228*, *229*, *230*, *232*
Goodwin PB, see Letham DS 85, 86, 122, *138*, 219, 224, *226*, *229*
Goodwin RH 150, *168*
Gordon ME, Letham DS, Parker CW 119, *132*
Gordon ME, Wilson MM, Parker CW, Letham DS 119, *132*
Gordon SA, Eib M 84, *133*
Gordon SA, see Dedolph RR 103, *130*
Gordon SA, see Naqvi SM 31, *73*, 82, 91, 103, 106, 114, 122, 123, *141*
Goren R, Bukovac MJ 28, *68*
Goren R, see Altman A 42, *62*
Goren R, see Goldschmidt EE 30, *67*
Goren R, see Ilan I 7, *19*
Göring H, see Köhler K-H 33, *70*
Gorter CJ 82, 125, *133*
Goto N, Esashi Y 11, *18*, 24, 31, 50, *68*
Gougler J, Evans ML 52, 54, *68*
Gould AR, Bayliss MW, Street HE 35, *68*
Gould AR, see Wang TL 175, *217*
Graebe JE, Ropers HJ *227*, *230*, *233*
Grahm L 102, *133*
Grahm L, Hertz CH 101, 102, *133*
Green C, see Rubenstein J 178, *215*
Green PB 47, *68*
Green PR 15, *18*
Greene DW 121, *133*
Greenhalgh WJ, Godley GL 200, *211*
Greenwood MS, Goldsmith MHM 91, 103, 111, 122, *133*
Greenwood MS, Hillman JR, Shaw S, Wilkins MB 118, 126, *133*
Greenwood MS, see Ross SD 189, *215*
Gregory FG 188, *211*
Gregory FG, Hancock CR 82, 89, 98, 99, 102, 113, 122, *133*
Gregory FG, Veale JA 15, *18*
Greyson RI, see Gee H 100, *132*
Griggs WH, see Gil GF 199, *211*
Grochowska M, see Thimann KV 13, *21*
Grochowska MJ 100, *133*
Grotha R 40, *68*
Grove MD, Spencer GF, Rohwedder WK, Mandava N, Worley JF, Warthen JD Jr , Steffens GL, Flippen-Anderson JL, Cook JC Jr 5, *18*
Gruen HE 59, *68*
Guern J, Dorée M, Sadorge P 119, *133*
Guern J, Sadorge P 119, *133*
Guern J, see Leguay JJ 39 *71*
Guern J, see Pilet PE 124, *142*
Guernsey FS, see Leopold AC 82, 123, *138*
Guha AK, Banerjee AB 60, *68*
Gunning BES, Pate JS 14, *18*
Gunning BES, Pate JS, Briarty LG 14, *19*
Gunning BES, see Pate JS 14, *21*
Gunning BES, see Wick SM 175, *217*
Guo-qing T, see Tsui C 57, *77*
Gurley LR, Walters RA, Hildebrand CE, Ratliff RL, Hohmann PG, Tobey RA 173, *211*
Gurley LR, see Hohmann P 173, *211*
Gustafson FG 198, 199, *211*
Guttenberg H v., Zetsche K 82, 91, *133*
Guttenberg H v., see Reiff B 82, 91, 114, 122, *142*
Guttman R 39, *68*
Guttman R, Back A 36, *68*

Haber AH 45, 47, *68*
Haccius B 177, *211*
Haccius BZ 177, *211*
Hagen GL, Marcus A 47, *68*
Halevy AH, see Mayak S 29, 30, *72*
Halevy AH, see Rudich J 194, *215*
Hall JF, see Elliott MC 41, *66*
Hall MA, see Staden J van 33, *77*
Hall OF, see Leopold AC 115, 123, *138*
Hall PL, Bandurski RS 101, 122, *133*
Hall RH *229*
Hall SM, Baker DA 121, *133*
Hall SM, Medlow GC 120, 121, *133*
Hall WC 10, *19*
Halliday MBW, Wangermann E 101, *133*
Halloran B, see Russo VEA 60, *75*
Halperin W, see Minocha SC 167, *169*, 181, *213*
Hamilton RH, see Liao SH 117, *139*
Hammond JBW, see Wood DA 60, *78*
Hamner CL, see Hamner KC 186, *211*
Hamner KC, Lyon CB, Hamner CL 186, *211*
Hamzi HQ, see Skoog F 38, *76*

Hancock CR, see Gregory FG 82, 89, 98, 99, 102, 113, 122, *133*
Hanney CEA, see Wareing PF 162, *171*
Han-ping L, see Tsui C 57, *77*
Hare RC 83, *133*
Harel S 122, *133*
Harper R, see Cleland RE 102, *130*
Harper SHT, see Lynch JM 60, *71*
Harrison MA, see Walton DC 33, *78*
Hartnett CM, see Tramontano WA 175, *217*
Hartung W 92, 98, 100, 119, *133*
Hartung W, Behl R 92, 118, 126, *133*, *134*
Hartung W, Phillips IDJ 92, 104, 118, 125, 126, *134*
Hartung W, see Phillips IDJ 92, 124, *142*
Harvey BMR, see Reid DM 34, *75*, 120, *143*
Hasenstein K-H 96, 107, 116, *134*
Hasenstein K-H, Kaldewey H 96, 107, 116, *134*
Hashem MA, see Mohsen AF 58, *73*
Hashizume H 189, *211*
Havsankova J, see Roth J 8, 14, *21*
Hayashi F, Naito R, Bukovac MJ, Sell HM 13, *19*
Hayes AB, Lippincott JA 55, *68*
Haystee R van, see Cherry JH 189, *209*
Hayward HE 160, *168*
Heald JK, see Little CHA 163, *169*
Hebard FV, see Dayanadan P 48, 55, *65*
Hejmova L, see Chvojka L 119, *129*
Hejnowicz A, Tomaszewski M 164, *168*
Hejnowicz Z 244, 248, 252, *260*
Hejnowicz Z, Romberger JA 252, *260*
Helgeson JP, see Skoog F 38, *76*
Helm J 179, *211*
Hemberg T 27, 48, 49, *68*, 151, *168*
Hemberg T, Larsson U 24, 27, 50, *68*
Hemberg T, see Jindal KK 27, *69*
Hemming HG, see Brian PW 48, 56, *64*, 225, *234*
Hendricks SB, see Taylorson RB 206, *216*, *230*, *231* *233*
Henson IE, Wareing PF 33, *68*, *234*
Henson IE, Wheeler CT 120, *134*
Henson IE, see Wareing PF 33, 54, *78*
Hepler PK, see Torrey JG 165, *171*
Hermann RK, see Lavender DP 120, *138*
Hertel R 82, 84, 94, 103, 107, 113, 114, *134*
Hertel R, Evans ML, Leopold AC, Sell HM 84, 123, 124, *134*
Hertel R, Flory R 90, 92, 95, 96, 103, 107, 111, 112, 113, 122, *134*, 248, 251, *260*
Hertel R, Leopold AC 82, 84, 85, 94, 96, 103, 107, 112, 113, 114, 115, 118, 122, *134*
Hertel R, Thomson KS, Russo VEA 14, *19*
Hertel R, Thomson K-St, Russo VEA 116, *134*
Hertel R, see Dohrmann U 116, *131*
Hertel R, see Filner B 102, *131*
Hertel R, see Jacobs M 117, *135*
Hertel R, see Lembi CA 85, 116, *138*
Hertel R, see Ouitrakul R 114, *141*
Hertel R, see Ray PM 117, *142*
Hertel R, see Rayle DL 92, 96, 103, 107, *142*
Hertel R, see Thomson K-S 116, *145*, 249, *261*
Hertz CH 102, *134*
Hertz CH, see Grahm L 101, 102, *133*
Hertz CH, see Woodcock AER 102, *147*
Heslop-Harrison J 4, *19*, 190, 192, *211*
Hess T, Sachs T 164, *168*
Hewett EW 121, *134*
Hewett EW, Wareing PF 33, *68*, 120, 121, 122, *134*
Hewett EW, see Alvim R 120, 121, 122, *126*
Hewett EW, see Horgan R 120, 122, *134*
Heyn ANJ 16, *19*
Hield HZ, see Stewart WS 199, *216*
Higgins TJV, see Letham DS 85, 86, 122, *138*, 219, 224, *226*, *229*
Higinbotham M 102, *134*
Higinbotham N, see Cleland RE 102, *130*
Hild V, see Schurzmann M 54, *75*
Hildebrand CE, see Gurley LR 173, *211*
Hildebrandt AC, see Vasil IK 180, *217*
Hill AC, see White KL 25, *78*
Hill TA 172, *211*
Hiller LK, Kelly WC, Powell LE *233*
Hillman JR 12, *19*, *226*
Hillman JR, see Bridges IG 118, *128*
Hillman JR, see Greenwood MS 118, 126, *133*
Hillman JR, see Hocking TJ 119, *134*
Hillman JR, see Philipson JJ 27, 48, *74*
Hillman JR, see Weyers JDB *231*
Hillman SK, Phillips IDJ 91, *134*
Hillman SK, see Hoad GV 100, 101, 121, *134*
Hillman WS, see Galston AW 83, *132*
Hink J, see Šebánek J 124, *144*
Hiraki Y, Ota Y 34, *68*
Hiron RWP, Wright STC 33, *68*
Hiron RWP, see Wright STC 33, *79*
Hirshberg G, see Sabnis DD 117, *143*, 160, *170*

Ho THD 9, *19*
Ho THD, Nolan RC, Shute DE 14, *19*
Ho THD, Shih S, Kleinhofs A 14, *19*
Hoad GV 13, *19*, 120, 121, *134*, *233*
Hoad GV, Bowen MR 13, *19*, 121, *134*
Hoad GV, Hillman SK, Wareing PF 100, 101, 121, *134*
Hoad GV, see Bowen MR 120, *128*
Hocking TJ, Hillman JR, Wilkins MB 119, *134*
Hofinger M, see Darimont E 54, 55, *65*
Hohmann P, Robey RA, Gurley LR 173, *211*
Hohmann PG, see Gurley LR 173, *211*
Hollen LB, see Blakely LM 52, *64*
Hollis CA, Tepper HB 100, *134*
Honma S, see Phatak SC 195, *214*
Horgan JM, see Horgan R 120, 122, *134*
Horgan JM, see Purse JG 120, *142*
Horgan R, Hewett EW, Purse JG, Horgan JM, Wareing PF 120, 122, *134*
Horgan R, see Lorenzi R 7, *20*
Horgan R, see Purse JG 120, *142*
Horgan R, see Wareing PF 33, 54, *78*
Horton RF, see Osborne DJ 92, 101, 122, 123, 124, *141*
Horton RF, see Rauser WE 53, *75*
Houck DF, LaMotte CE 155, 156, 158, 159, *169*
Howard A, Pelc SR 35, *68*
Howlett FS 194, *211*, *212*
Hsiao TC *233*
Hsu CL, see Stewart JM *76*
Hsu F 177, *212*
Huber B, Schmidt E, Jahnel H 121, *135*
Hughes EWD, Street HE 45, *68*
Hugon E, see Pilet PE 124, *142*
Huisinga B 28, 31, *68*
Hui-ying C, see Tsui C 57, *77*
Humphries EC, Wheeler AW 16, *19*
Humphries EC, Wheller AW 55, *68*
Hunt ER Jr, see Biro RL 34, *64*
Hunt NH, see Johnson LP 173, *212*
Hurst V, see Jacobs WP 7, *19*, 152, *169*
Huxter TJ, Reid DM, Thorpe TA 32, *69*

Ikuma H, see Adams PA 49, *62*
Ilag L, Curtis RW 53, 60, *69*
Ilan I, Goren R 7, *19*
Imaseki H, Kondo K, Watanabe A 29, *69*
Imaseki H, Pjon C-J 50, *69*
Imaseki H, Pjon C-J, Furuya M 32, *69*
Imaseki H, see Kondo K 24, 27, 29, *70*
Imaseki H, see Sakai S 29, *75*
Imber D, see Tal M 121, *145*, 223, 225, *231*, *235*
Ingersoll RB, Smith OE 91, 125, *135*
Inoue K, see Asakawa Y 119, *127*
Itai C, Ben-Zioni A 120, 121, *135*
Itai C, Ben-Zioni A, Ordin L 120, 121, *135*
Itai C, Richmond A, Vaadia Y 120, *135*
Itai C, Vaadia Y 33, *69*, 120, 121, *135*
Itai C, see Sitton D 121, *144*
Itai C, see Tal M 121, *145*
Ivanova I, see Kudrov T *226*
Ivanova OA, see Romanova LV 34, *75*
Iversen T-H, see Aasheim T *126*
Ivonis IY 172, *212*
Iwami S, Masuda Y 104, 117, *135*
Iyengar DSP, see Schildknecht H 5, *21*

Jablonski JR, Skoog F 166, *169*
Jackson DI 199, *212*
Jackson GAD 199, 200, *212*
Jackson GAD, Blundell JB 199, *212*
Jackson MB, see Musgrave A 24, 50, *73*
Jacobs M, Gilbert SF 85, 115, *135*
Jacobs M, Hertel R 117, *135*
Jacobs M, Ray PM 102, *135*
Jacobs WP 11, 12, *19*, 82, 83, 85, 86, 99, 101, 111, 115, 122, 123, 124, 126, *135*, 149, 150, 151, 152, 153, 154, 157, 158, 164, 165, 166, *169*, *226*
Jacobs WP, Case DB 11, *19*, 164, *169*
Jacobs WP, Danielson J, Hurst V, Adams P 7, *19*, 152, *169*
Jacobs WP, De Muth PJ 126, *136*
Jacobs WP, Kaldewey H 124, *136*
Jacobs WP, McCready CC 91, 118, 122, *136*, 153, 158, *169*
Jacobs WP, McCready CC, Osborne DJ 101, *136*
Jacobs WP, Morrow IB 159, 166, *169*, 179, *212*
Jacobs WP, Pruett P 91, 118, 124, 126, *136*
Jacobs WP, see Aloni R 152, 155, 166, *168*
Jacobs WP, see Chang YP 164, *168*
Jacobs WP, see Kaldewey H 101, *137*
Jacobs WP, see Kirk SC 91, 125, *137*
Jacobs WP, see Koevenig JL 122, *137*
Jacobs WP, see Kruszewski SP 125, *138*
Jacobs WP, see LaMotte CE 154, 155, 156, 157, 166 *169*
Jacobs WP, see McCready CC 84, 91, 96, 103, 107, 122, 123, *139*, 152, *169*
Jacobs WP, see Sabnis DD 117, *143*, 160, *170*
Jacobs WP, see Scott TK 101, 122, *144*, 151, 153, 159, *170*
Jacobs WP, see Smith CW 82, 91, 103, 111, 112, 115, 122, 123, *145*, 164, *170*

Jacobs WP, see Thompson NP 152, 153, 154, 157, 158, 166, *171*
Jacobs WP, see Veen H 91, 112, 122, 123, 124, *146*
Jacobs WP, see Werblin TP 112, 122, 123, *147*
Jacobs WP, see Wetmore RH 101, *147*
Jacobsen JV, see Letham DS *229*
Jacobson A, Corcoran MR 26, *69*
Jacqmard A 40, *69*
Jaffe LF 15, *19*, 174, 175, *212*
Jaffe MJ, see Biro RL 34, *64*
Jaffee M 34, *69*
Jahnel H, see Huber B 121, *135*
Jamin D, see Stebbins GL 45, *76*
Janes HW, Loercher L, Frenkel C 32, *69*
Janistyn B 173, *212*
Jarvis WR, see Matson DT 6, *20*
Jeffs RA, Northcote DH 166, *169*
Jennings RC 59, *69*
Jensen WA, see Cass DD 174, *209*
Jensen WA, see Schulz SR 176, *215*
Jindal KK, Hemberg T 27, *69*
Johansen O, see Brown BT 112, *128*
John WW, see Lau O-L 25, 29, *70*, *71*
Johnson AL, see Beyer EM 112, *127*
Johnson LP, McLeod JK, Parker CW, Letham DS, Hunt NH 173, *212*
Johnsson A 102, *136*
Jones CM 199, *212*
Jones JF, see Kende H 61 *70*
Jones KL 194, *212*
Jones OP 121, 122, *136*
Jones OP, Lacey HJ 120, *136*
Jones R 40, 49, *69*
Jones RA, Kaufman PB 45, 49, *69*
Jones RA, see Kaufman PB 23, 24, 49, *70*
Jones RL 206, *212*, *227*, *230*, *231*, *234*
Jones RL, Lang A 7, *19*, 32, *69*, 124, *136*
Jones RL, see Durnam DJ 47, 49, *66*
Jones RL, see Phillips IDJ 120, *142*
Jones RL, see Stuart DA 49, *76*
Jones RL, see Taiz L 206, *216*
Jordan WR, Skoog F 27, *69*
Jordan WR, see Davenport TL 97, 101, *130*
Jost L 159, *169*
Jouanneau JP 36, 37, 39, *69*
Jouanneau JP, Tandeau De Marsac N 37, *69*, 173, *212*
Jouanneau JP, see Peaud-Lenoël C 36, 39,, *74*
Juniper BE *231*
Jura P, see Stebbins GL 45, *76*
Jursic F, see Andreae WA 24, 29, 53, *62*

Kadir GO, see Morris DA 84, 100, 101, 113, 118, *140*
Kadyrov ChSh, see Kefeli VI *231*
Kahn AA, Tao KL *232*
Kaiser S, see Albaum HG 115, *126*
Kaji J, see Asakawa Y 119,, *127*
Kaldewey H 82, 83, 86, 87, 88, 89, 90, 91, 92, 93, 98, 99, 102, 103, 106, 107, 108, 109, 110, 122, *136*
Kaldewey H, Ginkel U 91, *136*
Kaldewey H, Ginkel U, Wawczyniak G 92, 93, 94, 103, *137*
Kaldewey H, Ginkel U, Karmann R, Paland I 91, 93, 113, *137*
Kaldewey H, Ginkel U, Lehmann I, Seiwert R 91, *137*
Kaldewey H, Jacobs WP 101, *137*
Kaldewey H, Kraus L 94, 100, 103, 113, *137*
Kaldewey H, Vardar Y *226*
Kaldewey H, see Hasenstein K-H 96, 107, 116, *134*
Kaldewey H, see Jacobs WP 124, *136*
Kamerbeek GA, see Swart A 60, *76*
Kamisaka S 49, *69*
Kamisaka S, Sano H, Katsumi M, Masuda Y 49, *69*
Kamisaka S, see Shibata K 5, *21*
Kang BG, Burg SP 41, *69*
Kang BG, Newcomb W, Burg SP 29, *69*
Kang BG, Ray PM 32, *69*
Karanov Ed, see Kudrov T *226*
Karmann R, see Kaldewey H 91, 93, 113, *137*
Karmoker JL, Steninck RFM van *231*
Karmoker JL, Van Steveninck RFM *233*
Karn J, see Goldsmith MHM 100, 101, 117, *132*
Karnosky DP 182, *212*
Katekar GF 112, *137*
Katekar GF, Geissler AE 112, *137*
Katekar GF, see Brown BT 112, *128*
Kato J 124, *137*
Kato J, see Katsumi M 49, *70*
Kato J, see Ueda J 5, *22*
Kato K, see Nishi A 39, *73*
Katsumi M, Kazama H 25, 49, *70*
Katsumi M, Purves WK, Phinney BO, Kato J 49, *70*
Katsumi M, see Kamisaka S 49, *69*
Katsumi M, see Kazama H 27, 48, 49, *70*
Kaufman PB, Jones RA 23, 24, 49, *70*
Kaufman PB, Petering LB, Adams PA 27, 40, 45, 48, 49, *70*
Kaufman PB, see Adams PA 49, *62*
Kaufman PB, see Dayanadan P 48, 55, *65*
Kaufman PB, see Jones RA 45, 49, *69*
Kaufman PB, see Soni SL 49, *76*
Kawase M, see Leopold AC 14, *20*, 56, *71*
Kazama H, Katsumi M 27, 48, 49, *70*

Kazama H, see Katsumi M 25, 49, *70*
Kedar N, see Elassar G 199, *210*
Kedar N, see Rudich J 194, *215*
Kefeli VI, Kadyrov ChSh *231*
Kefford NP, Rijven MGC 14, *19*
Kefford NP, see Bottomley W 164, 166, *168*
Keitt GW, Baker RA 112, *137*
Kelly WC, see Hiller LK *233*
Kende H 13, *19*, 120, *137*
Kende H, Acaster MA, Jones JF, Metraux J-P 61, *70*
Kende H, Gardner GM 117, *137*
Kende H, Lang A 9, 15, *19*
Kende H, Sitton D 120, *137*
Kende H, see Cummins WR 119, *130*
Kende H, see Musgrave A 32, *73*
Kende H, see Sitton D 121, *144*
Kendrick RE, see Cooke RJ 31, *65*
Kentzer T, Libbert E 124, *137*
Kentzer T, Synak R, Burkiewicz K, Banas A 59, *70*
Kerridge GH, see Skene KGM 34, *76*, 121, *144*
Kessell RHJ, Carr AH 182, *212*
Khaleafa AF, see Mohsen AF 58, *73*
Khan AA 206, *212*
Khan AA, Downing RD 26, *70*
Khan AA, Tao K-L 206, *212*, *228*, *229*, *230*, *233*, *234*
Khan AA, see Braun J 207, *209*
Khan AA, see Kopcewicz J 51, *70*
Khazhakyan KhK, see Chailakhyan MKh 119, *129*
Khlopenkova LP, see Chailakhyan MKh 119, *129*
Kidwai P, see Robards AW 164, *170*
King RW 177, *212*
Kirk SC, Jacobs WP 91, 125, *137*
Kirkpatrick JW, see Constabel F 41, *65*
Kirschner H, Sachs T, Fahn A 252, *260*
Klämbt HD 121, *137*
Klebe JF, see Bünning E 84, *129*
Klebs G 186, 187, 188, *212*
Kleczdowski K, see Bralczyk J 59, *64*
Klein RM, see Gentile AC 60, *67*
Kleinhofs A, see Ho THD 14, *19*
Kluge M, Reinhard E, Ziegler H 121, *137*
Knox RB, Evans LT 189, *212*
Koevenig JL 124, *137*
Koevenig JL, Jacobs WP 122, *137*
Koevenig JL, Sillix D 122, 123, *137*
Kögl F, Elema J 27, *70*
Köhler D 40, *70*
Köhler K-H, Dörfler M, Göring H 33, *70*
Konar A, see MacKenzie IA 38, *72*
Kondo K, Watanable A, Imaseki H 24, 27, 29, *70*
Kondo K, see Imaseki H 29, *69*
Kondo N, Fujii T, Yamaki T 31, *70*
Kondo N, see Suzuki T 54, *76*
Kondo N, see Watanabe T 58, *78*
Konings H 100, 118, *138*
Konings H, Gayadin AP 118, *138*
Kopcewicz J, Khan AA 51, *70*
Kormanik PP, see Brown CL 15, *18*
Kotob MA, Schwabe WW 199, *212*
Kowalik H, see Dohrmann U 116, *131*
Kozlowski TT 16, *19*, 101, *138*
Kraus EJ, Kraybill HR 15, *19*, 187, 188, *212*
Kraus L, see Kaldewey H 94, 100, 103, 113, *137*
Kraybill HR, see Kraus EJ 15, *19*, 187, 188, *212*
Krelle E, Libbert E 112, *138*
Kriedemann PE, Loveys BR, Fuller GL, Leopold AC 33, *70*
Kriedemann PE, see Leopold AC 196, *213*, *226*
Kriedemann PE, see Loveys BR 6, *20*, 32, *71*
Krishnamorthy HN, see Nanda KK 190, *214*
Krizek DT, Mandava NB 193, *213*
Krul WR 92, 103, *138*
Krul WR, Colclasure GC 91, *138*
Kruszewski SP, Jacobs WP 125, *138*
Ku HS, Suge H, Rappaport L, Pratt HK 24, 50, *70*
Kubota T, see Shibata K 5, *21*
Kudrov T, Ivanova I, Karanov Ed *226*
Kuehn GD, see Atmar VT 61, *62*
Kulaeva OM, Tsibulya LV 57, *70*
Kulaeva ON 121, *138*
Kumamoto J, see Shindy WW 119, *144*
Kumar D, Wareing PF 183, *213*
Kumor S, Datta KS, Nauda KK *230*
Kundu KK, Audus LJ 54, *70*
Kunishi AT, see Lieberman M 10, *20*, 29, *71*, 221, *234*
Kuo CG, see Pharis RP 189, *214*
Kurosawa E 225, *234*
Kurtz E Jr, see Nitsch JP 194, *214*
Kurz WGW, see Constabel F 41, *65*
Kuse G 82, 112, *138*
Küster E 17, *19*
Kuzmanoff KM, see Mulkey TJ 51, 53, *73*

Lacey HJ, see Fulford RM 100, 101, *132*
Lacey HJ, see Jones OP 120, *136*
Lado P, see Marré E 102, *139*
Lagerstedt HB, Langston RG 27, *70*, 119, 124, *138*

Lahib G, see Takebe I 180, *216*
Laibach F 101, *138*
Laloraya MM, see Banerji D 56, *63*
Laloraya MM, see Narain A 56, *73*
Lam SL, see Leopold AC 91, 92, 96, 122, 123, *138*
Lamb CJ 117, *138*
LaMotte CE, Jacobs WP 154, 155, 156, 157, 166, *169*
LaMotte CE, see Houck DF 155, 156, 158, 159, *169*
Lance-Nougarède A, see Nitsch JP 39, 43, *73*
Lang A 9, *19*, 57, *70*
Lang A, Chailakhyan MK, Frolova IA 188, *213*
Lang A, Chailakhyan MKh, Frolova IA 5, *19*
Lang A, see Atsmon D 120, *127*
Lang A, see Jones RL 7, *19*, 32, *69*, 124, *136*
Lang A, see Kende H 9, 15, *19*
Lang A, see Michniewicz M 190, *213*
Langenauer HD, see Webster PL 38, *78*
Langston RG, see Lagerstedt HB 27, *70*, 119, 124, *138*
Larsen P 97, *138*
Larsen P, see El-Antably HMM 54, *66*
Larsson U, see Hemberg T 24, 27, 50, *68*
Lata K, see Nanda KK 190, *214*
Lau O-L, John WW, Yang SF 25, 29, *70, 71*
Lau O-L, Yang SF 27, 29, *70*
Lavender DP, Sweet GB, Zaerr JB, Hermann RK 120, *138*
Lavender DP, see Sweet GB 120, *145*
Lawson VR, Weintraub RL 31, *71*
Le Page-Degivry MT 34, *71*
Le Roith D, see Roth J 8, 14, *21*
Leavitt RA, see Ries SK 5, *21*
Lee BO 61, *71*
Lee TT 27, *71*
Lefebvre M-F, see Chenou E 100, 117, *129*
Leguay JJ, Guern J 39, *71*
Lehmann I, see Kaldewey H 91, *137*
Lembi CA, Morré DJ, Thomson KSt, Hertel R 85, 116, *138*
Lenton JR, Bowen MR, Saunders PF 120, *138*
Lenton JR, Perry VM, Saunders PF 6, *19*
Leonard NJ, see Skoog F 38, *76*
Leopold AC 9, 10, 12, 14, *19, 20,* 80, 82, 83, 86, 112, 113, 125, *138*, 251, 252, *260*
Leopold AC, Fuente RK de la 114, 122, 123, *138*
Leopold AC, Guernsey FS 82, 123, *138*
Leopold AC, Hall OF 115, 123, *138*
Leopold AC, Kawase M 14, *20*, 56, *71*
Leopold AC, Kriedemann PE 196, *213*, *226*
Leopold AC, Lam SL 91, 92, 96, 122, 123, *138*
Leopold AC, see Christie AE 82, 85, 112, 113, 114, 115, *129*
Leopold AC, see Fuente RK de la 11, *18*, 90, 91, 92, 93, 94, 96, 103, 105, 107, 114, 115, 123, *130,* 131, 247, *260*
Leopold AC, see Hertel R 82, 84, 85, 94, 96, 103, 107, 112, 113, 114, 115, 118, 122, 123, 124, *134*
Leopold AC, see Kriedemann PE 33, *70*
Leopold AC, see Loveys BR 6, *20*, 32, *71*
Leopold AC, see Müller K 14, *20*
Leopold AC, see Niedergang-Kamien E 82, 112, 114, *141*
Leopold AC, see Noodén LD 6, 12, 16, *20*, 101, *141*, *228*, *229*, *230*, *231*, *232*
Leopold AC, see Rubinstein B 101, *143*
Leopold AC, see Scott PC 10, *21*
Leopold AC, see Thomson KSt 116, *145*
Lepp LW, Peel AJ 100, 101, 119, *138*
Lerner HR, Mayer AM, Evenari M 206, *213*
Lesniak MA, see Roth J 8, 14, *21*
Letham DS 5, 9, 10, *20*, 24, 56, 57, *71*, 164, *169*, *227*, *233*, *234*
Letham DS, Goodwin PB, Higgins TJV 85, 86, 122, *138*, 219, 224, *226*
Letham DS, Higgins TJV, Goodwin PB, Jacobsen JV *229*
Letham DS, see Goodwin PB *227*, *228*, *229*, *230*, *232*
Letham DS, see Gordon ME 119, *132*
Letham DS, see Johnson LP 173, *212*
Liao SH, Hamilton RH 117, *139*
Libbenga KR, Torrey JG 43, *71*
Libbert E, Gerdes I 124 *139*
Libbert E, Risch H 224, *234*
Libbert E, see Kentzer T 124, *137*
Libbert E, see Krelle E 112, *138*
Lieberman M *232*
Lieberman M, Kunishi AT 10, *20*, 29, *71*, 221, *234*
Lieberman M, see Chalutz E 60, *65*
Lieberman M, see Fuchs Y 10, *18*, 29, *67*, 195, *211*
Lieberman M, see Spalding DH 60, *76*
Liese W 255, *260*
Light EN, see Atsmon D 120, *127*
Lindoo SJ, see Noodén LD 16, *20*
Ling E, see Musgrave A 24, 50, *73*
Linskens HF, see Barendse GWM 198, *208*
Linthilhac PM, Vesecky TB 15, *20*
Lintilhac PM 46, *71*
Lippincott BB, Lippincott JA 55, *71*
Lippincott JA, see Hayes AB 55, *68*

Lippincott JA, see Lippincott BB 55, *71*
Lis E, see Antoszewski R 100, *126*
List A Jr 51, 52, *71*
Little CHA, Goldsmith MHM 114, *139*
Little CHA, Heald JK, Browning G 163, *169*
Little ECS, Blackman GE 101, *139*
Liu PBW, Loy JB 40, *71*
Liverman J, see Nitsch JP 194, *214*
Livne A, Vaadia Y 33, *71*
Lockhart JA 25, *71*
Lockhart JA, see Chin TY 119, *129*
Loeb J 4, *20*, 81, *139*
Loeffler JE, Overbeek J van 121, *139*
Loeffler JE, see Overbeek J van 56, *77*
Loehwing WF 194, *213*
Loeppky RN, see Skoog F 38, *76*
Loercher L, see Janes HW 32, *69*
Lonski J, see Abeles FB 32, *62*
Loomis RS, Torrey JG 165, *169*
Loomis RS, see Radin JW 53, *74*, 124, 126, *142*, 165, *170*
Loomis RS, see Torrey JG 165, *171*
Lorenzi R, Bennici A, Cionini P, Alpi A, D'Amato F 177, *213*
Lorenzi R, Horgan R, Wareing PF 7, *20*
Lorenzi R, see Alpi A 177, *208*
Lorenzi R, see Ceccarelli N 177, *209*
Loveys BR 11, *20*, 32, *71*
Loveys BR, Leopold AC, Kriedemann PE 6, *20*, 32, *71*
Loveys BR, Wareing PF 28, 29, 32, *71*
Loveys BR, see Kriedemann PE 33, *70*
Low VHK 23, 49, *71*
Loy JB 40, 47, *71*
Loy JB, Pollard JE 50, *71*
Loy JB, see Liu PBW 40, *71*
Lückel H, see Dörffling K 91, 100, 119, 125, *131*
Luckwill LC, Weaver P, MacMillan J 203, *213*
Luckwill LC, Whyte P 120, 121, *139*
Lukasik S 200, *213*
Lund EJ 83, 112, *139*
Lutz DH, see Gaither DH 23, 54, *67*
Lynch JM, Harper SHT 60, *71*
Lynch RS, see Addicott FT 16, *17*
Lyndon RF 40, 45, *71*
Lynn DG, see Tramontano WA 175, *217*
Lyon CB, see Hamner KC 186, *211*
Lyon CJ 56, *72*, 111, *139*
Lyon JL, see Addicott FT 23, *62*, 206, *208*

Macdonald PDM, see Barlow PW 38, *63*
MacKenzie IA, Konar A, Street HE 38, *72*
MacLeod RD 35, 37, 38, *72*
MacLeod RD, Davidson O 40, *72*
MacLeod RD, see Davidson D 40, *65*
MacMillan J, see Dennis FG Jr 207, *210*
MacMillan J, see Frydman VM 202, *211*
MacMillan J, see Luckwill LC 203, *213*
Mai G 101, 118, 122, *139*
Mainland CM, Eck P 199, *213*
Major DJ, see Rood S 194, *215*
Malik CP, Mehan M 23, 57, *72*
Mandava N, see Grove MD 5, *18*
Mandava NB 5, *20*
Mandava NB, Thompson MJ, Meudt WJ 5, *20*
Mandava NB, see Krizek DT 193, *213*
Mann LK 194, *213*
Manos GE 52, *72*
Manos PJ, Goldthwaite J 50, *72*
Manuel J, see Wareing PF 28, 34, *78*
Maravolo NC 111, *139*
Marcus A, see Hagen GL 47, *68*
Marigo G, Boudet AM 112, *139*
Marinos NG 50, *72*
Marlangeon RC 199, *213*
Marré E, Lado P, Ferroni A, Ballarin Denti A 102, *139*
Martin GC, see Dennis FG Jr 207, *210*
Martin GC, see Gil GF 199, *211*
Martin HV, Elliott MC, Wangermann E, Pilet PE 100, *139*
Martin MH, see Goldsmith MHM 96, 107, 116, *132*
Martin N, see Wilkins MB 114, *147*
Mason GM, see Zweig G 119, *148*
Mason MIR, see Overbeek J van 56, *77*
Masuda Y, Shimoda C, Takao N, Yanagishima N 59, *72*
Masuda Y, Yamamoto R 48, *72*
Masuda Y, see Iwami S 104, 117, *135*
Masuda Y, see Kamisaka S 49, *69*
Mathur RS, see Teaotia SS 199, *216*
Mathur SN, see Sorokin HP 165, *170*
Matson DT, Jarvis WR 6, *20*
Matsubara S 57, *72*
Matsuo E, Minohara Y, Takanashi S 194, *213*
Mau SC, see Davies FS 13, *18*
Mayak S, Dilley DR 29, *72*
Mayak S, Halevy AH 29, 30, *72*
Mayer AM, Poljakoff-Mayber A 205, *213*
Mayer AM, Shain Y 12, 16, *20*
Mayer AM, see Lerner HR 206, *213*
McAlpine RG, see Brown CL 15, *18*
McArthur ICS, Steeves TA 179, *213*

McBride R, Evans ML 29, 52, 53, *72*
McBride R, see Basler E 100, *127*
McComb AJ 119, *139*
McCorquodale I, see Bowen MR 104, 117, 118, 125, 126, *128*
McCorquodale I, see Wilkins MB 103, *147*
McCready CC 86, 87, 91, 114, 122, *139*
McCready CC, Jacobs WP 84, 91, 96, 103, 107, 122, 123, *139*, 152, *169*
McCready CC, see Jacobs WP 91, 101, 118, 122, *136*, 153, 158, *169*
McCready CC, see Osborne DJ 124, *141*
McGlasson WB, Wade DL, Adato I *228*, *230*, *231*
McGlasson WB, Wade DL, Adato I *232*
McIllrath WJ, see Smith HJ 193, *216*
McLeod JK, see Johnson LP 173, *212*
McMullan EE 189, *213*
McMullan EE, see Puritch GS 189, *215*
McNitt RE, see Shen-Miller J 47, *75*
McWha J 204, *213*
Meagher MD, see Puritch GS 189, *215*
Medlow GC, see Hall SM 120, 121, *133*
Meeuse BJD, see Chen J 5, *18*
Mehan M, see Malik CP 23, 57, *72*
Melchers G, see Takebe I 180, *216*
Melkonian M, Weber A 58, 59, *72*
Menary RC, Staden J van 121, *140*
Menary RC, see Staden J van 121, *146*
Menhenett R, Wareing PF 120, 121, *140*
Menzer U, see Dörffling K 91, 100, 119, 125, *131*
Mertz D 52, 53, *72*
Metraux J-P, see Kende H 61, *70*
Metwalli A, see Mohsen AF 58, *73*
Metzger JD, Zeevaart JAD 32, *72*
Meudt WJ, see Mandava NB 5, *20*
Meyer MM Jr, see Chin T-Y 42, *65*
Meyer RF, Boyer JS 47 *72*
Michael G, see Wagner H 121, *146*
Michniewicz M, Lang A 190, *213*
Miginiac E, Sossountzov L, Dugue N 98, 100, 117, 118, *140*
Milborrow BV 23, 24, 27, 30, 33, *72*, 121, 125, *140*, 206, *213*, *227*, *231*, *233*, *234*, 249, *260*
Milborrow BV, Robinson DR 33, *72*
Miller CO 202, *213*
Miller CO, see Skoog F 6, 10, *21*, 172, 181, 202, *216*
Miller JH, see Edwards ME 41, *66*
Miller JH, see Miller PM 57, *72*
Miller PM, Sweet HC, Miller JH 57, *72*
Milthorpe FL, Newton P 47, *73*
Miners J, see Phillips IDJ 5, 6, *21*, 32, *74*
Minina EG, Tylkina LG 194, *213*
Minocha SC 38, 42, *73*
Minocha SC, Halperin W 167, *169*, 181, *213*
Minohara Y, see Matsuo E 194, *213*
Mitchell EK, Davies PJ 104, 118, *140*
Mitchell EK, see Davies PJ 118, *130*
Mitchell JW, Skaggs DP, Anderson WP 197, *213*
Mitchinson GJ 116, *140*
Mitchison JM, Creanor J 36, *73*
Mittelheuser CJ, Steveninck RFM van 119, *140*
Moeschler H, see Schildknecht H 5, *21*
Mohr H, Noble A 32, *73*
Mohsen AF, Khaleafa AF, Hashem MA, Metwalli A 58, *73*
Molisch H 4, 15, *20*
Moloney RJ, see Dhaliwal AS 58, *66*
Mondal MH 49, *73*
Monselise SP, see Goldschmidt EE 26, *67*
Montague MH, see Adams PA 49, *62*
Montoya E, see Barea JM 36, 59, *63*
Monula MAK 177, *213*
Moore TC 5, *20*, *213*, *226*
Moore TC, Ecklund PR 195, *214*
Morath M 102, *140*
Morgan DG 112, *140*
Morgan DG, Söding H 112, *140*
Morgan PW *232*
Morgan PW, Beyer E, Gausman HW 28, *73*
Morgan PW, Gausman HW 11, *20*, 28, *73*
Morgan PW, see Beyer EM 11, *17*, 28, *64*
Morgan PW, see Beyer EM Jr 97, 107, *127*
Morgan PW, see Davenport TL 97, 101, *130*
Morré DJ, see Lembi CA 85, 116, *138*
Morris DA 100, 101, 122, *140*
Morris DA, Briant RE, Thomson PG 51, *73*, 100, 117, *140*
Morris DA, Kadir GO 84, 101, 118, *140*
Morris DA, Kadir GO, Barry AJ 100, 101, 113, *140*
Morris DA, Thomas AG 84, 100, 101, 104, 117, 118, *140*
Morris DA, Winfield PJ 100, 119, *140*
Morris DA, see Eliezer J 84, 100, 101, 114, 115, 116, *131*
Morris DA, see Elisezer J *227*
Morris DA, see Rowntree RA 100, 117, *143*, *228*
Morris GH, see Brown HT 4, *18*
Morris RO, Zaerr JB, Chapman RW 120, *140*
Morrow IB, see Jacobs WP 159, 166, *169*, 179, *212*

Most BH, see Scott TK 122, 123, 124, *144*
Mothes K 14, *20*, 124, *140*
Mothes K, Engelbrecht L 124, *140*
Mozes R, Altman A 119, *140*
Mozes R, see Altman A 119, *126*
Muir RM, Chang KD 31, *73*
Muir RM, see DeGroote D 55, *66*
Muir RM, see Sastry KSK 27, *75*
Mulkey TJ, Evans ML, Kuzmanoff KM 51, 53, *73*
Mulkey TJ, Kuzmanoff KM, Evans ML 51, 53, *73*
Müller K, Leopold AC 14, *20*
Müller S, see Thomson K-S 249, *261*
Müller S, see Thomson KSt 116, *145*
Mulligan DR, Patrick JW *230*
Mullins MG, see Osborne DJ 92, 114, *141*
Munro M, Crick FHC 248, *260*
Murakami Y 120, *140*
Murashige T, Nakano R 183, *214*
Murashige T, Skoog F 182, 192, *214*
Murofushi N, see Yamane H 202, *218*
Murphy GJP 117, *140*
Murray PB, see Acton GJ 31, 32, *62*
Musgrave A, Jackson MB, Ling E 24, 50, *73*
Musgrave A, Kende H 32, *73*
Muto S, see Abe H 58, *62*
Myers RM 101, *141*

Nadeau R, Rappaport L, Stolp CF 28, *73*
Nadeau R, see Stolp CF 28, *76*
Nagao M, see Ohwaki Y 118, 125, 126, *141*
Nagao M, see Wada S 9, *22*
Nagao MA, see Rubinstein B *229*
Nagl W 36, 40, 42, 43, *73*, 177, *214*
Nagl W, Rücker W 43, *73*
Nagl W, see Przybyllok T 177, *214*
Nair PKR, see Randhawa GS 199, *215*
Naito R, see Hayashi F 13, *19*
Nakano R, see Murashige T 183, *214*
Nakatani S, see Shimo S 59, *76*
Namekawa K, see Yamaki T 39, *79*
Nanda KK, Krishnamorthy HN, Toky KL, Lata K 190, *214*
Naqvi SM 84, 91, 106, 122, *141*
Naqvi SM, Dedolph RR, Gordon SA 103, 114, *141*
Naqvi SM, Gordon SA 31, *73*, 82, 91, 106, 114, 122, 123, *141*
Narain A, Laloraya MM 56, *73*
Narayanaswamy S 181, 182, *214*
Nash LJ, see Wilkins MB 92, 124, *147*
Nauda KK, see Kumor S *230*
Navarro E, see Barea JM 36, 59, *63*
Naylor AW 188, 192, 193, *214*
Naylor AW, see Stewart JW 205, *216*
Naylor JM, Simpson GM 10, *20*
Negi SS, Olmo HP 43, *73*, 195, *214*
Neidle EK 193, *214*
Nestler HA, see Albaum HG 115, *126*
Neville P, see Nozeran R 179, *214*
Newcomb W, see Kang BG 29, *69*
Newcomb W, see Syono K 43, *76*
Newman IA 82, 83, 87, 90, 93, 94, 95, 97, 99, 102, 103, 106, 109, 111, 112, *141*, 248, 251, *260*
Newton P, see Milthorpe FL 47, *73*
Newton RJ 42, 54, 56, *73*
Nickell LG 17, *20*, 172, 173, 200, *214*
Niedergang-Kamien E, Leopold AC 82, 112, 114, *141*
Niedergang-Kamien E, Skoog F 82, 112, *141*
Nielsen N 81, *141*
Nightingale GT 192, *214*
Nishi A, Kato K, Takahashi M, Yoshida R 39, *73*
Nishi A, Sugano N 45, *73*
Nishi A, see Yoshida R 120, *148*
Nishinari N, Yamaki T 39, *73*
Nitsch C 191, *214*
Nitsch C, Nitsch JP 190, 191, *214*
Nitsch C, see Nitsch JP 120, 121, *141*, 163, 164, *169*
Nitsch JP 121, *141*, 195, 196, 198, *214*
Nitsch JP, Kurtz E Jr, Liverman J, Went FW 194, *214*
Nitsch JP, Lance-Nougarède A 39, 43, *73*
Nitsch JP, Nitsch C 120, 121, *141*, 163, 164, *169*
Nitsch JP, see Bose TK 195, *209*
Nitsch JP, see Nitsch C 190, 191, *214*
Nitsch JP, see Vardjan M 10, *22*
Nix LE, Wodzicki TJ 247, 253, *260*
Noble A, see Mohr H 32 *73*
Nolan RC, see Ho THD 14, *19*
Noodén LD 7, 15, 16, *20*
Noodén LD, Leopold AC 6, 12, 16, *20*, 101, *141*, *228*, *229*, *230*, *231*, *232*
Noodén LD, Lindoo SJ 16, *20*
Noodén LD, Weber JA 5, 6, *20*
Nooden LD, see Davies FS 13, *18*
Northcote DH, see Jeffs RA 166, *169*
Northcote DH, see Roberts K 36, *75*
Nowacki J, Bandurski RS 8, *20*, 101, 122, *141*
Nozeran R, Bancilhon L, Neville P 179, *214*
Nyeki J 199, *214*

Odato I, see McGlasson WB *232*
Odhnoff C 165, *169*
Ohlrogge JB, Garcia-Martinez JL, Adams D, Rappaport L 11, *21*
Ohwaki Y, Tsurumi S, Nagao M 118, 125, 126, *141*
Ohwaki Y, see Tsurumi S 103, 104, 117, 118, *145*
Okerse R, Galston AW 25, *74*
Olatoye ST, see Staden J van 33, *77*
Olmo HP, see Negi SS 43, *73*, 195, *214*
Olson RA, see Du Buy HG 114, *131*
Ooi SC, see Thomas RL 199, *216*
Oostveldt P van, Goethem G van, Parijs R van 47, *77*
Oostveldt P van, see Boeken G 40, 47, *64*
Ordin L, see Itai C 120, 121, *135*
Oritani T, see Yoshida R 120, *148*
Osborn M, see Wick SM 175, *217*
Osborne DJ 58, *74*, 92, 112, 114, 122, 123, *141*, *232*
Osborne DJ, Black MK 124, *141*
Osborne DJ, Horton RF, Black MK 92, 101, 122, 123, 124, *141*
Osborne DJ, McCready CC 124, *141*
Osborne DJ, Mullins MG 92, 114, *141*
Osborne DJ, see Black MK 122, 124, *127*, 164, *168*
Osborne DJ, see Jacobs WP 101, *136*
Osborne DJ, see Sargent JA 24, *75*
Osborne DS 220, *228*, *231*, *233*, *234*
Osharova LM, see Gamburg KZ 39, *67*
Osmond G, see Skene KGM 121, *145*
Ota Y, see Hiraki Y 34 *68*
Ouitrakul R, Hertel R 114, *141*
Ouitrakul R, see Rayle DL 92, 96, 103, 107, *142*
Overbeek J van, Conklin ME, Blakeslee AF 200, *217*
Overbeek J van, Loeffler JE, Mason MIR 56, *77*
Overbeek J van, see Loeffler JE 121, *139*
Overbeek J van, see Weaver RJ 199, *217*

Paál A 81, *142*
Paddock EF, see Padmanabhan V 183, *214*
Padmanabhan V, Paddock EF, Sharp WR 183, *214*
Paland I, see Kaldewey H 91, 93, 113, *137*
Paleg LG 52, *74*
Paleg LG, see Cohen D 124, *130*
Palevitch D, see Elassar G 199, *210*
Palmer JH 24, 55, 56, *74*
Palomares A, see Barea JM 36, 59, *63*
Pandey IC, see Teaotia SS 199, *216*
Parijs R van, see Oostveldt P van 47, *77*
Parkash V, see Rijven AHGC 56, *75*
Parker CW, see Gordon ME 119, *132*
Parker CW, see Johnson LP 173, *212*
Parmentier A, see El Tannir Lomba J 190, *210*
Patching JW, see Considine PJ 60, *65*
Pate JS, Gunning BES 14, *21*
Pate JS, see Gunning BES 14, *18*, *19*
Patrick JW *228*
Patrick JW, Woodley DJ 103, *142*
Patrick JW, see Mulligan DR *230*
Pavlidis T 248, *260*
Peaud-Lenoël C 35, 37, *74*
Peaud-Lenoël C, Jouanneau JP 36, 39, *74*
Pedersen M 59, *74*
Peel AJ, see Field RJ 100, 119, *131*
Peel AJ, see Lepp LW 100, 101, 119, *138*
Pegg GF 60, *74*
Pelc SR, see Howard A 35, *68*
Penny D, see Penny P 220, *235*
Penny P, Penny D 220 *235*
Pernet JJ, Pilet PE 51, *74*, 118, *142*
Perry VM, see Lenton JR 6, *19*
Petering LB, see Kaufman PB 27, 40, 45, 48, 49, *70*
Peterson RL 165, *170*
Peyton C, see Corgan JN 32, *65*
Phan CT *232*
Pharis RP, Kuo CG 189, *214*
Pharis RP, see Durley RC 23, *66*
Pharis RP, see Rood S 194, *215*
Pharis RP, see Ross SD 189, *215*
Phatak SC, Wittwer SH, Honma S, Bukovac MJ 195, *214*
Philipson JJ 189, *214*
Philipson JJ, Hillman JR, Wilkins MB 27, 48, *74*
Phillips DA 42, *74*
Phillips DA, Cleland CF 121, 122, *142*
Phillips HL Jr, Torrey JG 38, *74*
Phillips IDJ 15, *21*, 34, *74*, *229*
Phillips IDJ, Hartung W 92, 124, *142*
Phillips IDJ, Jones RL 120, *142*
Phillips IDJ, Miners J, Roddick JG 5, 6, *21*, 32, *74*
Phillips IDJ, Vlitos AJ, Cutler H 5, *21*
Phillips IDJ, see Hartung W 92, 104, 118, 125, 126, *134*
Phillips IDJ, see Hillman SK 91, *134*
Phillips IDJ, see Wareing PF 190, *217*, *227*
Phillips R, see Rubenstein J 178, *215*
Phinney BO 23, *74*
Phinney BO, West CA 49, *74*
Phinney BO, see Katsumi M 49, *70*

Pickard BG, Thimann KV 84, *142*
Pieniazek J 42, *74*
Pilet PE 5, *21*, 24, 54, *74*, 82, 84, 91, 94, 107, 118, 122, 124, 125, *142*, *227*
Pilet PE, Guern J, Hugon E 124, *142*
Pilet PE, see Martin HV 100, *139*
Pilet PE, see Pernet JJ 51, *74*, 118, *142*
Pilet PE, see Rivier L 51, *75*, 118, *143*
Pjon C-J, see Imaseki H 32, 50, *69*
Plant Growth Substances 1973 *227*
Pluijm JE van der 174, *217*
Polevoy VV 251, *260*
Poljakoff-Mayber A, see Mayer AM 205, *213*
Pollard JE, see Loy JB 50, *71*
Pool RM, see Weaver RJ 199, *217*
Poole RJ, Thimann KV 115, *142*
Porandowski J, Rakowski K, Wodzicki TJ 245, *260*
Porandowski J, see Wodzicki TJ 245, *261*
Powell LE 7, *21*
Powell LE, see Hiller LK *233*
Pratt HK, Goeschl JD 41, *74*, *232*
Pratt HK, see Goeschl JD 32, *67*
Pratt HK, see Ku HS 24, 50, *70*
Preiss A, see Dathe W 200, *210*
Primer PE, see Crane JC 199, *210*
Prins H, see Cleland RE 102, *130*
Pritchard JB, see Scott TK 100, *144*
Provasoli L 59, *74*
Provasoli L, Carlucci AF 58, 59, *74*
Pruett P, see Jacobs WP 91, 118, 124, 126, *136*
Przybyllok T, Nagl W 177, *214*
Puritch GS, McMullan EE, Meagher MD, Simmons CS 189, *215*
Purse JG 121, *142*
Purse JG, Horgan R, Horgan JM, Wareing PF 120, *142*
Purse JG, see Horgan R 120, 122, *134*
Purves WK, see Cleland R 27, 28, 48, 49, *65*
Purves WK, see Katsumi M 49, *70*

Quatrano RS 175, *215*
Québédeaux B, Sweetser P, Rowell J 177, *215*
Quinlan JD, see Fulford RM 100, 101, *132*

Radin JW, Loomis RS 53, *74*, 124, 126, *142*, 165, *170*
Radley M 200, 201, 204, *215*, *228*, *230*, *231*
Raghavan V 176, *215*
Railton ID, Wareing PF 28, 32, *74*
Railton ID, see Durley RC 23, *66*
Railton ID, see Reid DM 28, *75*
Rakowski K, see Porandowski J 245, *260*
Rakowski K, see Wodzicki TJ 245, *261*
Ram G, see Balasimha D 24, *63*
Randhawa GS, Dhuria HS, Nair PKR 199, *215*
Rappaport L 41, *75*
Rappaport L, see Davies LJ 119, *130*
Rappaport L, see Ku HS 24, 50, *70*
Rappaport L, see Nadeau R 28, *73*
Rappaport L, see Ohlrogge JB 11, *21*
Rappaport L, see Stolp CF 28, *76*
Raschke K *233*
Raschke K, Zeevaart JAF 7, *21*
Raschke K, see Cummins WR 119, *130*
Rasmussen HP, Bukovac MJ 101, *142*
Ratliff RL, see Gurley LR 173, *211*
Rauser WE, Horton RF 53, *75*
Raven JA 85, 102, 115, *142*
Raven JA, see Smith FA 102, *145*
Ray PM 83, *142*
Ray PM, Dohrmann U, Hertel R 117, *142*
Ray PM, see Cande WZ 113, 114, *129*
Ray PM, see Goldsmith MHM 96, 114, 115, *132*
Ray PM, see Jacobs M 102, *135*
Ray PM, see Kang BG 32 *69*
Rayle DL, Ouitrakul R, Hertel R 92, 96, 103, 107, *142*
Rayle DL, see Adams PA 49, *62*
Rayle DL, see Cleland R 27, 28, 48, 49, *65*
Rebeiz CA, Crane JC 199, *215*
Rehm MM, Cline MG 23, 48, *75*
Reid DM, Burrows WJ 120, *143*
Reid DM, Carr DJ 120, *143*
Reid DM, Crozier A 34, *75*, 120, *143*
Reid DM, Crozier A, Harvey BMR 34, *75*, 120, *143*
Reid DM, Railton ID 28, *75*
Reid DM, see Carr DJ 13, *18*, 120, 121, *129*
Reid DM, see Crozier A 119, *130*
Reid DM, see Huxter TJ 32, *69*
Reid SM, see Bentley-Mowat JA 58, 59, *63*
Reiff B, Guttenberg H v. 82, 91, 114, 122, *142*
Reinders-Gouwentak CA 162, *170*
Reinert J, Bajaj YPS 181, *215*
Reinert J, Bajaj YPS, Zbell B 181, 200, *215*
Reinert J, see Tazawa M 182, *216*
Reinhard E, see Kluge M 121, *137*
Reinhold L 115, *143*, *228*, *233*, *234*
Reinholt L *232*
Reisener HJ 84, *143*
Reisener HJ, Simon H 84, *143*

Reisener HJ, see Bünning E 84, *129*
Richmond A, see Itai C 120, *135*
Richmond A, see Sitton D 120, *144*
Richmond AE, see Back A 26, *63*
Rier JP, see Beslow DT 153, *168*
Rier JP, see Wetmore RH 11, *22*, 166, *171*
Ries SK, Wert V, Sweeley CC, Leavitt RA 5, *21*
Rijven AHGC, Parkash V 56, *75*
Rijven AHGC, see Zwar JA 112, *148*
Rijven MGC, see Kefford NP 14, *19*
Riov JW, see Goldschmidt EE 26, *67*
Risch H, see Libbert E 224, *234*
Rivier L, Pilet PE 51, *75*, 118, *143*
Robards AW, Davidson E, Kidwai P 164, *170*
Roberts K, Northcote DH 36, *75*
Roberts LW 35, 36, *75*, 165, *170*, 181, *215*
Roberts LW, see Zobel RW 42, *79*
Robertson J, see Berrie AMM 34, *64*
Robey RA, see Hohmann P 173, *211*
Robinson BJ, Forman M, Addicott FT 117, *143*
Robinson DK, see Atkin RK 34, *62*, 120, *127*
Robinson DR, see Milborrow BV 33, *72*
Robinson JB, see Cohen D 124, *130*
Robinson PM, see Thomas TH 28, *77*
Rodaway SJ, see Blakely LM 52, *64*
Roddick JG, see Phillips IDJ 5, 6, *21*, 32, *74*
Rodrigues Pereira AS, see Barendse GWM 198, *208*
Rogozińska JH 164, *170*
Rohwedder WK, see Grove MD 5, *18*
Romanova LV, Ivanova OA, Tkachenko VN 34, *75*
Romberger JA, see Hejnowicz Z 252, *260*
Romberger JA, see Zajaczkowski S 245, *262*
Rönsch H, see Dathe W 200, *210*
Rood S, Pharis RP, Major DJ 194, *215*
Ropers HJ, see Graebe JE *227*, *230*, *233*
Rosenberry TL, Bernhard SA 25, *75*
Rosenzweig JL, see Roth J 8, 14, *21*
Ross JD, Bradbeer JW 34, *75*
Ross SD 189, *215*
Ross SD, Greenwood MS 189, *215*
Ross SD, Pharis RP 189, *215*
Roth J, Le Roith D, Shiloach J, Rosenzweig JL, Lesniak MA, Havsankova J 8, 14, *21*
Rowan KS, see Shepherd KR 163, *170*
Rowell J, see Québédeaux B 177, *215*
Rowntree RA, Morris DA 100, 117, *143*, *228*
Rubenstein J, Phillips R, Green C, Gengenbach B 178, *215*
Rubery PH 115, 117, *143*
Rubery PH, Sheldrake AR 85, 115, *143*
Rubery PH, see Davies PJ 115, *130*
Rubinstein B, Leopold AC 101, *143*
Rubinstein B, Nagao MA *229*
Rubinstein B, see Abeles FB 29, *62*
Rücker W 172, 184, 185, 186, 193, *215*
Rücker W, see Nagl W 43, *73*
Rudich J, Halevy AH, Kedar N 194, *215*
Rudich J, Sell HM, Baker LR 124, *143*
Rudich J, see Elassar G 199, *210*
Rudnicki R, Suszka B 207, *215*
Rummens FHA 26, *75*
Rünger W *215*
Russo VEA 60, 61, *75*
Russo VEA, Halloran B, Gallori E 60, *75*
Russo VEA, see Hertel R 14, *19*, 116, *134*

Sabnis DD, Hirshberg G, Jacobs WP 117, *143*, 160, *170*
Sachar RC 199, *215*
Sachs J 80, *143*, 187, *215*
Sachs T 154, 160, 167, *170*, 252, *260*
Sachs T, Thimann KV 124, *143*
Sachs T, see Aloni R 160 *168*
Sachs T, see Benayoun J 153, 158, *168*
Sachs T, see Hess T 164, *168*
Sachs T, see Kirschner H 252, *260*
Sachs T, see Thimann KV 225, *235*
Sachs TA *229*
Sadorge P, see Guern J 119, *133*
Sakai S, Imaseki H 29, *75*
Salminen SO, Gagianas AA, Berg AR 25, *75*
Saltman P, see Conrad H 59, *65*
Sano H, see Kamisaka S 49, *69*
Sargent JA, Atack AV, Osborne DJ 24, *75*
Sarma CM 56, *75*
Sasse WH, see Brown BT 112, *128*
Sasse WHF, see Brown BT 112, *128*
Sastry KSK, Muir RM 27, *75*
Sato R, see Abe H 58, *62*
Saunders P *229*, *230*
Saunders PF, see Alvim R 120, 121, 122, *126*
Saunders PF, see Cooke RJ 31, *65*
Saunders PF, see Lenton JR 6, *19*, 120, *138*
Saunders PF, see Wareing PF *231*, *233*, *234*
Schade W, see Dathe W 200, *210*
Schaffner JH 194, *215*
Schildknecht H, Iyengar DSP 5, *21*

Schildknecht H, Tausher B, Moeschler H, Edelmann J 5, *21*
Schmid R 255, *260*
Schmidt E, see Huber B 121, *135*
Schmitt MR, see Evans ML 28, *67*
Schneider EA, Wightman F *228*, *229*, *233*, *234*
Schneider EA, see Wightman F 125, *147*
Schnepf E 174, *215*
Schrank AR 83, 112, *143*
Schrank AR, see Backus GE 101, *127*
Schreiber K, see Dathe W 200, *210*
Schreiber K, see Sembdner G 119, 120, *144*
Schulz SR, Jensen WA 176, *215*
Schulze A, see Bandurski RS 31, *63*
Schurzmann M, Hild V 54, *75*
Schwabe WW 192, *215*
Schwabe WW, see Eeuwens CJ 177, 200, 202, 203, 204, *210*, *227*, *228*, *229*, *230*
Schwabe WW, see Kotob MA 199, *212*
Schwabe WW, see Smith DJ 121, *145*
Scott BIH 83, 102, 103, 112, *143*
Scott PC, Leopold AC 10, *21*
Scott TK 23, 51, 52, 54, *75*, 86, 125, *143*, *227*, *228*, *229*, *233*
Scott TK, Briggs WR 7, *21*, 31, *75*, 83, 91, 98, 99, 109, 122, *143*
Scott TK, Jacobs WP 101, 122, *144*, 151, 153, 159 *170*
Scott TK, Most BH 122, 123, 124, *144*
Scott TK, Pritchard JB 100, *144*
Scott TK, Wilkins MB 91, 125, *144*
Scott TK, see Batra MW 51, *63*, 86, 100, 101, 115, 118, *127*
Scott TK, see Edwards KL 51, *66*
Scott TK, see Wilkins MB 114, 125, *147*
Seagull RW, see Wick SM 175, *217*
Šebánek J, Hink J 124, *144*
Seigel N, see Vanderhoef LN 24, 27, 50, *78*
Seiwert R, see Kaldewey H 91, *137*
Sell HM, see Hayashi F 13, *19*
Sell HM, see Hertel R 84, 123, 124, *134*
Sell HM, see Rudich J 124, *143*
Sell HM, see Wittwer SH 199, *218*
Sembdner G, Weiland J, Aurich D, Schreiber K 119, 120, *144*
Sembdner G, see Dathe W 200, *210*
Sequeira L, Steeves TA 225, *235*
Seth A, Wareing PF 14, *21*
Seth AK, Davies CR, Wareing PF 11, *21*, 30, *75*, 124, *144*
Seth AK, see Thomas RL 199, *216*
Setterfield G 40, *75*
Setterfield G, see Wightman F *227*
Seubert E 81, *144*
Shain Y, see Mayer AM 12, 16, *20*
Shantz EM, see Stewart FC 6, *21*
Sharma PK, Baruah P 39 *75*
Sharp WR, see Padmanabhan V 183, *214*
Shaw S, Gardener G, Wilkins MB 98, *144*
Shaw S, Wilkins MB 91, 103, 104, 118, 125, 126, *144*
Shaw S, see Greenwood MS 118, 126, *133*
Shaw S, see Wilkins MB 114, *147*
Sheldrake AR 84, 91, 103, 104, 105, 112, 118, 120, 122, 123, *144*, 163, *170*, *228*
Sheldrake AR, see Rubery PH 85, 115, *143*
Shen-Miller J 94, 97, 103, 107, 108, 109, 111, *144*, 248, *261*
Shen-Miller J, McNitt RE, Wojciechowski M 47, *75*
Shepherd KR, Rowan KS 163, *170*
Shibata K, Kubota T, Kamisaka S 5, *21*
Shih S, see Ho THD 14, *19*
Shiloach J, see Roth J 8, 14, *21*
Shimo S, Nakatani S 59, *76*
Shimoda C, Yanagishima N 59, *76*
Shimoda C, see Masuda Y 59, *72*
Shimojo E, see Yamaki T 39, *79*
Shindy WW, Asmundson CM, Smith OE, Kumamoto J 119, *144*
Shininger TL 40, *76*, 164, 165, *170*, 178, *215*
Short KC, Tepfer DA, Fosket DE 36, *76*
Short KC, Torrey JG 38, *76*
Short KC, see Fosket DE 37, 43, *67*
Shropshire W, see Craker LE 32, *65*
Shu-huan L, see Tsui C 57, *77*
Shute DE, see Ho THD 14, *19*
Sillix D, see Koevenig JL 122, 123, *137*
Simard A 37, *76*, 173, *216*
Simmons CS, see Puritch GS 189, *215*
Simon H, see Bünning E 84, *129*
Simon H, see Reisener HJ 84, *143*
Simon M, see Evans ML 28, *67*
Simon S 149, *170*
Simpson GM, see Naylor JM 10, *20*
Simpson SF 37, *76*
Sims WL, Gledhill BL 194, *216*
Singh RN, see Chacko EK 199, *209*
Sinnott EW, Bloch R 149, *170*
Sircar SM, Chakraverty R 160, *170*
Sitton D, Itai C, Kende H 121, *144*

Sitton D, Richmond A, Vaadia Y 120, *144*
Sitton D, see Kende H 120, *137*
Skaggs DP, see Mitchell JW 197, *213*
Skene KGM 30, 38, 54, *76*, 120, 121, *144*
Skene KGM, Antcliff AJ 121, *144*
Skene KGM, Kerridge GH 34, *76*, 121, *144*
Skene KGM, Osmond G 121, *145*
Skene KGM, see Carr DJ 120, *129*
Skoog F *227*
Skoog F, Hamzi HQ, Sweykowska AM, Leonard NJ, Carraway KL, Fujii T, Helgeson JP, Loeppky RN 38, *76*
Skoog F, Miller CO 6, 10, *21*, 172, 181, 202, *216*
Skoog F, Tsui C 10, *21*
Skoog F, see Dravnieks DE 27, *66*
Skoog F, see Jablonski JR 166, *169*
Skoog F, see Jordan WR 27, *69*
Skoog F, see Murashige T 182, 192, *214*
Skoog F, see Niedergang-Kamien E 82, 112, *141*
Šmid N, Vardjan M 121, *145*
Smith AM 60, *76*
Smith AR, see Staden J van *228*
Smith CW, Jacobs WP 82, 91, 103, 111, 112, 115, 122, 123, *145*, 164, *170*
Smith DJ, Schwabe WW 121, *145*
Smith FA, Raven JA 102, *145*
Smith HJ, McIllrath WJ, Bogorad L 193, *216*
Smith OE, see Ingersoll RB 91, 125, *135*
Smith OE, see Shindy WW 119, *144*
Snow R 4, *21*
Söding H 81, 89, *145*, 163, 164, *170*, 180, *216*
Söding H, see Morgan DG 112, *140*
Sondheimer E, Galson EC, Tinelli E, Walton DC 207, *216*
Sondheimer E, Tzou DS, Galson EC 206, *216*
Soni SL, Kaufman PB 49 *76*
Sorokin HP, Mathur SN, Thimann KV 165, *170*
Sorokin S, see Wetmore RH 181, *217*
Sossountzov L, see Chenou E 100, 117, *129*
Sossountzov L, see Miginiac E 98, 100, 117, 118, *140*
Spalding DH, Lieberman M 60, *76*
Spencer GF, see Grove MD 5, *18*
Spencer MS *232*
Staaden J van, see Webb DP 223, *230*, *231*, *235*
Staden J van 121, 122, *146*
Staden J van, Bornman CH 23, 56, *77*
Staden J van, Breen CM 58, *77*
Staden J van, Davey JE 120, 121, *146*
Staden J van, Dimalla GGA 121, *146*
Staden J van, Menary RC 121, *146*
Staden J van, Olatoye ST, Hall MA 33, *77*
Staden J van, Smith AR *228*
Staden J van, Wareing PF 33, *77*
Staden J van, see Davey JE 7, *18*, 121, 122, *130*
Staden J van, see Dimalla GG 28, *66*
Staden J van, see Menary RC 121, *140*
Staden J van, see Webb DP *229*
Stahl CA, see Vanderhoef LN 24, 27, 48, 50, *78*
Stant MY 160, *170*
Starck Z, see Wodzicki TJ 245, *261*
Stark P 81, *145*
Stebbins GL 45, 46, *76*
Stebbins GL, Suryakant SS, Jamin D, Jura P 45, *76*
Steele Ch, see Filner B 102, *131*
Steeves TA, Briggs WR 91, *145*, 160, *170*
Steeves TA, Sussex IM 9, *21*, 179, *216*
Steeves TA, see Briggs WR 91, *128*
Steeves TA, see McArthur ICS 179, *213*
Steeves TA, see Sequeira L 225, *235*
Steeves TA, see Thair BW 252, *261*
Steffens GL, see Grove MD 5, *18*
Steninck RFM van, see Karmoker JL *231*
Steveninck RFM van, see Mittelheuser CJ 119, *140*
Stewart FC, Shantz EM 6 *21*
Stewart JM, Hsu CL *76*
Stewart JW, Naylor AW 205, *216*
Stewart WS, Hield HZ 199, *216*
Stoddart JL 190, *216*
Stohr J, see Chvojka L 119, *129*
Stolp CF, Nadeau R, Rappaport L 28, *76*
Stolp CF, see Nadeau R 28, *73*
Stowe B, Yamaki T 12, *21*
Street HE 51, *76*, 167, *170*
Street HE, see Butcher DN 52, *64*
Street HE, see Gould AR 35, *68*
Street HE, see Hughes EWD 45, *68*
Street HE, see MacKenzie IA 38, *72*
Street HE, see Wang TL 175, *217*
Strogonov BP, see Butenko RG 182, *209*
Stuart DA, Jones RL 49, *76*
Stuart DA, see Durnam DJ 47, 49, *66*
Stumm C, see Croes AF 43, *65*
Sugano N, see Nishi A 45, *73*
Suge H 24, 50, *76*
Suge H, see Ku HS 24, 50, *70*
Suryakant SS, see Stebbins GL 45, *76*
Sussex IM 176, *216*

Sussex IM, see Briggs WR 91, *128*
Sussex IM, see Steeves TA 9, *21*, 179, *216*
Sussex IM, see Wochok ZS 122, *147*
Sussex IM, see Yeung EC 177, *218*
Sussman MR, Goldsmith MHM 117, *145*
Suszka B, see Rudnicki R 207, *215*
Suttle JC, Zeevaart JAD *234*
Suzuki T, Kondo N, Fujii T 54, *76*
Svarinskaya RA, Gavrilova NS 40, *76*
Svensson SB 52, 53, 54, 55, *76*
Swart A, Kamerbeek GA 60, *76*
Sweeley CC, see Ries SK 5, *21*
Sweet GB, Zaerr JB, Lavender DP 120, *145*
Sweet GB, see Lavender DP 120, *138*
Sweet HC, see Miller PM 57, *72*
Sweetser P, see Québédeaux B 177, *215*
Sweetser PB, see Beyer EM 112, *127*
Sweykowska AM, see Skoog F 38, *76*
Symposia of Society for Experimental Biology XXXI *227*
Synak R, see Kentzer T 59, *70*
Syono K, Newcomb W, Torrey JG 43, *76*

Taiz L, Jones RL 206, *216*
Takahashi K 23, 24, 25, 50, *77*
Takahashi M, see Nishi A 39, *73*
Takahashi N, see Yamane H 202, *218*
Takanashi S, see Matsuo E 194, *213*
Takao N, see Masuda Y 59, *72*
Takebe I, Lahib G, Melchers G 180, *216*
Tal M, Imber D, Erez A, Epstein E 223, 225, *231*, *235*
Tal M, Imber D, Itai C 121, *145*
Tamari K, see Asakawa Y 119, *127*
Tanada T 26, 32, *77*
Tandeau De Marsac N, see Jouanneau JP 37, *69*, 173, *212*
Tao K-L, see Khan AA 206, *212*, *228*, *229*, *230*, *232*, *233*, *234*
Tarbox AW, see Davies PJ 51, *65*, 100, 118, *130*
Tausher B, see Schildknecht H 5, *21*
Tavares JE, see Thomson KSt 116, *145*
Taylor IEP, Wilkinson AJ 58, *77*
Taylorson RB, Hendricks SB 206, *216*, *230*, *231*, *233*
Tazawa M, Reinert J 182, *216*
Teaotia SS, Pandey IC, Mathur RS 199, *216*
Teltscherova L 28, 30, *77*
Tepfer DA, see Short KC 36, *76*
Tepfer M, see Adams PA 49, *62*
Tepper H, see Brossard D 100, *128*
Tepper HB, Brossard D 118, *145*
Tepper HB, see Hollis CA 100, *134*
Tewari MN, see Balasimha D 24, *63*
Thair BW, Steeves TA 252, *261*
Thimann KV 4, 5, 9, 16, *21*, 51, *77*, 86, *145*, *227*, 249, 252, *261*
Thimann KV, Beth K 58 *77*
Thimann KV, Grochowska M 13, *21*
Thimann KV, Sachs T 225, *235*
Thimann KV, Wardlaw IF 91, *145*
Thimann KV, see Gepstein S 32, *67*
Thimann KV, see Gillespie B 84, *132*
Thimann KV, see Goldsmith MHM 84, 91, 94, *132*
Thimann KV, see Pickard BG 84, *142*
Thimann KV, see Poole RJ 115, *142*
Thimann KV, see Sachs T 124, *143*
Thimann KV, see Sorokin HP 165, *170*
Thimann KV, see Thornton RM 90, 91, 103, 107, 111, *145*
Thimann KV, see Went F 4, *22*
Thimann KV, see Wickson M 10, *22*
Thimann KV, see Wightman F 125, *147*
Thomas AG, see Morris DA 84, 100, 101, 104, 117, 118, *140*
Thomas RL, Seth AK, Chau KW, Ooi SC 199, *216*
Thomas TH, Wareing PF, Robinson PM 28, *77*
Thomas TH, Wurr DCE 34, *77*
Thompson M, see Cleland R 27, 28, 48, 49, *65*
Thompson MJ, see Mandava NB 5, *20*
Thompson NP 154, 155, 156, 157, 158, 166, *171*
Thompson NP, Jacobs WP 152, 153, 154, 157, 158, 166, *171*
Thompson PA 198, 199, *216*
Thomson K-S, Hertel R, Müller S 249, *261*
Thomson KS, see Hertel R 14, *19*
Thomson KSt, Hertel R, Müller S, Tavares JE 116, *145*
Thomson KSt, Leopold AC 116, *145*
Thomson K-St, see Hertel R 116, *134*
Thomson KSt, see Lembi CA 85, 116, *138*
Thomson PG, see Morris DA 51, *73*, 100, 117, *140*
Thornton RM, Thimann KV 90, 91, 103, 107, 111, *145*
Thorpe TA, see Huxter TJ 32, *69*
Throneberry GO, see Atmar VT 61, *62*
Tiedjens VA 194, *216*

Tietz A 32, *77*
Tillberg E 31, 32, *77*
Tinelli E, see Sondheimer E 207, *216*
Ting IP, see Beasley CA 57, *63*
Ting IP, see Dhindsa RS 57, *66*
Tkachenko VN, see Romanova LV 34, *75*
Tobey RA, see Gurley LR 173, *211*
Toky KL, see Nanda KK 190, *214*
Tomaszewski M, see Hejnowicz A 164, *168*
Tompsett PB 189, *216*
Tompsett PB, Fletcher AM 189, *216*
Tompsett PB, Fletcher AM, Arnold GM 189, *216*
Torrey JG 38, 43, *77*, 165, *171*, 183, *216*, *228*, *229*, *231*, *233*, *234*
Torrey JG, Clarkson DT *227*
Torrey JG, Fosket DE, Hepler PK 165, *171*
Torrey JG, Loomis RS 165, *171*
Torrey JG, see Bonnett HT Jr 82, 92, 125, *128*
Torrey JG, see Libbenga KR 43, *71*
Torrey JG, see Loomis RS 165, *169*
Torrey JG, see Phillips HL Jr 38, *74*
Torrey JG, see Short KC 38, *76*
Torrey JG, see Syono K 43, *76*
Tramontano WA, Hartnett CM, Lynn DG, Evans LS 175, *217*
Tramontano WA, see Evans LS 5, *18*
Tran Thanh Van K, see Cousson A 191, 192, *210*
Trelease RN, see Choinski JS Jr 177, *209*
Trewavas A 35, *77*
Trewavas AJ 15, *21*, *22*, 206, 207, *217*
Trip P, see Goldsmith MHM 100, 101, 117, *132*
Tronchet A 56, *77*
Tsibulya LV, see Kulaeva OM 57, *70*
Tsivion Y, see Zamski E 118, *148*
Tsuboi M, Yanagishima N 43, *77*
Tsui C, Guo-qing T, Hui-ying C, Yan-ru S, Han-ping L, Zhe T, Shu-huan L, Xianzhang L 57, *77*
Tsui C, see Skoog F 10, *21*
Tsurumi S, Ohwaki Y 103, 104, 117, 118, *145*
Tsurumi S, Wada S 101, 118, *145*
Tsurumi S, see Ohwaki Y 118, 125, 126, *141*
Tukey HB, Young JO 196, 197, *217*
Tylkina LG, see Minina EG 194, *213*
Tzou DS, see Sondheimer E 206, *216*

Uchiyama RM, see Abe H 58, *62*
Ueda J, Kato J 5, *22*
Ueda M, Bandurski RS 101, 122, *145*

Vaadia Y, see Itai C 33, *69*, 120, 121, *135*
Vaadia Y, see Livne A 33, *71*
Vaadia Y, see Sitton D 120, *144*
Valdovinos JG, see Ernest LC 28, *66*
Van Steveninck RFM, see Karmoker JL *233*
Vance BD, see Chapman RW 58, *65*
Vanderhoef LN, Briggs WR 9, *22*, 31, *78*
Vanderhoef LN, Stahl CA 48, *78*
Vanderhoef LN, Stahl CA, Seigel N, Zeigler R 24, 27, 50, *78*
Van't Hof J 35, 36, 37, 38, 39, *77*, *78*
Van't Hof J, Bjerknes CA 37, *78*
Van't Hof J, see Evans LS 42, 43, *66*
Vardar Y 82, 86, 103, 113, *146*, *227*
Vardar Y, see Kaldewey H *226*
Vardjan M, Nitsch JP 10, *22*
Vardjan M, see Šmid N 121, *145*
Varner JE 206, *217*
Vasil IK 180, *217*
Vasil IK, Hildebrandt AC 180, *217*
Vasseur J, see Bourbouloux A 84, 98, 100, 117, 118, *128*
Veale JA, see Gregory FG 15, *18*
Veen H 92, 117, 123, 125, *146*
Veen H, Frissel MJ 116, 125, *146*
Veen H, Jacobs WP 91, 112, 122, 123, 124, *146*
Veldstra H, see Westringh C van de 115, *146*
Venis MA 9, 14, *22*, 117, *146*
Venis MA, see Andreae WA 24, 29, 53, *62*
Venis MA, see Batt S 116, 117, *127*
Vesecky TB, see Linthilhac PM 15, *20*
Vesper MJ, see Evans ML 28, *67*
Vince-Prue D 189, *217*, 223, *234*, *235*
Vlitos AJ, see Black M 31, *64*
Vlitos AJ, see Phillips IDJ 5, *21*
Vonk CR 121, *146*

Wada M, Furuya M 46, *78*
Wada S 9, *22*
Wada S, Nagao M 9, *22*
Wada S, see Tsurumi S 101, 118, *145*
Wade DL, see McGlasson WB *228*, *230*, *231*, *232*
Wagner H, Michael G 121, *146*
Wagner J 58, *78*
Wain RL, see Wilkins H 54, *78*
Walker GWR, Dietrich J 43, *78*
Walters RA, see Gurley LR 173, *211*
Walton DC 11, 13, *22*
Walton DC, Harrison MA, Cote P 33, *78*

Walton DC, see Sondheimer E 207, *216*
Wang TL, Everett NP, Gould AR, Street HE 175, *217*
Wangermann E 84, 91, 92, 113, 117, 118, 122, 123, *146*, 159, 160, *171*
Wangermann E, Withers LA 84, 104, 112, 113, 117, 118, 122, *146*
Wangermann E, see Fry SC 111, 122, *131*
Wangermann E, see Halliday MBW 101, *133*
Wangermann E, see Martin HV 100, *139*
Wardell WL 173, *217*
Wardlaw CW 40, *78*, 179, *217*
Wardlaw IF, see Thimann KV 91, *145*
Wardrop AB 255, 258, *261*
Wareing PF 4, *22*, 162, *171*, 205, *217*
Wareing PF, Good JEG, Manuel J 28, 34, *78*
Wareing PF, Hanney CEA, Digby J 162, *171*
Wareing PF, Horgan R, Henson IE, Davis W 33, 54 *78*
Wareing PF, Phillips IDJ 190, *217*, *227*
Wareing PF, Saunders PF *231*, *233*, *234*
Wareing PF, see Bowen MR 100, 119, *128*
Wareing PF, see Digby J 46, *66*, 163, 164, *168*, 180, *210*
Wareing PF, see Frankland B 7, *18*, 28, *67*
Wareing PF, see Henson IE 33, *68*, *234*
Wareing PF, see Hewett EW 33, *68*, 120, 121, 122, *134*
Wareing PF, see Hoad GV 100, 101, 121, *134*
Wareing PF, see Horgan R 120, 122, *134*
Wareing PF, see Kumar D 183, *213*
Wareing PF, see Lorenzi R 7, *20*
Wareing PF, see Loveys BR 28, 29, 32, *71*
Wareing PF, see Menhenett R 120, 121, *140*
Wareing PF, see Purse JG 120, *142*
Wareing PF, see Railton ID 28, 32, *74*
Wareing PF, see Seth A 14, *21*
Wareing PF, see Seth AK 11, *21*, 30, *75*, 124, *144*
Wareing PF, see Staden J van 33, *77*
Wareing PF, see Thomas TH 28, *77*
Wareing PF, see Webb DP 206, *217*, 223, *229*, *230*, *231*, *235*
Wareing PF, see Wooley DJ 119, *148*
Wareing PF, see Woolley DJ 30, 32, 33, *79*
Wareing PF, see Zamski E 100, 117, 119, *148*
Warren GF, see Brantley BB 194, *209*
Warthen JD Jr , see Grove MD 5, *18*
Wasilewska-Dabrowska LD, see Bralczyk J 59, *64*
Watanabe A, see Imaseki H 29, *69*
Watanable A, see Kondo K 24, 27, 29, *70*
Watanabe T, Kondo N 58, *78*
Wawczyniak G, see Kaldewey H 92, 93, 94, 103, *137*
Weaver P, see Luckwill LC 203, *213*
Weaver RJ 17, *22*, 199, 200, *217*, *227*
Weaver RJ, Overbeek J van, Pool RM 199, *217*
Webb DP, Staaden J van, Wareing PF 223, *230*, *231*, *235*
Webb DP, Staden J van, Wareing PF *229*
Webb DP, Wareing PF 206, *217*
Weber A, see Melkonian M 58, 59, *72*
Weber JA, see Noodén LD 5, 6, *20*
Weber K, see Wick SM 175, *217*
Webster J, Wilkins MB 98, *147*
Webster PL 40, *78*
Webster PL, Langenauer HD 38, *78*
Weij HG van der 81, 82, 83, 87, 88, 89, 90, 94, 105, 106, 111, 112, 114, 118, 122, *145*, *146*
Weiland J, see Sembdner G 119, 120, *144*
Weintraub RL, see Lawson VR 31, *71*
Weis A 122, *147*
Weis JS, see Fox JE 124, *131*
Wellensiek SJ 195, *217*
Weller LE, see Wittwer SH 199, *218*
Went F, Thimann KV 4, *22*
Went FW 7, *22*, 81, 83, 88, 112, 122, *147*, 187, *217*
Went FW, White R 115, 123, *147*
Went FW, see Nitsch JP 194, *214*
Werblin TP, Jacobs WP 112, 122, 123, *147*
Wert V, see Ries SK 5, *21*
West CA, see Phinney BO 49, *74*
Westergaard M 194, *217*
Weston GD, see Elliott MC 41, *66*
Westringh C van de, Veldstra H 115, *146*
Wetmore RH, Jacobs WP 101, *147*
Wetmore RH, Rier JP 11, *22*, 166, *171*
Wetmore RH, Sorokin S 181, *217*
Wetmore RH, see Briggs WR 91, *128*
Wetmore RH, see Brown CL 99, 113, 114, *129*
Weyers JDB, Hillman JR *231*
Weygand F, see Bünning E 84, *129*
Wheeler AW 202, 204, *217*
Wheeler AW, see Humphries EC 16, *19*
Wheeler CT, see Henson IE 120, *134*
Wheller AW, see Humphries EC 55, *68*
White KL, Hill AC, Bennett JH 25, *78*
White PR 182, *217*
White R, see Went FW 115, 123, *147*
Whitehouse RL, Zalik S 100, 101, 114, *147*

Whittington WJ 51, *78*
Whyte P, see Luckwill LC 120, 121, *139*
Whyte P, see Wilkins MB 114, *147*
Wick SM, Seagull RW, Osborn M, Weber K, Gunning BES 175, *217*
Wickson M, Thimann KV 10, *22*
Wielgat B, see Bralczyk J 59, *64*
Wiesner J 4, *22*
Wightman F, Schneider EA, Thimann KV 125, *147*
Wightman F, Setterfield G *227*
Wightman F, Thimann KV 125, *147*
Wightman F, see Schneider EA *228*, *229*, *233*, *234*
Wilkins H, Wain RL 54, *78*
Wilkins MB *227*, *233*, *234*
Wilkins MB, Cane AR 91, 102, 103, *147*
Wilkins MB, Cane AR, McCorquodale I 103, *147*
Wilkins MB, Martin N 114, *147*
Wilkins MB, Nash LJ 92, 124, *147*
Wilkins MB, Scott TK 114, 125, *147*
Wilkins MB, Shaw S 114 *147*
Wilkins MB, Whyte P 114, *147*
Wilkins MB, Woodcock AER 102, *147*
Wilkins MB, see Batt S 116, 117, *127*
Wilkins MB, see Bowen MR 104, 117, 118, 125, 126, *128*
Wilkins MB, see Bridges IG 118, *128*
Wilkins MB, see Gibbons GSB 54, *67*
Wilkins MB, see Goldsmith MHM 84, *132*
Wilkins MB, see Greenwood MS 118, 126, *133*
Wilkins MB, see Hocking TJ 119, *134*
Wilkins MB, see Philipson JJ 27, 48, *74*
Wilkins MB, see Scott TK 91, 125, *144*
Wilkins MB, see Shaw S 91, 98, 103, 104, 118, 125, 126, *144*
Wilkins MB, see Webster J 98, *147*
Wilkins MB, see Woodcock AER 102, 114, *147*, *148*
Wilkinson AJ, see Taylor IEP 58, *77*
Williams PM, see Arias I 34, *62*
Williams RF 193, *217*
Wilson BF, Wodzicki TJ, Zahner R 254, *261*
Wilson MM, see Gordon ME 119, *132*
Winfield PJ, see Morris DA 100, 119, *140*
Winfree AT 248, *261*
Winter A 13, *22*, 112, 113, *147*
Withers LA, see Wangermann E 84, 104, 112, 113, 117, 118, 122, *146*
Wittwer SH 172, *218*
Wittwer SH, Bukovac MJ 190, *218*
Wittwer SH, Bukovac MJ, Sell HM, Weller LE 199, *218*
Wittwer SH, see Phatak SC 195, *214*
Wochok ZS, Sussex IM 122, *147*
Wodzicki AB, see Wodzicki TJ 245, 246, 247, 248, 250, 255, *261*
Wodzicki T 245, *261*
Wodzicki TJ 163, *171*, 245, 246, 247, 254, 255, *261*
Wodzicki TJ, Brown CL 255, *261*
Wodzicki TJ, Rakowski K, Starck Z, Porandowski J, Zajaczkowski S 245, *261*
Wodzicki TJ, Wodzicki AB 245, 246, 247, 250, 255, *261*
Wodzicki TJ, Wodzicki AB, Zajaczkowski S 246, 248, 250, *261*
Wodzicki TJ, Zajaczkowski S 245, 246, *261*
Wodzicki TJ, see Brown CL 245, *260*
Wodzicki TJ, see Nix LE 247, 253, *260*
Wodzicki TJ, see Porandowski J 245, *260*
Wodzicki TJ, see Wilson BF 254, *261*
Wodzicki TJ, see Zajaczkowski S 164, *171*, 245, 246, 248, 253, 254, 257, 258, 259, *262*
Wojciechowski M, see Shen-Miller J 47, *75*
Wolpert L 244, 245, *262*
Wood DA, Hammond JBW 60, *78*
Wood NL, Berliner MD 58, *78*
Woodcock AER, Hertz CH 102, *147*
Woodcock AER, Wilkins MB 102, 114, *147*, *148*
Woodcock AER, see Wilkins MB 102, *147*
Woodley DJ, see Patrick JW 103, *142*
Wooley DJ, Wareing PF 119, *148*
Woolhouse HW, see Beever JE 120, *127*
Woolley DJ, Wareing PF 30, 32, 33, *79*
Worley JF, see Grove MD 5, *18*
Wright RD, Aung LH 56, *79*
Wright STC 11, *22*, *231*, *232*, *233*
Wright STC, Hiron RWP 33, *79*
Wright STC, see Hiron RWP 33, *68*
Wurr DCE, see Thomas TH 34, *77*

Xhaufflaire A, see Gaspar Th 24, 27, 52, 53, 54, *67*
Xian-zhang L, see Tsui C 57, *77*

Yamaguchi S, see Zweig G 119, *148*
Yamaki T, Fujii T 100, *148*
Yamaki T, Shimojo E, Yoshizawa K, Namekawa K 39, *79*
Yamaki T, see Kondo N 31, *70*
Yamaki T, see Nishinari N 39, *73*
Yamaki T, see Stowe B 12, *21*
Yamamoto R, see Masuda Y 48, *72*

Yamane H, Murofushi N, Takahashi N 202, *218*
Yanagishima N 59, *79*
Yanagishima N, see Masuda Y 59, *72*
Yanagishima N, see Shimoda C 59, *76*
Yanagishima N, see Tsuboi M 43, *77*
Yang SF 29, *79*, *232*
Yang SF, Adams DO *232*
Yang SF, see Adams DO 29, *62*
Yang SF, see Bradford KJ 8, *17*, 220, *234*
Yang SF, see Lau O-L 25, 27, 29, *70*, *71*
Yang SF, see Yu Y-B 29 *79*
Yan-ru S, see Tsui C 57, *77*
Yeoman MM, Brown R 46, *79*
Yeoman MM, see Aitchison PA 37, *62*
Yeomans LM, Audus LJ 82, 91, 125, *148*
Yeung EC, Sussex IM 177, *218*
Yoshida R, Oritani T 120, *148*
Yoshida R, Oritani T, Nishi A 120, *148*
Yoshida R, see Nishi A 39, *73*
Yoshizawa K, see Yamaki T 39, *79*
Young H, see Davison RM 13, *18*, 120, *130*
Young JO, see Tukey HB 196, 197, *217*
Yu Y-B, Adams DO, Yang SF 29, *79*

Zaerr JB 122, *148*
Zaerr JB, see Lavender DP 120, *138*
Zaerr JB, see Morris RO 120, *140*
Zaerr JB, see Sweet GB 120, *145*
Zahner R, see Wilson BF 254, *261*
Zajaczkowski S 163, 164, *171*, 245, 246, *262*
Zajaczkowski S, Romberger JA 245, *262*
Zajaczkowski S, Wodzicki TJ 164, *171*, 245, 246, 248, 253, 254, 257, 258, 259, *262*
Zajaczkowski S, Wodzicki TJ, Bruinsma J 246, *262*
Zajaczkowski S, see Wodzicki TJ 245, 246, 248, 250, *261*
Zakrzewski J 246, 247, *262*
Zalik S, see Fletcher RA 31, *67*, 101, *131*
Zalik S, see Whitehouse RL 100, 101, 114, *147*
Zamski E, Tsivion Y 118 *148*
Zamski E, Wareing PF 100, 117, 119, *148*
Zbell B, see Reinert J 181, 200, *215*
Zeevaart JAD 32, 33, *79*, 121, *148*, 221, 223, *229*, *230*, *232*, *234*, *235*
Zeevaart JAD, see Metzger JD 32, *72*
Zeevaart JAD, see Suttle JC *234*
Zeevaart JAF, see Raschke K 7, *21*
Zeevart JAD 192, *218*
Zeigler R, see Vanderhoef LN 24, 27, 50, *78*
Zetsche K, see Guttenberg H v. 82, 91, *133*
Zhe T, see Tsui C 57, *77*
Ziegler H 86, 114, *148*
Ziegler H, see Kluge M 121, *137*
Zimmerman DC, Coudron CA 5, *22*
Zobel RW, Roberts LW 42, *79*
Zwar JA, Rijven AHGC 112, *148*
Zwar JA, see Bottomley W 164, 166, *168*
Zweig G, Yamaguchi S, Mason GM 119, *148*

Subject Index

ABA see Abscisic acid
Abbreviations of chemicals
  ABA = Abscisic acid
  ACC = l-aminocyclopropane-l-carboxylic acid
  AMP = Cyclic adenosine monophosphate
  ATP = Adenosine triphosphate
  BA = Benzyladenine
  BAP = 6-Benzylaminopurine
  CCC = 2-Chlorethyltrimethylammonium chloride
    = Chlorocholine chloride
  2,4-D = 2,4-dichlorophenoxyacetic acid
  DNA = Deoxyribonucleic acid
  DNP = 2,4-dinitrophenol
  GA = Gibberellin
  GA_3 = Gibberellic acid
  IAA = Indoleacetic acid
  IBA = Indolebutyric acid
  NAA = Naphthalene acetic acid
  NPA = Naphthylphthalamic acid
  RNA = Ribonucleic acid
  SAM = S-adenosylmethionine
  2,3,5-T = 2,3,5-trichlorophenoxyacetic acid
  TIBA = Triiodobenzoic acid
Abscisic acid (ABA) 5, 7, 25, 54, 97
  algae 58
  and auxin
    gravitropic responses 222
    phototropic responses 222
  basipetal efflux of auxin 250
  bound
    seeds, stratification 34
  callus formation 42
  cell
    biosynthesis 30
    cycle 42
    division 38
    elongation 49, 50
    ethylene enhancement 30
    ethylene production 29
    expansion 24
  cold 34
  cotton fiber growth 57
  effect on indoleacetic acid (IAA) transport 164
  endogenous, hormonal modification 29
  flowering in vitro 191
  frond growth 56
  fruits 34
  and gibberellins (GA) export of 92
  gravitropic responses 221
  growth of the seed 177
  ethylene, cell elongation 23
  inhibition by
    ethylene production 27
    indoleacetic acid (IAA)-induced 27
  intact plant 219
  light 32
  movement of
    gibberellic acid (GA_3) 97
    kinetin 97
  parthenocarpy, cytokinins 200
  pollen-tube elongation 57
  polyploidy 42, 43
  role(s) of 226
  root
    caps 54
    elongation 54
    growth 54
    modulation 42
  rooting 42
  seed(s) 206
    development 204
    dormancy, other inhibitors 207
    germination inhibitors, coumarin 206
    translocation, basipetal 249
  vivipary 204
  water balance 221
  xylem 164
Abscisic acid (ABA) antagonism
  leaves, chlorophyll-sparing effect of kinetin 26
Abscisic acid (ABA) biosynthesis
  in leaves 33
  water stress 33
Abscisic acid (ABA) content
  photoperiod 32
  water stress 33
Abscisic acid (ABA) enhancement, GA_1 28
Abscisic acid (ABA) inhibition, auxin 27
Abscisic acid (ABA) – like inhibitor
  gibberellin (GA) 30
  kinetin 30
  suppression of 30

Abscisins, translocation of 124
Abscission 6, 10, 12, 14, 152, 180
ACC see 1-aminocyclopropane-1-carboxylic acid
Acer 34, 122, 180, 247
Acer pseudoplatanus 39, 206, 223, 246
Acer saccharinum 207
Acer saccharum 223
Acetabularia 58, 175
ACTH 8
Acropetal displacement, inhibitory effect of cytoplasm 114
Acropetal movement, by diffusion 114
Actinomycin-D 97
Adenosine triphosphate (ATP)-interactions, in auxin transport 91
S-adenosylmethionine (SAM)-aminocyclopropylcarboxylic acid (ACC) 29
Adiantum 46
Adventious rooting 221
Agaricus bisporus (mushroom) 60
Agaricus campestris 60, 61
Ageing effect, in relation to auxin transport 92
Agriculture, chemicals 173
Air 174
Alaria esculenta 58
Algae 57, 61, 187
  abscisic acid (ABA) 58
  auxin effects on 58
  cell division 58
  cell expansion 57, 58
  cytokinin effects on 59
  ethylene 58
  gibberellin (GA) effects on 59
  indoleacetic acid (IAA) effects on 58
Allium cepa 39
Amaranthus 33
1-aminocyclopropane-1-carboxylic acid (ACC) 8, 220, 222
  conversion of
    flooding 33
    ethylene production 33
Ambrosia trifida 194
Amo 1618, root-elongation, 2-chlorethyltrimethylammonium chloride, chlorocholine chloride (CCC) 53
α-amylase 9
Ananas sativus L. (pineapple) 191, 196
"Angker" 191
Antagonism
  and negative synergism 26
  synergism, in hormone interaction 25
Antagonistic actions 221
Antiflorigen 5
Antipodal cells 175, 224
Aphids, phloem feeding 117
Apical dominance 7, 10, 13, 15, 224, 258
Apple 13, 189, 190, 196, 200, 203
  cerab 98
Ascorbic acid 154
Aspartic acid and indoleacetic acid (IAA) 83
Aspen, abscisic acid (ABA) – like inhibitor, indoleacetic acid (IAA) maintenance 29
ATP see adenosine triphosphate
Autolysis, of protoplast 254, 255
Auxin(s) 118, 149, 221
  and abscisic acid (ABA) 222
  activity in aphid honeydew 101
  activity of polar transport 84
  basipetal efflux 250
  basipetal polarity 155
  bioassay of 150, 246
  biosynthesis
    abscisic acid (ABA) inhibitor 27
    gibberellin (GA) increase 27
  biphasic response, cytokinin interactions 48
  bound 27
  cambial area
    extraction 163
    diffusion 163
  cell
    division 39, 40
    elongation, H^+ ion secretion 48
  constant delivery of 91
  cytokinin 38
  dependence, uptake, pH 115
  in the developing embryo 177
  diffusible 7, 150
    profile 109
    identification of 151
  diffusion, polar, chemiosmotic 85
  efflux, oscillation 249
  elongation in roots, promotion and H^+ efflux 51
  embryogensis 182
  endogenous, exact substitution 151
  ethylene biosynthesis 53
  export, oscillations of 90
  extractable 7
  flooding 34
  flower development
    levels of carbohydrate 192
    other plant hormones 192
  flowering in vitro 191
  free, kinetin increase 27
  fruits
    and GA_4 200
    and Ga_7 200
    and gibberellin (GA) 200
    and gibberellic acid (GA_3), cambial activity 180
  gibberellin (GA), cell elongation 23

gibberellin (GA) regulated growth, suppression of 48
  and gravitropic responses 222
growth hormone 5
hyponasty 55
inhibition 123
intact plants 84
  abscisic acid (ABA) 219
  cytokinins 219
  ethylene 219
  gibberellin (GA) 219
interactions 180
  and kinetin 182
increase of content 27
leaf 55
leaf cell expansion 55
light 31
mitosis 40
mitotic activity 39
mobile, variations in density 103
movement
  applied to loci of natural production 100
  applied to mature leaves 101
  kinetin increases 27
  polarity 82
petiole growth 55
and phototropic responses 222
in plant morphogenesis 246
polyploidy 43
polarity 253
polar transport 244, 248, 249
polar transport ocillatory system 249, 251
polar transport velocity, of signal propagation 251
in pollen 198
red light 31
replacement, cambial activity 180
role(s) of 226
root 33, 40, 50, 53
  elongation 51, 53
in roots 51
in seeds
  bound, turnover 204
  growth 203
short-term application of 94
source, young leaves 179
stimulus incorporation of ^{3}H-Thymidine 173
synthesis
  in expanded leaves 221
  in short meristems 221
transit, across injured cells 96
transport 81, 84, 85
  actinomycin-D 97
  applied to unfolded leaves 101
  basipetal polarity 94
  channels 103
  conveyor belt model 112
  decline 91
  dependence on temperature 98
  early experiments 81
  faster and slower component of 90
  in intact plants 84
  intensity 98
  interactions
    cycloheximide 97
    2,4-dinitrophenol (DNP) 97
    Endothal 97
    KCN 97
    morphactin 97
    1-N-naphtylphthalamic acid (NPA) 97
    Triiodobenzoic acid (TIBA) 97, 98
  leaves 55
  oscillations 103
  oxygen dependence supply 113
  pathways, in shoots 118
  polar 84, 85
  roots 125
  and sieve tubes 84
  stream
    fast fraction 103
    slower fraction 103
  system, specificity 123
  tissue types 105
  uptake, pH dependence 115
  velocity 82, 98
  wave 257
    deviation of the vector 256
    organizing center 258
    positional information 254
    propagation, vectors 252, 254
    three-dimensional 253
    vector(s) 255, 256
    vector field(s)
      microfibrils 255
      positional information 252
      woody shoots 98
Auxin antagonists, root growth 51
Auxin biosynthesis, feedback inhibition, red light effect 29
Auxin conjunction, ethylene 29
Auxin degradation 27
Auxin deprivation, cell division 39
Auxin destruction, ethylene 29
Auxin effects
  on algae 58
  on fungi 59
Auxin enhancement
  ethylene 29
  ethylene production 29
Auxin-induced, cell elongation 48
Auxin-induced ethylene production, kinetin enhancement 25
Auxin levels, miotic activity 39

Auxin production, red light 31
Auxin-promoted elongation, gibberellic acid (GA_3) enhancement 25
Auxin transport, red light 31
Auxin uptake, red light 31
Avena 27, 45, 48, 49, 50, 55, 59, 88, 98, 101, 103, 106, 107, 111, 114, 118, 122, 124, 246, 248
  coleoptiles 93, 94, 97
    profiles of mobile ^{14}C 99
  curvature test 81
Avena sativa 204
Avena test 246
Axis fixation, cytoskeleton 175
Azalea 190

BAP see 6-benzylaminopurine
Barley 9, 12, 28, 31, 40, 45, 206
Basipetal efflux of auxin, abscisic acid (ABA) 249
Basipetal translocation
  of abscisic acid (ABA) 249
  of Triiodobenzoic acid (TIBA) 249
Bean 11, 31, 32, 50, 56, 57, 196
Bear Hybrid (Corn) 54
Benzyladenine (BA) 124
  callus formation 42
  ethylene production
    cell elongation 50
    enhancement of 50
^{14}C-benzylaminopurine 30
6-benzylaminopurine (BAP) 57
Binding sites 116
  on the plasma membrane 85
Bioassay
  auxin 246
  cytokinins, from cambium 164
Bioelectric theory 85
Birch 28, 32
Black currant 190
Bleeding sap, grape
  2-chlorethyltrimethylammonium Chloride, chlorocholine chloride (CCC) 30
  cytokinins 30
Blueberry 200
Blue green algae 58
Blue light, ethylene production, enhancement of 32
Boehmeria 160
Bolting juvenile characteristics 222
Branching 221, 258
Brassinolide, flower development 192
  light period 193
  macronutrients 193
Brassins 5
Brown algae 58
Bryophyllum cuttings 81
Bud(s) 7, 14, 30, 92, 155, 161, 222, 258
  cytokinins 4
  indoleacetic acid (IAA) 30

Calcium pump 175
Callitriche 50
Callus
  from guard cells 181
  mesophyll cells 181
  microsporocytes 181
  pollen 181
  tissue, differentiation, culture media 181
  xylogenesis 181
Callus formation
  abscisic acid (ABA) 42
  benzyladenine 42
  indoleacetic acid (IAA) 42
Calorigen 5
Calotropis procera 57
Cambial
  activity 162, 221
  fusiform cells, length of 247
  initial cells 256
  initials 252
Cambium 150, 162
  development 153
Cancer cells, indoleacetic acid (IAA) 39
Carnation 29
Carriers 113
Carrot cambium 164
Carrot cells 39, 41
CCC see 2-Chloroethyltrimethylammonium chloride, Chlorocholine chloride
Cell(s)
  cycle 35, 172, 173, 183
    abscisic acid (ABA) 42
    cytokinin
      control 36
      root meristems 37
    gibberellin (GA) 40
  differentiation 152, 172, 175
  division 162
    abscisic acid (ABA) 38, 42
    algae 58
    arrest 35
    auxin 39
      root 40
    cytokinin 36, 49
      inhibition of 38
    ethylene 41
    and geotropically stimulated roots 47
    gibberellin (GA) 40
      abscisic acid (ABA), inhibition 49
    kinetin 45, 47
    megaspore mother cells, cytokinin 43
    meiotic 43
    microscope mother cells, ethylene 44

orientation, light 46
plane 44, 46
root cap 38
root meristems
indoleacetic acid (IAA) 39
kinetin 39
synchronization, of auxin deprivation 39
synchronous, cytokinin, induction of 37
synchrony, ethylene 41
in tissue culture, cytokinin 47
trigonelline 42
water stress 47
elongation
abscisic acid (ABA) 50
auxin induced, abscisic acid (ABA) inhibition of 48
benzyladenine (BA) 50
ethylene 50
gibberellin (GA) promotion, dwarf plants 49
gibberellin (GA) stimulation 27
inhibition of
abscisic acid (ABA) 49
cytokinin 49
ethylene 49
kinetin 47
in roots
stimulation 52
auxin 52
expansion
direction of 45
epidermal, indoleacetic acid (IAA) 45
gibberellin (GA) 49
hormonal regulation, stem cells, auxin 48
in lower plants, chemical regulation 57
relationship of cell multiplication 46
in tissue culture, cytokinin 47
water stress 47
free 176
loosening, gibberellin (GA) enhancement 49
plate 175
orientation 45
single 175
wall
extensibility 28
formation 173
gibberellin (GA) enhancement 49
Cell division 6, 222
algae 58
cytokinins 177
embryonic development 177
Cell elongation
abscisic acid (ABA) 23
auxin 23
ethylene 23
gibberellin (GA) 23
Cell enlargement 6
Cell expansion
abscisic acid (ABA) 24
algae 58
cytokinin 24
Cell oscillations, period of 251
Centrifugation studies, inhibitory effect 114
Chara 40
Charophyceae 58
Chemiosmotic polar difussion, theory 115
Chemotropic factor gradients 174
Chenopodium, buds, indoleacetic acid (IAA) 30
Chenopodium rubrum 28, 30
Cherry 196, 200
sour 197
Chicory 10
Chlorocholine chloride (CCC) 190
flowering
angosperms 190
gymnosperms 190
2-Chloroethyltrimethylammonium chloride, chlorocholine chloride (CCC) 30, 53, 190
cytokinins 30
root elongation, Amo 1618 53
Chlorophyceae 58
Chlorophyll degradation 26
Chlorophyll retention 26
Chlorophyll-sparing effect of kinetin
leaves 26
abscisic acid (ABA) antagonism 26
Cholorella 59
Chromatin 173
Chrysanthemum maximum (LDP) 223
Chrysanthemum morifolium (SDP) 223
Cichorium intybus 191
^{14}C-indoleacetic acid (IAA) pulse migration
interaction with ethylene with water stress 97
velocity of 97
Citrullus 125
hypocotyls 93
Citrus fruits 30
Cladophora 175
Cleome 122–124
Cocklebur 50, 188
Coconut milk 201
Cold
abscisic acid (ABA) 34
gibberellin (GA) biosynthesis 34
gibberellin (GA) production 34
Coleoptile(s) 24, 28
elongation
gibberellin (GA) 26
tannins 26
regeneration of physiological tip, abscisic acid (ABA) and indoleacetic acid (IAA) 27

Coleus 46, 56, 101, 103, 106, 111, 116, 118, 122–126, 150–154, 158–160
Coleus blumei 27
Conchocelis 59
Convolvulus 125
Corn 7, 9, 34, 41, 47, 51, 53–55, 94, 96, 173, 194
  Bear Hybrid 54
  Golden Cross Bantam 54
  Velox 54
Corn coleoptiles 108, 111, 114
Corynebacterium fasciens 225
Cotton 57
Cotton fibers
  abscisic acid (ABA) 57
  gibberellin (GA) 57
  indoleacetic acid (IAA) 57
  kinetin 57
Cotyledon(s) 24, 176
Cotyledonary differentiation 177
Cotyledon factor and gibberellin (GA) promoted elongation 49
Cotyledon growth
  cytokinin 56
  gibberellin (GA) 56
Coumarin 206
Cowpea 28
Crepis parviflora 190
Cucumber 28, 47, 49
  hypocotyl 25
Cucurbita 117
Cucurbitaceae 194
Cucurbita pepo 194
Cucumis 124, 158
Cucumis melo 194
Cucumis sativus 194
Culture(s)
  cell arrest 175
  cell differentiation
    cytogenetics 183
    genetic uniformity 183
  organ, defined culture medium 181
  single cell, polarity 175
  stem segments 245
  suspension, cytokinin, cell cycle 175
  tissue
    defined culture medium 181
    development of flowers 191
    differentiation
      cytogenetics 183
      genetic uniformity 183
    hormonal effects 165
    initiation of flowers 191
Cupressaceae 189
Currant, black 190
Cyanophyceae 58
Cyclic adenosine monophosphate (AMP) 10, 173
Cycloheximide 97
Cyctochalasin B 114
Cytokinesis 173
Cytokinin(s) 92, 158, 221
  cell
    division 36, 38, 43, 47, 50, 177
    elongation 49
    expansion 47
  2-chlorethyltrimethylammonium chloride, cholorcholine chloride (CCC) 30
  cotyledon 56
  in embryo development 177
  endomitosis 43
  endoreduplication 37
  flowering in vitro 191
  gibberellin (GA) 28
  indoleacetic acid (IAA) 30
  intact plant 219
  kinetin, attraction center 124
  leaf growth 56
  liquid endosperm, other growth substances
    gibberellin (GA) 201
    indoleacetic acid (IAA) 201
  light, increase 33
  meristem activity 54
  movement
    gibberellin (GA) reduction of 30
    indoleacetic acid (IAA) radiation of 30
    Pisum 126
    Raphanus 126
  pollen tube elongation 57
  polyploidy 43
  ratio with auxin, endomitotic cycling 43
  roles of 226
  root(s) 34, 54
    growth 54
    meristems 37, 38
    nodule 43
    quiescent center 38
  secondary xylem cells 164
  seeds 202
    germination
      overcoming dormancy 206
      time sequence 206
    gibberellin (GA)
      endosperm volume 203
      growth of seed 202
  synthesis, root tips 221
  translocation of 124
Cytokinin biosynthesis, roots 34
Cytokinin content
  photoperiod 32
  light, decrease of 33
Cytokinin effects
  on algae 59

on fungi 60
Cytokinin-like materials, light 33
Cytokinin production
gibberellin (GA) 30
quiescent center 38
Cytokinin stimulation, ethylene production 29
Cytoplasmic streaming 14

2,4,-D see 2,4-dichlorophenoxyacetic acid
two-dimensional growth 46
Datura 201
Degradation, hormone 9, 11
Deoxyribonucleic acid (DNA) content, mitotic cycle time 37
Deoxyribonucleic acid (DNA) synthesis
inhibition, ethylene 41
kinetin 37
naphthalene acetic acid (NAA) 39
tobacco cells 37
Dependence on pH, auxin, uptake 115
Desication
cytokinin, reduction, roots 33
ethylene production, increase in 33
Development
cambium 153
embryo, liquid endosperm, composition 201
role of endosperm 201
floral 187
flower 186, 188, 192, 193
fruit, set 186, 195
organ 172
polarity 111
seed 186, 200
in shoots 159
vascular strand 179
2,4-dichlorobenzyltributyl phosphonium chloride (Phosphon-D), flowering
angosperms 190
gymnosperms 189, 190
2,4-dichlorophenoxyacetic acid (2,4-D) 91
Differentiation
cell 149, 172
fiber 149, 160
gibberellic acid (GA_3) 162
indoleacetic acid (IAA) 162
leaf primordia procambial stage 179
leaves 178
root 178
sieve-tube(s) 149
tissue culture 166
stem 178
tissue, callus, culture media 181
tracheary cells 149
tracheid
abscisic acid (ABA) 181
flower primordia 188
gibberellic acid (GA_3) 181
vascular bundle, auxin 179
xylem 11
Diffusible auxin 159
Diffusible hormone 13
Diffusion
accelerated 97, 116
acropetal movement 114
chemiosmotic 115
polar 116
proton and emf dependent 116
translocation 13
Digitalis, stamen 122
Digitalis purpurea 186
Dihydroconiferol 5
Diploidia natelensis 60
Division figure, orientation, flower primordia 186
DNA see deoxyribonceic acid
DNP see 2,4-dinitrophenol
Domain pattern, vector fields 252
Dormancy 5–7, 10, 16, 222
bud 173
seed 173
abscisic acid (ABA) 207
other inhibitors 207
Douglas fir 189
DPX 1840 123
Duckweed 56

Ecklonia 58, 59, 222
Ecological niche(s) 224
Ectoparpus 59
Electric potential gradient 175
wave method 101
Elongation 12
cell
abscisic acid (ABA) 50
ethylene 50
root 50, 52
auxin 51
root cell, indoleacetic acid (IAA) 51
stimulation of kinetin 50
Embryo
cotyledonary stage
zeatin-glucoside 177
zeatin riboside 177
in culture 176
development, cytokinins 177
GA_1 177
gibberellic acid (GA_3) 177
gibberellin (GA) 177
globular 176
growth, metabolic inhibitors 205
phenolic compounds 206
heart shaped
2-isopentenyladenine 177
zeatin 177

Embryo, heart shaped
    octant 176
  proembryo 176
  torpedo stage 176
Embryogenesis
  appearance of buds 182
  auxin 182
  kinetin 182
  light 182
  morphogenetic potential 183
  ontogenetic development, culture medium 182
  pH 182
  photoperiod 182
  regeneration 183
  rooting 182
  submergence of a callus 182
  sugar concentration 182
  temperature 182
Embryogeny 174
Embryoids 180
Endogenous auxin, replacement 152
Endomitosis, differentiation, cytokinin 43
Endorphin 8
Endothal 97
English ivy 42
Enteromorpha prolifera 59
Enzyme, biosynthesis
  cytoninin 26
  ethylene 26
  gibberellic acid (GA_3) 26
Epidermal hairs 57
Epinasty, petiole growth 56
Error function curves 116
Ether narcosis 82
Ethylene 5
  abscisic acid (ABA), cell elongation 23
  algae 58
  autocatalytic production 29
  biosynthesis pathway 29
  cell
    division 41, 43–44
    elongation 49, 50
    deoxyribonucleic acid (DNA) synthesis 41
  effects on
    auxin and production of ethylene 222
    flowering 222
    fruiting 222
    seed germination 222
    senescence 222
    vegetative growth 222
  flowering in vitro 191
  intact plant 219
  modification of
    auxin conjugation 28
    auxin destruction 28
    uptake of auxin 28
  nyctinasty 32
  petiole growth 56
  production
    enhancement by auxin
      flowers 29
      fruits 99
      leaves 29
      roots 29
    indoleacetic acid (IAA) induced, inhibition by abscisic acid (ABA) 27
  reduction of 28
  root elongation 50
Ethylene effects, on fungi 60
Ethylene enhancement, abscisic acid (ABA) 30
Ethylene formation 57
Ethylene production 53
  abscisic acid (ABA) 29
  auxin, kinetin enhancement 25
  auxin enhancement 29
  benzyladenine 50
  blue light 32
  cytokinin stimulation 29
  desiccation 33
  far red light 32
  flooding, enhancement, conversion of 1-aminocyclopropane-1-carboxylic acid (ACC) 33
  gibberellin (GA) 29
  light 32
  mechanical stimulation 34
  phytochrome 31
  red light 32
  roots, auxin enhancement 53
  in roots 53
Etioplasts 31
Euglena 59

Far red light, ethylene production, enhancement of 32
Fern 57, 187
  gametophytes 41, 46
  osmunda 160
Fiber
  differentiation 149, 160
  phloem 150, 151
Fig 200
Flooding
  auxin, increase in roots 34
  ethylene production, enhancement, conversion of 1-aminocyclopropane-1-carboxylic acid (ACC) 33
  gibberellin (GA), decrease in 34
Floral
  induction
    photoperiodism 192
    plant hormones 187

initiation 188
primordia 186
Florigen 5
gibberellin (GA), short-day plants 190
Flowers
development 186–188, 192
iron
cytochromes 193
feredoxin 193
nitrate reductase 193
sulfate reductase 193
levels of carbohydrate 192
light period, brassinolide 193
macronutrients, brassinolide 193
male flower, cytokinins 195
nitrogen levels 193
plant hormones, auxin 193
ethylene 29
fertilization 186
formation on root tissues 191
hormone synthesis 220
initiation 30
and flower promoting factors 188
and gene activation 189
and the gibberellin (GA) 189
indolebutyric acid (IBA) 192
and inhibitors 188
and natural synthetic growth regulators 189
primordia 188
Flower induction 15
Flowering 174, 222
biochemical balance, environmental stimuli, vernalization 187
explant tissue 191
long-day plant 223
long-day species, rosette forms 190
mineral nutrients 187
physical and chemical conditions 191
physiological balance 187
ratio between the carbohydrate and nitrogen 188
in stem segments 191
Fraxinus americana 207
Free indoleacetic acid (IAA), cytokinin increase 29
Fritillaria 103, 107–109, 111
flower stalks 92, 99
fruit stalks 122
Fritillaria melegris L. 89, 110
Fritschiella 59
Frond growth, abscisic acid (ABA) 56
Fruit(s) 4
abscisic acid (ABA) 34
appearance 200
auxin 200
6-benzyladenine 200
cell division
embryo 196
gibberellin (GA) 196–197
pericap 196
cell enlargement
embryo 196
gibberellin (GA) 196–197
pericap 196
color 174
development 174, 186
plant hormones 200
drop 174
embryo 196
ethylene 29
GA_4 200
GA_7 200
gibberellin (GA) 200
growth 196
set 195, 222
hormone synthesis 220
indolebutyric acid (IBA) 198
integuments 196
1-naphthaleneacetic acid (NAA) 198
2-naphtoxyacetic acid 198
nucellus 196
ovary 196
parthenocarpic 200
pollination 198
production
growth regulation 195
hormones 195
ripening 6, 16
seed 4, 196
senescence 26
set 174, 186, 195
shape, gibberellin (GA) 200
thinning, growth regulators 200
zygote 196
Fuchsia 190
Fucus 175
Fucus vesiculosus 59
Fungi
auxin effects on 59
cytokinin effects on 61
ethylene effects on 60
Fusarium oxysporum 60
Fusiform initials 252

GA see gibberellin
GA_3 see gibberellic acid
Galls 17
Gametophytes
fern 41
protonema stage, light 46
Genetisic acid 5
Geotropic responses 256
Geotropism 9

Germinating seeds, hormone, synthesis 220
Germination 4
  seed(s), overcoming dormancy
    cytokinins 206
    gibberellin (GA) 206
  seed dormancy, inhibitors, endogenous
    causes 205
  time sequence 206
Germination inhibitors, seeds
  abscisic acid (ABA) 206
  coumarin 206
Geum Chiloense 179
Gibberella fujikuroi 60
Gibberellic acid (GA_3) 154, 158–159, 160, 186, 249
  and auxin, cambial activity 180
  differentiation 162, 181
  embryo 177
  enhancement of, auxin promoted elongation 25
  and indoleacetic acid (IAA) 185
  interactions 180, 226
  intact plants 161
  and kinetin 100
    movement
      abscisic acid (ABA) 97
      basipetal 126
      kinetin 97
  role(s) of
    auxin 226
    cytokinin 226
    ethylene 226
  secondary xylem cells 164
  translocation of 91
Gibberellin (GA) 5, 53, 222, 223
  abscisic acid (ABA) and hinderance of export of 92
  abscisic acid (ABA) - like inhibitors 29, 30
  auxin 198
  and auxin translocation
    and abscisic acid (ABA) 97
    and kinetin 97
  bioassay 27
  cell
    cycle 40
    division 49
    elongation 23
    expansion 49
  content 28
  cotton fibers growth 57
  cotyledon growth 56
  cytokinin
    endosperm volume 203
    production 30
  in the developing embryo 177
  embryo 177
  ethylene production 29
  flooding 34
  florigen 190
  flowering
    angosperms 190
    in gymnosperms 189
    long-day species 190
  flowering in vitro 191
  fruit
    auxins 200
    GA_4 200
    GA_7 200
    plant hormones 200
  fruit growth
    cell division 196
    cell enlargement 196
    embryo 196
    pericarp 196
  guard cell mother cells 45
  increase, cytokinins 28
  indirect flower initiation 190
  intact plant 219
  leaf growth 56
  light 32
  long-day biennial plants 189
  movement, polarization, of indoleacetic acid (IAA) 124
  petiole growth 56
  pollen 198
  pollen tube elongation 57
  polyploidy 43
  rhizome growth 185
  root elongation 52
  seeds, germination, overcoming dormancy 206
  short-day plants 190
  stolons, indoleacetic acid (IAA) 185
  synthesis 13
  time sequence 206
  translocation of 124
  triiodobenzoic acid (TIBA)
    interaction 124
Gibberellin (GA) biosynthesis
  cell division 41
  cold 34
  roots 34
  seeds 34
  tuber 34
Gibberellin (GA) content, photoperiod 32
Gibberellin (GA) effects
  on algae 59
  on fungi 60
Gibberellin (GA) increase auxin 27
Gibberellin (GA) synthesis, roots 52
L-glutamic acid 159
Golden Cross Bantum (Corn) 54

Gossypium 103, 125
  petioles 97
  seedlings 94
Gradients 175
  chemotropic factor 174
  electric 15
  electric potential 175
  external 175
Grain(s) 4, 45
Grape 43, 190, 200
  bleeding sap
    cytokinins 30
    2-Chlorethyltrimethylammonium chloride, Chlorocholine chloride (CCC) 30
  Thompson Seedless, cytokinin biosynthesis, retardation of roots 34
Grapefruit 176
Grasses 48
Gravimorphism 258
Gravitropic responses, abscisic acid (ABA) 221, 222
  auxin 222
Gravity 175, 258
Green algae 58
Griffithsia 175
Ground meristem 176
Growth
  embryos
    liquid endosperm, composition 201
    metabolic inhibitors, phenolic compounds 206
    role of endosperm 201
  hormones 81
  promoting substance 81
  regulation 81
  substance, first isolation of 81
Growth hormone
  auxin 5
Growth regulators 173, 174
Growth substances 172
Growth suppressants 178
Guard cell mother cells
  gibberellin (GA) 45
  indoleacetic acid (IAA) 45
Gymnophorus arachis 122

Helianthus 55, 56, 91, 93, 103, 106, 118, 124
  epicotyls 122
  hypocotyls 94
Helianthus annuus L. 105
Hemp 160
Herbicide(s) 174
Hibiscus exculentus 154
Hiearchial control of plant morphogenesis 259
Histone 173
Hook closure 32
Hormonal controls 11, 15
  effects
    vectoral nature 208
  factors
    interplay 183
  interactions 226
Hormone(s)
  action 172
  application modes 100
  attachment 6
  binding 9
  bioassays 12
  circulation within the vascular system 119
  comparmentation 11
  concentration effects 10
  concepts 8
  correlations 13
  defined 4
  degradation 9, 11
  differentiation of vascular tissue 149
  diffusible 13
  effects at the molecular level 172
  evolution 20
  exchange between phloem and xylem 119
  exogenously applied 224
  genetics of developmental regulation 173
  growth 81
  identification 224
  induced secretion of H^+ ions 102
  interaction 172
  messenger systems 6
  movement
    early call for 81
    general remarks 86
    history of research 80
    indirect evidence for 80
    pathways 117
    in sieve tubes 114
    terminology of 86, 87
    transport equation of 87
    vascular systems 119
  multiple 10
  nonfunctional 7
  opposing effects 10
  quantification 224
  ratios 10
  receptors 14
  receptor sites 8
  regulation
    of growth 221
    of maturation 221
    of physical homeostasis, to cellular differentiation biochemical modulations 221
    of responses to the external environment 221
    remote 14
  regulatory sites 9

Hormone(s)
  second messengers 9
  sequential regulation 11
  synergism 180
  synthesis
    anthers 220
    embryos 220
    in flowers 220
    in fruits 220
    in germinating seeds 220
    growing 220
    ovaries 220
    pollen tubes 220
  synthetic, exact substitution 152
  theory 81
  translocation 8
    direct evidence for 81
    polarity 94
    root 125
    shoot sections 122
  transport 10, 13
    1-aminocyclopropane-1-carboxylic acid (ACC) 220
    in phloem 220
    roots, channels 118
    xylem 220
  triangle method 186
    determination of concentrations of growth substances, induction of differentiation of roots, buds or callus 186
  trigonellin 175
  two-way interchange 100
    xylem and extra cambial tissue 100
      auxin 100
      gibberellic acid (CA_3) 100
      kinetin 100
  types 5
  and wounding 149
Hormone interaction
  antagonism in 25
  definition of 24
  synergism 25
Hormone sensitivity 35
H^+ secretion 251
Human cancer tissue 39
Hydrodictyon 115
Hypnea 59
Hyponasty, auxin 55
Hypocotyl 176

IAA see indoleacetic acid
IBA see indolebutyric acid
Inactivation, loss of transport sites 92
Incompatibilities 178
Indoleacetic acid (IAA) 157–159, 161
  after applications, phototropically stimulated oat coleoptiles 102
  aspartic acid 83
  bioassay 159
  callus formation 42
  cancer cells 39
  cell (s)
    epidermal 45
    guard cell mother cell 45
  cell division 38
  convenience to indoleacetylaspartic acid 27
  cotton fibers growth 57
  cytokinins 30
  differentiation 162
  embryo development 177
  gibberellic acid (GA_3) 183
  and gibberellins (GA), movement, polarization 124
  inhibition of lateral bud growth 183
  light 31
  liquid endosperms, growth substances
    cytokinins 201
    gibberellins (GA) 201
  mitosis 40
  movement
    acropetal 151
    basipetal 151
    fluctuations of 90
    in intact plants 98
    regeneration 153
    replacement 153
    in roots 51
    saturation 153
    xylogenic action 154
  one-dimensional growth 46
  pollen tube elongation 57
  red-light 31
  replacement, provascular tissue 179
  rhizome growth 185
  root 40
    cell
      elongation 50, 51
      meristems 39
  stolons 185
  suppressing effect on cytokinin content, mediated by gibberellin (GA) 30
  synthesis, root 51
  transport, rhythmic fluctuations 111
Indoleacetic acid (IAA) biosynthesis, regulation by indoleacetic acid (IAA) 28
Indoleacetic acid (IAA) effects on algae 58
Indoleacetic acid (IAA) – induced ethylene, production, inhibition by abscisic acid (ABA) 27
Indoleacetic acid (IAA)-oxidase 83
Indoleacetic acid (IAA)-oxidase activity, kinetin 55
Indoleacetylaspartic acid, indoleacetic acid (IAA) 27

Indole-3-acetic acid (IAA) 151
Indolebutyric acid (IBA), translocation of 91
Induction, vascular bundle, auxin 179
Inhibitors
  endogenous causes 205
  root 50
  seed dormancy, germination 204
Inhibitory effect
  acropetal displacement, of cytoplasm 114
  by centrifugation 114
Insulin 8
Intact plant
  abscisic acid (ABA) 219
  auxin 219
  cytokinins 219
  ethylene 219
  gibberellin (GA) 219
Interaction(s)
  auxin 180
  in auxin transport 91
  in the developing embryo
    auxin 177
    gibberellin (GA) 177
  with ethylene 97, 226
  gibberellic acid (GA_3) 226
    cambial activity 180
  with gibberellin (GA) 124
  hormonal 226
  in hormone translocation 86
  triiodobenzoic acid (TIBA) 124
  with water stress 97
Intercept method 81
Internode sections 122
Ion
  currents, development gradients 174
  pumps, distribution and rhizoid formation 175
N^6-isopentenyl-adenine 158

Japanese holly 56
Jerusalem artichoke 42, 46

Kalanchoe blossfeldiana 193
Kinetin 97, 124, 158
  abscisic acid (ABA) – like inhibitors 30
  auxin 182
  cell
    division 38, 45, 47
    elongation 47
  cotton fiber growth 57
  differentiation 162
  deoxyribonucleic acid (DNA) synthesis 37
  elongation 50
  embryogenesis 182
  and gibberellic acid (GA_3) 97
  increase of auxin content 27
  indoleacetic acid (IAA)-oxidase activity 55
  movement 97, 100
  rhizome growth 185
  root(s) 55
    elongation 55
    meristems 39
  translocation
    and abscisic acid (ABA) 91, 97
    and gibberellic acid (GA_3) 91, 97
  transport of 92
Kinetin enhancement of auxin, induced ethylene production 25
Kinetin increases, auxin, movement 27
Koch's postulates 7

Lactuca dentata 223
Lactuca sativa 223
Lactuca scariola 223
Larix 247
Larix decidua Mill. 246
Leaf, leaves 84
  abscisic acid (ABA) biosynthesis 33
  auxin transport 55
  chlorophyll-sparing effect of kinetin, abscisic acid (ABA) antagonism 26
  detached 26
  discs 24
  differentiation 16
  ethylene 29
  gamma-irradiated 45
  hyponasty auxin 55
  unrolling 28
  wheat, ethylene production, cytokinin stimulation 29
  wilting 220
Leaf cell expansion
  auxin 55
  6-benzylaminopurine (BAP) 57
Leaf growth
  cytokinin 56
  gibberellin (GA) 56
Leaf primordia 176
  in regulating vascular differentiation, procambial stage 179
Leaves (see Leaf)
Lemna gibba (LDP) 223
Lemna minor 42, 54
Lemna paucicostata (SDP) 223
Lemna perpusilla (SDP) 223
Lens 107, 122, 124, 125
  stem sections 94
Lentil 55
Lettuce 27, 31, 32, 41, 49
Light
  abscisic acid (ABA) increases 32
  cell division 46
  cytokinin content 33
  cytokinin-like materials 33

Light
  ethylene production 32
  extension inhibition 31
  gibberellin (GA) reduction 32
  glucosylation of abscisic acid (ABA) 31
  hypocotyl elongation 32
  indoleacetic acid (IAA)
    bound 31
    free 31
  mitosis 47
  root elongation, gibberellin (GA) reversal of inhibition 53
  seed germination, increase in cytokinin 33
Lilium 122
Lily 57, 198
Liquid endosperm, growth substances
  cytokinins 201
  gibberellins (GA) 201
  indoleacetic acid (IAA) 201
Liverworts 57
Lolium temulentum 189, 190
Long-day flowering plants
  Chrysanthemum maximum 223
  Lemna gibba 223
  Nicotiana tabacum var. Havana A 223
  Nicotiana tabacum sylvestris 223
Long-day plants, flowering, rosette forms 190
Lower plants 57
Luffa acutangula 195
Lupinus albus 32
Lychnis dioica 194
Lycopersicon 125
Lycopersicon esculentum 194
Lycoricidinol 93
Lyngbya 58
Lysimachia punctata 179

Maize 28
Malus stems 96, 122
Mango 200
Mathematical model for polarity 115
Mechanical stimulation, ethylene production 34
Mechanical stress 46
Megagametophyte 174, 175
Megasporangium 175
Meiosis, spore production 43
Melons 205
Membrane fractions 117
Mercurialis annua 195
Meristem(s)
  apical
    root 35
      indoleacetic acid (IAA) 40
  root
    auxin 45
    differentiation 35
  terminal 179
Meristem activity
  cytokinin 54
Meristemoids
  guard cells 178
  root-hairs 178
  stomata 178
  vascular strands 178
Metabolic poisons 114
3-methyleneoxindole, root elongation 52
Methyljasmonate 5
N-methylnicotinic acid, G_2 factor 42
Micrasterias thomasiana 58
Microcapillary injections 98
Microfibrils and auxin waves 255
  vector fields 255
Microphyle 175
Microtubule
  band 175
  oventation of the spindle 175
Migration of indoleacetic acid (IAA), speed, variation of 97
Mitochondria 114
Mitosis 173
  auxin 40
  antheridial filaments 40
  light 47
  protein synthesis dependence 37
  suppression of indoleacetic acid (IAA), root 40
Mitotic activity
  auxins 39
  auxin levels 39
  naphthalene acetic acid (NAA) 39
Mitotic index, gibberellin (GA) 40
Mobilization 14
Monocotyledons 122
Morphactin 97
Morphogenic maps and positional information 252
Mosses 57
Movement
  abscisic acid (ABA) 97
    Phaseolus 126
  auxin
    applied to loci of natural production 100
    applied to mature leaves 101
    kinetin increases 27
    polarity 82
  basipetal, gibberellic acid (GA_3) 97, 126
  cytokinins
    Pisum roots 126
    Raphanus roots 126
  gibberellin (GA), polarization, of indoleacetic acid (IAA) 124
  hormone
    pathways 117

sieve tubes 119
vascular system 119
indoleacetic acid (IAA) 151, 152, 158
acropetal 151
in intact plants 98
polarity, regeneration 152
saturation 153
from storage tissues 101
of plant hormones, modern period of research 83
in roots
abscisic acid (ABA) 126
cytokinins 126
gibberellic acid (GA_3) 126
thiamine 126
mRNA 176
Multiphasic efflux 103
Mung bean(s) 27, 32, 42
hypocotyls 25, 29
Mushroom (*agaricus bisporous*) 60
Mutants
and abscisic acid (ABA) 225
and gibberellin (GA) 225
temperature sensitive 225
Myo-inositol ester 8

NAA see naphthalene acetic acid
Naphthalene acetic acid (NAA) 107, 154
deoxyribonucleic acid (DNA) 39
fruit growth 198
mitotic activity 39
translocation of 91
1-N-naphtylphthalamic acid (NPA) 112, 123
Narcissus
stamen 122
styles 122
Nasturitium 30
Neurospora crassa 61
Nicotiana 33, 112, 118, 122, 123
internodes 105
Nicotiana rustica 33
Nicotiana tabacum 181, 191
Nicotiana tabacum var. Maryland mammouth (SDP) 223
Nicotiana tabacum var. Havana A (LDP) 223
Nicotiana tabacum sylvestris (LDP) 223
Nicotinic acid 154
Nitrogen
bud formation 186
root development 186
Nitzschia putrida 58
Nonhormonal correlative controls 15
Nuclear division 172

Oat coleoptiles 102, 108, 111
Oenothera biennis 223
Oenothera lamarkiana 223
Okra 154
Omphalia flavida 225
One-dimensional growth, indoleacetic acid (IAA) 46
Onion 40, 46
Orange 29, 176
Orchid 56
Organ
culture, defined culture medium 181
development 172
distal 155
indoleacetic acid (IAA) placement 157
proximal 155
tissue, defined culture medium 181
Organogenesis 15, 181
light intensity 182
pH 182
photoperiod 182
sugar concentration 182
temperature 182
Organ polarity 256
Orientational domain 252
Oscillation(s)
of auxin efflux 249
of electric potential 111
of export rate 111
Oscillators, cellular 257
Oscillatory phenomena 248
Oscillatory system, in polar transport, of auxin 249, 251
Ovary growth 198
Ovule 175
Oxygen, auxin transport dependence 98

Paramecium caudatum 36
Parthenocarpy
abscisic acid (ABA) 200
cytokinins 200
Parthenocissus 115
Pea 9, 28, 29, 31, 32, 39, 41, 42, 47, 56, 92, 109, 160, 196, 204
Peach 196
Peanut 154
Pear 189, 190
Penicillium degitatum 60
Perilla 193
Petiole growth
auxin 55
epinasty 56
ethylene, epinasty 56
gibberellin (GA) 56
Petroglossum 58
Petunia hybrida 198
pH, dependence of auxin uptake 115
Phalophyceae 58
Phaseic acid 32

Phaseolus 56, 59, 98, 103, 107, 117, 119, 123, 124, 159, 178
  hypocotyls 122
  petioles 122
    ageing effect in relation to auxin transport 92
Phaseolus coccineus 177
Phaseolus vulgaris 29, 43, 202
Phloem 13
  aphids 117
  elements 176
  fibers 159, 161
Photoinhibition, ethylene production 32
Photoperidism 187
Photoperiod
  abscisic acid (ABA) content 32
  cytokinin content 32
  gibberellin (GA) content 32
  Phototropic responses
  abscisic acid (ABA) 222
  auxin 222
Phototropism 9
Phycomyces 60, 61
Phyllotaxy 179, 186, 208
Phytochrome
  cytokinin levels 33
  ethylene production 32
  gibberellin (GA) metabolism 31
Phytochrome activity in the algae 57
Picea 247
Picea abies (L.) Karst 246
Pinaceae 189
Pine 247
Pineapple (*ananas sativus* L.) 191, 196
D-pinitol 5
Pinus
  embryonic axes 103
  seedlings 122
Pinus silvestris L. 246
Pinus taeda 189
Pisum 101, 103, 111, 115, 117–119, 125, 126
  internodes 93
  seedlings 99
Pisum arvense 203
Pisum sativum (pea) 202, 204
Pith 166
Plant
  productivity 173
  resistance 174
Plant growth substances 4
Plant hormones 174
  mechanism of action 208
Plantlets in culture
  auxin 181
  kinetin 182
Plant morphogenesis and auxin waves 246
Plant pathology
  and cytokinins 225
  and disease of plants 225
  and gibberellin (GA) 225
  and pathogen host and hormonal interactions 225
Plasmodesmata 225
Plectonema 58
Plum 196
Plumbago indica 191
Plumbago indica L. 191
Polarity
  auxin
    movement 82
    transport 253
  axial 245
  basipetal 149
  cell wall growth 175
  degree of 123
  distribution of cellular constituients 174
  loss 256
  mathematical model for 115
  membrane 175
  organ 256
  of plant body 252
  of plant organ 256
  spindle orientation, influx of ions 175
  of translocation, hormone 94
  vesicle secretion 175
Polar secretion theory 112
Polar transport 13
  of auxin 244, 245, 248, 249, 251
  oscillatory system 249
  velocity of signal propagation 249, 251
Pollen
  auxin 198
  gibberellin (GA) 198
Pollen tube 174
Pollen tube elongation
  abscisic acid (ABA) 57
  cytokinin 57
  ethylene formation 57
  gibberellin (GA) 57
  indoleacetic acid (IAA) 57
Pollution 174
Polyploidy
  abscisic acid (ABA) 42, 43
  auxin 43
  cytokinins 43
  gibberellin (GA) 43
Poplar 33
Population 173
Populus 164, 180
Porphyra tenera 59
Positional information 244, 245
  auxin wave(s) and vector fields 252–255, 258

morphogenic maps 252
Potato 32, 34
Potato (*solanum andigena*) 28
Procambial system 176
Procambium 176
Production of ethylene 222
Proembryo, radial symmentry 176
Propagation
auxin wave(s) 254
vector(s) 254
Propagation of plants
test tube
bud formation 182
callus formation 182
root induction 182
seedling growth 182
Protoderm 176
Protoplast autolysis 254, 255
Prunus acida 196
Prunus persica 207
Pulse experiments
modifications of 94
transport 107
Puromycin 97
Pyrus communis 207

Quercus robur L. 246
Quiescent center, cytokinin production 38

Radish 165
Ranunculus sceleratus 50
Raphanus 124, 126
Raspberry 195
Red light
auxin
elongation 31
production 31
transport 31
uptake 31
ethylene production, inhibition of 32
gibberellin (GA)-like material 31
indoleacetic acid (IAA), extractable 31
Regeneration
dicotyledonous plants 181
indoleacetic acid (IAA) 153
monocotyledonous plants 181
protoplast 180
replacement 153
shoots 149
sieve tubes 149, 157
from single cells 180
sugar 153
tracheary cells 149, 160
Regulation, hormone
of growth 221
of maturation 221
of physiological homeostasis to cellular differentiation biochemical modulations 221
remote 14
of responses to the external environment 221
synthesis 223
transport 223
Regulators, growth 173, 174
Relationship of cell multiplication, cell expansion 46
Retardation of auxin destruction 27
Rhizoid formation 175
Rhizome growth 185
Rhodophyceae 58
Rice 49, 50
Ricinus communis 179
Riella heliophylla 40
Robinia 180
Role(s) of
abscisic acid (ABA) 226
auxin 226
cytokinin 226
ethylene 226
gibberellic acid (GA_3) 226
Root(s)
abscisic acid (ABA) modulation 42
auxin 33, 40
induction of hydrogen ion secretion 53
bean 165
cambial activity, cytokinins 165
cytokinin 33, 54
cytokinin biosynthesis 34
cytokinin production 30
differentiation 165, 178
elongation 23
auxin 51
ethylene 29
ethylene production, auxin enhancement 53
formation of ethylene, promotion by kinetin 55
gibberellin (GA) production 34
gibberellin (GA) synthesis 52
hormone
translocation 125
transport channels 118
indoleacetic acid (IAA), synthesis 51
indoleacetic acid (IAA) oxidase activity, kinetin 55
inhibitors 50
kinetin 55
meristem, apical 35
meristem activity, cytokinin 54
pea 165
Phaseolus 126
quiescent center, cytokinins 38
transition region 165

Root(s)
  Vicia 125
  Zea 126
Root caps
  abscisic acid (ABA) 54
    root growth 54
Root cell elongation
  indoleacetic acid (IAA)
    inhibition of 51
    stimulation of 51
Root elongation
  abscisic acid (ABA) 554
  auxin
    ethylene biosynthesis, action of inhibitors 53
    inhibitory action 53
  2-Chlorethyltrimethylammonium chloride, Chlorocholine chloride (CCC), Amo 1618 53
  ethylene 53
  gibberellin (GA) 52
  kinetin 55
  3-methyleneoxindole 52
Root growth
  abscisic acid (ABA) 54
  auxin antagonists, enhancement of 51
  cytokinin 54
  root cap 54
  H^+ secretion 54
Root hair, production 178
Rooting
  abscisic acid (ABA) 42
  embryogenesis 182
Root meristems
  auxin 38
  cycle 37
  cytokinin 38
  indoleacetic acid (IAA) 39
  kinetin 39
Root nodule, formation, cytokinin 43
Root tip, surface charge
  abscisic acid (ABA) 26
  auxin 26
Rosa sheradii 200
Rose petals 29, 30
Rudbeckia bicolor 40
Rumex obtusifolius 33
Rye 44

Saccharomyces cerevisiae 59
Saccharum 122–124
S-adenosylmethionine (SAM) synthesis 29
Salix 119
Saxifraga sarmentosa 116, 118
  stolons of 97
Scenedesmus 58
Scots pine 164
Second messenger, transport
  of Ca^{+2} 220
  of ethylene 220
  of gibberellin (GA) 220–221
  of indoleacetic acid (IAA) 220
Secretion of H^+, root growth 54
Secretion theory of polar auxin transport 85
Seed(s) 4, 13, 34
  abscisic acid (ABA) 204
    bound 34
  auxins
    bound, turnover 204
    growth 203
  cytokinin, endosperm volume 203
  development 186, 200, 204
  dormancy
    growth inhibitors 205
    inhibitors, germination, endogenous causes 205
  germination
    overcoming dormancy 206
      cytokinins 206
      gibberellin (GA) 206
    time sequence 206
  germination inhibitors
    abscisic acid (ABA) 206
    coumarin 206
  gibberellin (GA) 202
  gibberellin (GA) biosynthesis 34
Seed germination 33, 222
Selaginella 122
Senescence 4, 7, 15, 16, 222
Sex expression 174
Shoot
  apex 278
  growth 7
  transition region 165
Short-day flowering, of long-day plants 223
Short-day plants
  Chrysanthemum morifolium 223
  Lemna paucicostata 223
  Lemna perpusilla 223
  Nicotiana tabacum var. Maryland mammouth 223
  florigen 190
  gibberellin (GA) 190
Sieve-tube(s)
  aphids 101
  auxin transport 84
  differentiation 149
  regeneration 149, 157
Sieve-tube elements 160
  regeneration 154
Silene armeria 190
Sinapis alba 189, 190
Site of auxin transport 85
Sitka spruce 189

Size organ plant 174
control of 174
Solanum andigena (potato) 28, 30, 32, 33
Solanum tuberosum 28
Somatic hybridization, bioengineering 183
Soybean 37, 47, 50
hypocotyl elongation 27
interaction of indoleacetic acid (IAA) and cytokinin 27
Soybean callus 42
Soybean cells 36, 41, 175
Spinach 32
Spindle, orientation, flower primordia 186
Spirodela 56
Spoilage, post harvest 174
Stem
differentiation 178
growth 6
Stem segments
culture 245
Stomata, closure 222
Storage organs 222
Stratification, bound abscisic acid (ABA) 34
Strawberry 195, 196, 198
Streptocarpus nobilis 191
Sugars, interactions in auxin transport 91
Sunflower 33, 34, 92, 106, 107, 208
Sucrose
increased indoleacetic acid (IAA) transport 164
xylem cells 164
Suspensor 176
Sycamore 36
Sycamore cells 41
Synergism
antagonism, in hormone interaction 25
definition, interdependent interaction 25
hormones 180
Synergistic effects 221
Synergistic inhibitory interaction, equation for 25
Synthesis
and abscisic acid (ABA) 220
of auxin 223
expanding leaves 221
shoot meristems 221
of cytokinins 220
root tips 221
gibberellin (GA) 13, 220
indoleacetic acid (IAA) 220
regulation in
root cap 220
wilting leaves 220
Syringa 190

2,3,5-T see 2,3,5-trichlorophenoxyacetic acid
Taxodiaceae 189
Taxus baccata 34
Terminal meristem, in regulating vascular differentiation 179
Temperature
cold, abscisic acid (ABA), reduction, fruits 34
gibberellin (GA) biosynthesis, enhancement
seeds 34
tuber 34
gibberellin (GA) production, inhibition
roots 34
stratification, bound abscisic acid (ABA), increase in seeds 34
Thermodormancy
abscisic acid (ABA) levels 34
Thiamine 5, 126, 147
translocation of 124
Thompson Seedless grapes
cytokinin biosynthesis 34
H-thymidine 173
TIBA see triiodobenzoic acid
Tissue
callus
differentiation, culture media 181
xylogenesis 181
culture(s)
defined culture medium 181
development of flowers 191
differentiation
cytogenetics 183
genetic uniformity 183
initiation of flowers 191
leaves 183
growth 172
organ, defined culture medium 181
Tissue culture, cytokinin 47
Tobacco 47, 104, 166, 223
callus 10, 27, 32, 165, 173, 181
Tobacco cells 36, 37, 39, 41
Tomato 34, 45, 183, 194–196, 200, 205
Totipotency 180
concept 181
Tracheary cells
differentiation 149
regeneration 149, 160
Tradescantia paludosa 43
Transfer cells 14
Translocation
abscisic acid (ABA) 97
abscisins 124
auxin 97
basipetal
abscisic acid (ABA) 249
triiodolbenzoic acid (TIBA) 249
cytokinins 124
diffusion 14
gibberellin (GA) 97, 124

Translocation
  hormone 8
    direct evidence of 81
    polarity 94
    root 125
    shoot section 127
  indolebutyric acid (IBA) 91
  kinetin 95
  naphthalene acetic acid (NAA) 91
  thiamine, within the transpiration stream 119
Trans-membrane potentials 102
Transpiration 13
Transport
  auxin 81, 85, 123
    applied to unfolded leaves 101
    basipetal occurrence 122
    conveyor belt model 112
    dependence on temperature 98
    intensity 98
    leaves 55
    oscillations 103
    oxygen supply 113
    pathway in shoots 118
    polar 13, 84, 85
    roots 125
    sieve tubes 84
    system specificity 123
    tissue types 105
    velocity 98
    woody shoots 98
  capacity 87
  channels, for auxin 103
  equation 87, 90
  hormone 10, 13
    1-aminocyclopropane-1-carboxylic acid (ACC)
      in phloem 220
      in xylem 220
  regulation 223
  roots, channels 118
  of second messenger
    Ca^{+2} 220
    ethylene 220
    gibberellin (GA) 221
    indoleacetic acid (IAA) 220
  indoleacetic acid (IAA), rhythmic fluctuations 111
  mechanisms, models 111
  polar 13
  pulse experiments 107
  roots, velocity 102
  shoots, velocity 102
  sites, generation 105
  symplast 114
  system
    flexibility 109
    loadings 105
  velocity 82, 99
    auxin 83
    fluctuation(s) 108
    relationships between donor concentrations 106
    temperature dependence 82
Triacontanol 5
2,3,5-trichlorophenoxyacetic acid (2,3,5-T) 92
Trigonella foenum-graecum 42
Trigonelline, cell division 42
Triiodobenzoic acid (TIBA) 4, 27, 28, 33, 48, 63
  inhibitor 249
  translocation, basipetal 249
Triticum 118
Triticum aestivum 202
Tropisms 18
Tuber
  dormancy 222
  formation 221
  gibberellin (GA) biosynthesis 34
Tulipa 103
Tumors 9
Two-dimensional growth 46

Ulothrix 59
Ulva 59
Ulva fasciata 58
Ulva lactuca 59
Uptake, auxin dependence 115
Uptake of auxin, ethylene 26

Vascular development, spatial arrangement 178, 179
Vascular strand 149
Vectors
  of auxin, wave 255, 256
  critical, angle 255
  propagation 252, 254
  deviation, of auxin waves 256
  fields
    auxin wave
      positional information 252, 255
    domain pattern 252
    domain pattern 252
    positional information 258
    three-dimensional 253, 255
  of gravity 257
Velocity
  of auxin transport 83
  polar transport, of auxin 249, 251
  of signal propagation 251
  validity of, estimations 106
Velox (Corn) 54
Verticillium-albo-atrum 61
Vescile secretion 175

Vicia 40, 118
  roots 103, 125
Vicia faba 35, 40
Vitis 163
Vivipary, abscisic acid (ABA) 204

Water 174
Water balance
  and abscisic acid (ABA) 221
  and ethylene 221
Watermelon 48, 40
  dwarf 40
Water stress 11, 93, 97
  abscisic acid (ABA) biosynthesis
    enhancement of, in leaves 33
  abscisic acid (ABA) content
    enhancement of, in roots 33
  cell
    division 47
    expansion 47
Wheat 11, 14, 29, 31, 47, 52, 53, 55, 202, 204
Wound hormone 5, 9
Wuchshormone 81

Xanthium 33, 55, 122–124, 193
Xanthium pennsylvanium 188
Xanthium strumarium 193
Xanthophyceae 58
Xylem 149, 159
  differentiation 11

Yeast 36, 39, 43, 59, 61

Zea 38, 88, 101, 106–108, 115, 117, 118, 124
  coleoptiles 91, 122, 248
  mesocotyls 124
  roots 126
Zea mays 165, 173, 194, 202, 204
Zeatin 5, 158, 250
Zeatin riboside 158
Zygote(s), apolar 175

Encyclopedia of Plant Physiology

New Series

Editors: A. Pirson, M. H. Zimmermann

Volume 1

Transport in Plants I

Phloem Transport
Editors: M. H. Zimmermann, J. A. Milburn
With contributions by numerous experts
1975. 93 figures. XIX, 535 pages
ISBN 3-540-07314-0

Volume 2 A

Transport in Plants II

Cells
Editors: U. Lüttge, M. G. Pitman
With contributions by numerous experts
1976. 97 figures, 64 tables. XVI, 419 pages
ISBN 3-540-07452-X

Volume 2 B

Transport in Plants II

Tissues and Organs
Editors: U. Lüttge, M. G. Pitman
With contributions by numerous experts
1976. 129 figures, 45 tables. XII, 475 pages
ISBN 3-540-07453-8

Volume 3

Transport in Plants III

Intracellular Interactions and Transport Processes
Editors: C. R. Stocking, U. Heber
With contributions by numerous experts
1976. 123 figures. XXII, 517 pages
ISBN 3-540-07818-5

Volume 4

Physiological Plant Pathology

Editors: R. Heitefuss, P. H. Williams
With contributions by numerous experts
1976. 92 figures. XX, 890 pages
ISBN 3-540-07557-7

Volume 6

Photosynthesis II

Photosynthetic Carbon Metabolism and Related Processes
Editors: M. Gibbs, E. Latzko
With contributions by numerous experts
1979. 75 figures, 27 tables. XX, 578 pages
ISBN 3-540-09288-9

Volume 7

Physiology of Movements

Editors: W. Haupt, M. E. Feinleib
With contributions by numerous experts
1979. 185 figures, 19 tables. XVII, 731 pages
ISBN 3-540-08776-1

Volume 8

Secondary Plant Products

Editors: E. A. Bell, B. V. Charlwood
With contributions by numerous experts
1980. 176 figures, 44 tables and numerous schemes and formulas. XVI, 674 pages
ISBN 3-540-09461-X

Volume 9

Hormonal Regulation of Development I

Molecular Aspects of Plant Hormones
Editor: J. MacMillan
With contributions by numerous experts
1980. 126 figures. XVII, 681 pages
ISBN 3-540-10161-6

Volume 10

Hormonal Regulation of Development II

From the Cell to the Whole Plant
Editor: T. K. Scott
With contributions by numerous experts
ISBN 3-540-10196-9
In preparation

Springer-Verlag
Berlin
Heidelberg
New York
Tokyo

Volume 11

Hormonal Regulation of Development III

Role of Environmental Factors
Editors: A. Pirson, M. H. Zimmermann
With contributions by numerous experts
ISBN 3-540-10197-7
In preparation

Physiological Plant Ecology

Editors of Volume 12 A - D: O. L. Lange, P. S. Nobel, C. B. Osmond, H. Ziegler

Volume 12 A

Responses to the Physical Environment

1981. 110 figures. XV, 625 pages
ISBN 3-540-10763-0

Volume 12 B

Water Relations and Carbon Assimilation

1982. 153 figures. XI, 747 pages
ISBN 3-540-10906-4

Volume 12 C

Responses to the Chemical and Biological Environment

1983. 104 figures. XI, 799 pages
ISBN 3-540-10907-2

Volume 12 D

Ecosystem Processes: Mineral Cycling, Productivity and Man's Influence

1983. 61 figures. XI, 644 pages
ISBN 3-540-10908-0

Volume 13 A

Plant Carbohydrates I

Intracellular Carbohydrates
Editors: F. A. Loewus, W. Tanner
1982. 103 figures. XXII, 918 pages
ISBN 3-540-11060-7

Volume 13 B

Plant Carbohydrates II

Extracellular Carbohydrates
Editors: W. Tanner, F. A. Loewus
With contributions by numerous experts
1981. 124 figures. XXII, 769 pages
ISBN 3-540-11007-0

Volume 14 A

Nucleic Acids and Proteins in Plants I

Structure, Biochemistry and Physiology of Proteins
Editors: D. Boulter, B. Parthier
With contributions by numerous experts
1982. 135 figures. XX, 768 pages
ISBN 3-540-11008-9

Volume 14 B

Nucleic Acids and Proteins in Plants II

Structure, Biochemistry and Physiology of Nucleic Acids
Editors: B. Parthier, D. Boulter
With contributions by numerous experts
1982. 173 figures. XVIII, 774 pages
ISBN 3-540-11140-9

Volume 15

Inorganic Plant Nutrition

Editors: A. Läuchli, R. L. Bieleski
With contributions by numerous experts
1983. 131 figures. Approx. 900 pages
(In 2 volumes. Not available separately).
ISBN 3-540-12103-X

Volume 16 (Part A + B)

Photomorphogenesis

Editors: W. Shropshire, Jr., H. Mohr
With contributions by numerous experts
1983. 173 figures. XXX, 832 pages
(In 2 parts, not available separately)
ISBN 3-540-12143-9

Springer-Verlag
Berlin
Heidelberg
New York
Tokyo

www.ingramcontent.com/pod-product-compliance
Lightning Source LLC
Chambersburg PA
CBHW061159240926
51846CB00006B/1065
* 9 7 8 3 6 4 2 6 7 7 3 3 5 *